Handbook of
Neurochemistry

SECOND EDITION

Volume 2
EXPERIMENTAL
NEUROCHEMISTRY

Handbook of
Neurochemistry
SECOND EDITION

Edited by Abel Lajtha
Center for Neurochemistry, Ward's Island, New York

Handbook of
Neurochemistry

SECOND EDITION

Volume 2
EXPERIMENTAL
NEUROCHEMISTRY

Edited by
Abel Lajtha
Center for Neurochemistry
Ward's Island, New York

PLENUM PRESS · NEW YORK AND LONDON

Library of Congress Cataloging in Publication Data

Main entry under title:

Handbook of neurochemistry.

  Includes bibliographical references and index.
  Contents: v. 1. Chemical and cellular architecture—v. 2. Experimental neuro-
chemistry.
  1. Neurochemistry—Handbooks, manuals, etc. I. Lajtha, Abel. [DNLM: 1.
Neurochemistry. WL 104 H434]
  QP356.3.H36 1982   612′.814   82-493
ISBN 978-1-4684-4210-6   ISBN 978-1-4684-4208-3 (eBook)
DOI 10.1007/978-1-4684-4208-3

© 1982 Plenum Press, New York
A Division of Plenum Publishing Corporation
233 Spring Street, New York, N.Y. 10013
Softcover reprint of the hardcover 1st edition 1982

Contributors

K. Adriaenssens, Provincial Instituut voor Hygiëne, Antwerp, Belgium

Glen B. Baker, Neurochemical Research Unit, Department of Psychiatry, University of Alberta, Edmonton, Alberta T6G 2N8, Canada

Nicole Baumann, INSERM and CNRS Laboratory of Neurochemistry, Salpetriere Hospital, Paris 13, France

E. D. Bird, McLean Hospital, Harvard Medical School, Belmont, Massachusetts 02178

Alan A. Boulton, Psychiatric Research Division, University Hospital, Saskatoon, Saskatchewan S7N OXO, Canada

Jonathan D. Brodie, Department of Psychiatry, New York University School of Medicine, New York, New York 10016

N. Chamoles, Laboratorio de Neuroquimica, Clinica del Sol, Buenos Aires, Argentina

Jørgen Clausen, Neurochemical Institute, Copenhagen, Denmark, and Institute of Biology and Chemistry, University of Roskilde, DK 4000 Roskilde, Denmark

Thomas B. Cooper, Analytical Psychopharmacology Laboratory, Rockland Research Institute, Orangeburg, New York 10962

Ronald T. Coutts, Neurochemical Research Unit, Faculty of Pharmacy and Pharmaceutical Sciences, University of Alberta, Edmonton, Alberta T6G 2N8, Canada

David A. Durden, Psychiatric Research Division, University Hospital, Saskatoon, Saskatchewan S7N OXO, Canada

David D. Gilboe, Departments of Neurosurgery and Physiology, University of Wisconsin Center for Health Sciences, Madison, Wisconsin 53706

Fritz A. Henn, University of Iowa, Department of Psychiatry, College of Medicine, Iowa City, Iowa 52242. Present address: Long Island Research Institute, State University of New York, Stony Brook, New York 11794.

Suella W. Henn, University of Iowa, Department of Psychiatry, College of Medicine, Iowa City, Iowa 52242. Present address: Long Island Research Institute, State University of New York, Stony Brook, New York 11794.

R. Humbel, Centre Hospitalier de Luxembourg, Luxembourg

L. L. Iversen, Neurochemical Pharmacology Unit, Medical Research Council Centre, Medical School, Cambridge CB2 2QH, England

G. Jean Kant, Department of Medical Neurosciences, Walter Reed Army Institute of Research, Walter Reed Army Medical Center, Washington, D.C. 20012

Barry B. Kaplan, Department of Cell Biology and Anatomy, Cornell University Medical College, New York, New York 10021

Katrina L. Kelner, Department of Cell Biology, Baylor College of Medicine, Texas Medical Center, Houston, Texas 77030

Francois Lachapelle, INSERM and CNRS Laboratory of Neurochemistry, Salpetriere Hospital, Paris, France

Robert H. Lenox, Neuroscience Research Unit, Department of Psychiatry, University of Vermont, Burlington, Vermont 05405

A. Lowenthal, Laboratory of Neurochemistry, Born Bunge Foundation, Universitaire Instelling Antwerpen, Antwerp, Belgium

James L. Meyerhoff, Department of Medical Neurosciences, Walter Reed Army Institute of Research, Walter Reed Army Medical Center, Washington, D.C. 20012

Volker Neuhoff, Forschungsstelle Neurochemie, Max-Planck-Institut für Experimentelle Medizin, Göttingen, Federal Republic of Germany

Jose M. Palacios, Department of Neuroscience, The Johns Hopkins University, School of Medicine, Baltimore, Maryland 21205. Present address: Sandoz Ltd., Preclinical Research, CH-4002 Basel, Switzerland.

Ernest J. Peck, Jr., Department of Cell Biology, Baylor College of Medicine, Texas Medical Center, Houston, Texas 77030

Stephen R. Philips, Psychiatric Research Division, University Hospital, Saskatoon, Saskatchewan S7N OXO, Canada

John Rotrosen, Psychiatry Service, Veterans Administration Medical Center, New York, New York 10010

Stanley Stein, Roche Institute of Molecular Biology, Nutley, New Jersey 07110

N. M. van Gelder, Centre de Recherche en Sciences Neurologiques, Département de Physiologie, Faculté de Médicine, Université de Montréal, Montréal, Québec H3C 3J7, Canada

Nora Volkow, Department of Psychiatry, New York University School of Medicine, New York, New York 10016

James K. Wamsley, Departments of Psychiatry and Anatomy, University of Utah, Salt Lake City, Utah 84132

David L. Wilson, Department of Physiology and Biophysics, University of Miami, School of Medicine, Miami, Florida 33101

Preface

The second volume of the *Handbook* does not parallel any volume of the first edition; it is one more sign, or reflection, of the expansion of the field. By emphasizing the experimental approach, it illustrates the tools that have recently become available for investigating the nervous system. Also, perhaps even more than other volumes, it illustrates the multidisciplinary nature of the field, requiring multidisciplinary methodology. It is now recognized that the availability of methodology is often the rate-limiting determinant of studies and that improvements or innovations in instrumentation can open up new avenues. A new improved method, although opening up new possibilities and being crucial to making advances, is only a tool whose use will determine its usefulness. If we do not recognize its possibilities, its use will be limited; if we do not recognize its limitations, it will mislead us. It is the possibilities and limitations and the results obtained that are illustrated here.

As with the other volumes of this *Handbook*, many more chapters could be included, and each of the present chapters could have been expanded severalfold. Some topics that could be included in this volume will be dealt with in later chapters; some are better discussed in texts of other disciplines. The purpose was not to include so much detail that further literature searches would not be necessary—it was to evaluate the approaches, the limitations, the possibilities, and then to indicate where further details can be found. The fact that the details of most of the approaches described here were not available when the first edition was written is but another illustration of the rapid and exciting development of neurochemistry.

Abel Lajtha

Contents

Chapter 2

Receptor Mapping by Histochemistry

> *James K. Wamsley and Jose M. Palacios*

Chapter 3

Receptor Measurement

> *Ernest J. Peck, Jr. and Katrina L. Kelner*

Chapter 4

Rapid Enzyme Inactivation

> *Robert H. Lenox, G. Jean Kant, and James L. Meyerhoff*

Chapter 5

Radioenzymatic Analyses

Stephen R. Philips

Chapter 6

Two-Dimensional Polyacrylamide Gel Electrophoresis of Proteins

David L. Wilson

Chapter 7

The Identification of Subcellular Fractions of the CNS

Suella W. Henn and Fritz A. Henn

Chapter 8

Cell Isolation

Jørgen Clausen

Chapter 9

Principles of Compartmentation

   N. M. van Gelder

Chapter 10

Diagnosis of Hereditary Neurological Metabolic Diseases

   A. Lowenthal, N. Chamoles, K. Adriaenssens, and R. Humbel

Chapter 11

Human Brain Postmortem Studies of Neurotransmitters and Related Markers

E. D. Bird and L. L. Iversen

Chapter 12

Neurological Mutants

Nicole Baumann and François Lachappelle

Chapter 13

Analytical Aspects of the Pharmacokinetics of Psychotrophic Drugs

   Thomas B. Cooper

Chapter 14

Perfusion of the Isolated Brain

   David D. Gilboe

Chapter 15

Principles and Application of PET in Neuroscience

Jonathan D. Brodie, Nora Volkow, and John Rotrosen

Chapter 16

Selected Micromethods for Use in Neurochemistry

Volker Neuhoff

Chapter 17

Mass Spectrometric Analysis of Some Neurotransmitters and Their Precursors and Metabolites

David A. Durden and Alan A. Boulton

Chapter 18

Gas Chromatography

   Ronald T. Coutts and Glen B. Baker

Chapter 19

High-Performance Liquid Chromatography

   Stanley Stein

RNA–DNA Hybridization
Analysis of Gene Expression

Barry B. Kaplan

1. INTRODUCTION

During the last decade, significant insight into the regulation of eukaryotic gene expression has been achieved, in part from studies made possible by RNA–DNA hybridization. These studies take advantage of the ability of RNA to form stable complexes with denatured strands of complementary DNA, enabling the investigator to relate genomic DNA to the immediate products of gene expression. Nucleic acid hybridization is a research tool that has proven fundamental to the major advances in molecular biology and provides a framework for recent recombinant DNA technology. Indeed, quantitative hybridization analysis has become to the molecular biologist what enzyme assays are to the enzymologist or protein biochemist.

To date, RNA–DNA hybridization analysis has been most often used to (1) elucidate the nature and fraction of the genome transcribed, (2) determine the diversity of various RNA populations and/or concentrations of specific mRNAs within a given cell compartment, and (3) characterize alterations in gene expression during differentiation, development, regeneration, or in response to a variety of environmental stimuli.

In this chapter, the principles of RNA-driven hybridization reactions will be discussed and recent results obtained by hybridization analysis reviewed. Special emphasis will be placed on the application of hybridization technology to subjects of neurobiological interest. Excluded from this review will be analysis of data derived from DNA-driven hybridization reactions (reviewed by Flint,[1] Lewin[2]) and detailed discussion of the experimental protocols involved in hybridization assays (see refs. 3,4).

Barry B. Kaplan • Department of Cell Biology and Anatomy, Cornell University Medical College, New York, New York 10021.

2. *PRINCIPLES OF RNA-DRIVEN HYBRIDIZATION REACTIONS*

Generally, the hybridization of radiolabeled DNA to RNA is carried out in vast RNA excess (RNA:DNA mass ratios 10^2–10^4:1) so that the concentration of RNA at the start of the reaction remains essentially unchanged throughout. Therefore, the rate at which the DNA tracer enters into duplex formation with RNA follows pseudo-first-order kinetics and can be described as follows:

$$dD/dt = -kR_0D \qquad [1]$$

where D represents the concentration of DNA remaining single stranded (unreacted) at time t, R_0 the initial (and final) RNA concentration, and k the observed rate constant for RNA–DNA hybrid formation. Integration of equation 1 from the initial condition of $t = 0$ and $D = D_0$ yields:

$$\ln(D/D_0) = -kR_0t \qquad [2]$$

$$D/D_0 = \exp(-kR_0t) \qquad [3]$$

where D/D_0 is the fraction of DNA tracer remaining unreacted. As is customary, t is expressed in seconds, and R_0 in moles of nucleotide per liter, yielding a k with the units liter mol^{-1} sec^{-1}.

Ideally, the rate of the reaction, expressed as a function of R_0t, is inversely proportional to the sequence complexity of RNA.[5,6] Therefore, the greater the RNA complexity, the slower the reaction, and the longer it takes to reach completion.

A useful expression relating the reaction rate constant and RNA complexity is the $R_0t_{1/2}$, the value at which half the hybridization reaction is complete. Thus, when $D = D_0/2$ and R_0t is defined as $R_0t_{1/2}$, equation 2 can be written

$$\ln(D_0/2/D_0) = -kR_0t_{1/2} \qquad [4]$$

$$k = \ln 2/R_0t_{1/2} \qquad [5]$$

Thus, the pseudo-first-order rate constant of a single kinetic component can be approximated by determining the R_0t value at which half of the input DNA enters RNA–DNA hybrids. Whereas the hybridization rate constant is inversely proportional to the sequence complexity of the RNA, the $R_0t_{1/2}$ value is directly related to RNA complexity. Most simply defined, the RNA complexity is the equivalent amount of single-copy DNA homologous to the transcribed (hybridizing) RNA and is expressed as a percentage of the DNA complexity or as an equivalent number of nucleotides. The number of different RNA sequences comprising the population can be approximated by dividing the RNA complexity by the experimentally determined number average size (in nucleotides) of the RNA molecules comprising the population. The RNA size is most often estimated by polyacrylamide gel electrophoresis or density gradient centrifugation using denaturing conditions to eliminate intermolecular aggregation or RNA secondary structure. Size determinations have also been made by direct visualization using electron microscopy.

Knowledge of the complexity of an RNA population and its hybridization rate constant enables one to estimate the fraction of the total RNA that drives the reaction. From this value, the number of RNA copies per gram of tissue or cell (copy frequency) can be determined. The fraction of the total RNA input serving as driver is calculated by

$$D_f = k_{\mathrm{obs}}/k_{\mathrm{exp}} \qquad [6]$$

where D_f is the driver fraction, k_{obs} the measured pseudo-first-order rate constant, and k_{exp} the constant expected if all input RNA were driving the reaction.[7] Here, the k_{exp} of the experimental RNA is approximated with reference to a suitable kinetic standard by the expression

$$G_2 = (k_1/k_2)G_1 \qquad [7]$$

where G_1 and k_1 are the complexity and hybridization rate constant of the RNA standard, respectively, and G_2 and k_2 are the equivalent parameters for the experimental RNA. Useful standards are the renaturation of the *Escherichia coli* genome (4.2×10^6 nucleotide pairs, $k = 0.25$ liter mol^{-1} sec^{-1}) or the hybridization of RNA enzymatically synthesized from ϕX174 DNA with radiolabeled ϕXDNA. For this calculation, the RNA complexity is taken as 5.4×10^3 nucleotides (nt), and $k = 170$ liter mol^{-1} sec^{-1}.[8] Using this approach, the driver fraction and copy frequency of various brain RNA populations have been estimated (for discussion see Section 8.1.2).

3. VARIABLES AFFECTING HYBRIDIZATION

The proposed rate-limiting step in the hybridization reaction is a nucleation event caused by the collision and formation of a few correct base pairs.[9] The subsequent "zippering" reaction is presumably fast and results in the formation of a stable complex derived from cooperative "stacking free energy." Hydrogen bonds, although probably important for the specificity of the reaction, are thought to play only a minor part in the total stability of the duplex. Studies performed with synthetic polynucleotides or nucleic acids sheared to defined lengths suggest that, depending on G + C content, the minimum size for a stable complex is 10–20 nucleotides (reviewed by McCarthy and Church).[10] Somewhat larger lengths (25–40 nt) are necessary to form stable duplexes with oligonucleotides obtained from DNA of higher organisms.

In addition to the concentration and sequence complexity of the reacting

RNA, there are several other factors that determine the rate of the hybridization reaction. The major variables are briefly summarized below.

3.1. Temperature

Not unexpectedly, the rate of the reaction approaches zero at incubation temperatures approaching the melting temperature (T_m) of the DNA probe. The relationship between hybridization rate and incubation temperature (T_i) is best described by a bell-shaped curve with a broad optimum 20–30°C below the T_m of the DNA.[11,12] The T_m of the DNA is itself influenced by the G + C content, but these effects are relatively small and are usually ignored.

3.2. Ionic Strength and pH

The effect of salt concentration on the rate of DNA renaturation has been investigated. At low salt concentrations (<0.4 M Na$^+$), the rate is proportional to the cube of the cation concentration.[9] The effect of salt on the reaction rate is diminished significantly between 0.4 M and 1.0 M Na$^+$ yet still increases the rate twofold. Britten *et al.*[4] have developed a useful table which facilitates comparison of DNA renaturation rate constants obtained at different salt concentrations. In practice, kinetic data obtained at elevated ionic strengths are usually expressed as the values they would exhibit under standard conditions of 0.18 M Na$^+$ (equivalent C_0t).

It bears mention that the relationship between hybridization rate and Na$^+$ concentration was established empirically from DNA renaturation studies. Recent evidence indicates, however, that the salt dependence of RNA-driven hybridization reactions is 2.5-fold less than that of DNA renaturation.[13] The reason for this difference is uncertain, but caution should be used in applying correction schedules derived from DNA annealing studies to kinetic data obtained from RNA-driven hybridization experiments.

Investigation of the effect of pH on DNA annealing (and presumably RNA–DNA hybridization) revealed little effect on reaction rates in the range at which hybridization experiments are normally conducted (pH 6–8).[9,10]

3.3. Nucleic Acid Fragment Length

The rate of DNA annealing is markedly influenced by the length of the reacting DNA fragments. In DNA renaturation experiments, the second-order rate constant is directly proportional to the square root of the single strand length.[9,14,15] This also appears to be the case for reactions between RNA fragments and longer DNA strands present in stoichiometric amounts.[16] In RNA-driven hybridization reactions, the situation is less clear. In these experiments, the radiolabeled DNA tracer is usually short (200–400 nt) relative to the excess RNA driving the reaction. Under these circumstances, the rate of hybridization, as monitored by hydroxyapatite chromatography, was inversely proportional to RNA driver length.[17] This result is unexpected and represents a departure from standard DNA renaturation reactions. Using an

S_1 nuclease assay to determine the rate constant of RNA-driven reactions containing complementary DNA, no difference in reaction rate was observed when DNA tracer length was varied between 200 and 1400 nt (W. E. Hahn, personal communication). Therefore, the effect of fragment length on the kinetics of RNA-driven hybridization reactions may depend, in part, on both the type of DNA used as probe and the method of assay.

3.4. Viscosity

Since hybridization is dependent on intermolecular collision between complementary nucleic acid fragments, it is reasonable to assume that the rate is influenced by the solution viscosity. Rates of hybridization performed under optimal conditions are demonstrated to be inversely proportional to viscosity (reviewed by Wetmur[18]). Therefore, alterations in the viscosity of reactions caused by varying amounts of RNA, especially at high concentrations (15–20 mg per ml), could significantly affect both the rate of the reaction and the complexity estimate. This effect would be most marked with complex RNA populations (e.g., brain RNA). In practice, however, little allowance has been made for the effect of viscosity on rates of reaction.

4. SPECIFICITY OF HYBRIDIZATION

During hybrid formation, base pairs could differ from perfect complementarity at randomly scattered individual bases or in short regions interspersed with well-matched segments. The extent to which imperfectly base-paired structures are discriminated against is dependent on the reaction conditions and is proportional to the T_i and inversely proportional to salt concentration. Stringent hybridization reactions are generally conducted in 0.18 M Na^+ at 60°C or 0.6 M Na^+ at 70°C and permit approximately 10% base-pair mismatch.

Base-pair mismatch reduces the thermal stability of hybrids. The amount of noncomplementarity contained in hybrids can be estimated by determination of the mean thermal denaturation temperature (T_m). Measurements obtained from model systems indicate that 1.0–1.5% base-pair mismatch reduces the hybrid T_m approximately 1°C.[4,19] Therefore, comparison of the T_m of RNA–DNA hybrids to that of an appropriate standard (e.g., native DNA sheared to a similar fragment length)[20] permits the routine quantitative assessment of hybrid fidelity.

5. USE OF ORGANIC REAGENTS

Hybridization reactions containing complex RNA populations often require prolonged incubation times to reach completion (apparent saturation). Lengthy exposure of nucleic acids to elevated temperatures (60–70°C), however, results in considerable depurination and strand scission. In an effort to reduce thermal degradation, hybridization has been conducted in solutions containing organic reagents such as formamide, dimethylsulfoxide, and urea. These agents disrupt hydrogen bonds and decrease the stability of DNA duplexes. The effects of formamide on the stability and renaturation kinetics of DNA were systemically studied by Hutton.[21] Increasing the concentration of formamide from 0 to 50% lowers the T_m of DNA by 0.6°C/percent formamide at Na$^+$ concentrations ranging from 0.035 M to 0.88 M. Thus, the T_m of rat DNA (89°C in 0.18 M Na$^+$) would be reduced to 59°C in 50% formamide [89°C − 0.6(50)]. Given that optimal hybridization rates are achieved at approximately 25°C below T_m, reactions containing 50% formamide could be incubated at 34°C (59°C–25°C).

In principle, maximal hybridization rates and acceptable specificity could be achieved by judicious selection of the reaction temperature and concentrations of salt and formamide. Conditions for RNA–DNA hybridization in formamide at reduced temperatures (25–45°C) have been described.[22,23] However, formamide has the negative effect of increasing the viscosity of the reaction mixture, reducing the hybridization rate two- to fourfold compared to aqueous salt solutions without formamide.[21,23] For this reason, the use of organic reagents has been limited despite the apparent advantage that reduced incubation temperatures might afford.

6. HYBRIDIZATION ASSAYS

The fraction of labeled DNA tracer remaining unreacted (D/D_0 in equation 3) following hybridization is generally determined by either hydroxyapatite chromatography or S$_1$ nuclease assay. The principles involved in these standard assays are described below.

6.1. Hydroxyapatite Chromatography

Hydroxyapatite (HAP), a calcium phosphate complex, binds double-stranded nucleic acids while excluding or releasing single-stranded fragments.[24–26] The main factor involved in the absorption of nucleic acids to HAP is the interaction between phosphate groups of the nucleic acid and calcium on the surface of the HAP crystals. Therefore, fractionation of single- and double-stranded DNA on HAP is based, in part, on differences in the density and distribution of phosphate groups in the nucleic acid fragments.

To determine hybridization values, reaction mixtures are passed over HAP in 0.12 M Na-phosphate buffer (0.18 M Na$^+$) at 60°C. Under these conditions, duplexes bind to HAP, while single-stranded material passes through the column. Hybrids are recovered from HAP with 0.4 M Na-phosphate or by elevating the column temperature to 97°C. Measurement of the radioactivity in column-bound and unbound fractions yields the total amount of labeled DNA in the sample and the DNA fraction in RNA–DNA hybrids. Self-renaturation of the DNA tracer (DNA–DNA hybrids) in hybridization mixtures ("blank values") is monitored by a low-salt RNase procedure in which RNA–DNA hybrids are eliminated while DNA duplexes remain intact.[27] Blank values are also obtained by assay of reaction mixtures in which the RNA is destroyed by RNase treatment or alkaline hydrolysis prior to hybridization.[28]

Under standard conditions, RNA–DNA hybrids less than 40–50 base pairs in length do not bind efficiently to HAP. Thus, the use of short DNA probes (100–150 nt) can lead to significant underestimates of hybridization. Conversely, use of long DNA tracers may yield overestimates of hybridization since partial duplex structures will also bind to HAP. In our hands, DNA tracers of about 250–350 nt seem to yield the most reliable and consistent hybridization values. The advantages provided by HAP include the ease with which one can (1) maintain specific criteria with regard to temperature and salt concentration, (2) recover, intact, both unreacted and hybridized DNA tracer, and (3) monitor the T_m of the hybrids formed.

6.2. S_1 Nuclease Assay

In this procedure, unreacted DNA tracer is hydrolyzed by exposure to a single-strand-specific endonuclease from *Aspergillus oryzae*.[29] The specificity of S_1 nuclease for single-stranded DNA has been rigorously demonstrated.[30–32] Enzyme preparations containing little contaminating double-strand activity are now commercially available. Hybridization reaction mixtures are adjusted to the appropriate concentrations of salt (0.15–0.3 M Na^+), zinc (1 mM), and pH (4.5) required for optimal enzyme activity,[33,34] incubated at 37°–45°C, and S_1-resistant nucleic acid precipitated with trichloroacetic acid. Alternatively, nuclease-resistant hybrids can be bound to DEAE-cellulose filters[35,36] which quantitatively bind fragments as small as 30 nucleotides. An important feature of the filter assay is that nuclease-digested DNA can be efficiently washed from the filters so that background levels of less than 1.0% of the input DNA tracer can be achieved. In contrast, acid precipitation procedures usually provide background levels of 5–10%. The ability to achieve low background is especially important in studies using single-copy DNA probes (Section 7.1) where often less than 10% of the input DNA is reacted at saturation. A second advantage of the DEAE-cellulose filter assay is that hybrids can be recovered intact from the filters for further analysis.[35]

6.3. Comparison of Methods

A subtle but fundamental difference exists between the two hybridization assays. Whereas the S_1 nuclease procedure measures the fraction of DNA in complete duplex structures, HAP measures the fraction of DNA that is totally single-stranded. Complexity estimates of brain RNA obtained by HAP or S_1 nuclease are shown in Table I. Consistent with these findings, Colman *et al.*[37] obtained hybridization values for brain poly(A+) mRNA using S_1 nuclease that were 80–85% those determined by HAP. Therefore, under the appropriate conditions, estimates of RNA complexity derived from these methods are in close agreement. Nevertheless, the S_1 nuclease assay appears the method of choice, in part because it is more rigorous and less time consuming. It should

Table I
Complexity of RNA from Mouse Brain Assayed by HAP or S_1 Nuclease[a]

RNA	Methods of assay[b]	Percent scDNA as hybrid	Complexity[c] (nt)
Poly(A+) mRNA	S_1 nuclease	3.9	1.25×10^8
Poly(A+) mRNA	HAP	4.1	1.33×10^8
Poly(A+) mRNA	HAP	3.6	1.17×10^8
Total nuclear RNA	S_1 nuclease	18.0	5.85×10^8
Total nuclear RNA	HAP	19.0	6.17×10^8
Total nuclear RNA	HAP	19.9	6.46×10^8

[a] Reproduced from Hahn *et al.*[166]
[b] Close agreement between HAP and S_1 nuclease assay of RNA–DNA hybrids does not necessarily imply complete base pairing of the hybridized DNA as measured by HAP chromatography. In the S_1 nuclease procedure, short duplexes, which are effectively measured, are bound with poor efficiency to HAP. In measurements by HAP chromatography, non-base-paired regions in the hybrid are included. These two aspects may be quantitatively similar, thus accounting for the close agreement of values obtained by the two methods.
[c] Complexity estimate calculated on the basis that the single-copy fraction of the mouse haploid genome is 3.25×10^9 nt.

be noted, however, that in contrast to the agreement obtained when estimating initial and terminal hybridization values, kinetic data obtained for DNA renaturation (second-order reaction) by the two methods differ significantly.[38,39]

7. HYBRIDIZATION PROBES

Two general approaches are used to determine the complexity of RNA populations. The first involves the use of single-copy genomic DNA (saturation hybridization), whereas the second employs complementary DNA copies of RNA populations or specific RNA species (cDNA analysis). Since both experimental approaches involve the hybridization of trace amounts of radiolabeled DNA to vast excesses of unlabeled RNA, the hybridization reactions follow pseudo-first-order kinetics (Section 2).

7.1. Saturation Hybridization

Unlike prokaryotic DNA, the genomes of higher organisms contain multiple copies of closely related DNA sequences (repetitive DNA). Depending on species, the repetitive DNA fraction can comprise 20–80% of the total genomic DNA.[40,41] Early DNA-driven hybridization experiments established that the majority of heterogeneous nuclear RNA (hnRNA)[42–45] and messenger RNA (mRNA)[46–48] was transcribed from DNA sequences found in one or a few copies per haploid genome (single-copy DNA). Therefore, direct estimates of RNA sequence complexity are routinely made using single-copy DNA (scDNA). Single-copy DNA is obtained by renaturing sheared genomic DNA to C_0t values at which the repetitive sequences have annealed but single-copy fragments remain single-stranded. Removal of double-stranded DNA (repetitive fraction) is then accomplished by HAP chromatography. A typical experimental protocol for the isolation of scDNA is illustrated in Fig. 1.

Genomic DNA or scDNA can now be radiolabeled *in vitro* to high specific activity by nick[49–51] or gap translation.[52] This is a significant advantage, making it easier to attain the necessary RNA excess to drive the hybridization reaction to completion. High RNA driver:DNA tracer ratios are especially important in investigations of complex RNA populations, since individual RNA species are often present at very low frequencies. Generally, nick translation provides DNA probes of greater length and reactivity and therefore appears to be the labeling method of choice.

Typically, hybridization data are expressed as the fraction of scDNA in RNA–DNA hybrids as a function of $R_0 t$, i.e., RNA $C_0 t$ (see Fig. 2). In practice, the RNA sequence complexity is determined directly from the amount of scDNA in hybrid form at the termination of the reaction (apparent saturation). For example, whole brain polyadenylated [poly(A+)] mRNA hybridizes to approximately 4.0% of the scDNA at saturation (Fig. 2). Assuming that gene transcription is asymmetric (involves only one DNA strand), the experimental value represents 8% of the available DNA tracer. Given that the complexity of the rat single-copy genome is about 1.96×10^9 nt (complexity of the rat haploid genome is 2.8×10^9 nt of which 70% is single-copy DNA),[53,54] the RNA complexity is 1.56×10^8 nt ($1.95 \times 10^9 \times 0.08$). Assuming that the average size of brain poly(A+) mRNA is 1500 nt,[55] the sequence complexity of this RNA population is the equivalent of 104,000 different mRNA sequences

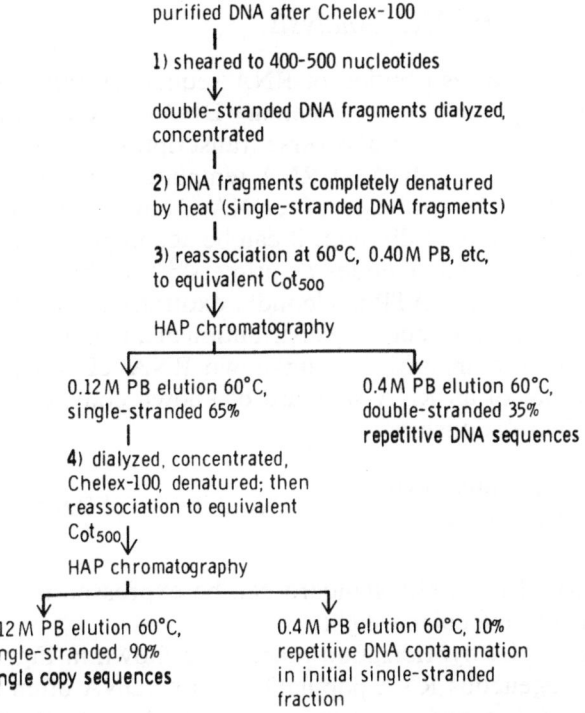

Fig. 1. A typical experimental protocol for the isolation of scDNA. Modified from Church.[3]

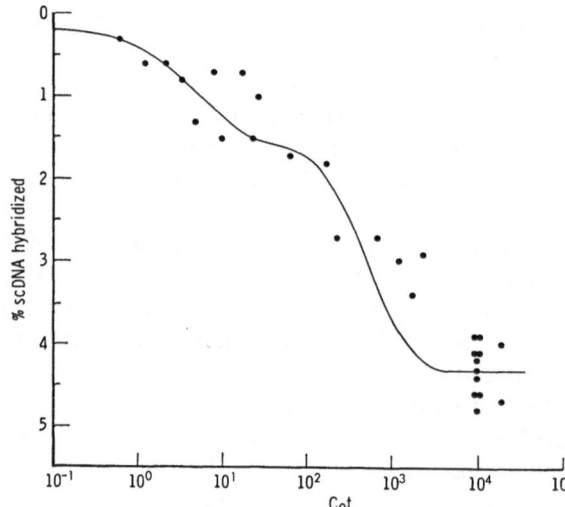

Fig. 2. Kinetics of hybridization of rat cerebellar poly(A+) mRNA to [³H]-labeled scDNA. Hybridization values were obtained by HAP chromatography, and the line through the data was graphed using a least-squares computer program.[67]

$(1.56 \times 10^{8}/1.5 \times 10^{3})$. Identical calculations have been used to estimate the sequence complexity of nuclear RNA.

7.2. Complementary DNA Analysis

The number and distribution of RNA sequences can also be determined by hybridizing radiolabeled complementary DNA (cDNA) with excess template RNA.[56] In this approach, a viral reverse transcriptase[57,58] is used to synthesize a cDNA probe from a poly(A+) RNA template *in vitro* using an oligo(dT) primer.[59–61] Although synthesis of cDNA from RNA lacking 3′-terminal poly(A) sequences is more difficult, it can be accomplished using a mixture of random oligodeoxyribonucleotides as primers[62,63] or by enzymatic polyadenylation using the enzyme ATP nucleotidyl exotransferase.[64–66]

The complexity and frequency distribution of an unknown RNA population are determined by comparing the unknown RNA–cDNA hybridization rate constant with that of an RNA standard of known complexity with its cDNA according to the relationship:

$$G = \frac{C_0 t_{1/2} \text{ unknown RNA}}{C_0 t_{1/2} \text{ standard RNA}} \times \text{(complexity of RNA standard)}$$

The complexity of the RNA standard can be expressed in either nucleotides or the equivalent molecular weight.

A typical cDNA hybridization reaction is shown in Fig. 3. Reaction of a complex, heterogeneous RNA population to its cDNA often takes place over 3–5 log units of $C_0 t$ and can contain several discrete kinetic components. According to ideal pseudo-first-order kinetics, the reaction of a single RNA component to its cDNA should reach completion within 1.5 decades of C_0t. Deviation from ideal kinetics indicates heterogeneity in the concentration of RNA sequences driving the reaction. Resolution of the reaction into its individual kinetic components is most often accomplished by computer analysis using a nonlinear least-squares program.[67] The presence of discrete abundance classes in poly(A+) mRNA is widespread[68–70] but by no means a universal occurrence. The concept of mRNA abundance classes has been critically reevaluated by Quinlan *et al.*[71]

Visual inspection of the data in Fig. 3 suggests the presence of three transitions or abundance classes. The fraction of cDNA in each of the transitions and its $C_0t_{1/2}$ value are presented in Table II. Given the average size and amount of mRNA per cell, the sequence complexity of each abundance class can be expressed in terms of the number of different RNA sequences per cell (Table II, column 6) and RNA copy frequency (Table II, footnote *f*).

As mentioned above, calculation of sequence complexity represented by each transition is accomplished with reference to a suitable kinetic standard. Therefore, the value obtained for the absolute number of sequences in each transition (Table II, Column 6) may be dependent on the standard selected (see, for example, Savage *et al.*[72]). However, both the fraction of cDNA represented in each transition and the relative number of sequences are independent of the standard chosen. Obviously, the accuracy of the RNA complexity estimate depends, in part, on the reliability of the kinetic standard. Since the size of the RNA (driver) and cDNA probe (tracer) may affect the reaction rate (see Section 3.3), it is preferable to select a similar sized RNA and cDNA as standard. It is also desirable to determine the hybridization kinetics of the standard under the same condition as the test system to avoid introducing errors stemming from normalization of the data.

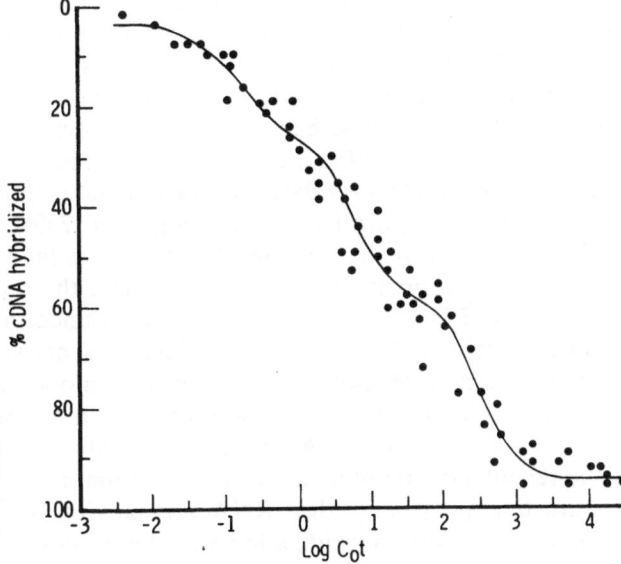

Fig. 3. Hybridization of rat liver cDNA and its poly(A+) mRNA template. The reactions were analyzed using S_1 nuclease, and solid line is a computer-generated nonlinear least-squares fit of the data. Reproduced from Savage *et al.*[72]

Table II

Frequency Classes in Rat Liver Poly(A +) mRNA[a]

Component	Percent cDNA hybridized[b]	C_0t_i Observed	C_0t_i Corrected[c]	Sequence complexity[d] (nt)	No. of sequences[e]	RNA copies/cell[f]
1	18.7	0.09	0.017	6.0×10^3	4	7400
2	36.6	3.7	1.35	4.4×10^5	293	200
3	44.6	246	110	3.5×10^7	23,000	3

[a] Obtained from a computer best-fit analysis of data shown in Fig. 3. From Savage *et al.*[72]

[b] Final hybridization value of the reaction is 94%. The fraction of cDNA in each transition is normalized for cDNA probe reactivity (% cDNA hybridized ÷ 0.94).

[c] Calculated from observed $C_0t_{1/2}$ assuming the component was pure and comprised all the reacting RNA sequences observed ($C_0t_{1/2}$ × fraction cDNA hybridized).

[d] Using ovalbumin mRNA as the complexity standard.[167] This mRNA has a complexity of 1,900 nt and hybridizes to its cDNA with a rate constant of 150 liter mol^{-1} sec^{-1} ($C_0t_{1/2} = 4.6 \times 10^{-3}$).

[e] Calculated assuming that the number average size rat poly(A +) mRNA is 1500 nt (complexity ÷ average size).

[f] The frequency of sequences in each abundance class is given by:

$$\frac{[\text{RNA content per cell } (\mu g)] \, [330 \times 10^6 \, \mu g \, (\text{mole of nucleotide})]}{6 \times 10^{23} \, (\text{Avogadro's number})}$$

Which is the number of nt per cell poly(A +) mRNA. This number is then multiplied by the fraction of DNA hybridized in each class and divided by the base sequence complexity. Here, the amount of rat liver poly(A +) mRNA/cell is taken 0.13 pg.[168]

7.3. Comparison of Experimental Approaches

The base sequence complexity of poly(A +) mRNA from several mammalian cell types and tissues as evaluated by saturation hybridization and cDNA kinetic analysis is given in Table III. In most cases, the complexity estimates obtained with cDNA are two- to fourfold less than values obtained by saturation hybridization. Although cDNA kinetic analysis yields invaluable data concerning the distribution and relative abundance of RNA sequences in the population, there are several drawbacks to using this technique for determining the overall complexity of the RNA population. First, there is still some uncertainty as to whether the AMV reverse transcriptase transcribes all mRNAs with equal efficiency.[73,74] Second, the cDNAs used in the analysis are often small (*ca.* 300 nt) relative to the template RNA and only partially reactive in RNA-driven hybridization reactions (*ca.* 75%). If the cDNA tracer is not fully reactive or if the complex RNA class is inadequately represented in the cDNA, it is difficult to accurately estimate the $C_0t_{1/2}$ of the complex, slowly hybridizing RNA component. This creates uncertainty in the final complexity values obtained. The above considerations notwithstanding, relatively good agreement in complexity estimates has been obtained with scDNA and cDNA measurements in instances where the length of the cDNA tracer approached that of the template RNA and cDNA reactivity was greater than 90%.[72,75,76] The recent agreement achieved between these two methods argues against the interesting possibility that RNA complexity is routinely overestimated by scDNA saturation hybridization as recently posited by Kiper.[77] The ability to synthesize full-length cDNA is apparently a function of the concentration of deoxyribonucleoside triphosphates,[78–80] Mg^{2+},[81] and reverse transcriptase[82] in the *in vitro* reaction mixture. Inclusion of sodium pyrophosphate in the reaction mixture is also reported to increase the yield of full-length cDNA copies,[83] ostensibly by inhibiting RNase present in the enzyme preparation.

Perhaps the most significant aspect of cDNA analysis is its application to the study of the expression and structure of specific eukaryotic genes. Specific mRNAs have been isolated by immunochemical[84–90] and/or physiocochemical procedures (see for example Rosen[91]; Ross[92]) and cDNA probes prepared. At the single-gene level, cDNA analysis has been used to study (1) the regulation of globin gene expression[93–95]; (2) hormone induction of egg white proteins[96–100]; (3) estrogen induction of vitellogenin mRNA[101] and uteroglobin mRNA synthesis[102]; (4) effect of gestation and lactation on the synthesis of mammary gland milk protein mRNAs[103] and human placental lactogen mRNA[104]; (5) expression of protamine mRNA during trout testis development[105]; and (6) synthesis of crystallin mRNA during lens induction in chick embryo.[106] The

Table III

Comparison of scDNA and cDNA Complexity Estimates[a]

| Species | RNA source | Number of different mRNAs | | References |
		scDNA	cDNA	
Rat	Glioma C-6	65,000	33,000	153, 169
	Neuroblastoma (NS20)	51,000	8,000	170–172
	Myoblasts		29,000	177
	Liver	50,000		173
		43,000	23,000	72, 175
			15,000	174
			12,000	168, 176
Mouse	Friend cells	32,000	12,000	178, 179
			8,000	180, 181
	Myoblasts	24,000	22,000	182
			17,000	181
	PCC3 cells[b]	12,000	10,000	183
			8,000	181
	Liver	43,000	39,000	76
			16,000	69
			10,000	68
	Brain	100,000	73,000	55, 75
			30,000	70
			19,000	69
			15,000	68
			12,000	184
Chicken	Myoblasts	24,000	22,000	182
	Liver	24,000	15,000	185
	Oviduct	21,000	19,000	100
			18,000	185

[a] Sequence complexity estimates converted to mRNA species assuming number average mRNA size 1500 nt.
[b] Embryonal carcinoma cell.

screening of cloned genomic DNA libraries for specific gene sequences also relies heavily on the synthesis of cDNA probes from specific mRNAs (for review see Carbon *et al.*[107]; Maniatis *et al.*[108]; Maniatis[109]).

8. COMPLEXITY OF GENE EXPRESSION IN BRAIN

8.1. Transcriptional Level

Heterogeneous nuclear RNA consists of two major populations of RNA, one containing a poly(adenylic acid) sequence approximately 100–200 nucleotides in length at the 3′-terminus[110–115] and a second RNA fraction that is nonadenylated. Polyadenylation is an early posttranscriptional event the biological function of which is still unclear. Regardless of the significance of the poly(A) sequence, it has proven a valuable tool in the isolation of many mRNA species and their precursors. Purification of poly(A+) RNA is routinely accomplished by affinity chromatography using oligo(dT)-cellulose[116] or poly(U)-Sepharose.[117,118] The sequence complexity of brain nuclear RNA and its subpopulations is considered below.

8.1.1. Total Nuclear RNA

Results of early saturation hybridization analysis of brain nuclear RNA from mouse[28,119,120] and rabbit[121] suggested that at least twice as much scDNA was transcribed in the brain as in other organs. For example, mouse brain nuclear RNA hybridized to approximately 10–12% of the scDNA, whereas liver and kidney nuclear RNA was complementary to 4–5%.[28] Recent estimates of brain nuclear RNA complexity obtained under more advantageous conditions are even larger than these pioneering studies indicated (Table IV). Assuming asymmetric transcription, 30–40% of the single-copy genome is transcribed in the brain, a value equivalent to 130,000–180,000 different hnRNA sequences averaging 4,500 nucleotides in length (Table IV, footnotes *b* and *c*). Presumably, the early hybridization studies underestimated brain nuclear RNA complexity because of (1) a selective loss of poly(A+) RNA sequences during RNA isolation, (2) suboptimal hybridization conditions which frequently led to a breakdown of RNA and DNA with premature termination of the reaction, and (3) difficulties in reaching maximum hybridization values in the presence of contaminating ribosomal RNA (rRNA).

The extent of similarity (sequence homology) in the RNA sequences transcribed in various tissues was estimated by hybridizing scDNA to mixtures of nuclear RNA from different organs. At saturation, hybridization of mixtures of brain and liver or brain and kidney RNA yields values equal to that obtained for brain alone.[122] Theoretically, if brain and liver contained totally nonoverlapping populations of RNA, the complexity values would be additive or equal to the sum of values obtained for each organ alone. Conversely, if all the RNA sequences transcribed in one organ appear in another, the hybridization values would not summate but would reflect the value of the organ with the higher-

Table IV
The Base Sequence Complexity of Whole-Brain RNAs

RNA source	Species	Fraction of scDNA hybridized (%)[a]	Sequence complexity (nt)[b]	Number of different sequences[c]	References
Total cellular RNA[f]	Mouse[e]	18.8	7.3×10^8	163,000	188
	Cat[e]	14.8	5.8×10^8	129,000	186
	Rat[e]	16.4	6.4×10^8	143,000	123
Nuclear RNA	Mouse[e]	17.2	6.5×10^8	144,000	36
	Mouse[d]	21.2	8.1×10^8	179,000	55
	Rat[e]	15.6	5.9×10^8	131,000	122
	Rat[e]	16.9	6.4×10^8	142,000	123
Total poly(A+) RNA	Mouse[d]	13.3	5.0×10^8	111,000	55
	Rat[d]	12.5	4.8×10^8	107,000	53
	Rat[d]	13.0	4.9×10^8	109,000	37
Poly(A+) mRNA	Mouse[d]	3.8	1.4×10^8	100,000	55
	Rat[e]	3.2	1.2×10^8	86,000	123
	Rat[d]	4.8	1.8×10^8	129,000	122
	Rat[d,e]	4.4	1.7×10^8	121,000	37
	Rat[d]	4.6	1.7×10^8	121,000	155
Poly(A−) mRNA	Mouse[d]	3.6	1.4×10^8	100,000	13
	Rat[d,e]	4.0	1.5×10^8	107,000	127

[a] Hybridization values given for each study are corrected for the reactivity of the [^{3}H]DNA tracer.
[b] Complexity estimates assume that DNA transcription is asymmetric and that the rodent single-copy DNA represents 70% of a haploid genome comprising 2.8×10^9 nucleotides.
[c] Values obtained considering the number average length of brain hnRNA and mRNA are 4500 and 1400 nucleotides, respectively.[55]
[d] Hybridization values determined by hydroxyapatite chromatography.
[e] Hybridization values estimated by S_1 nuclease assay.
[f] A computer extrapolation to saturation from hybridization kinetic data.

complexity RNA. These data suggest that most nuclear RNAs present in liver and kidney also occur in brain. It follows that gene transcripts required for brain-specific function are present in addition to an RNA population that is apparently shared by several less complex organs.

Brain nuclear RNA exists in different abundance classes. In an attempt to define the RNA frequency classes using saturation hybridization, Grouse et al.[123] described three major abundance classes. The most rapidly hybridizing component representing a significant portion of the RNA mass contributed only 6% of the total RNA diversity. The middle component, comprising RNA species present at much lower abundance, contained 27% of the total diversity, and the least abundant species contained the major portion of the nuclear RNA sequence complexity.

8.1.2. Polyadenylated Nuclear RNA

Poly(A+) hnRNA constitutes about 10–12% of the mass of nuclear RNA and has a sequence complexity of approximately 5.0×10^8 nt (Table IV), a value representing 65–75% of the total nuclear RNA complexity.[37,53,55] Ample evidence indicates that much of the hnRNA serves as a precursor to cytoplasmic messengers on average two to five times smaller than the nuclear transcript (reviewed by Monahan[124]; Darnell[125]; Tobin[126]). Recently, Hahn *et al.*,[75] using cDNA synthesized from poly(A +) RNA fragments cleaved from large poly(A +) hnRNAs, demonstrated that the vast majority of the sequences adjacent to the 3'-terminus of poly(A +) hnRNA appear in the poly(A +) mRNA population in brain.

The percentage of the nuclear RNA mass that drives the hybridization reaction has been estimated by comparing the observed pseudo-first-order rate constant (k_{obs}) with that expected (k_{exp}) if all RNA sequences in the population participated equally in the reaction (see Section 2, equation 6). In brain, early studies indicated that approximately 5% of the nuclear RNA[55] and 5–10% of the poly(A +) hnRNA mass[53] drives the hybridization reaction. Recent evidence suggests, however, that values derived using scDNA probes underestimate the fraction of total RNA that drives the hybridization reaction (for discussion, see Van Ness and Hahn[76]). For example, kinetic data obtained with scDNA indicate that approximately 5–10% of the poly(A +) mRNA drives the reaction, in contrast to a value of 40% obtained by cDNA analysis. This discrepancy results, in part, from a retardation of hybridization rate apparently caused by sequence discontinuity in scDNA and RNA driver (J. Van Ness and W. E. Hahn, unpublished observations): scDNA fragments (350 nt) are frequently a mixture of coding and noncoding (intervening) sequences. Although the mechanism whereby sequence discontinuity affects hybridization is unclear, driver values obtained with scDNA probes may misrepresent the driver fraction by as much as four- to eightfold.

8.2. Translational Level

8.2.1. Total Polysomal RNA

The complexity of whole-brain polysomal RNA has been compared to that of other organs by saturation hybridization.[127] At maximum hybridization, total-brain polysomal RNA was complementary to 8% of the scDNA, a finding consistent with the remarkable complexity manifest in brain nuclear RNA. Assuming asymmetric transcription, this value represents sufficient information to code for 190,000 different proteins averaging 45,000 daltons. By comparison, liver and kidney mRNA contained information to code for the equivalent of 57,000 and 39,000 different proteins, respectively. Similar findings were obtained in mouse brain[13] with special care given to obtain polysomes free of hnRNA contamination (see below).

8.2.2. Polyadenylated mRNA

The base sequence complexity of brain poly(A +) mRNA is summarized in Table IV. About half of the complexity of polysomal mRNA resides in the polyadenylated RNA fraction. Analysis of brain polysomal poly(A +) mRNA by DNA-excess hybridization indicates that the majority of brain mRNA is complementary to scDNA, with a small kinetic component complementary to repetitive DNA sequences.[128] The hybridization kinetics of poly(A+) mRNA to scDNA suggests that there are at least three RNA abundance classes[55,123] (also see Fig. 2). The most abundant mRNA sequences hybridize rapidly and contain about 10–20% of the total poly(A+) mRNA sequence complexity. These abundant poly(A+) mRNA sequences are believed present in most cells and include the so-called "housekeeping" genes that are requisite for basic function.[129]

The slower kinetic fractions include the intermediate and rare RNA abundance classes comprising 70–80% of the total sequence complexity. The rare RNA sequences may represent brain-cell-specific and/or region-specific gene transcripts.

Comparison of liver and kidney poly(A+) mRNA populations by cDNA kinetic analysis indicates that they also contain at least three major frequency classes.[69,72] Interestingly, there are significant tissue-specific differences in the complexity and composition of the most abundant class of messengers. For example, a liver cDNA probe enriched for rapidly hybridizing, abundant sequences cross hybridized with brain poly(A+) mRNA with kinetics indicative of a rare RNA sequence class. Thus, some gene transcripts present in liver in great abundance are also expressed in brain, but in greatly reduced amounts. Similar conclusions were reached in an independent cDNA analysis conducted by Young *et al.*[70] These findings are also consistent with the hybridization data of Chikaraishi and associates[122] who demonstrated that most nuclear RNA sequences present in liver were also transcribed in brain. Taken together, these studies indicate that the establishment of organ-specific function involves both quantitative and qualitative alterations in the pattern of gene expression.

8.2.3. Nonadenylated mRNA

In the past, it was sometimes assumed that the poly(A+) mRNA class represented the entire mRNA population. Recent data show, however, that poly(A−) mRNAs comprise as much as 30–50% of the mass of nascent mRNA.[130–132] Poly(A−) mRNA, like its adenylated counterpart, has an AU-rich base composition, 5'-terminal cap structure,[133,134] is released from polyribosomes by EDTA and puromycin,[130] is associated with protein,[135] and serves as a template for protein synthesis when assayed in a heterologous cell-free translation system.[136] The rates of metabolic turnover of these two classes of mRNA also appear similar. Cross-hybridization experiments in which DNA complementary to poly(A+) mRNA is reacted with excess polysomal poly(A−) RNA demonstrate that polysomal poly(A−) RNA does not arise from poly(A+) mRNA by nonadenylation, deadenylation, or random degradation.[130]

A highly diverse poly(A−) mRNA population exists in the brain. Hybridization of rat brain polysomal poly(A−) mRNA to scDNA yields complexity values equal to those obtained for poly(A+) mRNA.[13,127] Addition–hybridization experiments with polysomal poly(A−) RNA and poly(A+) mRNA clearly demonstrate little overlap in these RNA populations, i.e., that they contain few sequences in common. That little sequence homology exists between the two mRNA populations was also demonstrated by hybridizing cDNA synthesized from poly(A−) RNA to excess poly(A+) mRNA obtained from formaldehyde-fixed polysomes isolated by CsCl gradient centrifugation.[13] In support of the hybridization data, most translation products synthesized in a heterologous cell-free system derived from rabbit reticulocytes were different with respect to poly(A+) mRNA vs. poly(A−) mRNA templates, as analyzed by two-dimensional gel electrophoresis.[127] However, several proteins, the most abundant being actin, were present in the translaton products of each messenger population. That the poly(A+) and poly(A−) mRNA fractions code for qualitatively similar sets of abundant proteins was also observed in sea urchin embryos,[137] HeLa,[138] and mouse Friend cells.[139]

9. CELL AND REGIONAL DISTRIBUTION OF GENE TRANSCRIPTS

9.1. Cell Lines of Neuroectodermal Origin

The striking complexity of brain mRNA is generally interpreted as reflecting the heterogeneity of cell types in the tissue. The possibility arises, however, that neuroectodermal derivatives, like the brain itself, express an unusual amount of the haploid genome. As shown in Table V, nuclear poly(A+) RNA from neuroblastoma and glioma contains about 75% of the total complexity of brain nuclear RNA, whereas poly(A+) mRNA contains about 60% of that from whole brain. Complexity estimates for other somatic cell lines are also summarized in Table V. These results suggest that cultured cells of neural origin, like brain, express 1.5- to 3.0-fold more genetic information than other somatic cell types.

Given that these results are derived from transformed cells, the findings should be interpreted with some caution. Although transformation of cultured cells by chemical[140–142] or viral agents[143,144] does not result in major qualitative alterations in gene expression, elucidation of the diversity of gene activity in nerve cells will ultimately require the isolation of mRNA from a number of specific brain cell types.

9.2. Major Brain Regions

Comparisons have been made of the sequence complexity of nuclear RNA and total poly(A+) RNA from several brain regions differing markedly in cell composition, neuronal/glial ratio, structure, and function. In general, a surprising degree of homology (>80%) is observed in the nuclear RNA and poly(A+) RNA sequences present in cerebral cortex, cerebellum, hypothalamus, and hippocampus.[53,145] It is important to point out, however, that considering the resolution of the hybridization assay and complexity of the RNA populations involved, regional differences in the order of a few thousand se-

Table V
Sequence Complexity of Polyadenylated RNA from Cultured Mammalian Cells

Cell type	Species	scDNA hybridized (%)	Complexity (nt)	References
Nuclear poly(A+) RNA				
Neuroectodermal				
Neuroblastoma (B104)	Rat	8.7	3.1×10^8	53
Glioma (C6)	Rat	9.3	3.4×10^8	53
Other				
Friend cells (M2)	Mouse	5.4	1.9×10^8	178
PCC3 cells[a]	Mouse	2.5	0.9×10^8	183
Syrian embryo cells[b]	Hamster	2.8	1.0×10^8	141
Cytoplasmic poly(A+) mRNA				
Neuroectodermal				
Neuroblastoma (B104)	Rat	2.4	8.6×10^7	153
Neuroblastoma (NS20)	Mouse	2.1	7.6×10^7	170
Glioma (C6)	Rat	2.7	9.7×10^7	153
Other				
Friend cells (M2)	Mouse	1.3	4.7×10^7	179
PCC3 cells[a]	Mouse	0.5	1.8×10^7	183
Syrian embryo cells[b]	Hamster	0.7	2.5×10^7	141
Fibroblasts (PYAL/N)	Mouse	1.0	3.6×10^7	187

[a] Embryonal carcinoma cell.
[b] Complexity estimates calculated assuming hamster single-copy DNA is 4.16×10^9 nucleotides.

quences might go undetected. Therefore, the possibility that region-specific differences in gene transcription exist cannot be entirely eliminated.

Although gene transcription in different brain regions appears similar, posttranscriptional mechanisms could regulate region-specific gene expression by controlling the sequences that ultimately appear in the cytoplasm. There is now good evidence to indicate that posttranscriptional mechanisms are involved in the regulation of organ-specific gene expression in sea urchin embryos,[146,147] frog embryos,[148] and plants.[149] At present, there is little information on the regional compartmentalization of brain mRNA. Preliminary data indicate that poly(A+) and poly(A−) mRNA from rat cerebral cortex have a 10% greater complexity than that of cerebellum (B. B. Kaplan, S. L. Bernstein, and A.E. Gioio, unpublished results). This modest difference in hybridization values is equivalent to sufficient information to code for 5,000–10,000 different proteins, averaging 50,000 daltons. Clearly, more information is required on both the poly(A+) and poly(A−) mRNA populations before meaningful conclusions concerning the regional specificity of gene expression in brain are reached.

10. GENE EXPRESSION DURING THE ANIMAL LIFE-SPAN

In view of the importance of gene expression in cell differentiation and development, it is of obvious interest to examine brain RNA sequence complexity as a function of age. Conceivably, the diversity of genetic information expressed in brain could increase as the organ acquired adult structure and function. Given this situation, one might hope to identify genes whose transcription correlates with the development of region-specific structure and function. In addition, the effect of aging on gene expression in the brain bears on some fundamental issues in neurogerontology including several neuroendocrine theories of aging (reviewed by Finch[150,151]). Information concerning the effect of development and aging on gene activity in the CNS is summarized below.

10.1. Postnatal Development

Results of early studies of small mammals indicated that the fraction of the genome transcribed increased dramatically during postnatal development.[120,121,152] Recent studies using improved RNA extraction techniques and optimal hybridization conditions have confirmed these early observations. For example, age-related increases (*ca.* 15–20%) in whole-brain poly(A+) RNA occur in rat during the first 2 weeks after birth.[153] Marked increases in the sequence complexity of nonadenylated nuclear RNA and messenger RNA also occur in mouse brain during fetal and early postnatal development.[154] This is an important new finding, as these age-related alterations in gene expression occur in an RNA population largely specific to brain.

In contrast to the findings obtained with whole brain, the complexity of rat cerebellar poly(A+) RNA decreases (*ca.* 15–20%) during postnatal development.[155] This decrement occurs during the second and third week post-partum, when cerebellar cell proliferation and differentiation near completion.[156] Thus, there appear to be region-specific differences in the developmental program of gene expression in brain.

In addition to qualitative alterations in gene activity, development may also involve altered rates of mRNA production and processing, resulting in changes in the relative abundance of mRNA sequences. This situation appears to characterize the resting and proliferating states of cells in tissue culture. In this case, cells regulate their poly(A+) mRNA content by controlling the efficiency with which nuclear poly(A+) RNA is processed to cytoplasmic messengers.[157,158] Thus, future determination of brain mRNA frequency distribution by cDNA kinetic analysis might prove fruitful.

10.2. Aging Studies

In an extensive study of poly(A+) RNA from rat brain, no age changes were found in the yield or complexity of nuclear poly(A+) RNA or poly(A+) mRNA in two rat strains.[37] Data from addition–hybridization experiments suggest further that most poly(A+) mRNA sequences are common to all adult age groups. In accord with these findings is the study of Farquhar *et al.*,[159] who found no age-related differences in nuclear RNA complexity in cerebral cortex of female macaques.

Interestingly, the above studies revealed strain differences in RNA sequence complexity. Since strain differences in the complexity of the single-copy genome or in the extent of transcription seem unlikely, these results probably reflect differences in the relative concentration of the rare RNA sequences. Nonetheless, possible variation in brain RNA sequence complexity within a species suggests the need to use highly inbred animal models reared under standard conditions.

Although these studies indicate that no major age-related changes occur in the brain nuclear poly(A+) RNA or poly(A+) mRNA populations, they do not rule out differences in (1) a minority class of RNA sequences, (2) poly(A−) mRNA, or (3) RNA species in a minority population of brain cells. For example, two studies detected a 10–20% loss of cellular RNA (principally rRNA) in the striatum,[160,161] a brain region that contains about 2% of the brain's cells. Microspectrophotometric studies show 10–30% decreases of RNA in human cerebellar Purkinje cells[162,163] as well as in rodent neurons.[164,165] Perhaps hybridization analysis of brain regions or neurons showing reduced RNA content would reveal significant age-related alterations in gene expression.

11. CONCLUDING REMARKS

Recent application of RNA–DNA hybridization techniques to problems in neurobiology reveals a remarkable new property of mammalian brain, namely, that the brain expresses two- to fourfold more of its haploid genome than do other somatic tissues or organs. The sequence complexity of brain nuclear RNA and mRNA predicts the existence of 100,000–200,000 different brain polypeptides, greater than 95% of which, perhaps because of their relative scarcity, have not been distinguished by standard two-dimensional gel-electrophoretic analysis. The enormous complexity of brain mRNA bears on many fundamental issues in neurobiology including the ultimate function of this genetic information and the degree to which it is involved in the development and maintenance of specific connectivity in different brain cells and regions.

At the time of birth, the rodent brain has nearly achieved the adult inventory of polyadenylated nuclear and messenger RNA sequences; only modest changes are detectable at the whole-brain level or in the cerebellum despite this region's cellular and morphological immaturity at parturition. No subsequent changes in poly(A+) RNA complexity are observed throughout the remainder of the adult life-span. In contrast, major developmental alterations occur in the complexity of the nonadenylated nuclear and messenger RNA populations. These RNA sequences appear organ specific and may play an important role in specialized brain function. An important goal for future research is to define the degree to which environmental cues or experience affects the developmental program of gene expression in brain.

Studies of cultured cells of neuroectodermal origin suggest that both neurons and glia may transcribe an unusually large number of different mRNAs. The manner in which this large amount of genetic information is related to differentiated cell function(s) is unknown. A key question concerns the regional distribution of brain mRNA sequences. Although only limited data are available, most major brain regions appear to have poly(A +) RNA populations that are indistinguishable from each other or from whole brain. Perhaps future work will reveal significant region-specific differences in the poly(A −) RNA fractions.

Considering the enormous complexity of brain RNA and the limits in the resolution of RNA–DNA hybridization analysis, additional insights into the control of gene expression in brain will depend, in large measure, on the development of more specific hybridization probes. These probes will come in the near future from at least two avenues of approach: first, the purification of mRNAs for known brain-specific proteins and the isolation of the corresponding genes from an appropriate "shotgun-cloned" gene library; second, the identification and isolation of a population of DNA sequences transcribed specifically in brain or in a specific neural cell type. In either case, modern recombinant DNA technology will, no doubt, provide the tools with which to address questions of neurobiological import at the genetic level.

REFERENCES

1. Flint, S. J., 1980, *Genetic Engineering*, Volume 2 (J. K. Setlow and A. Hollander, eds.), Plenum Press, New York, pp. 47–82.
2. Lewin, B., 1980, *Gene Expression*, Volume 2, John Wiley & Sons, New York.
3. Church, R. B., 1973, *Molecular Techniques and Approaches in Developmental Biology* (M. J. Chrispeels, ed.), John Wiley & Sons, New York, pp. 223–300.
4. Britten, R. J., Graham, D. E., and Neufeld, B. R., 1974, *Methods Enzymol.* **29E:** 363–406.
5. Bishop, J. O., 1969, *Biochem. J.* **113:**805–811.
6. Birnstiel, M. L., Sells, B. H., and Purdom, I. F., 1972, *J. Mol. Biol.* **63:**21–39.
7. Hough, B. R., Smith, M. J., Britten, R. J., and Davidson, E. H., 1975, *Cell* **5:**291–299.
8. Galau, G. A., Britten, R. J., and Davidson, E. H., 1977, *Proc. Natl. Acad. Sci. U.S.A.* **74:**1020–1023.
9. Wetmur, J. G., and Davidson, N., 1968, *J. Mol. Biol.* **31:**349–370.
10. McCarthy, B. J., and Church, R. B., 1970, *Annu. Rev. Biochem.* **39:**131–150.
11. Marmur, J., Rownd, R., and Schildkraut, C. L., 1963, *Prog. Nucleic Acid Res.* **1:**231–300.
12. Nygaard, A. P., and Hald, B. D., 1964, *J. Mol. Biol.* **9:**125–142.
13. Van Ness, J., Maxwell, I. H., and Hahn, W. E., 1979, *Cell* **18:**1341–1349.
14. Wetmur, J. G., 1971, *Biopolymers* **10:**601–613.
15. Hinnebusch, A. G., Clark, V. E., and Klotz, L. C., 1978, *Biochemistry* **17:**1521–1529.
16. Hutton, J. R., and Wetmur, J. G., 1973, *J. Mol. Biol.* **77:**495–500.
17. Chamberlain, M. E., Galau, G. A., Britten, R. J., and Davidson, E. H., 1978, *Nucleic Acids Res.* **5:**2073–2094.
18. Wetmur, J. G., 1976, *Annu. Rev. Biophys. Bioeng.* **5:**337–361.
19. Laird, C. D., McConaughy, B. L., and McCarthy, B. J., 1969, *Nature* **224:**149–154.
20. Hayes, F. N., Lilly, E. H., Ratliff, R. L., Smith, D. A., and Williams, D. L., 1970, *Biopolymers* **9:**1105–1117.
21. Hutton, J. R., 1977, *Nucleic Acids Res.* **4:**3537–3555.
22. McConaughy, B. L., Laird, C. D., and McCarthy, B. J., 1969, *Biochemistry* **8:**3289–3294.
23. Schmeckpeper, B. J., and Smith, K. D., 1972, *Biochemistry* **11:**1319–1326.
24. Bernardi, G., 1971, *Methods Enzymol.* **21D:**95–139.

25. Wilson, D. A., and Thomas, C. A., 1973, *Biochim. Biophys. Acta* **331**:333–340.
26. Martinson, H. G., and Wagenaar, E. B., 1974, *Anal. Biochem.* **61**:144–154.
27. Galau, G. A., Britten, R. J., and Davidson, E. H., 1974, *Cell* **2**:9–21.
28. Hahn, W. E., and Laird, C. D., 1971, *Science* **173**:158–161.
29. Ando, T., 1966, *Biochim. Biophys. Acta* **114**:158–168.
30. Ghangas, G. S., and Wu, R., 1975, *J. Biol. Chem.* **250**:4601–4606.
31. Shenk, T. M., Rhoades, C., Rigby, P. W., and Berg, P., 1975, *Proc. Natl. Acad. Sci. U.S.A.* **72**:989–993.
32. Weigand, R. C., Godson, G. N., and Radding, C. M., 1975, *J. Biol. Chem.* **250**:8848–8855.
33. Sutton, W. D., 1971, *Biochim. Biophys. Acta* **240**:522–531.
34. Vogt, V. M., 1973, *Eur. J. Biochem.* **33**:192–200.
35. Salzberger, S., Levi, Z., Aboud, M., and Goldberger, A., 1977, *Biochemistry* **16**:25–29.
36. Maxwell, I. H., Van Ness, J., and Hahn, W., 1978, *Nucleic Acids Res.* **5**:2033–2038.
37. Colman, P. D., Kaplan, B. B., Osterburg, H. H., and Finch, C. E., 1980, *J. Neurochem.* **34**:335–345.
38. Morrow, J., 1974, Ph.D. dissertation, Stanford University, Stanford.
39. Smith, M. J., Britten, R. J., and Davidson, E. H., 1975, *Proc. Natl. Acad. Sci. U.S.A.* **72**:4805–4809.
40. Britten, R. J., and Kohne, D. E., 1968, *Science* **161**:529–540.
41. Davidson, E. H., and Britten, R. J., 1973, *Q. Rev. Biol.* **48**:565–613.
42. Perry, R. P., Greenberg, J. R., and Tartof, K. D., 1970, *Cold Spring Harbor Symp. Quant. Biol.* **35**:577–588.
43. Greenberg, J. R., and Perry, R. P., 1971, *J. Cell Biol.* **50**:774–786.
44. Smith, M. J., Hough, B. R., Chamberlin, M. E., and Davidson, E. H., 1974, *J. Mol. Biol.* **85**:103–126.
45. Holmes, D. S., and Bonner, J. B., 1974, *Biochemistry* **13**:841–848.
46. Campo, M. S., and Bishop, J. O., 1974, *J. Mol. Biol.* **90**:649–663.
47. Klein, W. H., Murphy, W., Attardi, G., Britten, R. J., and Davidson, E. H., 1974, *Proc. Natl. Acad. Sci. U.S.A.* **71**:1785–1789.
48. Spradling, A., Penman, S., Campo, M. S., and Bishop, J. O., 1974, *Cell* **3**:23–30.
49. Rigby, P. W. J., Dieckmann, M., Rhodes, C., and Berg, D., 1977, *J. Mol. Biol.* **113**:237–251.
50. Balman, A., and Birnie, G. D., 1979, *Biochim. Biophys. Acta* **561**:155–166.
51. Nordstrom, J. L., Roop, D. R., Tsai, M. J., and O'Malley, B. W., 1979, *Nature* **278**:328–331.
52. Galau, G. A., Klein, W. H., Davis, M. M., Wold, B. J., Britten, R. J., and Davidson, E. H., 1976, *Cell* **7**:487–505.
53. Kaplan, B. B., Schachter, B. S., Osterburg, H. H., deVellis, J. S., and Finch, C. E., 1978, *Biochemistry* **17**:5516–5524.
54. Pearson, W. R., Wu, J. R., and Bonner, J., 1978, *Biochemistry* **17**:51–59.
55. Bantle, J. A., and Hahn, W. E., 1976, *Cell* **8**:139–150.
56. Bishop, J. O., Morton, J. G., Rosbash, M., and Richardson, M., 1974, *Nature* **250**:199–203.
57. Baltimore, D., 1970, *Nature* **226**:1209–1211.
58. Temin, H., and Mizutani, S., 1970, *Nature* **226**:1211–1213.
59. Kacian, D. L., Spiegelman, S., Bank, A., Terada, M., Metafora, S., Dow, L., and Marks, P. A., 1972, *Nature (New Biol.)* **235**:167–169.
60. Ross, J., Aviv, H., Scolnick, E., and Leder, P., 1972, *Proc. Natl. Acad. Sci. U.S.A.* **69**:264–268.
61. Verma, I. M., Temple, G. F., Fan, H., and Baltimore, D., 1972, *Nature (New Biol.)* **235**:163–167.
62. Taylor, J. M., Illmensee, R., and Summer, J., 1976, *Biochim. Biophys. Acta* **442**:324–330.
63. Dudley, J. P., Birtel, J. S., Socher, S. H., and Riser, J. M., 1978, *J. Virol.* **28**:743–752.
64. Winters, M. A., and Edmonds, M., 1973, *J. Biol. Chem.* **248**:4763–4768.
65. Sippel, A. E., 1973, *Eur. J. Biochem.* **37**:31–40.
66. Taniguchi, T., Palmiero, M., and Weissman, C., 1978, *Nature* **274**:223–228.
67. Pearson, W. R., Davidson, E. H., and Britten, R. J., 1977, *Nucleic Acids Res.* **4**:1727–1737.
68. Ryffel, G. U., and McCarthy, B. J., 1975, *Biochemistry* **14**:1379–1384.
69. Hastie, N. D., and Bishop, J. O., 1976, *Cell* **9**:761–774.

70. Young, B. D., Birnie, G. D., and Paul, J., 1976, *Biochemistry* **15**:2823–2829.
71. Quinlan, T. J., Beeker, G. W., Cox, R. F., Elder, P. K., Moses, H. L., and Getz, M. J., 1978, *Nucleic Acids Res.* **5**:1611–1625.
72. Savage, M. J., Sala-Trepat, J. M., and Bonner, J., 1978, *Biochemistry* **17**:462–467.
73. Buell, G. N., Wicken, M. P., Farhang, P., and Schimke, R. T., 1978, *J. Biol. Chem.* **253**:2471–2482.
74. Meyuhas, O., and Perry, R. P., 1979, *Cell* **16**:139–148.
75. Hahn, W. E., Van Ness, J., and Maxwell, I. H., 1978, *Proc. Natl. Acad. Sci. U.S.A.* **75**:5544–5547.
76. Van Ness, J., and Hahn, W. E., 1980, *Nucleic Acids Res.* **8**:4259–4269.
77. Kiper, M., 1979, *Nature* **278**:279–280.
78. Collett, M. S., and Faras, A. J., 1975, *J. Virol.* **16**:1220–1228.
79. Efstratiadis, A., Maniatis, T., Fotis, C., Kafatos, A. J., and Vournakis, J. N., 1975, *Cell* **4**:367–378.
80. Weiss, G. B., Wilson, G. N., Steggles, A. W., and Anderson, W. F., 1976, *J. Biol. Chem.* **251**:3425–3431.
81. Rothenberg, E., and Baltimore, D., 1977, *J. Virol.* **21**:168–178.
82. Friedman, E. Y., and Rosbash, J., 1977, *Nucleic Acids Res.* **4**:3455–3471.
83. Kacian, D. C., and Myers, J. C., 1976, *Proc. Natl. Acad. Sci. U.S.A.* **73**:3408–3412.
84. Schapiro, D. J., Taylor, J. M., McKnight, G. S., Palacios, R., Gonzalez, C., Kiely, M. L., and Schimke, R. J., 1974, *J. Biol. Chem.* **249**:3665–3671.
85. Schnecter, I., 1974, *Biochemistry* **13**:1875–1885.
86. Groner, B., Hynes, N. E., Sippel, A. E., Jeeps, S., Nguyen-Huu, M. C., and Schutz, G., 1977, *J. Biol. Chem.* **252**:6666–6674.
87. Schutz, G., Kieval, S., Groner, B., Sippel, A. E. Kurtz, D. T., and Feigelson, P., 1977, *Nucleic Acids Res.* **4**:71–84.
88. Payvar, F., and Schimke, R. J., 1979, *Eur. J. Biochem.* **101**:271–282.
89. Maurer, R. A., 1980, *J. Biol. Chem.* **255**:854–859.
90. Shapiro, S. Z., and Young, J. R., 1981, *J. Biol. Chem.* **256**:1495–1498.
91. Rosen, J. M., 1976, *Biochemistry* **15**:5263–5271.
92. Ross, J. 1978, *J. Mol. Biol.* **119**:21–35.
93. Ross, J., Gielen, J., Packman, S., Ikawa, Y., and Leder, P., 1974, *J. Mol. Biol.* **87**:697–714.
94. Orken, S. H., and Swerdlow, P. S., 1977, *Proc. Natl. Acad. Sci. U.S.A.* **74**:2475–2479.
95. Lo, J. C., Aft, R., Ross, J., and Mueller, G. C., 1978, *Cell* **15**:447–453.
96. Cox, R. F., Haines, M. E., and Emtage, J. S., 1974, *Eur. J. Biochem.* **49**:225–236.
97. Harris, S. E., Rosen, J. M., Means, A. R., and O'Malley, B. W., 1975, *Biochemistry* **14**:2072–2080.
98. McKnight, G. S., Pennequin, P., and Schimke, R. J., 1975, *J. Biol. Chem.* **250**:8165–8170.
99. Palmiter, R. D., Moore, P. B., Mulvihill, E. R., Emtage, S., 1976, *Cell* **8**:557–572.
100. Hynes, N. E., Groner, B., Sippel, A. E., Nguyen-Huu, M. C., and Schutz, G., 1977, *Cell* **11**:923–932.
101. Schapiro, D. J., and Baker, H. J., 1977, *J. Biol. Chem.* **252**:8428–8434.
102. Arnemann, J., Heins, B., and Beato, M., 1979, *Eur. J. Biochem.* **99**:361–367.
103. Nakhasi, H. L., and Qasba, P. K., 1979, *J. Biol. Chem.* **254**:6016–6025.
104. McWilliams, D., Callahan, R. C., and Baime, I., 1977, *Proc. Natl. Acad. Sci. U.S.A.* **74**:1024–1027.
105. Iatrou, K., and Dixon, G. H., 1977, *Cell* **10**:433–441.
106. Shinohara, T., and Piatigorsky, J., 1976, *Proc. Natl. Acad. Sci. U.S.A.* **73**:2808–2812.
107. Carbon, J., Clarke, L., Ilgen, C., and Ratzkin, B., 1977, *Recombinant Molecules: Impact on Science and Society* (R. F. Beers and E. G. Bassatt, eds.), Raven Press, New York, pp. 335–378.
108. Maniatis, T., Hardison, R. C., Lacy, E., Lauer, J., O'Connell, C., Quon, D., Sim, G. K., and Efstratiadis, A., 1978, *Cell* **15**:687–701.
109. Maniatis, T., 1980, *Cell Biology: A Comprehensive Treatise*, Volume 3 (L. Goldstein and D. M. Prescott, eds.), Academic Press, New York, pp. 564–608.
110. Darnell, J. E., Philipson, L., Wall, R., and Adesnik, M., 1971, *Science* **174**:507–510.

111. Edmonds, H., Vaughan, M. H., and Nakazato, H., 1971, *Proc. Natl. Acad. Sci. U.S.A.* **68:**1336–1340.
112. Philipson, L., Wall, R., Glickman, G., and Darnell, J. E., 1971, *Proc. Natl. Acad. Sci. U.S.A.* **68:**2806–2809.
113. Mendecki, J., Lee, S. Y., and Brawerman, G., 1972, *Biochemistry* **11:**792–798.
114. DeLarco, J., Abramowitz, A., Bromwell, K., and Guroff, G., 1975, *J. Neurochem.* **24:**215–222.
115. Mahoney, J. B., and Brown, I. R., 1975, *J. Neurochem.* **25:**503–507.
116. Aviv, J., and Leder, P., 1972, *Proc. Natl. Acad. Sci. U.S.A.* **69:**1408–1412.
117. Lindberg, U., and Persson, T., 1972, *Eur. J. Biochem.* **31:**246–254.
118. Molloy, G. R., Jelinek, W., Salditt, J., and Darnell, J. E., 1974, *Cell* **1:**43–53.
119. Brown, I. R., and Church, R. B., 1971, *Biochem. Biophys. Res. Commun.* **42:**850–856.
120. Grouse, L., Chilton, M. D., and McCarthy, B. J., 1972, *Biochemistry* **11:**798–805.
121. Brown, I. R., and Church, R. B., 1972, *Dev. Biol.* **29:**73–84.
122. Chikaraishi, D. M., Deeb, S. S., and Sueoka, N., 1978, *Cell* **13:**111–120.
123. Grouse, L. D., Schrier, B. K., Bennett, E. L., Rosenzweig, M. R., and Nelson, P. G., 1978, *J. Neurochem.* **30:**191–203.
124. Monahan, J. J., 1978, *Int. Rev. Cytol.* **8:**229–290.
125. Darnell, J. E., 1979, *Prog. Nucleic Acid Res. Mol. Biol.* **22:**327–352.
126. Tobin, A. J., 1979, *Dev. Biol.* **68:**47–58.
127. Chikaraishi, D. M., 1979, *Biochemistry* **18:**3249–3256.
128. Heikkila, J. J., and Brown, I. R., 1977, *Biochim. Biophys. Acta* **474:**141–153.
129. Galau, G. A., Klein, W. H., Britten, R. J., and Davidson, E. H., 1977, *Arch. Biochem. Biophys.* **179:**584–599.
130. Milcarek, C., Price, R., and Penman, S., 1974, *Cell* **3:**1–10.
131. Nemer, M., Graham, M., and Dubroff, L. M., 1974, *J. Mol. Biol.* **89:**435–454.
132. Greenberg, J. R., 1976, *Biochemistry* **15:**3516–3522.
133. Surrey, S., and Nemer, M., 1976, *Cell* **9:**589–595.
134. Faust, J., Millward, S., Duchastel, A., and Fromson, D., 1976, *Cell* **9:**597–604.
135. Greenberg, J. R., 1976, *J. Mol. Biol.* **108:**403–416.
136. Kaufman, Y., Milcarek, C., Berissi, H., and Penman, S., 1977, *Proc. Natl. Acad. Sci. U.S.A* **74:**4801–4805.
137. Brandhorst, B. P., Verma, D. P. S., and Fromson, D., 1979, *Dev. Biol.* **71:**128–141.
138. Milcarek, C., 1979, *Eur. J. Biochem.* **102:**467–476.
139. Minty, A. J., and Gros, F., 1980, *J. Mol. Biol.* **139:**61–83.
140. Getz, M. J., Reiman, H. M., Siegel, G. P., Quinlan, T. J., Proper, J., Elder, P. K., and Moses, H. L., 1977, *Cell* **11:**909–921.
141. Moyzis, R. K., Grady, D. L., Li, D. W., Mirvis, S. E., and Ts'o, P. O. P., 1980, *Biochemistry* **19:**821–837.
142. Supowit, S. C., and Rosen, J. M., 1980, *Biochemistry* **19:**3452–3460.
143. Rolton, H. A., Birnie, G. D., and Paul, J., 1977, *Cell Differ.* **6:**25–39.
144. Williams, J. G., Hoffman, R., and Penman, S., 1977, *Cell* **11:**901–907.
145. Hahn, W. E., 1973, *Soc. Neurosci. Abstr.* **3:**139.
146. Wold, B. J., Klein, W. H., Hough-Evans, B. R., Britten, R. J., and Davidson, E. H., 1978, *Cell* **14:**941–950.
147. Shepherd, G. W., and Nemer, M., 1980, *Proc. Natl. Acad. Sci. U.S.A.* **77:**4653–4656.
148. Shepherd, G. W., and Flickinger, R., 1979, *Biochim. Biophys. Acta* **563:**413–421.
149. Kamalay, J. C., and Goldberg, R. B., 1980, *Cell* **19:**935–946.
150. Finch, C. E., 1976, *Q. Rev. Biol.* **51:**49–83.
151. Finch, C. E., 1979, *Fed. Proc.* **38:**178–183.
152. Cutler, R. G. , 1975, *Exp. Gerontol.* **10:**37–60.
153. Kaplan, B. B., and Finch, C. E., 1982, *Molecular Approaches to Neurobiology* (I. R. Brown, ed.), pp. 71–98. Academic Press, New York.
154. Hahn, W. E., and Chaudhari, N., 1981, *Trans. Am. Soc. Neurochem.* **12:**71.
155. Bernstein, S. L., Gioio, A. E., and Kaplan, B. B., 1980, *Soc. Neurosci. Abstr.* **10:**775.
156. Zagon, I. S., and McLaughlin, P. J., 1979, *Brain Res.* **170:**443–457.

157. Johnson, L. F., Williams, J. G., Abelson, H. T., Green, H., and Penman, S., 1975, *Cell* **4:**69–75.
158. Getz, M. J., Elder, P. K., Benz, E. W., Stephens, R. E., and Moses, H. L., 1976, *Cell* **7:**255–265.
159. Farquhar, M. N., Kosky, K. J., and Omenn, G. S., 1979, *Aging in Nonhuman Primates* (D. M. Bowden, ed.), Van Nostrand Reinhold, New York, pp. 71–79.
160. Chaconas, G., and Finch, C. E., 1973, *J. Neurochem.* **21:**1469–1473.
161. Shaskin, E. G., 1977, *J. Neurochem.* **28:**509–516.
162. Mann, D. M. A., Yates, P. O., and Stamp, J. E., 1978, *J. Neurol. Sci.* **37:**83–93.
163. Mann, D. M. A., and Yates, P. O., 1979, *Acta Neuropathol.* **47:**93–97.
164. Bohn, R. C., and Mitchell, R. B., 1977, *Exp. Neurol.* **57:**161–178.
165. Zs-Nagy, V., Bertoni-Freddari, C., Zs-Nagy, I., Pieri, C., and Giuli, C., 1977, *Gerontology* **23:**267–276.
166. Hahn, W. E., Van Ness, J., and Maxwell, I. H., 1980, *Nature* **286:**601.
167. Monahan, J. J., Harris, S. E., Woo, S. L. C., Robberson, D. L., and O'Malley, B. W., 1976, *Biochemistry* **15:**223–233.
168. Sippel, A. E., Hynes, N., Groner, B., and Schutz, G., 1977, *Eur. J. Biochem.* **77:**141–151.
169. Grouse, L. D., Lettendre, C., and Schrier, B. K., 1979, *J. Neurochem.* **33:**583–585.
170. Schrier, B. K., Zubairi, M. Y., Letendre, C. H., and Grouse, L. D., 1978, *Differentiation* **12:**23–30.
171. Grouse, L. D., Schrier, B. K., Letendre, C. H., Zubairi, M. Y., and Nelson, P. G., 1980, *J. Biol. Chem.* **255:**3871–3877.
172. Felsani, A., Berthelot, F., Gros, F., and Croizat, B., 1978, *Eur. J. Biochem.* **92:**569–577.
173. Wilkes, P. R., Birnie, G. D., and Paul, J., 1979, *Nucleic Acids Res.* **6:**2193–2207.
174. Colbert, D. A., Tedeschi, M. V., Atryzek, V., and Fausto, N., 1977, *Dev. Biol.* **59:**111–123.
175. Coupar, B. E. H., Davies, J. A., and Chesterton, C. J., 1978, *Eur. J. Biochem.* **84:**611–629.
176. Towle, H. C., Dillman, W. H., and Oppenheimer, J. H., 1979, *J. Biol. Chem.* **254:**2250–2257.
177. Leibovitch, M. P., Leibovitch, S. A., Harel, J., and Kruh, J., 1979, *Eur. J. Biochem.* **97:**321–326.
178. Kleiman, L., Birnie, G. D., Young, B. D., and Paul, J., 1977, *Biochemistry* **16:**1218–1223.
179. Birnie, G. D., Macphail, E., Young, B. D., Getz, M. J., and Paul, J., 1974, *Cell Differ.* **3:**221–231.
180. Minty, A. J., Birnie, G. D., and Paul, J., 1979, *Exp. Cell Res.* **115:**1–14.
181. Affara, N. A., Jacquet, M., Jakob, H., Jacob, F., and Gros, F., 1977, *Cell* **12:**509–520.
182. Paterson, B. M., and Bishop, J. O., 1977, *Cell* **12:**751–765.
183. Jacquet, M., Affara, N. A., Robert, B., Jakob, H., Jacob, F., and Gros, F., 1978, *Biochemistry* **17:**69–79.
184. Croizat, B., Berthelot, F., Felsani, A., and Gros, F., 1979, *FEBS Lett.* **103:**138–143.
185. Axel, R., Fiegelson, Pa., and Schutz, G., 1976, *Cell* **7:**247–254.
186. Grouse, L. D., Schrier, B. K., and Nelson, P. G., 1979, *Exp. Neurol.* **64:**354–364.
187. Grady, L. J., North, A. B., Campbell, W. P., 1978, *Nucleic Acids Res.* **5:**697–712.
188. Grouse, L. D., Nelson, P. G., Omenn, G. S., and Schrier, B. K., 1978, *Exp. Neurol.* **59:**470–478.

Receptor Mapping by Histochemistry

James K. Wamsley and Jose M. Palacios

1. INTRODUCTION

The ability to apply receptor binding techniques to the localization of neurotransmitter receptors at the microscopic level has opened new avenues of research into the sites of drug and neurotransmitter action. Originally, localizations of this nature required the availability of irreversibly or nearly irreversibly binding ligands specific for the receptor of interest. The development of new techniques allowing the discrete localization of binding sites for diffusible substances has been a major advance in this field and will undoubtedly prove to be a useful adjunct to future studies involving the analysis of neurotransmitter pathways in the nervous system.

1.1. Necessity of Localizing Receptors

Knowledge of receptor distributions in the central nervous system provides us with important scientific and clinical information on the sites of drug and neurotransmitter action. Receptor binding techniques allow the definition of a pharmacologically relevant receptor based on its interactions with various compounds. Once the receptor is defined, these techniques can be used to screen other pharmacological agents to find their potency at that receptor locus. These studies coupled with functional or biobehavioral investigations provide insights into which neurotransmitter systems are involved in producing certain effects. Thus, clinically important pharmacological agents and their interactions with known neurotransmitter receptor mechanisms can be studied (for examples see refs. 1–4).

By employing microscopic techniques for receptor localization, we can

James K. Wamsley • Departments of Psychiatry and Anatomy, University of Utah, Salt Lake City, Utah 84132. *Jose M. Palacios* • Department of Neuroscience, The Johns Hopkins University, School of Medicine, Baltimore, Maryland 21205. Present address: Sandoz Ltd., Preclinical Research, CH-4002 Basel, Switzerland.

define neuroanatomic areas that contain receptors specific for neurotransmitter agents or pharmacologically related compounds. In certain instances, where the anatomic loci that mediate various drug effects are known, we can attempt to correlate the presence of the specific receptor types with these nuclei. Whereas immunohistochemical studies of neurotransmitters or their biosynthetic enzymes allow the localization of the presynaptic element (that is, the neuronal type that contains that specific neurotransmitter), the localization of receptors defines the postsynaptic element (the neuron receiving the information) or, in some instances, the presence of presynaptic receptors which may limit the release of the neurotransmitter. Coupling of these two techniques allows the establishment of neurotransmitter-specific connections in the central nervous system (CNS). Receptor localizations provide knowledge of new areas potentially important in the mediation of other drug effects and lay the foundation for new investigations to probe these areas in an attempt to find the functional significance of the presence of receptors. Areas that overlap in the distribution of receptor types can then be studied to determine receptor interactions and their functional relationship in individual brain regions. Areas of receptor nonoverlap can provide us with the knowledge of the differential distribution of receptor subtypes or of whether receptors are "coupled" or "uncoupled" in different regions of the brain.

Binding studies have shown that receptor densities change in human disease states.[5] The ability to use microscopic techniques to localize receptors allows the determination of which nuclei in the brain show a given receptor deficit and how this might relate to the pathological changes taking place. Determination of other receptor types on those cells or receptor interactions that take place along the same pathway may suggest potential treatments for the alleviation of symptoms of certain psychiatric and neurological disorders.

1.2. Historical Background

Very few techniques for the microscopic visualization of receptors are currently available. Fluorescent probes have been employed for the purpose of identifying β-adrenergic receptor sites,[6–10] but it is still not clear exactly what is being labeled with this technique.[11,12] A biologically active fluorescent agonist and a rhodamine-conjugated enkephalin have been used in preliminary investigations to identify angiotensin II[13] and opioid[14] receptor sites, respectively. Other techniques involving the use of peroxidase have reportedly been used to identify prolactin receptors[15] and receptors for luteinizing hormone-releasing hormone[16] in the brain. These compounds do not allow one to directly determine if the altered compound is binding to the pharmacologically relevant receptor; for that purpose, the concomitant introduction of tritium along with the probe makes it possible to perform binding studies as well as localization studies using the same molecule.

Microdissection techniques have been used to identify specific nuclei that contain certain receptors. This method involves the isolation of brain areas from slices and the subsequent homogenization and binding assay of the isolated area. The latter technique is hampered only by the investigator's ability to separate brain regions cleanly prior to the performance of the receptor determination. Consistently successful studies of the microscopic localization of receptors have relied on the technique of autoradiography, since it affords the investigator the ability to study the characteristics and pharmacology of the receptor binding prior to the receptor localization. In this way, it can be demonstrated that the radioactive ligand is binding to the pharmacologically defined receptor, and the appropriate experimental conditions can be established to allow highly specific binding with minimal background or nonspecific binding. Data from autoradiographic studies that do not show such specificity[17–21] should remain questionable until the specific nature and pharmacological relevance of the binding can be demonstrated.

Several autoradiographic techniques have been devised to localize receptors. Some of these employ *in vivo* administration of diffusible ligands or the use of irreversibly binding ligands; these will be discussed in detail. It may also be possible, in some instances, to label certain receptors with radioactive peptides and then "cross link" the ligand to the receptor with a fixative.[22,23] None of these techniques, however, is so widely applicable as *in vitro* labeling methods that employ diffusible ligands to label tissue for apposition techniques of autoradiography.

1.2.1. Autoradiography

Autoradiography is a technique that involves the exposure of photographic emulsions by tissues containing a radioactive component. Substances conjugated to elements that emit β particles (usually tritium) are employed. Charged particles (β radiation) released from the radioactive substance alter the silver bromide crystals suspended in gelatin which make up the emulsion. Subsequent development of the emulsion reduces the "altered" silver bromide crystals into metallic silver and washes the rest of the emulsion away. Thus, the position of the tritiated substance in the tissue can be localized by observing the pattern of autoradiographic grains in the emulsion. This technique nas been employed in neuroscience for the localization of uptake sites of certain substances and their precursors but was not generally applicable for this purpose or for receptor localization until the advent of dry-mount techniques for autoradiography.[24]

1.2.2. Hormone Receptors

The feasibility of localizing the uptake sites of many different compounds was realized when the dry-mount technique was employed for the autoradiography of diffusible substances.[25] This method has been successfully employed for the localization of many types of hormone uptake sites in the brain including sites for estrogen, androgen, progesterone, glucocorticoids, and thyroid hormone (see ref. 26).

1.2.3. Neurotransmitter Receptor Binding

Development of techniques for the study of receptor binding has opened up new avenues of receptor research.[27] Pioneering work in this field has made it possible to examine receptor binding in brain membrane or tissue slice prep-

arations using ligands specific for muscarinic cholinergic receptors, opiate receptors, glycine receptors, GABA receptors, kainate receptors, β-adrenergic
receptors, α-adrenergic receptors, dopamine receptors, serotonin receptors,
benzodiazepine receptors, substance P receptors, cholecystokinin receptors,
and more recently adenosine receptors (for recent reviews see ref. 27–33).
Some of these and subsequent studies have identified specific ligands with an
affinity high enough to be used for autoradiographic localization of neurotransmitter receptors in the CNS.

1.3. Recent Developments in Receptor Autoradiography

1.3.1. Autoradiography of Diffusible Ligands Bound in Vivo

1.3.1a. Quinuclidinyl Benzilate. Quinuclidinyl benzilate (QNB) is a potent
muscarinic cholinergic antagonist which binds specifically and with high affinity
to muscarinic cholinergic receptors in the CNS.[34] Receptor-bound tritiated
QNB can be localized by binding assay after *in vivo* injection of the drug.[35]
Since the dissociation rate of the QNB–receptor complex is slow, QNB could
be injected into an animal, allowed to bind to its receptor, and then localized
microscopically using dry-mount autoradiographic techniques.[36,37] Approximately 60 to 75% of the bound radioactivity was blocked by pretreatment with
atropine, indicating that most of the binding sites are specific.

These studies demonstrated the feasibility of using autoradiographic techniques to localize neurotransmitter receptors in the CNS and provided the first
"look" at the light microscopic localization of muscarinic cholinergic receptors.

1.3.1b. Etorphine and Diprenorphine. Autoradiographic localization of
opiate receptors in the brain was first performed using tritiated diprenorphine
(a potent opiate antagonist) and tritiated etorphine (a potent opiate agonist)
injected systemically or intracerebroventricularly.[38,39] These initial studies
were followed by detailed mapping of opiate receptor distributions in all parts
of the brain and spinal cord.[40–42] In general, these studies showed a similar
regional distribution to that demonstrated by biochemical assay. Much of the
binding could be prevented by pretreatment with drugs known to be specific
for the opiate receptor, and this blocking action corresponded to the potency
of the unlabeled compound for the opiate receptor.[43] The binding was also
saturable and was mostly bound to particulate matter, indicating similarities
between the original homogenate binding studies used to describe the opiate
receptor and the characteristics of the autoradiographically localized opiate
receptor.

1.3.1c. Spiperone. In vivo administration of radiolabeled spiperone (an
antagonist of dopamine receptors) or pimozide has been shown to label dopamine receptors in the mouse CNS.[44,45] Labeling of dopamine receptors by
in vivo injection of spiperone has also been studied in the rat CNS[46] and has
led to the autoradiographic localization of dopamine receptors.[46–48]

Recent binding studies, however, have shown that spiperone binds to more sites than just the dopamine receptor.[2,49-53] This evidence suggests that spiperone labels dopamine receptors, a subtype of serotonin receptor (5HT-2 site), α receptors, and possibly a specific site of its own called a spirodecanone site. Palacios *et al.*[54] have made use of these properties of spiperone to autoradiographically localize multiple receptors in the brain using *in vitro* labeling techniques.

1.3.2. Autoradiographic Localization of Irreversibly Binding Ligands

1.3.2a. α-Bungarotoxin. The autoradiographic localization of nicotinic cholinergic receptors in the peripheral nervous system has been accomplished using the virtually irreversibly binding compound α-bungarotoxin (a neurotoxin derived from snake venom).[55-59] α-Bungarotoxin (α-BTX) has been used in binding studies to examine nicotinic cholinergic receptors in the brain.[60-65] This binding is apparently specific for the nicotinic cholinergic receptor,[65-67] although the presence of the neurotoxin in the CNS does not appear to interfere with synaptic transmission in certain instances.[68-69] The toxin has been used, however, in several autoradiographic studies to identify regions containing nicotinic cholinergic receptors in the CNS.[70-74]

Because of the very slow dissociation rate of α-BTX from the nicotinic cholinergic receptor, this binding will withstand the rigorous fixation necessary for ultrastructural localization of binding sites. This has made possible the electron microscopic localization of nicotinic cholinergic receptors by autoradiography.[70-75]

1.3.2b. Propylbenzilylcholine Mustard. Propylbenzilylcholine mustard (PrBCM) is a potent synthetic antagonist specific for the muscarinic cholinergic receptor and binds irreversibly to the receptor by means of a covalent bond.[76] The specific nature of this binding has made possible the localization of muscarinic cholinergic receptors at the light microscopic level after *in vitro* labeling of the tissue sections with tritiated PrBCM.[77-80] The pharmacological specificity and the binding kinetics were studied directly on the slide-mounted tissue sections. This allowed the establishment of binding conditions that gave high ratios of specific to nonspecific binding and, thus, more precise localization of the muscarinic cholinergic receptor.

Since PrBCM binds to the receptor irreversibly, it is also possible to localize muscarinic cholinergic receptors at the electron microscopic level using techniques of autoradiography.[81]

1.3.2c. Flunitrazepam. Tritiated flunitrazepam is a potent ligand for the benzodiazepine receptor, and its binding has been studied after *in vivo* injection.[82] Although this ligand will diffuse from the receptor, the binding can be made irreversible by exposure to ultraviolet light.[83,84] The distribution of these binding sites has been studied by electron microscopic autoradiography following *in vivo* injection of the drug and subsequent photoaffinity labeling. Demonstration of specific benzodiazepine receptor binding by this method was examined in brain homogenates and tissue slices. It has been possible to show some correlation between the ultrastructural localization of benzodiazepine receptors and terminals showing glutamic acid decarboxylase (GAD) immunoreactivity.[85]

1.3.3. Autoradiography of Diffusible Ligands Bound in Vitro

In order to localize binding sites for diffusible substances, it is necessary to circumvent the diffusion problems associated with coating the treated tissue with liquid emulsion. Dry-mount techniques have been developed and widely used after *in vivo* labeling of tissues,[25] but because of low-affinity binding or metabolic breakdown of many ligands, *in vivo* binding is not always practical. Apposition techniques of autoradiography have also been developed[86] in which a coverslip is coated with emulsion and allowed to dry before being brought into contact with the labeled tissue.

Recently, Young and Kuhar[87] devised a method for applying receptor binding techniques to slide-mounted tissue sections followed by the apposition of an emulsion-coated coverslip to autoradiographically localize drug- and neurotransmitter-binding sites in intact tissue sections. This method has made possible the microscopic localization of many different types of receptors, albeit not all, which was not possible with previously available techniques. Currently, this method has been applied to the localization of opiate receptors[87–90] (both μ and δ sites[91]), benzodiazepine receptors[92,93] (both BZ-1 and BZ-2 sites[94]) GABA receptors, [95,96] muscarinic cholinergic receptors[97] (both high- and low-affinity sites[98]), dopamine receptors,[54] serotonin receptors,[54,99,100] histamine H_1 receptors,[101,102] glycine receptors,[103] β-adrenergic receptors,[104] α_1- and α_2-adrenergic receptors,[105,106] neurotensin receptors,[107,108] insulin receptors,[109] cholecystokinin receptors,[110] and kainic acid receptors.[111]

In vitro labeling of slide-mounted tissue sections has several distinct advantages. Probably foremost is the ability to study the binding kinetics and pharmacology of the ligand directly on the slide-mounted tissue sections. This allows the application of precise binding conditions for the localization of binding sites using parameters that provide very high specific-to-nonspecific (signal-to-noise) ratios. Thus, investigators can provide evidence that they are labeling the pharmacologically relevant receptor and demonstrate its localization with a high degree of specificity and resolution.

2. TECHNIQUES OF LOCALIZING NEUROTRANSMITTER RECEPTORS BY AUTORADIOGRAPHY

2.1. In Vivo Labeling and Processing

A general outline of the steps involved in the technique of autoradiographic localization of receptors after *in vivo* application of the appropriate ligand is as follows:

1. Inject the ligand into the animal intravenously.
2. Wait an appropriate amount of time for the unbound radioactive ligand to move through the system leaving as much of the ligand bound to the receptor as possible.
3. Sacrifice the animal; remove and freeze the brain.
4. Section the tissue in a cryostat.
5. Mount the sections (in a photographic darkroom under safelight conditions) on slides coated with a dry layer of photographic emulsion.
6. Wait an appropriate amount of time for the radioactivity present in the tissue to form a latent image in the emulsion.
7. Develop the latent image in the emulsion.
8. Stain, dry, and coverslip the tissue section.
9. View the tissue and autoradiogram under a microscope using brightfield or darkfield illumination.

Some of the applications of this method are shown in Table I. To study receptors in this fashion, it is necessary first to choose a ligand that has a high affinity and will selectively label a specific type of receptor. Usually, the initial studies have been performed by *in vitro* application of the ligand in tissue homogenates followed by binding assay. Studies of dissociation, association, and saturation curves reveal the binding characteristics of the ligand in brain tissue. Scatchard analysis of the saturation data will indicate the affinity of the ligand for its receptor and the number of receptors present. Hill plots can be used to check for cooperativity. Displacement curves are then usually performed using a variety of ligands known to be specific or not specific for the receptor of interest to determine if the ligand shows the appropriate pharmacology.

If a ligand shows a high affinity for a specific receptor and its dissociation rate is slow, it may be a good candidate for *in vivo* testing. Previous studies using this technique have involved the injection of the radioactive ligand into an animal followed by dissection, homogenization, and determination of the radioactivity in the brain. This step is useful to study the parameters necessary for maximizing the specific binding in the tissue. In tests such as these, it is appropriate to administer different concentrations of the radioactive ligand and to wait different lengths of time before sacrifice in order to determine the optimal concentration and postinjection time needed to obtain a high level of binding in the brain. It is also necessary to determine that the radioactive ligand is not degradated by metabolic processes occurring in the tissue.

The specificity of any binding noted can be tested by injecting an unlabeled compound known to be specific for the receptor of interest followed by the radioactive ligand in order to block the concentration of radioactivity in the brain. Any specific binding should then be shown to be stereospecific and saturable. The binding should also show a regional distribution similar to that determined in homogenization binding studies. For instance, if we used a ligand to label muscarinic cholinergic receptors *in vivo*, it should be possible to dissect the caudate and find a high concentration of radioactivity and to dissect the cerebellum and find a low concentration of radioactivity, since this is the

Table I

In Vivo Labeling Parameters for Receptor Autoradiography

Ligand	Receptor	Species	Amount injected (μCi/g)	Survival time	Control ligand	Concentration	Pretreatment time	Ref.
[³H]QNB	Muscarinic cholinergic	Rat	0.8–1.0	1 hr	Atropine	60 mg/kg, i.m.	1 hr	36,37
[³H]Diprenorphine	Opiate	Rat	0.6–0.7	1 hr	(-)Levallorphan	5 mg/kg	—	38,40–43
[³H]Diprenorphine	Opiate	Rat	1.0	1 hr	(-)Levallorphan	5 mg/kg, i.v.	30 min	129
[³H]Diprenorphine	Opiate	Rat	0.4–0.5	1 hr	Diprenorphine	0.8–1.0 mg/kg	5 min	130
[³H]Diprenorphine	Opiate	Monkey	0.5	1 hr	—	—	—	89
[³H]Etorphine	Opiate	Mouse	2.0	5 min	Naltrexone	1 mg/kg	—	39
[³H]Etorphine	Opiate	Rat	0.1 or 0.6–0.7	15 min	Levallorphan	1 mg/kg	—	40–42
[³H]Spiperone	Dopamine	Rat	0.6–0.8	2 hr	(+)Butaclamol	5 mg/kg	0–20 min	46,131
[³H]Spiperone	Dopamine	Rat	0.5	2 hr	Pimozide	5 mg/kg	1 hr	48
[³H]Spiperone	Dopamine	Rat	0.5	2 hr	(+)Butaclamol	5 mg/kg i.p.	20 min	47

distribution shown in studies using brain homogenates.[34] It would also be helpful to show appropriate physiological or behavioral changes in the animal whose receptors are supposedly occupied by the radioactive ligand.

When most, if not all, of these conditions have been determined, the optimal conditions can be used to label the specific receptors with the radioactive ligand, and these receptors can then be localized by autoradiography. Autoradiographs from animals injected with the radioactive ligand can be compared to those from animals previously injected with an unlabeled compound to effectively block the binding of the radioactive ligand to the specific receptor. The first set of tissues would have autoradiographic grains associated with all sites, whereas the second set of tissues would show grain distributions only over nonspecific binding sites. By comparing autoradiographs acquired using these two conditions, sites of specific binding can be determined.

In vivo labeling methods have been used to ultrastructurally localize nicotinic cholinergic and benzodiazepine receptors using electron microscopic autoradiography.[75,83,84] Nicotinic cholinergic receptors were labeled *in vivo* with [^{125}I] α-BTX by intracerebroventricular injection. The animals were then perfused, and the brains removed and sectioned with a vibratome and processed for electron microscopy. Benzodiazepine receptors were labeled by intravenous injection of [^{3}H]flunitrazepam. The animals were then sacrificed, and the brains were removed and sliced. The sections were exposed to ultraviolet light for photoaffinity labeling and processed for electron microscopy.

In both cases, ultrathin sections of the labeled brain tissues were picked up on grids and coated with emulsion by the loop technique.[112] After an exposure period, the emulsion was developed and examined under a transmission electron microscope.

2.2. In Vitro Labeling and Processing

2.2.1. Irreversible Ligands

When radiolabeled ligands that bind irreversibly to a specific receptor are employed, the necessity of taking precautions to prevent diffusion of the ligand from the receptor are minimized. Thus, tissue sections can be labeled *in vitro* with the radioactive ligand and then dipped into liquid emulsion to coat the tissue section directly. It is still important to determine optimal binding conditions in order to maximize specific-to-nonspecific ratios (Table II). Most of these parameters have, again, been predetermined by *in vitro* binding studies performed in brain homogenates.

Labeling of brain nicotinic cholinergic receptors with α-bungarotoxin has been accomplished by both *in vitro* labeling of brain slices and intracerebroventricular injection of the toxin although the *in vitro* method appears to give more satisfactory results.[71] Autoradiographic experiments with this ligand are accomplished using cryostat sections of frozen brain tissues, and the incubations are performed on slide-mounted tissue sections. The tissue sections are immersed in buffered solutions containing the iodinated α-BTX for the duration of the incubation period at room temperature. Control sections (blanks),

Table II
Labeling Receptors in Vitro with Irreversibly Binding Ligands

Receptor	Ligand	Concentration	Incubation	Rinse time	Blanks	Reference
Nicotinic cholinergic	[^{125}I]α-BTX	3 nM	30–60 min	3 × 20 min	0.1 µM α-BTX 1 mM Nicotine	72
Nicotinic cholinergic	[^{125}I]α-BTX	0.5 or 5 nM	30 min	30 min	1 mM Tubocurarine 1 mM Nicotine 1 mM Atropine	71
Nicotinic cholinergic	[^{125}I]α-BTX	8 or 15 nM	40–60 min	4 × 5 min	1 mM Nicotine	73
Nicotinic cholinergic	[^{125}I]α-BTX	5 nM	60 min	6 × 10 min	0.1 mM Tubocurarine	70
Muscarinic cholinergic	[^{3}H]PrBCM	2.4 nM	15 min	5 × 15 min	1 µM Atropine	77

which indicate areas of nonspecific binding, are generated by preincubating slide-mounted tissue sections for 30–60 min in buffer containing the displacer (an unlabeled compound known to be specific for the receptor of interest). The displacer is present in such a concentration that it will effectively block the binding of the radioactive ligand to the specific receptor. After the preincubation, the slides are incubated in the solution containing the radioactive α-BTX but in the added presence of the displacer. The slide-mounted tissue sections are then rinsed in buffer, fixed in ethanol, rinsed in water, dried, and then coated with a thin layer of emulsion for autoradiography. The slides can then be stored at 4°C for an exposure period judged to be appropriate for the specific activity of the ligand and the amount of radioactive ligand bound. The emulsion is then developed, and the tissue stained and coverslipped.

Use of propylbenzilylcholine mustard (PrBCM) to label muscarinic receptors for autoradiography also involves the use of cryostat sections. The slide-mounted tissue sections are first fixed in buffer containing 1% glutaraldehyde for 15 min, rinsed, and then preincubated 15 min in buffer alone or in buffer containing atropine for the production of controls. The PrBCM is then added to the preincubation solutions; the slides are incubated (30°C) and then transferred to Carnoy's fixative solution for an additional 15 min. The sections are washed in absolute alcohol, dried, coated with emulsion (in the dark), and allowed to expose. After this period, the emulsion is developed, and the tissue sections are stained prior to microscopic examination.

It has been possible to use these ligands to ultrastructurally localize cholinergic receptors. Administration of α-BTX for this purpose has been accomplished after *in vivo* intracerebroventricular injection[75] or by incubating slices in buffer containing the radiolabeled toxin.[70] In both cases, the animals were perfused with fixative to preserve the ultrastructural morphology of the brain tissues. To label muscarinic receptors at the electron microscopic level with tritiated PrBCM, thick sections were incubated in the presence of the mustard and subsequently fixed to maintain cellular integrity.

Although electron microscopic autoradiography is not practical for extended mapping studies, it does provide investigators with an appreciation of the distribution of receptors in a particular area with a resolution not currently feasible with other techniques. Many of the grains are associated with synapses which can thus be identified as cholinergic. It is still difficult to ascertain which membrane is actually responsible for binding the radioactive ligand and thus generating the autoradiographic grain.

2.2.2. Reversible Ligands

2.2.2a. Coverslip Method. This recently devised technique for the localization of diffusible ligands has already been applied to the microscopic localization of a wide variety of neurotransmitter and drug receptors. The versatility of the technique and its applicability to different ligands and incubation situations have opened new avenues of receptor localization. Some of the utilizations of the method and the parameters employed are shown in Table III.

Table III

In Vitro Labeling Parameters for Receptor Autoradiography

Receptor	Ligand	Concentration	Incubation time, temp.	Rinse time	Displacer	Ref.
Opiate	[³H]Dihydromorphine	4 nM	60 min, rm. temp.	2 × 5 min	1 μM naloxone	87
	[³H]Diprenorphine	4 nM	10 min, rm. temp.	2 × 5 min	1 μM naloxone	87,89
	[³H]-D-Alanine² methioninamide⁵ enkephalin	4 nM	53 min, 4°C	2 × 5 min	1 μM naloxone	87
	[¹²⁵I]FK33-824	0.2 nM	40 min, rm. temp.	2 × 5 min	1 μM naloxone	90,91
	[³H]Naloxone	6 nM	40 min, rm. temp.	2 × 5 min	1 μM naloxone	90
	[¹²⁵I]-D-Alanine²-D-leucine⁵-enkephalin	0.2 nM	40 min, rm. temp.	2 × 5 min	1 nM FK33-824 (δ sites)	91
Glycine	[³H]Strychnine	4 nM	20 min, 4°C	5 min	10 mM glycine	103
β-Adrenergic	[³H]Dihydroalprenolol	2 nM	30 min, rm. temp.	2 × 10 min	0.1 μM zinterol (β₂ site) 0.1 mM zinterol 10 μM DL-propranalol	104
α₁-Adrenergic	[³H]WB-4101	1 nM	70 min, 4°C	5 + 20 min	100 μM norepinephrine	105,106
α₂-Adrenergic	[³H]Aminoclonidine	2.5 nM	60 min, 4°C	5 + 10 min	100 μM norepinephrine	105,106
Neurotensin	[³H]Neurotensin	4 nM	60 min, 4°C	2 × 5 min	5 μM neurotensin	107,108

Receptor	Ligand	Concentration	Incubation	Wash	Displacer	Ref
Insulin	[125I]Insulin	0.17 nM	60 min, rm. temp.	5 + 15 min	3.3 µM insulin	109
CCK	[125I]Cholecystokinin	50 pM	45 min, rm. temp.	30 min	0.1 µM cholecystokinin	110
Kainic acid	[3H]Kainic acid	100 nM	30 min, 0°C	—	100 µM kainic acid	111
GABA	[3H]Muscimol	5 nM	40 min, 4°C	1 min	200 µM GABA	95,96
Benzodiazepine	[3H]Flunitrazepam	1 nM	40 min, 4°C	2 min	1 µM clonazepam 1 µM diazepam 200 nM CL218,872 (BZ-1 sites)	92,93 94
Muscarinic cholinergic	[3H]Quinuclidinyl benzilate	1 nM	2 hr, rm. temp.	2 × 5 min	1 µM atropine	97
	[3H]-N-Methylscopolamine	1 nM	1 hr, rm. temp.	2 × 5 min	0.1 mM carbachol (high-affinity site) 1 µM atropine	98
Dopamine	[3H]Spiperone	0.4 nM	1 hr, rm. temp.	2 × 5 min	1 µM ADTN 0.4 µM haloperidol 0.1 µM spiperone	54
Serotonin	[3H]Spiperone	0.4 nM	1 hr, rm. temp.	2 × 5 min	0.3 µM cinancerin (5HT-1 site) 0.4 M haloperidol 0.1 µM spiperone	54
Serotonin	[3H]Serotonin	1.0 nM	1 hr, rm. temp.	2 × 5 min	1 µM LSD (5HT-2 site)	100
	[3H]LSD	3.1 nM	1 hr, rm. temp.	2 × 5 min	1 µM LSD	100
	[3H]LSD	6 nM	1 hr, rm. temp.	—	10 µM 5-HT	99
Histamine-H$_1$	[3H]Mepyramine	5 nM	40 min, 4 °C	10 min	2 µM triprolidine	101,102

Cryostat sections taken from unfixed or lightly fixed animal tissues are routinely used. The method is also applicable to the use of human postmortem tissue and thus could be used to study microscopic changes in receptor density occurring in different pathological states. These tissue sections are thaw mounted on slides and incubated in coplin jars containing the radioactive ligand in a buffering solution. The unbound and nonspecifically bound radioactive ligand is then rinsed off in buffer alone, sacrificing as little of the specific binding as possible.

Initially, the ligand concentration, incubation time, and rinse time are estimated from previous binding studies performed in homogenates. By varying these parameters in different sets of tissue sections, it is possible to generate dissociation, association, and saturation data by directly studying the binding of the radioactive ligand to the tissue sections. This is accomplished by incubating and rinsing the tissues and then wiping them off the slide with a glass microfiber filter disk. The disk can be immersed in a scintillation cocktail, and the amount of radioactivity remaining in each set of tissue sections can be determined by counting in a liquid scintillation counter. Each variable introduced in the incubation or rinsing procedure is also performed on adjacent tissue sections which are incubated under similar conditions but in the presence of excess displacer. Thus, plotting the data for various rinse times for example, will show how the total amount of binding is affected compared to nonspecific binding. This allows the investigator to precisely choose the incubation conditions that result in highly specific labeling of the receptor of interest. By placing different displacers in varying concentrations in separate incubation solutions, displacement curves can be generated to directly study the pharmacology of the binding to the tissue section. The latter step is important to provide convincing evidence that it is indeed the pharmacologically relevant receptor that is being labeled by this procedure.

After the optimal binding conditions are established, another set of slide-mounted tissue sections can be incubated using these conditions, and then, instead of being wiped from the slide, the tissues are dried by cool dry air blown over their surfaces. Once dry, the labeled tissue sections can be stored at 4°C in the presence of desiccant until coverslips are applied.

Thin flexible coverslips (25×77 mm, Corning #0, Corning, NY) are dipped (under safelight conditions) in Kodak NTB-3 (diluted 1:1 with water) emulsion and allowed to dry. The slide-mounted tissue sections are then allowed to come to room temperature, and the coverslips are glued to the frosted end of the slide in such a way that the emulsion coating on the long coverslip is directly over the labeled tissue section. The coverslip is covered with a piece of Teflon® and clamped tightly down against the tissue with a binder clip. These assemblies are stored at 4°C in light-tight boxes containing desiccant for varying amounts of time. After this exposure period, the binder clip and Teflon® are removed, and the coverslip is bent back away from the tissue-containing end of the slide and maintained in that position with a spacer bar. The latent image on the coverslip is developed, the tissue section is stained with pyronin-Y and dried, and the coverslip is reapposed to the tissue section and maintained there with permanent mounting fluid.

Slides prepared in this manner provide quantifiable grain densities immediately overlying the brain tissue areas where the radioactivity was bound. Bright field illumination can be used to define morphological areas of interest on the tissue section. Then, by switching to darkfield illumination, the investigator can observe the concentration of radioactive grains associated with that area and thus establish the apparent density of receptors associated with specific regions of brain.

2.2.2b. Tritium–Film Method. The introduction of a photographic film sensitive to tritium is an advance in convenience and speed of acquiring autoradiographs of slide-mounted tissue sections. Labeling of the tissue is performed in the same fashion as described in the previous section. The backs of the slides containing the labeled tissue sections are then taped (double-stick tape) to photographic mounting board and placed in the bottom of an X-ray film cassette. A sheet of the tritium film is then placed over the slides, another sheet of mounting board is placed on top, and the cassette cover is closed by putting pressure on the sandwiched layers inside. Since the film is highly sensitive to tritium, the autoradiographs are visible with less exposure time. The film is simply removed from the cassette, developed, and viewed under a standard light microscope. In this case, the tissue is separate from the autoradiograph, and, thus, interpretation of the anatomic locus of the grains is more subjective. Quite satisfactory results, however, have been obtained in preliminary studies using this technique.[113–115]

3. CNS RECEPTOR LOCALIZATION: FOCUS ON THE HIPPOCAMPAL FORMATION

To illustrate the type of information that can be gained from receptor autoradiography, we have compiled a detailed table listing the localizations of many types of receptors found in the hippocampal formation (Table IV), and we have included photomicrographs of some of these distributions (Figs. 1, 2). Evidence such as this, indicating the presence of specific receptor types in microscopic divisions of the hippocampal formation, can be used to substantiate and complement experimentation from other fields of scientific endeavor. For instance, GABA-containing cells in the hippocampal formation are thought to be interneurons that are widespread throughout the hippocampus and dentate gyrus (see refs. 116, 117 for review). Electrophysiological evidence on hippocampal slice preparations indicates that GABA can excite or inhibit pyramidal cell dendrites.[118]

Glutamic acid decarboxylase (GAD) immunoreactivity has demonstrated GABA-containing short-axon neurons in all layers of the hippocampal formation[119] which are not affected by lesions of afferent and efferent fibers. High-affinity GABA receptors are widespread in the hippocampal formation as well, with the highest densities concentrated in the molecular layer of the dentate gyrus and in laminae of CA-1 of the hippocampus. Thus, GABAergic interneurons may release their neurotransmitter onto the dendritic arboriza-

Table IV

Microscopic Distribution of Receptors in the Hippocampal Formation

Receptor	Ligand	Distribution	Density[a]	References
Muscarinic cholinergic	[³H]QNB	Hippocampus		36,37,77
		CA1		
		Stratum oriens	+++	
		Stratum radiatum	+++	
		Dentate Gyrus		
		Stratum moleculare	+++	
Muscarinic cholinergic (high-affinity site)	[³H]NMS + carbachol	Hippocampus		98
		CA1		
		Stratum oriens	+	
		Dentate Gyrus		
		Stratum moleculare	++	
Nicotinic cholinergic	[¹²⁵I]α-BTX	Hippocampus		71–73,75
		CA1–3		
		Outer portion of	+++	
		Stratum oriens		
		Stratum lacunosum	++	
		Dentate Gyrus		
		Polymorphic layer of hilus	+++	
Opiate	[³H]DHM	Hippocampus		23,39,42,88,1
	[³H]Naloxone	CA1–2		30
	[³H]Etorphine	Stratum pyramidale	+++	
	[³H]Diprenorphine	CA3		
		Stratum pyramidale	++	
		CA2		
		Stratum moleculare	+++	
		Dentate Gyrus		
		Stratum moleculare	++	
		Stratum granulare	++	
		Subiculum		
		(anterior)	+++	
		Presubiculum	+++	

Receptor	Ligand	Region	Density	Reference
Histamine-H$_1$	[³H]Mepyramine	Hippocampus		101
		CA3		
		Stratum oriens	+ + +	
		Stratum radiatum	+ + +	
		Dentate Gyrus	+ + +	
		Polymorphic layer of hilus		
		Subiculum	+ + +	
		(ventral)		
		Presubiculum	+ + +	
Benzodiazepine	[³H]Flunitrazepam	Hippocampus		93,94
		CA1		
		Stratum oriens	+ + +	
		Dentate gyrus	+ + +	
		Stratum moleculare (BZ-2 site)	+ + +	
β-Adrenergic	[³H]Dihydroalprenolol	Hippocampus		104
		CA1		
		Stratum oriens	+	
		Stratum radiatum	+	
		Dentate gyrus		
		Stratum moleculare	+ +	
Spiperone (spirodecanone site)	[³H]Spiperone	Hippocampus		
		Stratum pyramidal	+ + +	
		Subiculum	+ + +	
Serotonin	[³H]Serotonin [³H]LSD	Hippocampus		99,100
		CA1		
		Stratum oriens	+	
		Stratum radiatum (rostrally)	+	
		Stratum lacunosum moleculare	+	
		Dentate gyrus		
		Stratum moleculare	+ + +	
		Hilar region	+ + +	
		Prosubiculum	+ + +	
α-Adrenergic	[³H]WB-4101	Dentate Gyrus		105.106

(Continued)

Table IV (Continued)

Receptor	Ligand	Distribution	Density[a]	References
α₁ site		Stratum moleculare (caudally)	+	
α₂ site	[³H]Para-aminoclonidine	Hippocampus	+ +	105,106
		Stratum lacunosum moleculare	+ + +	
GABA	[³H]Muscimol	Hippocampus		95
		CA1		
		Stratum oriens	+ +	
		Stratum pyramidale	+ +	
		Stratum radiatum	+	
		Stratum lacunosum moleculare	+	
		Dentate gyrus		
		Stratum moleculare	+ +	
		Subiculum	+ +	
		Presubiculum	+ +	
Kainic acid	[³H]Kainic acid	Hippocampus CA3		
		Stratum incidium	+ + +	111

[a] Density: + + +, high; + +, intermediate; +, low.

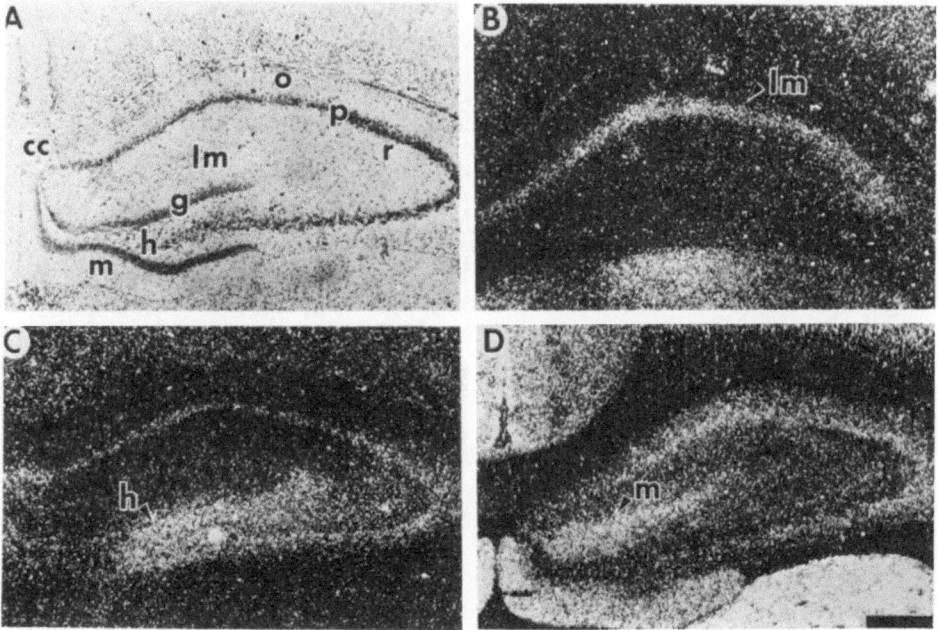

Fig. 1. Neurotransmitter receptors in the hippocampal formation. A. This is a brightfield photomicrograph of a section through the hippocampus stained with cresyl violet. cc, corpus callosum; o, stratum oriens; p, stratum pyramidal; r, stratum radiatum; lm, stratum lacunosum moleculare; g, granule cell layer of the dentate gyrus; h, hilar region of the dentate gyrus; m, molecular layer of the dentate gyrus. B. A darkfield photomicrograph of a section labeled with [³H]clonidine to localize α₂-adrenergic receptors. Under darkfield illumination, the autoradiographic grains appear on the coverslip as small white dots against a dark background. In this photomicrograph, a "streak" of autoradiographic grains appears over the stratum lacunosum moleculare (lm), indicating the presence of α₂ receptors in this area. The tissue immediately under the coverslip cannot be seen. C. Darkfield photomicrograph showing the distribution of histamine-H₁ receptors as identified by [³H]mepyramine binding. Most of the grains are confined to the hilar region of the dentate gyrus (h) extending into CA3. D. Darkfield photomicrograph demonstrating the autoradiographic grain distribution in the hippocampal formation after labeling with [³H]muscimol. In this case, the presence of grains indicates GABA receptors concentrated predominantly in the molecular layer of the dentate gyrus (m). Bar, 500 μm.

tions of the pyramidal and granule cells. These receptor localizations should prove valuable as "maps" for future electrophysiological studies in which GABA can be micropulsed in the dendritic fields of these cells.

Serotonin may inhibit hippocampal neurons after its release from fibers that originate in the raphe nuclei.[20] Serotonergic fibers have been shown to have a widespread distribution in the hippocampal formation.[121,122] High densities of serotonin receptors are found in CA-1 of the hippocampus and in the hilar region of the dentate gyrus. These receptor distributions correspond with areas that contain terminals from serotonin-containing fibers that project from the median raphe nucleus through supracallosal fibers.[122] Serotonin receptors are also found in the stratum lacunosum moleculare which, by fluorescence histochemical techniques,[123,124] has been shown to contain serotonin terminals.

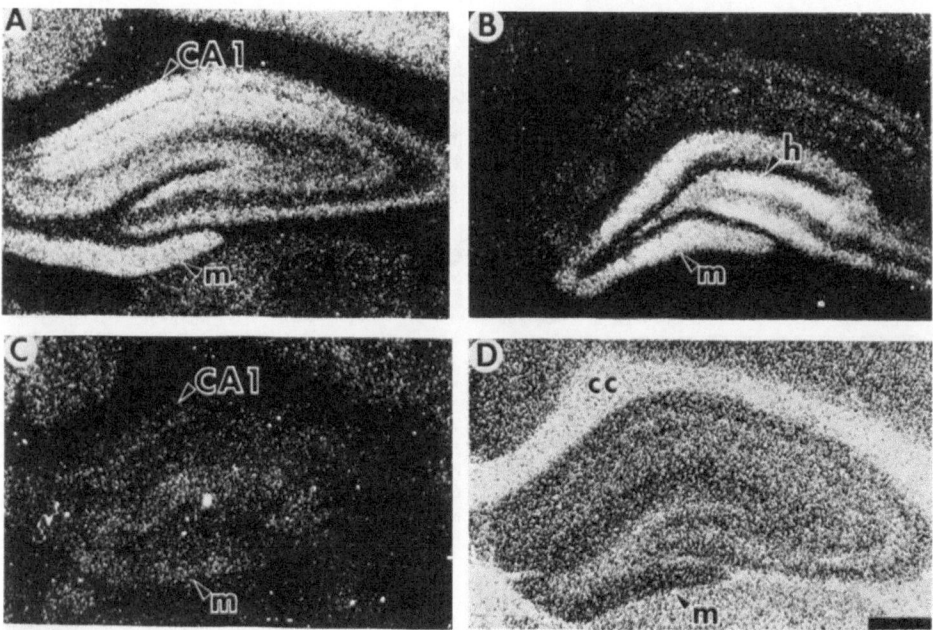

Fig. 2.   Neurotransmitter receptors in the hippocampal formation, A. This darkfield photomicrograph depicts the autoradiographic grains associated with muscarinic cholinergic receptors labeled with [³H]N-methylscopolamine. Above-background grain densities are seen in virtually all layers of the hippocampus and dentate gyrus (except cellular layers). The highest concentration of autoradiographic grains is found in the layers of CA1 in the hippocampus and in the molecular layer of the dentate gyrus (m). B. Darkfield photomicrograph showing the location of serotonin receptors in the hippocampal formation. These receptors were labeled with [³H]LSD and occur with the highest density in the molecular layer (m) and hilar region (h) of the dentate gyrus. C. The positions of β-adrenergic receptors are depicted in this darkfield photomicrograph after labeling of the tissue with [³H]dihydroalprenolol. The overall grain density associated with the hippocampal formation is very low and thus has affected the quality of this darkfield photomicrograph. It can be appreciated, however, that a small number of β receptors are found in CA1 of the hippocampus and in the molecular layer of the dentate gyrus (m). D. A brightfield photomicrograph of the grain distribution seen on tritium-sensitive film after exposure to a section labeled with [³H]flunitrazepam to demonstrate benzodiazepine receptors. Since this photomicrograph was taken under brightfield illumination, the autoradiographic grains appear black against a white background. There is no tissue present in the picture. Grain densities representing benzodiazepine receptors can be seen in virtually all layers of the hippocampus and dentate gyrus. Compare these grain concentrations with the background density seen associated with the corpus callosum (cc). The highest grain density is found in the molecular layer of the dentate gyrus (m). Bar, 500 μm.

Catecholamine-containing neurons have been localized in the hippocampal formation using fluorescence[125] and immunofluorescence[126] methods. These fibers originate from the locus coeruleus.[127] In the dentate gyrus, catecholamine-containing fibers are found concentrated in the hilar region, whereas adrenergic receptors are concentrated in the stratum moleculare (Table IV). The catecholamine-containing fibers do extend, however, into the stratum moleculare where they are oriented perpendicular to the surface. The hippocampus has catecholamine-containing fibers in the stratum lacunosum moleculare, arranged in an axonal plexus which correlates with the distribution of α_2-adrenergic receptors in this area.

Numerous studies have indicated the prominence and significance of cholinergic inputs to the hippocampus,[117] most of which originate in the septum. Attempts to block or mimic physiological responses of hippocampal neurons elicited with acetylcholine have indicated the presence of different populations of cholinoceptive neurons which are nicotinic or muscarinic.[128] Both nicotinic and muscarinic receptors have been localized in the hippocampal formation (Table IV). In the dentate gyrus, nicotinic cholinergic receptors are highly concentrated in the hilar region, whereas muscarinic cholinergic receptors are more numerous in the molecular layer. This would appear to be an excellent area for future studies to focus on in an attempt to define the differential effects of stimulation of these cholinergic receptors in the hippocampal formation.

It is apparent from the few examples outlined here that there are areas of overlap and areas of discrepancy when attempts are made to correlate neurotransmitter receptor density with the neuron population containing a specific neurotransmitter. Further studies will be necessary to resolve the reasons for these differences. It is possible to develop many different hypotheses that would explain the presence of receptors where no terminals have been localized that contain the indicated neurotransmitter. It is somewhat more disturbing and perplexing to find areas of supposed neurotransmitter-specific terminals where there are no identifiable receptors.

4. FUTURE CONSIDERATIONS

The ability to microscopically localize a wide variety of receptors using specific radioactive ligands is a major advance in receptology. Some types of receptors still cannot be localized by these techniques because of the low affinity of ligands specific for that receptor type. It is expected, however, that as new ligands with higher affinities are developed, receptor autoradiography will prove useful in the study of the binding distribution.

Development of methodology that better preserves the tissue morphology and increases the resolution of receptor autoradiography would be a welcomed improvement. The ease and quickness of generating autoradiographic results have been accelerated by the introduction of the tritium-sensitive film. It is now possible to quantitate autoradiographic grain concentrations using densitometry. Typical automated methods for measuring grain density are somewhat limited in their ability to resolve small areas; however, highly technical computer-based microdensitometers can approach the limits needed to resolve the autoradiographic grains associated with an individual cell. Computer-assisted microdensitometry can be gauged with appropriate standards to read femtomoles of receptors in individual brain areas. These densities can also be color coded for quick and easy determination of subtle differences between regions of the CNS.[113]

Electron microscopic localization of receptors would obviously increase the resolving power of receptor localization, but here we must question whether autoradiography is the method of choice. A single grain circles up through a monolayer of emulsion in a helical fashion, and we are confronted with the problem of calculating the probability of where the grain actually originated. This probability usually includes a synaptic cleft but involves the presynaptic element, the postsynaptic element, and any other cellular or subcellular membrane that may be nearby. The already excellent and highly developed emulsions will have to be further improved, or we will have to use other methods. Immunohistochemical localization of receptors using the peroxidase–antiperoxidase method is indicated here and awaits the isolation and purification of antibodies specific for different receptor types.

Thus, even though there remains the need for improvement, receptor autoradiography is providing us with information about the microscopic distribution of receptor types that was not previously attainable. Valuable information is being obtained about the sites of drug and neurotransmitter action and how these correlate with known effects or side effects of various compounds. The combination of receptor autoradiography with chemical and surgical lesions allows an investigator to pinpoint cell types that bear certain kinds of receptors and to determine how these might correlate with known transmitter inputs. Future studies such as these will undoubtedly improve our understanding of where different drugs exert their effects on the nervous system and perhaps even suggest other compounds to use in attempts to establish and manipulate functional neurotransmitter-specific connections in the CNS.

ACKNOWLEDGMENTS. The authors gratefully acknowledge the secretarial assistance of Jane Wamsley. We also wish to express our appreciation to Dr. Michael J. Kuhar whose help and guidance to us and many others has made possible much of the work cited in this chapter.

REFERENCES

1. Creese, I., and Snyder, S. H., 1977, *Nature* **270**:180–182.
2. Peroutka, S. J., U'Prichard, D. C., Greenberg, D. A., and Snyder, S. H., 1977, *Neuropharmacology* **16**:549–556.
3. Snyder, S. H., Greenberg, D., and Yamamura, H. I., 1974, *Arch. Gen. Psychiatry* **31**:58–61.
4. Snyder, S. H., and Yamamura, H. I., 1977, *Arch. Gen. Psychiatry* **34**:236–239.
5. Olsen, R. W., Reisine, T. D., and Yamamura, H. I., 1980, *Life Sci.* **27**:801–808.
6. Atlas, D., and Levitski, A., 1977, *Proc. Natl. Acad. Sci. U.S.A.* **74**:5290–5294.
7. Atlas, D., and Melamed, E., 1978, *Brain Res.* **150**:377–385.
8. Atlas, D., Teichberg, V. I., and Changeaux, J. P., 1977, *Brain Res.* **128**:532–536.
9. Melamed, E., Lahav, M., and Atlas, D., 1976, *Nature* **261**:420–422.
10. Melamed, E., Lahav, M., and Atlas, D., 1976, *Brain Res.* **116**:511–515.
11. Correa, F. M. A., Innis, R. B., Rouot, B., Pasternak, G. W., and Snyder, S. H., 1980, *Neurosci. Lett.* **16**:47–53.
12. Hess, A., 1979, *Brain Res.* **160**:533–538.
13. Landis, S., Phillips, M. I., Stamler, J. F., and Raizada, M. K., 1980, *Science* **210**:791–793.
14. Hazum, E., Chang, K. J., and Cuatrecasas, P., 1979, *Science* **206**:1077–1079.
15. Petrusz, P., 1975, *Anatomical Neuroendocrinology* (W. E. Stumpf and L. D. Grant, eds.), S. Karger, Basel, pp. 176–184.
16. Sternberger, L. A., 1978, *J. Histochem. Cytochem.* **26**:542–544.

17. Chan-Palay, V., 1978, *Proc. Natl. Acad. Sci. U.S.A.* **75:**1024–1028.
18. Chan-Palay, V., 1978, *Proc. Natl. Acad. Sci. U.S.A.* **75:**2516–2520.
19. Chan-Palay, V., and Palay, S. L., 1978, *Proc. Natl. Acad.Sci. U.S.A.* **75:**2977–2980.
20. Chan-Palay, V., Yonezawa, T., Yoshida, S., and Palay, S., 1978, *Proc. Natl. Acad. Sci. U.S.A.* **75:**6281–6284.
21. Yazulla, S., and Brecha, N., 1981, *Proc. Natl. Acad. Sci. U.S.A.* **78:**643–647.
22. Beaudet, A., Tremeau, O., Menez, A., and Droz, B., 1979, *C. R. Acad. Sci. [D] (Paris)* **289:**591–594.
23. Herkenham, M., and Pert, C. B., 1980, *Proc. Natl. Acad. Sci. U.S.A.* **77:**5532–5536.
24. Stumpf, W. E., and Roth, L. G., 1966, *J. Histochem. Cytochem.* **14:**274–287.
25. Roth, L. J., and Stumpf, W. E., 1969, *Autoradiography of Diffusible Substances*, Academic Press, New York.
26. Stumpf, W. E., and Grant, L. D. (eds.), 1975, *Anatomical Neuroendocrinology*, S. Karger, Basel.
27. Yamamura, H. I., Enna, S. J., and Kuhar, M. J., 1978, *Neurotransmitter Receptor Binding*, Raven Press, New York.
28. Enna, S. J., and Yamamura, H. I. (eds.), 1980, *Neurotransmitter Receptors*, Part 1, Chapman and Hall, London.
29. Hoffman, B. B., and Lefkowitz, R. J., 1980, *Annu. Rev. Pharmacol. Toxicol.* **20:**581–608.
30. Minneman, K. P., Pittman, R. N., and Molinoff, P. B., 1981, *Annu. Rev. Neurosci.* **4:**419–461.
31. Pepeu, G., Kuhar, M. J., and Enna, S. J. (eds.), 1980, *Receptors for Neurotransmitters and Peptide Hormones*, Raven Press, New York.
32. Williams, L. T., and Lefkowitz, R. J., 1978, *Receptor Binding, Studies in Adrenergic Pharmacology*, Raven Press, New York.
33. Yamamura, H. I., Olsen, R. W., and Usdin, E. (eds.), 1980, *Psychopharmacology and Biochemistry of Neurotransmitter Receptors,* Elsevier/North-Holland, Amsterdam.
34. Yamamura, H. I., and Snyder, S. H., 1974, *Proc. Natl. Acad. Sci. U.S.A.* **71:**1725–1729.
35. Yamamura, H. I., Kuhar, M. J., and Snyder, S. H., 1974, *Brain Res.* **80:**170–176.
36. Kuhar, M. J., and Yamamura, H. I., 1975, *Nature* **253:**560–561.
37. Kuhar, M. J., and Yamamura, H. I., 1976, *Brain Res.* **110:**229–243.
38. Pert, C. B., Kuhar, M. J., and Snyder, S. H., 1975, *Life Sci.* **16:**1849–1854.
39. Schubert, P., Hollt, V., and Herz, A., 1975, *Life Sci.* **16:**1855–1856.
40. Atweh, S. F., and Kuhar, M. J., 1977, *Brain Res.* **124:**53–67.
41. Atweh, S. F., and Kuhar, M. J., 1977, *Brain Res.* **129:**1–12.
42. Atweh, S. F., and Kuhar, M. J. 1977, *Brain Res.* **134:**393–405.
43. Pert, C. B., Kuhar, M. J.,and Snyder, S. H., 1976, *Proc. Natl. Acad. Sci. U.S.A.* **73:**3729–3733.
44. Baudry, M., Martres, M. P., and Schwartz, J. C., 1977, *Life Sci.* **21:**1163–1170.
45. Hollt, V., Czlonkowski, A., and Herz, A., 1977, *Brain Res.* **130:**176–183.
46. Kuhar, M. J., Murrin, L. C., Malouf, A. T., and Klemm, N., 1978, *Life Sci.* **22:**203–210.
47. Klemm, N., Murrin, L. C., and Kuhar, M. J., 1979, *Brain Res.* **169:**1–9.
48. Murrin, L. C., and Kuhar, M. J., 1979, *Brain Res.* **177:**279–285.
49. Creese, I., and Snyder, S. H., 1978, *Eur. J. Pharmacol.* **49:**201–202.
50. Howlett, D. R., Morris, H., and Nahorski, S. R., 1979, *Mol. Pharmacol.* **15:**506–514.
51. Leysen, J. E., Gommeren, W., and Laduron, P., 1978, *Biochem. Pharmacol.* **27:**307–316.
52. Leysen, J. E., Niemegeers, C. J. E., Tollenaere, J. P., and Laduron, P. M., 1978, *Nature* **272:**168–171.
53. Peroutka, S. J., and Snyder, S. H., 1979, *Mol. Pharmacol.* **16:**687–699.
54. Palacios, J. M., Niehoff, D. L., and Kuhar, M. J., 1981, *Brain Res.* **213:**277–289.
55. Barnard, E. A., Wieckowski, J., and Chiu, T. H., 1971, *Nature* **234:**207–209.
56. Bourgeois, J. P., Ryter, A., Menez, A., Fromageot, P., Boquet, P., and Changeux, J. P., 1972, *FEBS Lett.* **25:**127–133.
57. Fertuck, H. C., and Salpeter, M. M., 1974, *Proc. Natl. Acad. Sci. U.S.A.* **71:**1376–1378.
58. Hartzell, H. C., and Fambrough, D. M., 1973, *Dev. Biol.* **30:**153–165.
59. Porter, C. W., Chiu, T. H., Wiekowski, J., and Barnard, E. A., 1973, *Nature* **241:**3–7.
60. Eterovic, V. A., and Bennett, E. L., 1974, *Biochim. Biophys. Acta* **362:**346–355.

61. Moore, W. M., and Brady, R. N., 1976, *Biochim. Biophys. Acta* **444**:252–260.
62. Salvaterra, P. M., Mahler, H. R., and Moore, W. J., 1975, *J. Biol. Chem.* **250**:6469–6475.
63. Salvaterra, P. M., and Moore, W. J., 1973, *Biochem. Biophys. Res. Commun.* **55**:1311–1318.
64. Schleifer, L. S., and Eldefrawi, M. E., 1974, *Neuropharmacology* **13**:53–63.
65. Schmidt, J., 1977, *Mol. Pharmacol.* **13**:283–290.
66. Lowy, J., MacGregor, J., Rosenstone, J., and Schmidt, J., 1976, *Biochemistry* **15**:1522–1527.
67. Salvaterra, P. M., and Mahler, H. R., 1976, *J. Biol. Chem.* **251**:6327–6334.
68. Duggan, A. W., Hall, J. G., and Lee, C. Y., 1976, *Brain Res.* **107**:166–170.
69. Miledi, R., and Szczepaniak, A. C., 1975, *Proc. R. Soc. Lond. Biol.* **190**:267–274.
70. Arimatsu, Y., Seto, A., and Amano, T., 1978, *Brain Res.* **147**:165–169.
71. Hunt, S., and Schmidt, J., 1978, *Brain Res.* **157**:213–232.
72. Polz-Tejera, G., Schmidt, J., and Karten, H. J., 1975, *Nature* **258**:349–351.
73. Segal, M., Dudai, Y., and Amsterdam, A., 1978, *Brain Res.* **148**:105–119.
74. Silver, J., and Billiar, R. B., 1976, *J. Cell. Biol.* **71**:956–963.
75. Hunt, S. P., and Schmidt, J., 1978, *Brain Res.* **142**:152–159.
76. Burgen, A. S. V., Hiley, C. R., and Young, J. M., 1974, *Br. J. Pharmacol.* **51**:279–385.
77. Rotter, A., Birdsall, N. J. M., Burgen, A. S. V., Field, P. M., Hulme, E. C., and Raisman, G., 1979, *Brain Res. Rev.* **1**:141–165.
78. Rotter, A., Birdsall, N. J. M., Burgen, A. S. V., Field, P. M., Smolen, A., and Raisman, G., 1979, *Brain Res. Rev.* **1**:207–224.
79. Rotter, A., Birdsall, N. J. M., Field, P. M., and Raisman, G., 1979, *Brain Res. Rev.* **1**:167–183.
80. Rotter, A., Field, P. M., and Raisman, G., 1979, *Brain Res. Rev.* **1**:185–205.
81. Kuhar, M. J., Taylor, N., Wamsley, J. K., Hulme, E. C., and Birdsall, N. J. M., 1981, *Brain Res.* **216**:1–9.
82. Williamson, M. J., Paul, S. M., and Skolnick, P., 1978, *Nature* **275**:551–553.
83. Mohler, H., Battersby, M. K., and Richards, J. G., 1980, *Proc. Natl. Acad Sci. U.S.A.* **77**:1666–1670.
84. Mohler, H., and Richards, G., 1980, *Psychopharmacology and Biochemistry of Neurotransmitter Receptors* (H. I. Yamamura, R. W. Olsen, and E. Usdin, eds.), Elsevier/North-Holland, Amsterdam, pp. 649–654.
85. Mohler, H., Wu, J. Y., and Richards, J. G., 1981, *GABA and Benzodiazepine Receptors* (E. Costa, G. DiChiara, and G. L. Gessa, eds.), Raven Press, New York, pp. 139–146.
86. Roth, L. J., Diab, I. M., Watanabe, M., and Dinerstein, R., 1974, *Mol. Pharmacol.* **10**:986–998.
87. Young, W. S., III, and Kuhar, M. J., 1979, *Brain Res.* **179**:255–270.
88. Meibach, R. C., and Maayani, S., 1980, *Eur. J. Pharmacol.* **68**:175–179.
89. Wamsley, J. K., Zarbin, M. A., Young, W. S. III, and Kuhar, M. J., 1981, *Neuroscience* **7**:595–613.
90. Young, W. S. III, Wamsley, J. K., Zarbin, M. A., and Kuhar, M. J., 1980, *Science* **210**:76–78.
91. Goodman, R. R., Snyder, S. H., Kuhar, M. J., and Young, W. S. III, 1980, *Proc Natl. Acad. Sci. U.S.A.* **77**:6239–6243.
92. Young, W. S. III, and Kuhar, M. J., 1979, *Nature* **280**:393–395.
93. Young, W. S. III, and Kuhar, M. J., 1980, *J. Pharmacol. Exp. Ther.* **212**:337–346.
94. Young, W. S. III, Niehoff, D., Kuhar, M. J., Beer, B., and Lippa, A. S., 1981, *J. Pharmacol. Exp. Ther.* **216**:425–430.
95. Palacios, J. M., Wamsley, J. K., and Kuhar, M. J., 1981, *Brain Res.* **222**:285–307.
96. Palacios, J. M., Young, W. S. III, and Kuhar, M. J., 1980, *Proc. Natl. Acad. Sci. U.S.A.* **77**:670–674.
97. Wamsley, J. K., Lewis, M. S., Young, W. S. III, and Kuhar, M. J., 1981, *J. Neurosci.* **1**:176–191.
98. Wamsley, J. K., Zarbin, M. A., Birdsall, N. J. M., and Kuhar, M. J., 1980, *Brain Res.* **200**:1–12.
99. Meibach, R. C., Maayani, S., and Green, J. P., 1980, *Eur. J. Pharmacol.* **67**:371–382.
100. Young, W. S. III, and Kuhar, M. J., 1980, *Eur. J. Pharmacol.* **62**:237–239.
101. Palacios, J. M., Wamsley, J. K., and Kuhar, M. J., 1981, *Neuroscience* **6**:15–37.
102. Palacios, J. M., Young, W. S. III, and Kuhar, M. J., 1979, *Eur. J. Pharmacol.* **58**:295–304.

103. Zarbin, M. A., Wamsley, J. K., and Kuhar, M. J., 1981, *J. Neurosci.* **1**:532–547.
104. Palacios, J. M., and Kuhar, M. J., 1980, *Science* **208**:1378–1380.
105. Young, W. S. III, and Kuhar, M. J., 1979, *Eur. J. Pharmacol.* **59**:317–319.
106. Young, W. S. III, and Kuhar, M. J., 1980, *Proc. Natl. Acad. Sci. U.S.A.* **77**:1696–1700.
107. Young, W. S. III, and Kuhar, M. J., 1979, *Eur. J. Pharmacol.* **59**:161–163.
108. Young, W. S. III, and Kuhar, M. J., 1981, *Brain Res.* **206**:273–285.
109. Young, W. S. III, Kuhar, M. J., Roth, J., and Brownstein, M. J., 1980, *Neuropeptides* **1**:15–22.
110. Zarbin, M. A., Innis, R. B., Wamsley, J. K., Snyder, S. H., and Kuhar, M. J., 1981, *Eur. J. Pharmacol.* **71**:349–350.
111. Foster, A. C., Mena, E. E., Monaghan, D. T., and Cotman, C. W., 1981, *Nature* **289**:73–75.
112. Caro, L., and Van Tubergen, R. P., 1963, *J. Cell Biol.* **15**:173–188.
113. Palacios, J. M., Niehoff, D. L., and Kuhar, M. J., 1981, *Neurosci. Lett.* **25**:101–106.
114. Penny, J. B., Frey, K., and Young, A. B., 1981, *Eur. J. Pharmacol.* **72**:421–422.
115. Palacios, J. M., Unnerstall, J. R., Niehoff, D. L., and Kuhar, M. J., 1982, *J. Neurosci. Methods* (in press).
116. Fonnum, F., and Storm-Mathisen, J., 1978, *Handbook of Psychopharmacology*, Volume 9 (L. L. Iversen, S. D. Iversen, and S. H. Snyder, eds.), Plenum Press, New York, pp. 357–401.
117. Storm-Mathisen, J., 1978, *Functions of the Septo–Hippocampal System* (Ciba Symposium 58), Elsevier/North-Holland, Amsterdam, pp. 49–79.
118. Andersen, P., Bie, B., Ganes, T., and Mosfeldt Larsen, A., 1978, *Iontophoresis and Transmitter Mechanisms in the Mammalian Central Nervous System* (R. W. Ryall and J. S. Kelly, eds.), Elsevier/North Holland, Amsterdam, pp. 179–182.
119. Ribak, C. E., Vaughn, J. E., and Saito, K., 1979, *Brain Res.* **140**: 315–332.
120. Segal, M., 1975, *Brain Res.* **94**:115–131.
121. Azmitia, E. C., 1978, *Handbook of Psychopharmacology*, Volume 9 (L. L. Iversen, S. D. Iversen, and S. H. Snyder, eds.), Plenum Press, New York, pp. 233–314.
122. Azmitia, E. C., and Segal, M., 1978, *J. Comp. Neurol.* **179**:641–647.
123. Bjorkland, A., Nobin, A., and Stenevi, U., 1973, *Z. Zellforsch. Mikrosk. Anat.* **145**:479–501.
124. Fuxe, K., and Jonsson, G., 1974, *Adv. Biochem. Psychopharmacol.* **10**:1–12.
125. Blackstad, T. W., Fuxe, K., and Hokfelt, T., 1967, *Z. Zellforsch. Mikrosk. Anat.* **78**:463–473.
126. Swanson, L. W., and Hartman, B. K., 1975, *J. Comp. Neurol.* **163**:467–506.
127. Lindvall, O., and Bjorklund, A., 1978, *Handbook of Psychopharmacology*, Volume 9 (L. L. Iversen, S. D. Iversen, and S. H. Snyder, eds.), Plenum Press, New York, pp 139–231.
128. Segal, M., 1978, *Neuropharmacology* **17**:619–623.
129. Murrin, L. C., Coyle, J. T., and Kuhar, M. J., 1980, *Life Sci.* **27**:1175–1183.
130. Pearson, J., Brandeis, L., Simon, E., and Hiller, J., 1980, *Life Sci.* **26**:1047–1052.
131. Murrin, L. C., Gale, K., and Kuhar, M. J., 1979, *Eur. J. Pharmacol.* **60**:229–235.

Receptor Measurement

Ernest J. Peck, Jr. and Katrina L. Kelner

1. INTRODUCTION

Many cellular activities are controlled via signals from outside the cell. Examples of extracellular signals include hormones and neurotransmitters which act on target cells by way of receptors to induce specific responses. Cells unresponsive to a given signal are nontargets and generally possess few, if any, receptors for that signal. Receptors are macromolecules that may exist as free and soluble components of the cytoplasm, may reside within various intracellular compartments, or may exist primarily on the outer surface of plasma membranes. Receptors function via their capacity to recognize and bind specific signals, i.e., ligands. Implicit in receptor theory is the assumption that the occupied receptor, i.e., receptor–ligand complex, is responsible for the production of a biological response. Although the correlation of receptor occupancy with biological response is an important issue in receptor studies, the assessment of receptor–ligand interactions is the primary subject of this chapter. We have recently reviewed the receptor occupancy/response relationship for steroid[1] and neurotransmitter[2] systems.

2. CRITERIA FOR RECEPTORS

A set of criteria has been established for the demonstration of receptor systems. Wherever possible, these criteria should be met before the term "receptor" is applied to a binding system.

2.1. Finite Binding Capacity

Biological responses to extracellular signals are generally saturable phenomena. If formation of receptor–ligand complex is an obligatory step in the

Ernest J. Peck, Jr. and Katrina L. Kelner • Department of Cell Biology, Baylor College of Medicine, Texas Medical Center, Houston, Texas 77030.

production of a biological response, then the saturable nature of responses implies that the number of receptors per cell or per unit mass of tissue should be finite. To meet this criterion, the ligand-binding or receptive site under study must be saturable. This is demonstrated by exposing the receptive site source to various concentrations of radioactive ligand and subsequently measuring the amount of bound and/or free ligand at equilibrium. The attainment of a constant maximum binding value above a certain ligand concentration indicates that saturation has been achieved.

2.2. Appropriate Affinity

The affinity of a receptor for its ligand should be appropriate to the physiological concentration of ligand, i.e., that which exists *in vivo*. For example, signals that exist at low concentrations should have high-affinity receptors; otherwise, no complex would form, and no response could result. Conversely, low-affinity receptors are required when signals exist at high levels. However, it should be noted that both affinity and number of receptors determine the probability of complex formation. Thus, an increase in receptor number allows a decrease in receptor affinity with a similar probability of a biological response, assuming that response is proportional to the number of receptor–ligand complexes formed.

2.3. Ligand Specificity

Receptors should possess specificity for a given class of ligands. This selectivity enables a given target to respond to a specific signal without interference from other signals. Agonists and antagonists of the same class should compete for their receptor in a rank order that parallels their biological effectiveness. This principle may be employed to distinguish specific from nonspecific binding of ligands and to recognize multiple classes of receptor for a given signal. Although an important criterion, receptor sites do not display absolute specificity. The binding or receptive sites of many receptors have a limited capacity for discriminating between similar ligand structures. If the concentration of any compound greatly exceeds that of the physiological ligand, competition for the receptive site may occur.

2.4. Target Specificity

For many systems, only specific cell types respond to a given signal. If the response is receptor mediated, receptors should exist in or on these cell types but not others. Receptor systems for hormones generally meet this criterion. Thus, a limited number of tissues are stimulated by sex steroids, e.g., uterus, vagina, mammary gland, pituitary, and brain in the case of estrogens. In general, the density of receptor is much higher per unit mass of tissue in targets than in nontargets. However, caution must be exercised with this criterion. Targets such as the brain may have low receptor densities because of heterogeneous cell populations containing both target and nontarget cells.

2.5. Correlation with Biological Response

Implicit in all studies of macromolecules that bind effector ligands and meet the above criteria is the assumption that this binding results in a biological response. Thus, the binding of hormone to a putative receptor must precede or accompany a tissue response, and the extent of that response should relate to some function of receptor occupancy. This criterion, the demonstration of receptor-dependent response, is not often met in studies of receptor systems because it is the most difficult to establish, especially with neurotransmitters.

3. ANALYSIS OF SIMPLE AND COMPLEX BINDING SYSTEMS

The trials and tribulations of receptor study come with the analysis of binding data. For most receptor systems, the biochemist uses broken cell preparations or subcellular fractions. Often ligand binding is the only parameter available for the study of receptors. The theory and practice of such analyses will be the subject of this and subsequent sections.

3.1. Single-Component Systems

In most cases, receptors exist in the presence of other binding components which complicate the analysis of receptor binding parameters. However, for introduction and illustration, a system with only one binding or receptive site will be discussed initially. In such a system, the amount of receptor (R_t) can be determined *in vitro* by exposing replicates of a given receptor source (as a cytosol or membrane preparation) to varying ligand concentrations until near saturation of the receptor species is obtained. One then measures the amount of bound and/or free ligand at equilibrium for each concentration and relates these values to the affinity and capacity of the receptor system. The amount of bound ligand (RL) at equilibrium is related to free ligand (L), total receptor (R_t), and the dissociation constant (K_d) of the receptor–ligand complex by the following:

$$[RL] = ([R_t][L])/(K_d + [L])$$

This equation is equivalent to that which describes rapid equilibrium kinetics of enzyme systems and which is known as the Michaelis–Menten expression.[3] This expression applies equally well to receptor–ligand interactions at equilibrium.

The data from a saturation analysis (usually 10–90% saturation) can be employed to determine the number of binding sites (R_t, B_{max}, or n) and the K_d of the receptor–ligand complex. A number of mathematical manipulations of the equation above allow one to linearize binding data for the calculation of K_d and B_{max}. For instance, in a Scatchard analysis, one plots the ratio bound ligand/free ligand (y), as a function of bound ligand (x) at equilibrium.[4] The number of binding sites is given directly by $x_{y \to 0}$, whereas the K_d is given by the reciprocal of the slope. Alternatively, these parameters may be obtained by plotting the reciprocal of bound ligand concentration (1/[RL]) as a function of the reciprocal of free ligand concentration (1/[L]). This plot, analogous to the Lineweaver–Burk analysis of enzyme kinetics,[5] gives the number of sites (B_{max}) as the reciprocal of $y_{x \to 0}$, whereas $x_{y \to 0}$ equals $-1/K_d$. A third method is the direct linear plot,[6] a simple graphic procedure which requires no calculation and in which B_{max} and K_d are given directly from the intersection of a series of lines defined by [L] and [RL]. These methods will be illustrated for more complex systems in Section 3.2.1 below.

3.2. Complex Binding Systems

A simple system such as that described in the preceding section does not exist unless the receptor is pure and possesses only one class of active or receptive site. Usually other binding or receptive species are present in subcellular fractions and complicate the measurement of receptors. In this section, some of these complications are discussed together with how they can be manipulated such that receptor parameters can be ascertained.

3.2.1. Specific versus Nonspecific Binding

As discussed earlier, the binding of ligands by a receptor is selective and thus is defined as "specific." "Nonspecific" binding results from the interaction of ligand with sites that are of low affinity and high capacity relative to the receptor. The total bound ligand in a system composed of specific and nonspecific sites is the sum of that bound to receptor (RL) plus that bound to nonspecific sites (NL). The saturation analysis of such a mixed binding system is shown in Fig. 1A. Note that total binding (RL + NL) and nonspecific binding (NL) are nonsaturable, although specific binding, that is, the association of ligand with receptor (RL), is saturable.

The data from Fig. 1A are analyzed by the method of Scatchard in Fig. 1B. The [RL + NL]/[L] ratio is a curvilinear function of the amount of ligand bound, [RL + NL]. This curve is the vectorial sum of specific and nonspecific components, both of which are also plotted individually and appear as linear functions in Fig. 1B. The resolution of these components has proven a major difficulty for many receptor systems and will be discussed in detail below and in Section 3.2.3.

A direct assessment of nonspecific binding can sometimes be made by the inhibition of labeled ligand binding by nonlabeled competitive ligand. In such instances, the receptor source is exposed to multiple concentrations of radioactive ligand in the presence and absence of an excess of nonradioactive competitive ligand. Under these conditions, data such as that in Fig. 1A are obtained. The function designated as RL + NL is termed "total" binding and represents radiolabeled ligand that is bound to both receptor (or "specific") sites and nonspecific sites; thus, it contains both saturable and nonsaturable components. Nonspecific binding (NL) is the radioactive ligand bound in the

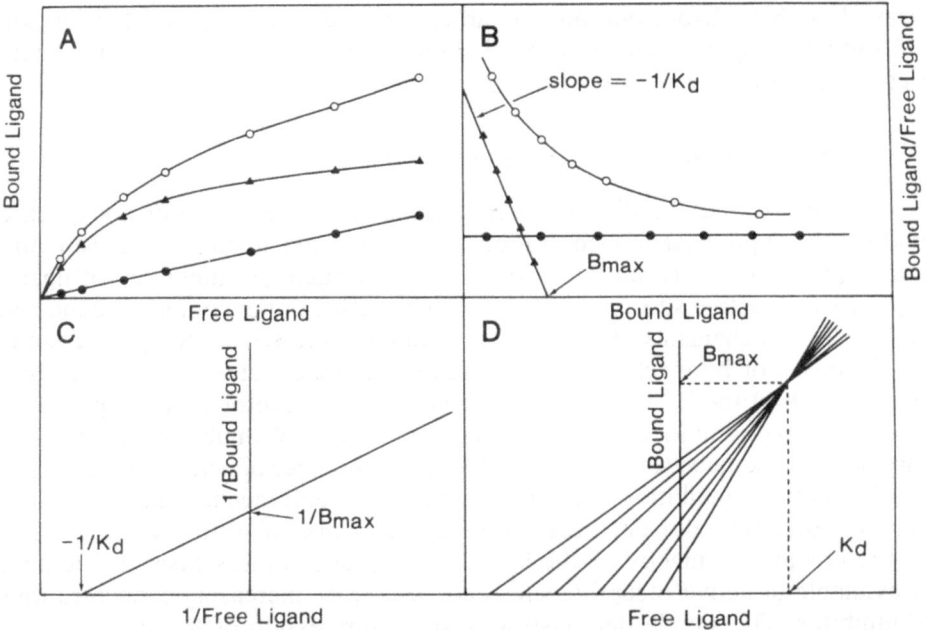

Fig. 1. Analysis of ligand binding to specific and nonspecific sites. A. Saturation analysis of total binding (○; labeled ligand only), nonspecific binding (●; labeled ligand plus excess unlabeled ligand), and specific binding (▲; total − nonspecific binding). B. Scatchard analysis of the theoretical data shown in A. C. Double-reciprocal analysis of the specific binding data calculated from A. D. Direct linear plot of the specific binding data calculated from A.

presence of excess unlabeled competitive ligand. The competing nonlabeled ligand occupies essentially all high-affinity receptive sites but does not interfere appreciably with the binding of labeled ligand to nonspecific sites. Receptor sites are estimated by subtracting NL from RL + NL. The number of receptive sites and their K_d can be determined by Scatchard, double-reciprocal, or direct linear analysis as shown in Figs. 1B, C, and D.

The use of inhibition to determine receptor binding parameters is based on the assumption that the nonlabeled ligand is a competitive inhibitor. If the nonlabeled ligand is identical to the radioactive ligand, this will be true. In some cases, however, it is necessary to use a nonidentical inhibitor, and the assumption of competitive inhibition should be verified. In addition, the use of competitive inhibition to determine receptor parameters is based on the assumption that nonspecific binding sites are of low affinity and high capacity relative to the receptor system. This has proven true for many receptor systems but should be validated by the demonstration of a straight line for nonspecific binding, as in Fig. 1A, or a horizontal line by Scatchard analysis (Fig. 1B).

For receptor assays, the use of the term nonspecific to describe nonreceptor binding is adequate. However, it should be realized that in practice, nonspecific binding is actually that binding that is nondisplaceable by a competitive ligand in the range of labeled and unlabeled ligand employed in the assay. For this reason, one must characterize the binding of labeled ligands and the nature of the inhibition of binding by unlabeled ligands before interpreting the results of such studies.

3.2.2. Competitive Inhibition of Ligand Binding

In order to use the displacement method for the estimate of nonspecific binding, the appropriate concentration of unlabeled ligand must be chosen, and displacement should result from competitive interactions. Inhibition of ligand binding by receptive sites may occur by competitive or noncompetitive means, that is, by mechanisms that involve mutually exclusive binding of ligands (competitive) or by mechanisms in which ligand inactivates (either reversibly or irreversibly) the ligand-binding capacity of the receptor (noncompetitive). In competitive inhibition, increasing concentrations of inhibitor alter the apparent K_d of the receptor but do not change the number of active binding sites. Noncompetitive inhibition may involve one of several mechanisms: the inhibitor may precipitate or denature the receptor or may bind to secondary sites on the receptor to inactivate the ligand-binding site. In this case, the number of available or active receptive sites decreases with increasing concentrations of inhibitor. The simple demonstration of a suppression of RL formation by the addition of an inhibitor does not establish a competitive mechanism. Double-reciprocal or Scatchard analyses of ligand binding in the presence of multiple concentrations of inhibitor should be employed to establish whether the inhibitor alters K_d (competitive) or R_t (noncompetitive).

The appropriate concentration of competitor to be used in measuring nonspecific binding can be determined by the ratio of the K_ds for receptor–ligand and receptor–competitor complexes. If these constants are the same, as when labeled and unlabeled ligand are identical, 10- and 100-fold excesses are sufficient to displace 91 and 99%, respectively, of labeled ligand from the high-affinity site. However, if the affinities of a receptive site for two dissimilar ligands are quite different, considerable molar excesses of one may be required to displace the other. In general, one should use the lowest concentration of unlabeled ligand consistent with displacement of 99% of "specific" ligand binding. One should avoid the use of excessive levels of competing ligand. Huge excesses of unlabeled ligand may result in the displacement of secondary, nonspecific sites and the artifactual production of multiple "specific" binding classes.

3.2.3. Methods for the Analysis of Nonspecific Binding

If nonspecific binding is small relative to specific binding, it can be ignored, and reasonable estimates of equilibrium or kinetic constants determined for a receptor system. However, in most receptor systems, the amount of nonspecific binding is significant with respect to total binding, and the probability of accurately estimating specific binding without appropriate correction is very low. A careful analysis of saturation in the presence and absence of competing ligand and/or the use of computer-assisted fitting procedures is necessary to establish the existence of receptor sites under these conditions. Methods for correcting binding data for nonspecific interactions are discussed below.

3.2.3a. Direct Subtraction. The simple subtraction method is most common for the estimation of specific binding in the presence of nonspecific binding. A series of incubation tubes containing only labeled ligand is employed together with a parallel set containing labeled ligand plus excess unlabeled ligand. The concentration of labeled ligand is the same for each pair of tubes. After the determination of bound radioactivity in all tubes, "nonspecific" binding is subtracted from total binding for each given pair. This manipulation is based on the assumption that nonspecific binding (NL) is the same in each of the pair of incubation tubes. The difference between the two values is often taken to be specific binding (RL) and used for the estimation of receptor binding parameters via such methods as are shown in Figs. 1B–D.

In fact, this method yields an erroneous estimate of specific binding because the values employed for nonspecific binding (NL) are not appropriate for the total-binding tubes (RL + NL). The actual nonspecific binding present in any total binding tube is proportional to the concentration of free radioactive ligand present in that tube. Free ligand is not the same in each tube of a pair, since free ligand is defined as total ligand added minus total ligand bound, and the value of total bound ligand is not the same in both tubes (RL + NL ≠ NL). Since free ligand at equilibrium is not the same for a given pair, nonspecific binding of radioactive ligand is, in fact, not the same for a given pair. However, under certain circumstances, these values may be the same. These are if (1) specific binding + nonspecific binding ~ specific binding; (2) specific binding + nonspecific binding ~ nonspecific binding; (3) an extremely small fraction of total ligand added is bound at equilibrium. If the first condition is true, one can ignore nonspecific binding, since it is negligible. Neither of the latter circumstances is desirable. If nonspecific binding is large relative to total binding, or if total binding represents an extremely low percentage of ligand bound, high levels of noise or scatter in saturation analyses will usually result. In such a circumstance, the calculation of equilibrium constants is compromised.

It should be noted that direct subtraction is the only possible method for determining specific binding in a one-point assay (that is, the use of a single saturating concentration of labeled ligand to measure total receptor number; see Section 3.2.5). Although not recommended, this procedure is necessary in some circumstances. To minimize the problems of the direct subtraction method, one should reduce nonspecific binding by purification of receptive site sources and/or extensive washing (if possible) after equilibration of ligand and receptor source.

3.2.3b. Linear Regression Estimate of Nonspecific Binding. The problems of direct subtraction are sometimes readily solved. One can use the values for nonspecific binding and free ligand as determined for tubes containing labeled plus unlabeled ligand to calculate a line by regression that defines nonspecific interactions as a function of free ligand concentration. The concentration of free ligand in total-binding tubes (total added minus total bound in tubes containing labeled ligand only) can then be used to calculate the nonspecific binding in those tubes via the line determined by regression. This value can then be subtracted from total binding, and the result defined as specific binding.

This method assumes that nonspecific binding is linear and thus that a regression analysis of nonspecific binding data will yield a good estimate of nonspecific binding as a function of free ligand. Scrutiny of nonspecific binding data from many experiments suggests that this assumption is not always valid; that is, a small saturable component exists in many assays of nonspecific binding, usually at lower concentrations of labeled ligand. Scatchard analysis of nonspecific binding ([NL]/[L] vs. [NL]) should result in a horizontal line if nonspecific binding is linear. However, in the assay of nuclear estrogen receptors and GABA receptors on junctional complexes (see Fig. 2), nonlinearities are apparent at low ligand concentrations. At higher concentrations of labeled ligand, [NL]/[L] versus [NL] approaches the horizontal. Clearly, the use of "nonspecific" binding data at lower concentrations in the regression analysis of [NL] versus [L] yields an erroneous estimate of the line.

Is this nonlinearity a property of nonspecific binding or a problem with the assay of nonspecific binding? In the former case, a regression analysis of nonspecific binding would not be valid; in the latter case, one might alleviate the problem. Figure 2 shows that this nonlinearity probably results from the competitive displacement employed to assess nonspecific binding. With successively higher concentrations of competing ligand, the saturable site can be eliminated almost completely. Thus, incomplete competition of receptor sites with unlabeled ligand is the probable source of this nonlinearity, even at 100-fold and 1000-fold excesses of unlabeled ligand. The linear regression method

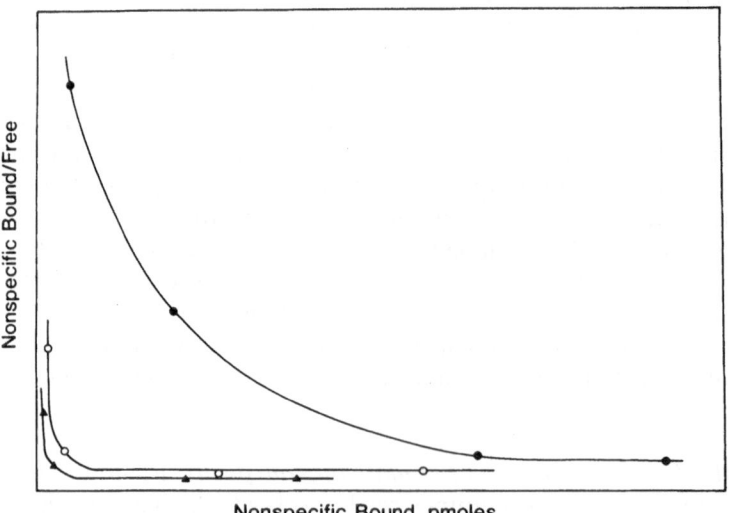

Fig. 2. Scatchard analysis of ligand binding in the presence of 10-, 100-, and 1000-fold excesses of unlabeled ligand. Junctional complexes were incubated with varying concentrations of [³H] muscimol in the presence of 10- (●) 100- (○), or 1000- (▲) fold excesses of unlabeled muscimol. Total bound/free was plotted as a function of total bound.

for the estimation of nonspecific binding cannot completely alleviate this problem. However, one can use values of [NL] versus [L] measured at high ligand concentrations to obtain a good estimate of the nonspecific parameter. This allows a reasonable determination of specific binding by difference.

One might consider the use of 10,000- or 100,000-fold molar excesses of unlabeled ligand on the assumption that, if a little is good, a lot is better. However, one should avoid the use of extraordinary excesses of unlabeled ligand, since nonspecific binding is displaced under these conditions. Such displacement can alter the equilibrium parameters calculated for specific sites and/or give curves suggestive of multiple sites. If nonspecific binding is large in a system, alternatives to "classic" procedures, such as those below, are necessary for the correction of nonspecific binding.

3.2.3c. Rosenthal Analysis. Data for total binding (RL + NL) can be corrected for nonspecific binding by the graphic method of Rosenthal[7] which resolves a curvilinear Scatchard plot ([RL + NL]/[L] vs. [RL + NL]) into two constitutive binding sites. The resolution employs vectors emanating from the origin and involves graphic iteration of any two sites. As reported, the method is time consuming and tedious. However, a modification of the Rosenthal method can be used for accurate and simple resolution of a single high-affinity binding site from nonspecific binding. For this analysis, nonspecific binding is first analyzed by the method of Scatchard ([NL]/[L] vs. [NL]). The horizontal component of this plot is subtracted by vector analysis from values of [RL + NL]/[L] versus [RL + NL]. This manipulation yields a corrected value for specific binding. Mathematically, this involves the transformation of rectangular coordinates for total and nonspecific binding into polar coordinates (length of vector and angle). Subtraction of vectors for nonspecific binding from vectors for total binding yields polar coordinates for specific binding at all concentrations of ligand. These values are converted to rectangular coordinates which represent [RL]/[L] versus [RL] and are analyzed via Scatchard. Since nonlinear nonspecific binding data are not utilized in this procedure, a reliable correction for nonspecific binding is obtained. This procedure is depicted in Fig. 3 and is used routinely in our laboratory for the resolution of saturation data. A simple program for use with either a programmable pocket calculator or a DEC-10 computer has been written and is available on request.

3.2.3d. Nonlinear Curve-Fitting Analyses. A more sophisticated, although conceptually similar, method of curve resolution is the nonlinear curve-fitting program, LIGAND, which was developed by Munson and Rodbard.[8] This program can separate specific from nonspecific binding parameters and, in addition, can resolve more complex binding systems. In this method, only primary data for total ligand added and total ligand bound are analyzed. No correction of the primary data is made for nonspecific binding.

This program can iteratively determine two or more fits to the data. One first assumes a single high-affinity receptive site plus a linear nonspecific binding model and allows the program to find the best fit under this assumption. Then one assumes two relatively high-affinity sides plus linear nonspecific

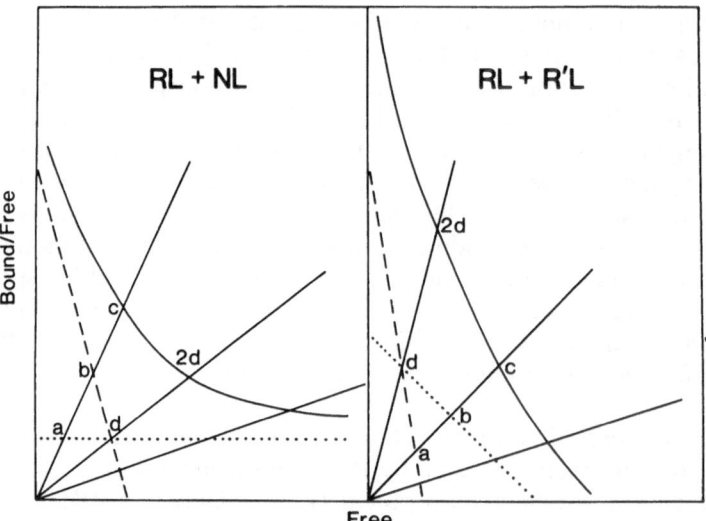

Fig. 3. Rosenthal resolution of ligand binding data. In the left panel is shown the resolution of nonspecific binding and a single saturable, specific site. Note that the vectors from the origin (0) to a and to b must sum to 0 to c; also, there exists a unique point where specific and nonspecific contributions are equal (2d on total bound curve). In the right panel, two saturable binding sites are resolved. In this case, the contribution by nonspecific binding has been corrected previously, and the resultant curve is resolved into two linear components.

binding and allows the computer to attempt to fit. Indicators of goodness of fit, including residual variance and a runs test, are calculated by LIGAND. One can then judge between alternative solutions. Data from several saturation analyses can be pooled using this program, as it also weights each saturation analysis to correct for differences between experiments in receptor number per tube.

The LIGAND program can readily resolve a single high-affinity binding site plus linear nonspecific binding. This is similar to the resolution performed with the modified Rosenthal. In addition, LIGAND has the capacity to resolve binding systems containing two relatively high-affinity binding sites plus a nonspecific site. This program represents a powerful method of data analysis in multicomponent binding systems (see below).

3.2.4. Multiple Specific Binding Components

In many receptor systems two or more binding species exist which may have high affinities for the same ligand. These mixtures of binding components are sometimes revealed in saturation curves; however, Scatchard plots of saturation analyses over a wide concentration range will usually demonstrate their presence. If a saturation analysis of a mixed binding system includes limited ligand concentrations, only a single component system may be apparent, and an improper estimate of K_d and B_{max} may result. Errors of this type are exaggerated when a saturable binding component of low affinity is in excess of a second site with a high affinity. In such cases, binding analyses employing low concentrations of ligand lead to gross overestimates of the number of a single class of sites and an underestimate of their affinity for ligand.

The ideal solution to the problem of mixed binding systems is the physical separation of components by purification. This allows the study of each as an isolated system. However, this is often not feasible because of limited sources of receptor. Rosenthal[7] and Feldman[9] have developed methods for resolving multiple binding components by the graphic analysis of curvilinear Scatchard plots. Simply stated, curved Scatchard plots are resolved by finding straight lines which, when summed point by point in a vectorial manner, reproduce the original curve. This procedure was discussed in Section 3.2.3c and is illustrated in Fig. 3. Note that two independent components must sum to the curve. The curve-fitting program, LIGAND, was also discussed for the analysis of nonspecific binding and is quite powerful at the resolution of multiple high-affinity systems. Theoretically, these methods could be employed to resolve any number of receptive species. However, the data from ligand-binding studies are usually limited and imprecisely determined. Consequently, the resolution of more than two specific components is improbable.

Analytical methods employing geometric or curve-fitting procedures are useful if extensive ligand-binding data are available. Sometimes this is not possible because of limitations in biological material. Other methods must be found to cope with the problem of multiple species of receptive sites. Differential inhibition of ligand binding is one method employed to measure a given component in a mixed system. The use of [³H]estradiol and diethylstilbestrol for the assay of estrogen receptors in the presence of α-fetoprotein is one example. α-Fetoprotein is present in large quantities in the neonatal rat and has an affinity for estradiol ($K_d \sim 10^{-9}$–10^{-10} M) similar to that of the estrogen receptor. The receptor is measured in the presence of α-fetoprotein by taking advantage of the fact that diethylstilbestrol binds with low affinity to α-fetoprotein but competes very effectively with [³H]estradiol for estrogen receptor binding sites. Thus, [³H]-ligand binding to receptor is displaced without altering that bound to α-fetoprotein.

As a second example, consider the binding of [³H]GABA to synaptic plasma membranes. Synaptic plasma membranes possess at least two binding sites for GABA, one coupled to a GABA transport mechanism and a second coupled to the receptor mechanism. Transport of GABA is competitively inhibited by nipecotic acid, although postsynaptic processes are not affected by this compound. On the other hand, muscimol is a potent GABA agonist but a poor substrate for the transport mechanism. Thus binding of [³H]GABA to postsynaptic receptors can be measured in the presence of neuronal transport sites by taking advantage of the differential pharmacological specificities of neuronal transport and postsynaptic systems.[2,10] Excess unlabeled nipecotic acid is employed to occupy all transport receptive sites while [³H]GABA in the presence or absence of unlabeled GABA (or a second competitive ligand such as muscimol) is used to assess nonspecific and total [³H]GABA binding to sites other than the transport site. Conversely, excess unlabeled muscimol may be employed to block all binding to postsynaptic receptors while variable

[³H]GABA plus or minus unlabeled GABA or nipecotic acid measures specific binding to transport receptive sites.[2] Given sufficient knowledge of the pharmacology of a mixed binding system, the experimentalist can use differential inhibition to measure multiple species individually.

3.2.5. The Use of One-Point Assays

Specific receptor sites should be demonstrated by their saturable nature, not by the use of "one-point" assays which employ a single concentration of labeled ligand plus or minus excess unlabeled ligand. In circumstances in which the tissue source of receptor and/or the number of receptive sites per unit of tissue are low, an investigator may be forced to employ "one-point" assays for the study of receptor dynamics in various physiological states. However, the receptor system must be characterized as to its saturable nature, and any extremes in receptor number must be confirmed by saturation analysis. Only in this manner can one be certain that changes in K_d, endogenous ligand, and/or number of receptive species do not produce artifacts in the receptor analysis.

3.2.6. Hooks and Curves in Scatchard Plots

When using graphic methods such as that of Scatchard to analyze ligand-binding data, one should remember the conditions under which such an analysis is valid. Values of bound and free ligand at equilibrium must be determined for a system in which receptor number is constant. The plot should be linear or readily resolved into linear components. The presence of convexities or concavities in Scatchard plots should not be taken as *a priori* evidence of cooperativity or multiple binding sites. Such nonlinearity simply indicates that the data do not fit a simple two-state model in which ligand and receptor exist at equilibrium as bound (occupied) or free. In addition to multistate systems, a number of methodological artifacts can lead to nonlinear Scatchard plots.

3.2.6a. High Concentrations of Ligand or Receptor Site. Although estimates of affinity by saturation analysis are valid when [L] or [R_t] is great compared to K_d, error in the estimate of bound or free ligand at extremes of either component compromises Scatchard analyses. Thus, when labeled ligand concentrations are high relative to K_d (i.e., only a small fraction, 0.01–0.5%, of the total ligand is bound), errors in the estimate of either bound or free ligand can be large. When receptor site concentration is high, an accurate estimate of free ligand is difficult. Over- or underestimates of either parameter, bound or free, are magnified by the ratio [RL]/[L] employed in Scatchard analysis; these may produce bizarre and deceiving nonlinearities.

3.2.6b. Nonequilibrium Conditions. Studies are often performed under conditions that are sufficient for equilibration of receptive site with ligand at saturating concentrations but insufficient at subsaturating concentrations of ligand. The resultant underestimate of "bound at equilibrium" at low ligand concentrations and the attendant overestimate of free produce a "hook" in the

Scatchard plot which is suggestive of "positive cooperativity." An increase in incubation time or temperature often eliminates such hooks.

3.2.6c. Instability of Ligand or Receptive Site. If labeled and unlabeled ligands differ significantly in chemical stability, there is no straightforward way to estimate [L] at equilibrium. Separation of total free ligand from bound ligand will not suffice, because apparent total free is composed of multiple ligand species. Such heterogeneity in ligand species can produce nonlinear Scatchard plots.

Additionally, receptor sites may show heterogeneity with respect to stability, a situation that results in a nonlinear Scatchard plot. For example, if receptors are stable when occupied but labile when unoccupied, hooks in Scatchard plots are likely. Thus, during equilibration with ligand, receptor sites at or near saturation are stable, whereas those exposed to subsaturating conditions constantly change in number. A variable receptor number does not allow a Scatchard analysis. This phenomenon probably has resulted in many reports of cooperativity in receptor systems.

3.2.6d. Dilution of Radioactivity. Many receptor preparations contain significant amounts of endogenous ligand. Levels of endogenous ligand may vary with physiological state, among treatment groups, and/or among subcellular fractions. Endogenous ligands dilute the specific activity of the labeled ligands employed in receptor assays. If the amount of endogenous ligand is similar to that of the labeled ligand in the binding analysis, then the specific activity of tracer in the low concentration range is markedly reduced relative to that at higher concentrations of labeled ligand. These effects may alter the shape and invariably alter the slope of a Scatchard plot. One can pretreat the receptor source, as with charcoal or multiple washes, to remove endogenous ligand. Alternatively, multiple saturation analyses at different incubation volumes may be employed to extrapolate apparent dissociation constants to infinite dilution. The latter is time consuming and expensive but necessary if levels of endogenous ligand cannot be reduced.

3.2.6e. Interacting Species. Scatchard plots are not linear in systems with multiple binding sites, ligand–ligand interactions, interactions between receptor sites, and cooperative states. Under such conditions, the interaction of ligand with receptor is not $R + S \rightleftharpoons RS$ and thus cannot be analyzed in a simple fasion. These are discussed by Hollenberg and Cuatrecasas[11] and by Rodbard and Feldman.[2] If faced with nonlinear Scatchard plots, one should first consider the many artifacts discussed above and then employ these sources for an analysis of interacting systems.

4. METHODS FOR RECEPTOR MEASUREMENT

A variety of procedures have evolved for the assay of receptors. These invariably include three discrete components: isolation of receptor source, equilibration of receptor source with labeled ligand, and the separation of bound from free ligand. The methods employed to measure a given receptor are determined by its subcellular localization, affinity for ligand, and abundance. Other factors of importance are the stability of the receptor and the physical nature of the ligand employed (hydrophobic vs. hydrophilic, charged vs. uncharged state, etc.). In the following discussion, two receptor systems will be used for illustration. The assay of cytoplasmic and nuclear estrogen receptors will be utilized to discuss soluble and partially immobilized receptors. The assay of GABA transport and postsynaptic receptive sites will illustrate the methods employed for membrane-bound receptors. Procedures other than those chosen as examples certainly exist and are used routinely for the assay of receptors. That they are not included herein does not imply that they are less desirable; rather, the number of examples is too large. Recent reviews and books should be consulted for additional methods.[1,11,13,14]

4.1. Steroid Receptors

After disruption of target tissues, steroid receptors are found primarily in two cellular compartments, the cytoplasm and nucleus. The number of receptors in the nuclear compartment reflects the previous exposure of the target to steroid and determines the extent of biological responses.[1] Thus, it is often important to measure both receptor states, cytoplasmic (primarily unoccupied by ligand) and nuclear (primarily occupied by ligand). Since steroids are hydrophobic, they interact readily with lipid and membrane fractions in a nonsaturable manner to produce high levels of nonspecific binding and noise in steroid receptor assays. For these reasons, a sucrose pad procedure has been developed to isolate cytosol and crude chromatin fractions that are relatively free of contaminating lipid. This procedure has been applied to estrogen receptors of the hypothalamus, pituitary, and uterus, as well as other steroid receptor systems.

4.1.1. Sucrose Pad/Exchange Procedure

Figure 4 illustrates the sucrose pad/exchange assay.[15] Target tissue is homogenized in ice-cold Tris–EDTA (TE) buffer (0.01 M Tris HCl, 1.5 mM EDTA, pH 7.4), and the homogenate diluted to an appropriate concentration (0.5 hypothalamus/ml; 0.2 pituitary/ml; 0.02 uterus/ml). Aliquots of 1 ml of homogenate are layered on 1.0-ml pads of 1.2 M sucrose in 12×75 mm plastic culture tubes and centrifuged at 6900 g for 45 min. Aliquots (875 μl) of the resultant supernatant are pooled and recentrifuged at 48,000 g for 20 min. This cytosol fraction is used for the assay of cytoplasmic receptors (see below). The sucrose pad is aspirated, and the crude chromatin pellet is used for the assay of nuclear receptor. Aliquots (800 μl) of the cytosol are incubated with various concentrations of [^{3}H]estradiol (0.05–5 nM) in TE buffer. Chromatin is resuspended in 250 μl of TE buffer containing 0.05–5 nM [^{3}H]estradiol. A parallel set of tubes containing cytosol or chromatin, [^{3}H]estradiol, and a 100-fold molar excess of diethylstilbestrol is employed to assess nonspecific binding. Incu-

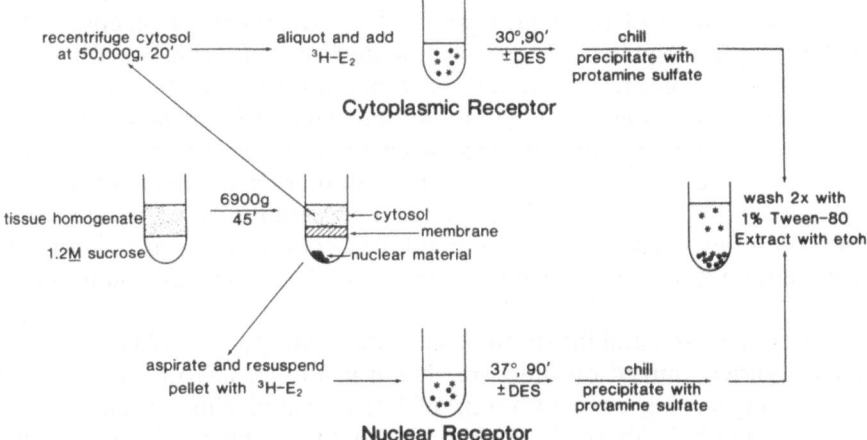

Fig. 4. Protocol for the assay of cytoplasmic and nuclear steroid receptors. This procedure has been designed to assess the distribution of steroid receptors between cytoplasmic and nuclear compartments. Particular advantages of this procedure include the separation of chromatin and membranous components in order to lower nonspecific binding in the assay of nuclear receptors and the precipitation of receptors by protamine sulfate, which insures complete recovery of receptor and also lowers nonspecific binding.

bation at 37°C for 90 min for nuclear fractions and at 30°C for 90 min for cytosol fractions insures complete equilibration of [³H]estradiol with the receptor species. Incubation is terminated by chilling the tubes in an ice-water bath for 5 min. Receptor–steroid complexes are separated from free steroid by precipitation with an equal volume of cold protamine sulfate solution (1 mg/ml) for 10 min. All manipulations after this point are at 0–4°C to avoid dissociation of receptor–ligand complexes. Samples are diluted with 3.5 (nuclear) or 2.5 (cytosol) ml wash buffer (TE buffer + 1% polysorbate 80; pH 7.4 at 0°C), vortexed, and centrifuged at 5000 g for 15 min. The supernatants are discarded, and the pellets washed with 2 ml wash buffer. After centrifugation, supernatants are discarded, and the pellets extracted with 1 ml 100% ethanol or scintillation fluid (overnight at room temperature, or 1–2 hr at 30°C). After decanting into scintillation vials, a second aliquot of ethanol or scintillation fluid is added to each tube; the tubes are incubated at room temperature for 1 hr and decanted into the same scintillation vials as after the previous extraction. The ethanol or toluene extracts are combined with toluene-based scintillation fluid, and radioactivity determined by scintillation spectrometry.

4.1.2. Alternative Procedures for Cytosol Receptors

If only the cytoplasmic receptor is of interest, one need not utilize the sucrose pad procedure. Simply homogenize the target tissue in question, centrifuge at high speed (48,000–100,000 g), and use aliquots of the clear supernatant for the assay of receptor.

In addition to protamine sulfate precipitation,[16] other methods for the separation of bound from free hormone in assays of soluble receptors include charcoal adsorption of unbound ligand[17,19] or adsorption of receptor–steroid complexes to glass pellets,[20] hydroxylapatite,[21] or ion-exchange filters.[22] Although these procedures are rapid and generally effective for the separation of free from bound steroid, none are without fault. Ion-exchange procedures are subject to influence by the ionic strength of buffers; hydroxylapatite adsorption is influenced by divalent cations and, more often than not, also gives a high level of nonspecific binding; charcoal procedures may strip steroid from receptor species.[23] Each should be employed cautiously, bearing in mind the dangers inherent in using nonequilibrium methods to measure equilibrium values.

For comparison and illustration, saturation data for cytoplasmic receptors of adult ovariectomized rat uteri are shown in Fig. 5. Cytosol from uteri was incubated in varying concentrations of [³H]estradiol plus or minus 100-fold excesses of diethylstilbestrol, and bound ligand was separated from free ligand by protamine sulfate precipitation (closed figures) or charcoal stripping (open figures). Note that nonspecific binding is considerably higher with the charcoal procedure. Since receptor levels per assay tube were relatively high in this assay, the final estimate of equilibrium parameters (K_d, B_{max}; see inset) were similar for both procedures; however, had receptor number been low relative to nonspecific binding (as for brain receptors), considerable variance would be expected between the two methods. Thus, one should select procedures that maximize the probability of measuring specific interactions in a given system.

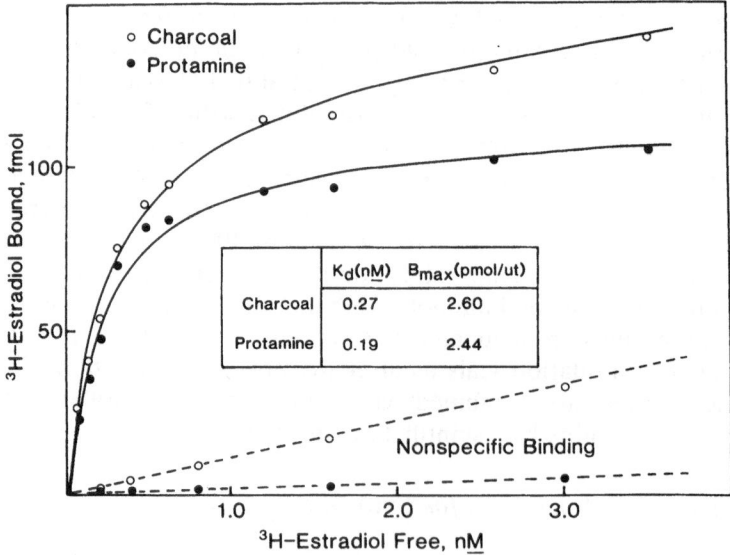

Fig. 5. Comparison of charcoal adsorption and protamine sulfate precipitation methods for the separation of bound and free steroid. High-speed cytosols from rat uteri were incubated with varying concentrations of [³H]estradiol in the presence (–––n) or absence (——) of excess unlabeled diethylstilbestrol, and bound and free ligand subsequently separated by charcoal adsorption (○) or protamine sulfate precipitation (●). Inset: Values for K_d and B_{max} as determined by Scatchard analysis of the data.

4.1.3. Alternative Procedures for Nuclear Receptors

Often one is concerned only with that receptor translocated to the nucleus after *in vivo* or *in vitro* steroid treatment. In such a circumstance, one can often simply homogenize the target tissue and isolate crude nuclear pellets for the assay of nuclear receptors. This is the basis for the original [^{3}H]steroid-exchange assay[24] and is still employed for various tissues and steroid systems in many laboratories. Details of this procedure and its modifications have been reviewed recently.[1] Major shortcomings of the original procedure include (1) dissociation of receptor from chromatin and its subsequent loss on washing and (2) high levels of nonspecific binding, especially in tissues such as brain with a high density of membrane lipid.

Nuclear receptors may also be extracted with high-ionic-strength media (e.g., 0.4 M KCl in TE buffer) and subsequently assayed as soluble entities.[25] This procedure has been employed to avoid the high nonspecific binding measured in crude nuclear pellets of brain.[26] In general, the method has not proven strictly quantitative. Depending on the target tissue in question, 10–40% of the receptor–steroid complexes in nuclear fractions may not be solubilized with high-ionic-strength media, the fractional solubilization varying with physiological state and the type of tissue.

4.1.4. Time and Temperature of Steroid Receptor Assays

For all receptor assays, procedures must be carried out at low temperature (0–4°C; usually with tubes in ice-water baths) except during equilibration with ligand. Equilibration with ligand at low temperatures generally results in the assay of unoccupied receptor sites; higher temperatures allow the measurement of both occupied and unoccupied sites. It is important to study the effect of various times and temperatures of equilibration to insure that maximum numbers of receptors are measured. Short time or low temperature may not allow complete equilibration of labeled ligand with all receptive sites; long times or elevated temperatures may lead to inactivation of receptor species. A compromise is often required to measure as many receptors as possible at equilibrium with ligand.

As one example, the measurement of the progesterone receptor by exchange differs from that of the estrogen receptor.[27] The differences stem from the labile nature of the mammalian progesterone receptor. Measurement of the progesterone receptor requires lower temperatures and longer periods for complete equilibration with ligand. These modifications permit exchange of labeled for unlabeled ligand to occur with minimal degradation of receptor.

4.2. Neurotransmitter Receptors

Receptors for neurotransmitters are enriched in preparations of synaptic plasma membranes or junctional complexes; however, they are found in many membrane fractions of the nervous system. As with steroid receptors, primary problems in their assay include isolation of a source enriched in receptor relative to nonspecific binding and the selection of an appropriate method for separating bound from free ligand. Many receptor studies have employed crude membrane preparations such as a lysed P_2 (crude synaptosomal–mitochondrial) fraction. Such membranes are invariably contaminated with endogonous ligands and multiple species of ligand-binding sites (enzymes, vesicles, glial elements, reticulum). As such, they are very poor for the study of postsynaptic receptors. A relatively clean preparation of synaptic plasma membrane (SPM) can be produced by simple flotation gradient centrifugation of such a mixture. The SPM is the usual starting material for our analysis of postsynaptic receptors.[2,10,28]

4.2.1. Isolation of Neuronal Membranes

Figure 6 depicts a typical approach to the assay of neurotransmitter receptor systems. Early steps in the procedure involve subcellular fractionation and membrane purification for the reduction of nonspecific binding and the removal of binding sites other than the receptive sites of interest. For instance, if a lysed P_2 or crude mitochondrial–synaptosomal fraction is employed to measure receptors, a variety of receptive sites coexist with postsynaptic receptors and may interfere with receptor analysis. Among these are membrane-bound enzymes, vesicular transport or binding systems, and transport systems of glial as well as of neuronal origin (see ref. 2 for extensive discussion). By employing a flotation gradient step,[25,26] reasonable pure synaptic plasma membrane (SPM) fractions are obtained; subsequent detergent treatment, as with 0.5% Triton X-100, and washing results in enriched junctional complex (JC)[27] or postsynaptic density fractions.[32] These fractions are enriched for many neurotransmitter receptors and allow their study in the absence of interfering extraneous sites.

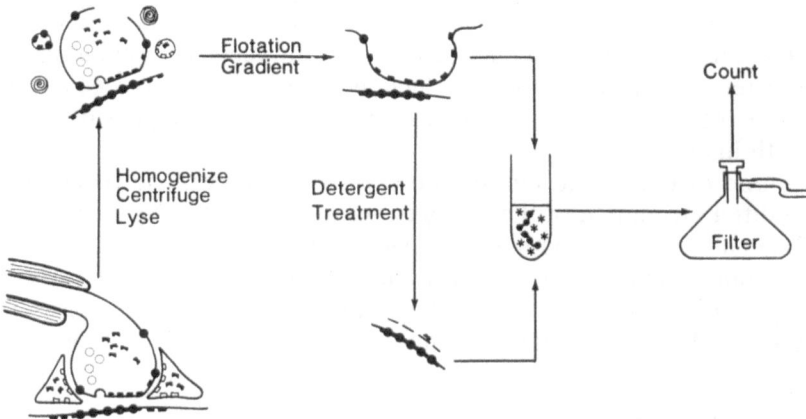

Fig. 6.   Scheme for the analysis of membrane-bound receptive sites. Neural tissue contains many different binding sites for neurotransmitters. For the study of ligand binding in neural systems, preliminary fractionation of membranes is advisable to avoid confusion with respect to the analysis of sites. Shown is one approach to this problem. Details are given in the text.

Briefly, the nervous tissue under study is homogenized in 10 volumes of 0–4°C isotonic sucrose (0.32 M) containing 10 μM $CaCl_2$, and the homogenate centrifuged at 1000 *g* for 10 min. The resultant supernatant is centrifuged at 12,000 *g* for 20 min, and the pellet resuspended in 5 mM Tris HCl, pH 8.6. The suspension is homogenized with a Tekmar Tissuemizer® or Beckman Poly-tron® and centrifuged at 48,000 *g* for 20 min. The pellet is resuspended in distilled water, centrifuged at 48,000 *g* for 20 min, and the resultant pellet, P_4, is stored frozen for future use. Pellets such as these are prepared routinely from bovine cerebral cortex and stored frozen for months. They represent a crude lysed and washed P_2 or synaptosomal–mitochondrial fraction.

To prepare SPM, the crude P_4 pellet is resuspended in 5 mM Tris, pH 8.6, and 2 M sucrose added to a final concentration of 1.2 M. The membrane suspension is overlaid in ultracentrifuge tubes with 0.9 and 0.3 M sucrose. Samples are centrifuged at 82,000 *g* for 120 min.[30] The interface between 1.2 and 0.9 M sucrose is taken as the purified SPM fraction. This can be resuspended to 2–4 mg protein/ml in 10 mM HEPES, pH 7.4, for subsequent binding experiments or can be used as the starting material for JC fractions.

To prepare JC, SPM are resuspended in 10 mM HEPES, pH 7.4, at about 0.5 mg protein/ml. Triton X-100 is added to a final concentration of 0.1–0.5%. After 30 min at 0–4°C, Triton-treated membranes are pelleted by centrifugation for 20 min at 48,000 *g*. Pellets are washed three times by resuspension in 10 mM HEPES, pH 7.4, and centrifugation at 48,000 *g*. The resulting fraction, enriched in washed JC, is resuspended in HEPES buffer at 1–3 mg protein/ml for subsequent binding studies.

4.2.2. Equilibrium Binding Studies

Equilibrium binding analysis of .GABA receptive sites on SPM or JC is conducted at 0–4°C (tubes in an ice-water bath). Reaction mixtures contain 10 mM HEPES, pH 7.4, plus or minus 100 mM NaCl. The NaCl is included if transport receptive sites are to be studied[10]; otherwise, receptor binding analyses are in the strict absence of sodium. Generally, [³H]GABA (5–500 nM) or [³H]muscimol (0.5–50 nM) is employed for the saturation analysis of the GABA postsynaptic receptor; for the assay of transport receptive sites.[³H]GABA (0.1–20 μM) or [³H]nipecotic acid (1.0–200 μM) is employed. Duplicate reaction mixtures contain excess unlabeled ligand to permit the correction for nonspecific binding. Reaction mixtures are incubated for 30 min or 1 hr at 0–4°C and terminated by the addition of 3 ml ice-cold 10 mM Tris and rapid filtration with suction through Whatman GF/C glass fiber filters. Filters are washed rapidly with two 3-ml portions of the same buffer, dried, and placed in scintillation vials. The elapsed time for the termination, filtration, and washing is 10–15 sec. The filters are covered with 0.5 ml of Bio-Solve® (Beckman, Fullerton, CA), and 4.5 ml of toluene-based scintillation fluid is added. These are maintained in the dark for 4–6 hr before radioactivity is measured by liquid scintillation spectrometry.

4.2.3. Alternative Procedures for Neurotransmitter Receptor Analysis

4.2.3a. Microcentrifugation. The filtration method described above is widely employed for the study of membrane-bound receptors.[33-37] The method is very simple and rapid. However, many investigators prefer microcentrifugation for the separation of bound from free ligand. The method was introduced by Rodbell[38] for the assay of glucagon receptors on liver plasma membranes and has been applied with modification to many membrane-bound receptor systems.[33-41] In the original procedure, membranes were suspended in 2.5% albumin, 1 mM EDTA, 20 mM Tris, pH 7.6, and labeled ligand at a final volume of 125 μl. After incubation for 15–30 min, duplicate 50-μl aliquots of incubation mixture were layered over 0.3 ml of 2.5% albumin in 20 mM Tris-HCl, pH 7.6, in Beckman microfuge tubes that were immersed in an ice-water bath. Tubes were centrifuged for 5 min at 5°C in a Beckman microfuge (10,000 g), the supernatant fluids aspirated with a #22 needle, and the wall and surface of the pellet washed once by the addition of 0.3 ml of 10% sucrose and subsequent aspiration. The tip of the microfuge tube was cut with a razor blade, and the radioactivity in the tip measured. Modifications of this procedure include the direct centrifugation of the incubation mixture without passage through an albumin pad and the introduction of the intact incubation tube into the liquid scintillation vials. Both modifications appear to increase background or blank values for ligand binding.

The microcentrifugation method has several disadvantages: (1) the determination of binding at short times of incubation is not possible; (2) high nonspecific binding is observed at high, saturating concentrations of labeled ligand; (3) ligand may dissociate from receptive sites if the albumin pad is employed; and (4) additional ligand may bind to receptive sites during centrifugation without the albumin pad, especially at low concentrations of ligand when receptor concentration is high. These conditions can produce dramatic changes in apparent equilibrium constants.

We have compared the centrifugation procedure with the filtration procedure to determine which is most reliable for the assay of high-affinity GABA receptive sites. The top panels in Fig. 7 illustrate saturation analyses for the binding of [3H]muscimol to SPM and JC preparations. These analyses were performed on a single membrane preparation using centrifugation (triangles) and filtration (circles) to separate free from bound ligand. For both membrane preparations, the ratio of specific to nonspecific binding is higher for the filtration procedure than for the centrifugation method. As discussed previously, high levels of nonspecific binding can lead to difficulties in the analysis of binding data. The bottom panels of Fig. 7 compare Scatchard analyses of the saturation data for [3H]muscimol to JC (upper right panel). Individual data points are represented, and duplicates are connected by a line; thus, the scatter of the data can be assessed. The fits of the data were derived via the Rosenthal procedure (Section 3.2.3c). Values of K_d of 4.1 and 3.9 nM were obtained by the centrifugation assay and filtration assay, respectively. Values of $B_{\max}$ were 2.1 pmol/mg protein and 1.4 pmol/mg protein, and the correlation coefficients were 0.77 and 0.96. Both procedures result in similar estimates of K_d and $B_{\max}$.

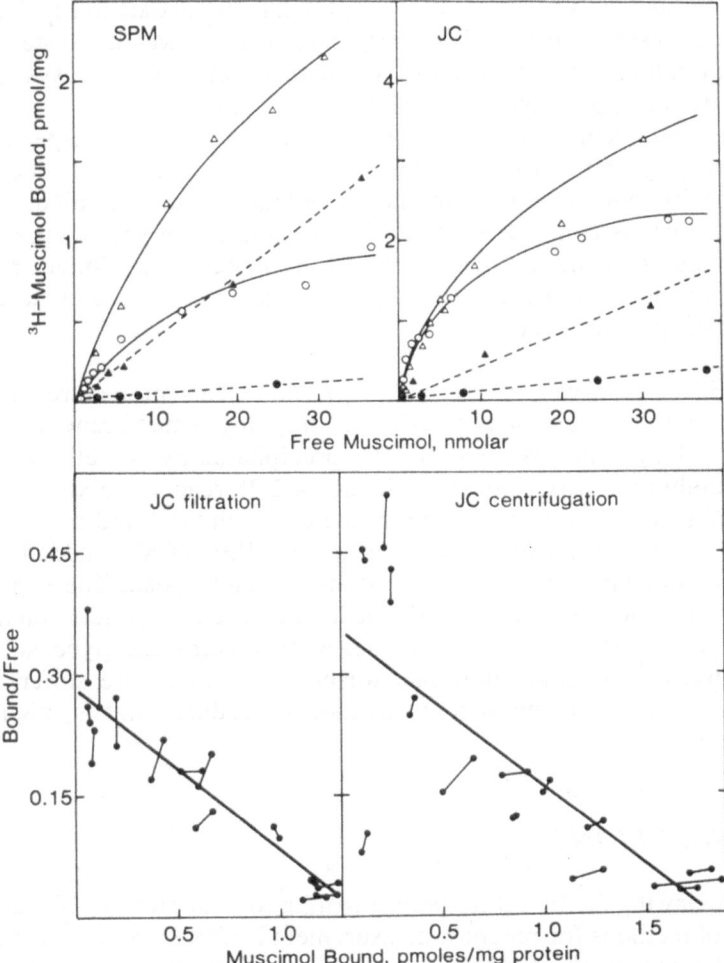

Fig. 7. Comparison of filtration and centrifugation as methods for the separation of bound from free ligand in membrane receptor systems. Upper panels: Synaptic plasma membranes (SPM) or junctional complex fractions (JC) were incubated with varying concentrations of [³H]muscimol in the presence (– – –) or absence (———) of unlabeled muscimol, and bound and free ligand subsequently separated by filtration (●) or centrifugation (▲). Lower panels: The data for binding of [³H]muscimol to JC was corrected for nonspecific binding by Rosenthal resolution, and individual data points plotted according to Scatchard. Duplicate points are connected by lines to assess the scatter of the data. The left lower panel shows the Scatchard analysis of data obtained by filtration, and the right panel is of data obtained by centrifugation.

However, the goodness of fit is appreciably diminished by the high nonspecific binding in the centrifugation method.

If the data in the lower right panel were fit by eye, two binding components might be extrapolated from the centrifugation data. However, the variance around the single-component, straight-line fit indicates that this is not the case. That is, as many data points are found below the fitted line as above, and a simple runs test suggests random error about the line. The nonlinear curve-fitting program LIGAND (Section 3.2.3d) readily fit all data in Fig. 7 to a single component system but failed in every case to fit a two-site system. Thus, a misinterpretation with respect to the existence of a two-site system can be avoided by the appropriate assay and analysis of data.

These data demonstrate a consistently higher level of nonspecific binding when centrifugation is used to separate bound from free ligand. The high level of nonspecific binding compromises the estimate of specific binding and may lead to erroneous conclusions regarding the number of species present in the membrane system. In contrast, the filtration procedure minimizes nonspecific binding, avoids entrapment of labeled ligand, and thus provides a more accurate estimate of specific binding.

4.2.3b. Polyethylene Glycol Precipitation. This procedure also usually employs centrifugation and is based on the trapping of membranes or solubilized receptors along with bovine γ-globulin precipitated by polyethylene glycol.[42] After equilibration with ligand (see Section 4.2.2), reaction mixtures are diluted with 2 ml of 0.1% bovine γ-globulin (ice-cold solution) and mixed, and 1 ml of 20% (w/v) polyethylene glycol (Carbowax® PEG 6000) is added. Tubes are vortexed and centrifuged at 4°C for 10 min at high speed. The supernatant is removed, and the radioactivity in the pellet measured. Depending on the affinity of receptor for ligand, additional washes with polyethylene glycol solution may be employed.[43] Without additional washes, this procedure suffers the same problem as the direct microcentrifugation procedure, that is, high levels of nonspecific binding.

5. CONCLUSIONS

In reviewing the literature before writing this chapter, we realized that the number of methods for receptor measurement is almost as large as the number of investigators in the field of receptorology. For this reason, we have purposefully kept the discussion of procedural matters to a minimum. The measure of any receptor depends on its unique physical properties and subcellular localization. Instead of cataloging many procedures, we have discussed a few with which we are familiar. Other specific procedures can be obtained from the plethora of books and reviews on the subject. We have devoted a large portion of this chapter to a discussion of the interpretation and misinterpretation of binding data. In our opinion (and that of others[44,45]), the mis- or overinterpretation of binding data is a serious problem in the field of receptor analysis. If this chapter assists in reducing any of the faults of this field, it will have served its purpose.

REFERENCES

1. Clark, J. H., and Peck, E. J., Jr., 1979, *Female Sex Steroids: Receptors and Function*, Springer-Verlag, New York.

2. Peck, E. J., Jr., and Lester, B. R., 1980, *The Cell Surface and Neuronal Function* (C. W. Cotman, G. Poste, and G. L. Nicolson, eds.), North-Holland, Amsterdam, New York, pp. 405–453.
3. Michaelis, L., and Menten, M. L., 1913, *Biochem. Z.* **49**:333–369.
4. Scatchard, G., 1949, *Ann. N.Y. Acad. Sci.* **51**:660–672.
5. Lineweaver, H., and Burk, D., 1934, *J. Am. Chem. Soc.* **56**:658–666.
6. Eisenthal, R., and Cornish-Bowdin, A., 1974, *Biochem. J.* **139**:715–720.
7. Rosenthal, H., 1967, *Anal. Biochem.* **20**:525–532.
8. Munson, P. J., and Rodbard, D., 1980, *Anal. Biochem.* **107**:220–239.
9. Feldman, H. A., 1972, *Anal. Biochem.* **48**:317–338.
10. Lester, B. R., Miller, A. L., and Peck, E. J., Jr., 1981, *J. Neurochem.* **36**:154–164.
11. Hollenberg, M. D., and Cuatrecasas, P., 1976, *Methods in Cancer Research*, Volume XII (H. Busch, ed.), Academic Press, New York, pp. 317–366.
12. Rodbard, D., and Feldman, H. A., 1975, *Methods Enzymol.* **36**:3–6.
13. O'Brien, R. D. (ed.), 1979, *The Receptors*, Volume 1, Plenum Press, New York.
14. Birnbaumer, L., and O'Malley, B. W. (eds.), 1978, *Receptors and Hormone Action*, Volumes II, III, Academic Press, New York.
15. Kelner, K. L., Miller, A. L., and Peck, E. J., Jr., 1980, *J. Recept. Res.* **1**:215–237.
16. Steggles, A. W., and King, R. J. B., 1970, *Biochem. J.* **118**:695–701.
17. Korenman, S. G., Perrin, L. E., and McCallum, T. P., 1969, *J. Clin. Endocrinol.* **29**:879–883.
18. Mester, J., Robertson, D. M., Feherty, P., and Kellie, A. E., 1970, *Biochem. J.* **120**:831–836.
19. Sanborn, B. M., Rao, B. R., and Korenman, S. G., 1971, *Biochemistry* **10**:4955–4961.
20. Clark, J. H., and Gorski, J., 1969, *Biochim. Biophys. Acta* **192**:508–515.
21. Erdos, T., Best-Belpomme, M., and Bessada, R., 1970, *Anal. Biochem.* **37**:244–252.
22. Santi, D. V., Sibley, C. H., Perriard, E. R., Tomkins, G., and Baxter, J. D., 1973, *Biochemistry* **12**:2412–2416.
23. Peck, E. J., Jr., and Clark, J. H., 1977, *Endocrinology* **101**:1034–1043.
24. Anderson, J. N., Clark, J. H., and Peck, E. J., Jr., 1972, *Biochem. J.* **126**:561–567.
25. Zava, D. T., Harrington, N. Y., and McGuire, W. L., 1976, *Biochemistry* **15**:4292–4297.
26. Roy, E. J., and McEwen, B. S., 1977, *Steroids* **30**:657–669.
27. Hsueh, A. J., Peck, E. J., Jr., and Clark, J. H., 1974, *Steroids* **24**:599–611.
28. Lester, B. R., and Peck, E. J., Jr., 1979, *Brain Res.* **161**:79–97.
29. Cotman, C. W., and Mathews, D. A., 1971, *Biochim. Biophys. Acta* **249**:380–394.
30. Jones, D. H., and Matus, A. I., 1974, *Biochim. Biophys. Acta* **356**:276–287.
31. Cotman, C. W., Banker, G., Churchill, L., and Taylor, D., 1974, *J. Cell Biol.* **63**:441–455.
32. Cohen, R. S., Blomberg, F., Berzins, K., and Siekevitz, P., 1977, *J. Cell Biol.* **74**:181–203.
33. Cuatrecasas, P., 1971, *Proc. Natl. Acad. Sci. U.S.A.* **68**:1264–1268.
34. Goldfine, I. D., Roth, J., and Birnbaumer, L., 1972, *J. Biol. Chem.* **247**:1211–1218.
35. Posner, B. I., Kelly, P. A., Shiu, R. P. C., and Friesen, H. G., 1974, *Endocrinology* **95**:521–531.
36. Brown, E. M., Aurbach, G. D., Hauser, D., and Troxier, F., 1976, *J. Biol. Chem.* **251**:1232–1238.
37. Birnbaumer, L., Pohl, S. L., and Kauman, A. J., 1974, *Adv. Cyclic Nucleotide Res.* **4**:239–281.
38. Rodbell, M., Krans, H. M. J., Pohl, S. L., and Birnbaumer, L., 1971, *J. Biol. Chem.* **246**:1861–1871.
39. Zukin, S. R., Young, A. B., and Synder, S. H., 1974, *Proc. Natl. Acad. Sci. U.S.A.* **71**:4802–4807.
40. Enna, S. J., and Snyder, S. H., 1975, *Brain Res.* **100**:81–97.
41. Beaumont, K., Chilton, W. S., Yamamura, H. I., and Enna, S. J., 1978, *Brain Res.* **148**:153–162.
42. Cuatrecasas, P., 1972, *Proc. Natl. Acad. Sci. U.S.A.* **69**:318–322.
43. Iyengar, R., Abramowitz, J., Bordelon-Riser, M., Blume, A. J., and Birnbaumer, L., 1980, *J. Biol. Chem.* **255**:10312–10321.
44. Hollenberg, M. D., and Cuatrecasas, P., 1979, *The Receptors* (R. D. O'Brien, ed.), Volume 1, Plenum Press, New York, pp. 193–214.
45. Munck, A., 1976, *Receptors and Mechanism of Action of Steroid Hormones* (J. R. Pasqualini, ed.), Volume 1, Marcel Dekker, New York, pp. 1–40.

Rapid Enzyme Inactivation

Robert H. Lenox, G. Jean Kant, and James L. Meyerhoff

1. INTRODUCTION

The brain functions at a high metabolic level and is dependent on the maintenance of energy stores that require constant access to oxygen and glucose via cerebral blood flow. The enzyme activities of individual metabolic pathways and relative concentrations of intermediary metabolites can be significantly altered by changes in regional blood flow in the brain. In order to determine endogenous metabolite levels, cessation of metabolic events should ideally occur *in situ* instantaneously, since any finite amount of time provides relative periods of ischemia during which artifactual changes in metabolite concentrations can occur. Efforts to measure *in vivo* levels of intermediary metabolites, high-energy phosphates, cyclic nucleotides, and amino acids in brain tissue require rapid enzyme inactivation to stop metabolism.

Enzymes can be inactivated by physiochemical events that denature proteins, i.e., alterations in temperature, pH, etc. Currently, only methods involving rapid temperature changes are adequate to prepare brain tissue for accurate determination of *in vivo* levels of high-energy metabolites in brain. This chapter describes a number of temperature-dependent methods used to inactivate enzymes rapidly in order to assay levels of metabolites in brain tissue. Each of the methods described has its advantages as well as certain limitations. The decision as to which technique to employ requires consideration of both the substrates to be measured and the purpose of the investigation.

This material has been reviewed by the Walter Reed Army Institute of Research, and there is no objection to its presentation and/or publication. The opinions or assertions contained herein are the private views of the authors and are not to be construed as official or as reflecting the views of the Department of the Army or the Department of Defense.

Robert H. Lenox • Neuroscience Research Unit, Department of Psychiatry, University of Vermont, Burlington, Vermont 05405. *G. Jean Kant and James L. Meyerhoff* • Department of Medical Neurosciences, Walter Reed Army Institute of Research, Walter Reed Army Medical Center, Washington, D.C. 20012.

2. FREEZING METHODS

2.1. Brain in Situ

Quick freezing of the brain *in situ* was the first method used to rapidly stop metabolic processes and permit estimation of *in vivo* levels of labile metabolites.[1-4] Current techniques include whole-body immersion into freezing media, decapitation into freezing media, funnel freezing, and cryoplate freezing. Different coolants including liquid air, liquid nitrogen, chilled isopentane, and Freon® 12 have been used in these methods. Some reports have suggested advantages of using one coolant over another, but metabolite concentrations are generally equivalent regardless of the coolant used.[5]

2.1.1. Immersion Techniques

In these methods, either the intact animal or the decapitated head is plunged into a container of coolant (usually liquid nitrogen or Freon® 12). These techniques are inexpensive and do not require elaborate equipment, surgical preparation, or anesthesia. The procedures are limited by the rate of freezing, which is relatively slow, by rapid hypoxic metabolic changes occurring during the latency to freezing, and by the difficulty of regional dissection of the frozen brain.

The problem of greatest concern is that significant changes in the levels of metabolites can take place during the time it takes for the brain temperature to fall to 0°C.[5] As shown in Fig. 1, deeper regions of the brain require more time to freeze. When decapitated rat heads were placed in Freon® 12, rat cerebral cortex required 30 sec to fall to 0°C, whereas the hypothalamus required 90 sec to fall to 0°C; in the mouse, cerebral cortex was frozen after 6 sec, and the hypothalamus after 30 sec.[6] Clearly, the relatively slow rate of

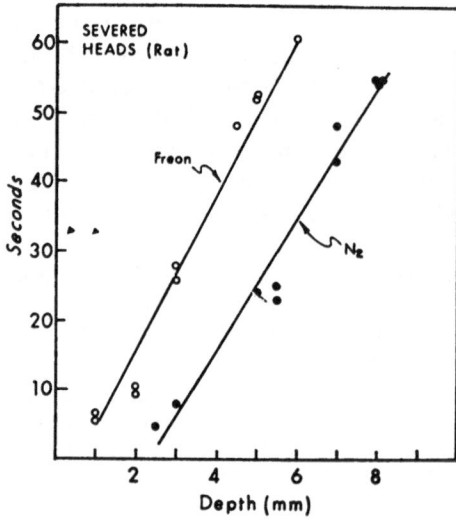

Fig. 1.   Time required to cool rat forebrain to 0°C after immersion of the severed head in coolant. Temperature was measured using a chromel–alumel thermocouple inserted into the brain prior to immersion. (Adapted with permission from Ferrendelli *et al.*[5])

inactivation achieved by plunging the head or intact animal into coolant will be a significant source of artifact when the metabolite of interest is labile and responds to hypoxia. The artifact will be greater in larger animals (rats vs. mice) and in deeper regions of the brain.

Although the rate of brain freezing in decapitated heads is slightly faster than in intact animals, Ferrendelli *et al.*[5] concluded that measurements of ATP, Phosphocreatine (P-creatine), and lactate in animals frozen intact more closely approximated *in vivo* levels (Table I). It was suggested that this difference might be caused by the disruption of blood flow and possible neuronal stimulation caused by decapitation. Other studies support the conclusion that immersion of the intact animal is a better method of fixation than immersion of the decapitated head for estimation of *in vivo* levels of metabolites.[7,8] Studies examining protein phosphorylation patterns in brain membranes also have found differences between animals sacrificed by decapitation into liquid nitrogen and animals frozen intact.[9]

Aprison and his colleagues described a modification of the immersion method that permits regional dissection of the brain.[10,11] In the near-freezing method, rats are immersed intact in liquid nitrogen for approximately 10 sec and then decapitated. Immersion time was selected so that no part of the brain fell below 0°C to avoid freeze-thawing conditions. The decapitated heads are placed in a specially constructed cold box at 0°C, and the brain is dissected using gloves attached to holes in the box. This method has been used to measure neurotransmitters in small pieces of tissue. Although this method retards postmortem artifact, deep regions of the brain require more than 10 min to cool to 0°C.[10] Therefore, only substances whose levels change slowly can be measured by this method.

2.1.2. Funnel Freezing

In order to avoid artifacts caused by hypoxia during the freezing process, a funnel freezing method has been developed that provides for blood flow and oxygenation to the deeper regions of the brain while the superficial regions of the brain are being frozen.[12,13] Animals are anesthetized and artificially ventilated to maintain normal blood gases. Physiological parameters can be monitored. After a midline incision, a funnel is fitted over the skull and secured with sutures; liquid nitrogen is poured through the funnel until a large drop in blood pressure occurs indicating that brainstem structures have frozen (approximately 2 min for rat brain). The head is then immersed in liquid nitrogen, and dissection is performed.

Although circulation is cut off from some regions (e.g., parietal cortex) because of early freezing of the vascular supply to the region, circulation is maintained in most regions until the tissue is frozen.[13]

This method appears to yield good estimates of *in vivo* levels of high-energy phosphates which are often used as criteria for the quality of fixation (Table I). Other advantages include the ability to use this method in larger animals such as cats.[14] The major disadvantages of the method are the requirement for anesthesia which can affect brain chemistry in itself,[15] the re-

Table I

Brain Intermediary Metabolites: Methods of Enzyme Inactivation

Method	Animal	Weight (g)	Brain region	Metabolite (µmol/g wet weight)						References
				P-Cr[c]	ATP	ADP	AMP	Lactate	Pyruvate	
Freezing[a]										
Freon®, severed head	Mouse	20	Cerebral cortex	2.43	2.36	0.91	0.21	2.26	—	3
N$_2$, severed head	Mouse	22	Cerebral cortex	2.69	3.01	—	—	1.77	—	5
N$_2$, intact animal	Mouse	22	Cerebral cortex	4.30	2.97	—	—	1.36	—	5
N$_2$, intact animal	Rat	225	Cerebral cortex	3.71	2.62	—	—	2.47	—	5
N$_2$, severed head	Rat	225	Whole brain	1.41	1.79	—	—	3.16	—	7
N$_2$, intact animal	Rat	225	Whole brain	3.40	2.30	—	—	1.90	—	7
Funnel freezing	Rat	350	Cerebral cortex	4.71	3.00	0.26	0.04	1.64	—	13
Freeze clamping (0.5 mm)[b]	Rat	200	Cerebral cortex	3.79	3.32	0.40	0.04	0.72	0.08	19
Freeze clamping (5.0 mm)[b]	Rat	200	Cerebral cortex	1.04	2.11	0.93	0.49	3.61	0.06	19
Brain chopping	Rat	100	Cerebral cortex	3.16	1.90	—	—	2.48	—	20
Freeze blowing	Rat	225	Forebrain	4.05	2.45	0.56	0.04	1.23	0.09	17
Microwave (2450 MHz)										
1.2 kW (14.0 sec)	Rat	225	Forebrain	1.69	1.69	1.32	0.40	1.71	0.06	17
1.2 kW (3.5 sec)	Rat	275	Cerebral cortex	2.97	1.33	0.91	0.44	2.01	0.09	25
2.0 kW (0.5 sec)	Mouse	50	Forebrain	4.00	2.20	—	—	—	—	26
2.0 kW (2.0 sec)	Rat	175	Forebrain	2.40	1.50	—	—	1.20	0.07	26
6.0 kW (0.3 sec)	Mouse	25	Cerebral cortex	3.23	2.54	0.65	0.06	1.39	—	27
6.0 kW (0.6 sec)	Rat	80	Cerebral cortex	3.71	2.64	0.36	0.06	1.49	0.10	28
Microwave (915 MHz)										
17.0 kW (1.0 sec)	Rat	275	Cerebral cortex	3.99	2.70	1.08	0.11	0.91	0.21	24

[a] Freezing medium: N$_2$, liquid nitrogen.
[b] Distance from surface to midpoint of brain slice (mm); 0.5mm slice was nearest to the cooling block. Adapted from ref. 29.
[c] P-creatine.

gional variation in freezing time, and the considerable time and effort expended per animal sacrificed.

2.1.3. Cryoplate

Skinner *et al.*[16] described a novel *in situ* freezing technique. Cryoplates, made from stainless steel hypodermic needle tubing, are surgically implanted on the surface of the cortex and maintained as a chronic preparation. Coolant is circulated through the plate by attached polyethylene tubes. Cortical samples (approximately 1 mm thick) can be frozen in milliseconds and then extracted by pulling briskly on the assembly, freeing both the plate and attached brain tissue. The method has the advantages of very rapid freezing and minimal observable behavioral stress at the time of sampling. The rat is unrestrained. Cyclic AMP levels using this method have been reported to be higher in rat parietal cortex than those found with other fixation methods.

Although reported in rat preparations, the method could be used in larger animals. Limitations of the method are that only cortical areas can be sampled, the method requires that the cryoplate be in contact with the brain surface for some time prior to sampling, and the method requires considerable time and effort per sample obtained. This method appears more useful as a comparative technique with other fixation methods than as a routine method of tissue fixation for neurochemical analyses.

2.2. Brain ex Situ

2.2.1. Freeze Blowing

The freeze blowing technique described by Veech *et al.*[17] offers an extremely rapid method of sacrifice which preserves high brain levels of phosphorylated intermediary metabolites and low levels of lactate, consistent with relatively little postmortem anoxic artifact. This technique employs an apparatus that immobilizes a rat by clamping the maxilla, mandible, and incisors. Sacrifice is accomplished by pneumatically driving two hollow steel probes into the cranial cavity; a jet of air at 25 lb/in.² of pressure enters through one probe and forces the supratentorial portion of the brain out via the second probe into a hollow aluminum disk, which has been precooled with liquid nitrogen. Thus, in less than 1 sec, the animal is sacrificed, and a sample of brain weighing approximately 1 g is removed and converted to a wafer of brain tissue frozen at approximately $-190°C$.

The advantages of the freeze blowing technique stem from its rapidity, permitting sacrifice with less biochemical evidence of anoxic artifact than many other techniques.[17] The freeze blowing technique preserves high levels of ATP and pyruvate (Table I). An additional advantage is the lack of a requirement to anesthetize the animal. The disadvantages of the technique include the requirement for immobilization of the animal, the loss of brain morphology, and the loss of all brain regions caudal to the superior colliculus (cerebellum, pons, and medulla). If desired, the technique may be modified to obtain posterior fossa structures, but only at the expense of losing structures rostral to the tentorium.

2.2.2. Guillotine Freeze Clamping

Quistorff described a mechanical device that rapidly sacrifices a rat, removes a coronal slice of the rat's head and brain, and then freezes the slice between precooled aluminum blocks (Fig. 2).[18] In this technique, rats enter a tube and are positioned so that the head is located at a gap in the tube which permits a pair of rotating blades to slice a coronal section of the head. The blades are activated by a pneumatically driven piston, and sectioning is accomplished within 30 msec. A second piston then moves two aluminum blocks, precooled in liquid nitrogen, into position anterior and posterior to the slice. A third piston presses the aluminum blocks together, compressing the tissue and initiating the freezing process within 80 msec of sacrifice. Slice thickness can be varied from 4.5 mm to 13 mm. During freeze clamping, these slices are compressed to 2.5 mm or 10 mm, respectively. A buzz saw mounted in a refrigerated glove box ($-30°C$) is used to cut 1-mm slices of the sample for biochemical analysis. Some anatomic features are macroscopically observable in the slices, and, therefore, a degree of regional sampling is possible.

Guillotine freeze clamping affords the advantages of rapid sacrifice and enzyme inactivation without a requirement for anesthesia. Although some compression of the slice occurs during freeze clamping, a degree of brain regional sampling is possible. Values for intermediary metabolites measured in tissue within 0.5 mm of the freezing surface compare favorably with those seen after freeze blowing (Table I).

The disadvantages of guillotine freeze clamping seem to be related to the thickness (10 mm) of the slice obtained and the delay in freezing the tissue in the middle of the slice (25 sec to reach 0°C). Quistorff has demonstrated decreases in ATP and P-creatine as well as increases in lactate with increasing distance from the freezing surface of the slice (Table I).[19] These changes are consistent with anoxic metabolic artifact. The authors indicate that the technique does not require the animal to be immobilized. Careful placement of the animal, however, does require restriction of movement, a procedure to which the authors suggest the animals can be habituated.

2.2.3. Brain Chopping

A third *ex situ* method, which has been called simply "brain chopping,"[20] utilizes a multibladed guillotine which sacrifices a rat by cutting the head into several coronal slices which drop into a pan containing isopentane chilled with liquid nitrogen. This method has the advantage of requiring neither anesthesia nor immobilization, but without the latter, it is unclear whether slices obtained would be anatomically comparable between rats. Although the chopping procedure is completed in 0.1 sec, the freezing requires 5.0 sec, and the method

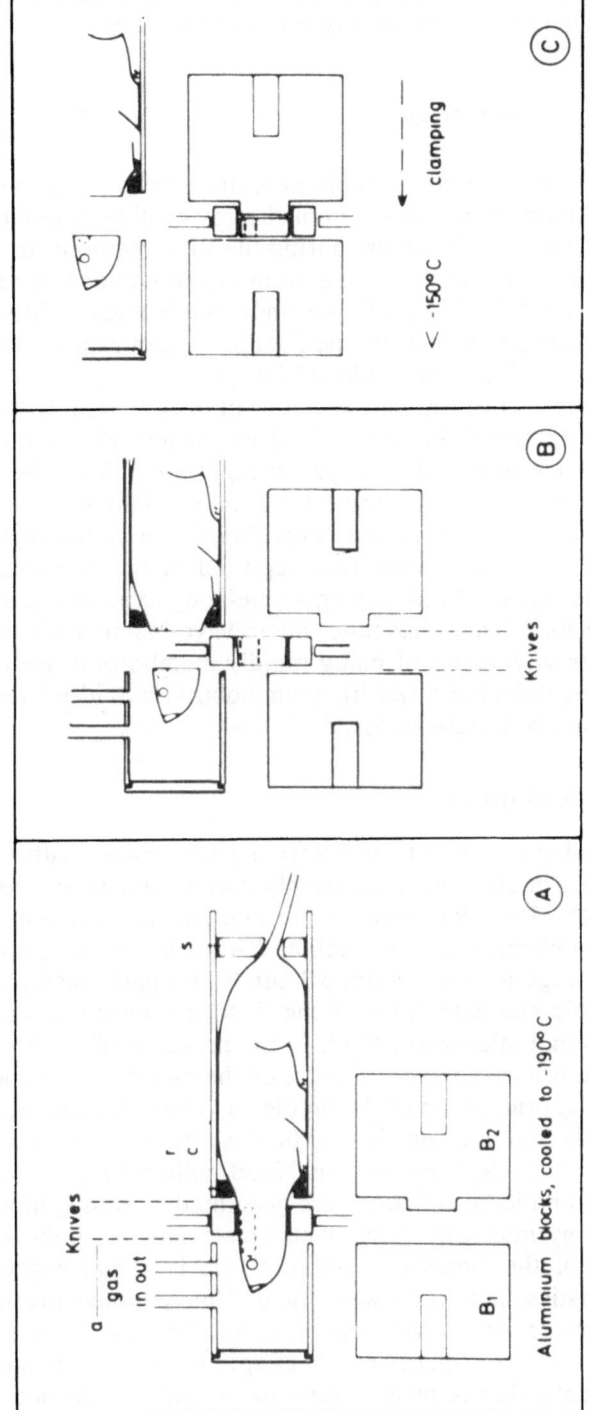

Fig. 2. Brain sampling by guillotine freeze clamping. (Adapted with permission from Quistorff.[19])

has the disadvantage of producing lower levels of ATP and P-creatine and higher levels of lactate than does freeze blowing (Table I).

3. MICROWAVE METHODS

Most enzymes denature irreversibly at temperatures in the range of 55 to 90°C. Enzyme inactivation can be accomplished by heating a brain homogenate at 90 to 100°C for 10 min. However, during the time required for sacrifice of the animal, removal of the brain, preparation of the homogenate and heating of the homogenate, metabolism continues under anaerobic conditions and produces artifactual changes in relative metabolite concentrations dependent on the activity of the individual metabolic pathways.

High-power microwave irradiation can be utilized to rapidly heat the brain *in situ*, thereby both sacrificing the animal and inactivating enzymes within seconds. This methodology offers a promising approach to the problem of measuring *in vivo* levels of heat-stable metabolites. This technique has generated a great deal of interest among neurochemists, and the technology has steadily improved since Stavinoha first reported microwave inactivation of brain acetylcholinesterase (AChE) using whole-body exposure in a microwave oven system in 1970.[21] Since that time, microwave inactivation has been utilized to measure *in vivo* levels of many rapidly metabolized, heat-stable substrates including acetylcholine (ACh), γ-aminobutyric acid (GABA), cAMP, cGMP, and intermediary metabolites.[22]

3.1. Microwave Radiation

Microwave radiation consists of electromagnetic waves with a wavelength between 1 cm and 1 m. The most commonly used commercial frequencies are 2450 MHz and 915 MHz. Microwaves are nonionizing radiation and, unlike higher-frequency ionizing radiation such as X-rays and γ rays, possess insufficient quantum energy to break chemical bonds in organic material. The electric-dipole and ionic characteristics of the dielectric materials determine the heating properties in a microwave field. In the presence of a microwave field, the electric dipole orients in the direction of the electric field and induces a dipole moment. The orientation of the dipole will change, and ionic movement will occur as the direction of the electric field is altered. Since the microwave electric field at 2450 MHz reverses itself 2450 million times a second, rapid dipole oscillations and ionic collisions are generated, resulting in a conversion of the kinetic energy into heat. Since water molecules are polarized (possess permanent dipoles), the water content of tissue is a major determinant of heating characteristics in a microwave field. Conventional heating methods heat from the surface, transfering the heat into the tissue by principles of conduction or convection. Microwave heating, on the other hand, penetrates the tissue to a depth that is proportional to its wavelength and heats more rapidly and uniformly than conduction or convection.

3.2. *Microwave Inactivation Systems*

The first microwave inactivation systems utilized commercial 1.25 kW microwave ovens at a frequency of 2450 MHz. Animals were placed in an oven cavity and exposed to whole-body irradiation for as long as 30 sec. In later systems, the head of the animal (rat, mouse) was inserted into a waveguide applicator, focusing the radiation directly into the head in order to increase the brain temperature more rapidly.

The major components of a current microwave inactivation system consist of (1) a microwave source, generally a magnetron or klystron tube generating the microwave fields; (2) a waveguide, a rectangular metal channel with dimensions of the wavelength through which the field propagates; (3) a tuner, a modified waveguide that permits optimal electrical coupling to the load; (4) an applicator, a section of waveguide modified as an exposure chamber; (5) a load, the animal in a plastic restraining tube. The waveguide applicator shown in Fig.3 was developed in our laboratory and is modified to optimize the coupling between the rat brain and the microwave field. In this case, the longitudinal axis of the rat head is perpendicular to the E field generated within the applicator. Other methods have utilized applicators in which the E field is parallel to the longitudinal axis of the head. A microwave inactivation system currently in use in our laboratory is shown in Fig. 4. This system permits optimal coupling of each animal by adjustment of the tuner for optimal impedance match using a low-power (10 mW) signal. This system also incorporates power-leveling and modified timing circuitry with recordings of net power received by each animal. Commercial models currently available utilize 2450 MHz systems with waveguide applicators similar to the ones noted above.

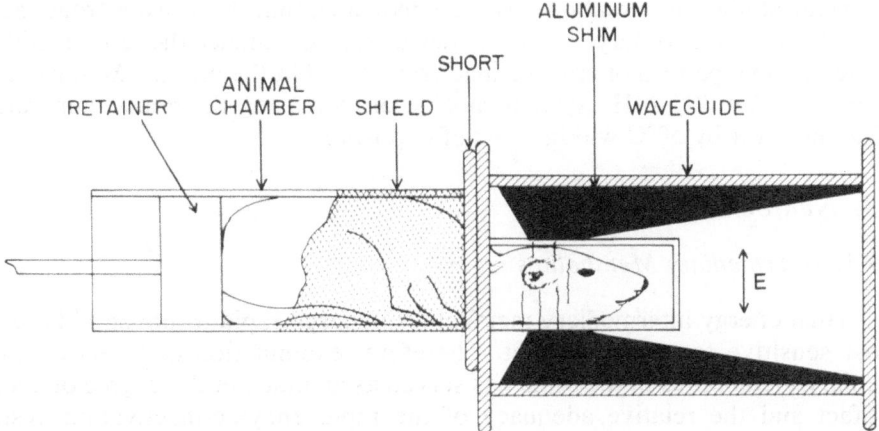

Fig. 3. Waveguide applicator. The rat is placed in an offset plexiglass restraining tube which is inserted into a shorting end plate of the waveguide perpendicular to the microwave field (*E*). Tapered aluminum plates attached to the broad faces of the waveguide serve to concentrate the energy into the rat head and improve the uniformity of field distribution over the longitudinal extent of the brain.

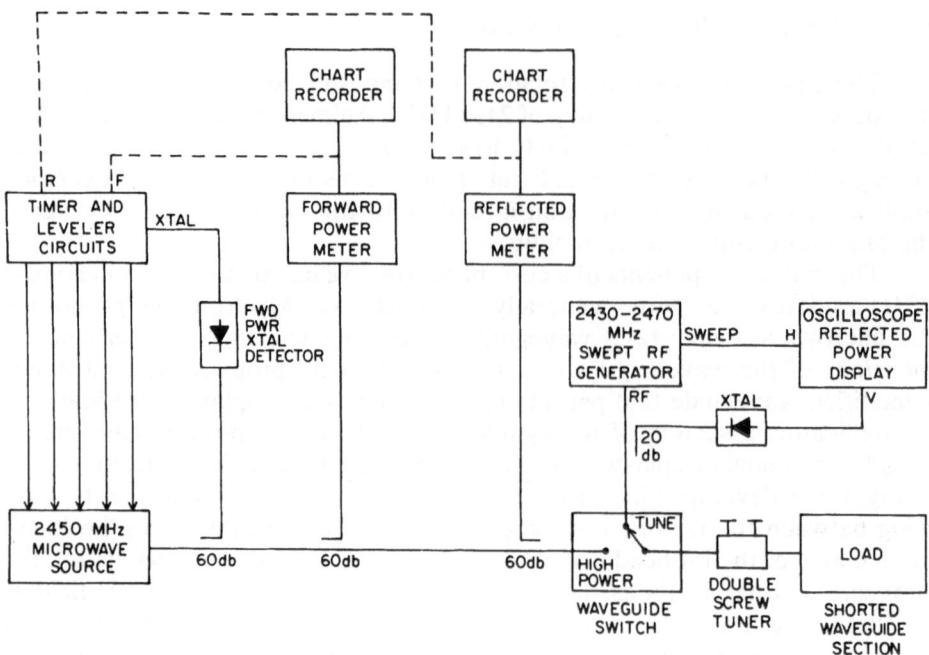

Fig. 4.   Block diagram of the microwave inactivation system. Features of this system permit in-line impedance matching at low power through a waveguide switch, continuous monitoring and recording of net microwave power absorbed by the load, a capacity for high-resolution exposure timing, and leveling of the output power.

Microwave inactivation systems using 915 and 986 MHz have also been reported. Studies in our laboratory have indicated that these lower frequencies may be preferable for larger rodents such as rats to enhance the reproducibility of the regional pattern of enzyme inactivation within the brain.[23] Medina *et al.* report a 17-kW 915-MHz system capable of increasing the brain temperature of an adult rat by 50°C within 1 sec of exposure.[24]

3.3. Neurochemical Studies

3.3.1. Intermediary Metabolites

High-energy intermediary metabolites undergo rapid turnover and are the most sensitive to anoxic artifact. Therefore, examination of levels of intermediary metabolites in the brain has served as an index of the degree of anoxic artifact and the relative adequacy of the rapid enzyme-inactivation system utilized. The relative levels of P-creatine, ATP, ADP, AMP, lactate, and pyruvate are altered during ischemia to the brain. During the anoxic period, there is a rapid decrease in P-creatine, a slower decrease in pyruvate and ATP, a rapid increase in AMP and lactate, and little change in levels of ADP.[3] Table I compares different methods of brain fixation for the determination of metabolite levels.

Only the higher-power microwave systems, i.e., 6 kW, 2450 MHz and 17 kW, 915 MHz, are capable of achieving values of intermediary metabolites in rat brain comparable to those observed using the most rapid freezing techniques, i.e., freeze blowing. It is also apparent that as the size of the animal increases, there is a requirement for increased power to achieve comparable levels.

Levels of intermediary metabolites have been examined as a function of the depth of tissue sampled from the surface of cortex following different methods of enzyme inactivation.[29] Since freezing occurs via conduction, deeper areas of the brain demonstrate more anoxic artifacts in levels of high-energy phosphates, particularly P-creatine. On the other hand, brain exposed to microwave showed significantly less variability as a function of depth of tissue sampled, even in slower microwave systems.

3.3.2. Cyclic AMP and Cyclic GMP

Cyclic nucleotides also rapidly turn over in brain, and determination of *in vivo* levels of cAMP and cGMP requires rapid inactivation of the enzymes adenylate cyclase (AC), guanylate cyclase (GC), and phosphodiesterase (PDE)[30] as well as cyclic nucleotide binding protein.[7]

Schmidt *et al.* were the first to report the use of microwave inactivation (1.25 kW, whole body) for the determination of cAMP.[31] With the improved microwave technology now available, microwave inactivation is the current method of choice for the measurement of cAMP in rodent brain. An excellent review of this subject has recently appeared.[32]

Several studies have reported elevations in cAMP levels in brain following decapitation or exposure to anoxia.[30,31,33] Studies in rat brain following decapitation indicate significant regional variability in cAMP elevation, with levels in the cerebellum increasing five- to tenfold within 90 sec while cortical levels remain relatively constant.[31,33] Animals exposed to anoxia without decapitation, on the other hand, demonstrate relatively comparable rates and extent of cAMP elevation in the same two brain regions.[32] The investigators have suggested that the difference might be caused by central activation and spinal shock following severing of the brainstem.

As shown in Fig. 5, sacrifice by slower whole-body microwave systems which required 30 sec or more exposure to adequately inactivate AC and PDE resulted in relatively high levels of cAMP with marked regional variability in cAMP levels. The resultant regional variability of cAMP levels under these conditions reflects a constellation of factors including nonuniform energy distribution throughout the brain, differential regional concentration of AC and PDE, and regional responses to varying degrees of anoxia.

It should be noted, however, that levels of cAMP in the cerebral cortex were not dependent on the rapidity of the microwave system utilized. Lust *et al.*[8] have similarly reported that in comparing both slower and rapid microwave systems, levels of cAMP from mouse cortex were consistently low, even in the deeper layers, and did not respond with the anoxic elevations observed using freezing methods of brain fixation (Fig. 6). Studies by Passonneau *et al.*[29]

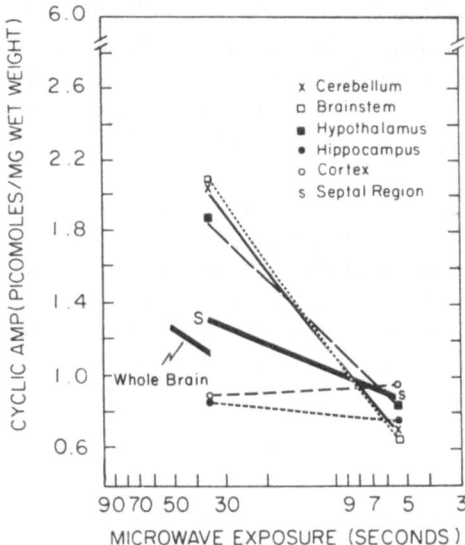

Fig. 5. Comparison of cAMP concentrations in rat brain regions determined with two different microwave inactivation systems. The 35-sec exposure duration was required for adequate regional enzyme inactivation using the earlier whole-body-oven cavity system, whereas the 5-sec duration was required with the head waveguide applicator system. Levels of cAMP in many brain regions were decreased as the inactivation time was reduced. Regions of the brain, most notably the hippocampus and cortex, remained relatively independent of changes in the inactivation time required in these two microwave systems. (Adapted with permission from Lenox *et al.*[63])

have clearly demonstrated that under similarly slow microwave inactivation conditions, significant anoxic artifacts do occur in levels of other intermediary metabolites. Thus, levels of cAMP in cortex from animals exposed to microwave cannot be utilized as an indicator of the degree of anoxia.

As microwave inactivation systems increased their power and began utilizing waveguide applicators to focus the microwave field onto the head of the animal, decreasing the required exposure time, it became apparent that concentrations of cAMP were more constant throughout the brain and were lower than previously reported. Studies using current, more rapid microwave systems generally report cAMP levels for all brain regions less than 1.0 pmol cAMP/ mg wet wt.

Jones and Stavinoha[32] have reported that regional cAMP levels in brain remain relatively unchanged in mice exposed to a range of more rapid microwave conditions, from 0.44 kW for 4 sec to 7 kW for 250 msec. It appears that cAMP in brain regions can be reliably measured in animals exposed to microwave inactivation systems requiring several seconds in spite of the fact that significant anoxic artifactual changes occur in levels of intermediary metabolites. Studies in our laboratories have utilized such a system (2.5 kW, 5 sec) to examine pharmacological and physiological characteristics of regional brain cAMP in the rat.[34–39]

Cyclic GMP in brain appears to decrease on exposure to anoxia or decapitation.[30] Jones and Stavinoha[32] report a regional variability in the postmortem decrease of cGMP following circulatory arrest, with levels remaining generally unchanged at 30 sec and decreasing at 120 sec. Levels of cGMP throughout the brain vary over a tenfold range from region to region, with the cerebellum having the highest concentrations.[37]

Studies in our laboratory have demonstrated that immobilization stress increases pituitary levels of cAMP[36] and that locomotor activity elevates levels of cGMP in the cerebellum.[39] Since microwave inactivation systems require restraint of the animal prior to placement into the waveguide applicator, attention must also be paid to these factors in the interpretation of cyclic nucleotide data obtained using rapid microwave inactivation systems.

3.3.3. Acetylcholine

Postmortem changes in the concentration of ACh are known to occur in brain.[40] The turnover of ACh in brain is very rapid, and there are high concentrations of the enzymes involved in its metabolism. Stavinoha and colleagues[43] first reported the use of microwave inactivation for the determination of ACh in rodent brain. As shown in Table II, levels of ACh are significantly lower in animals sacrificed by decapitation or whole-body exposure to low-power microwave inactivation than in those exposed to high-power focused microwave irradiation of the head. Data shown in Table II also compare re-

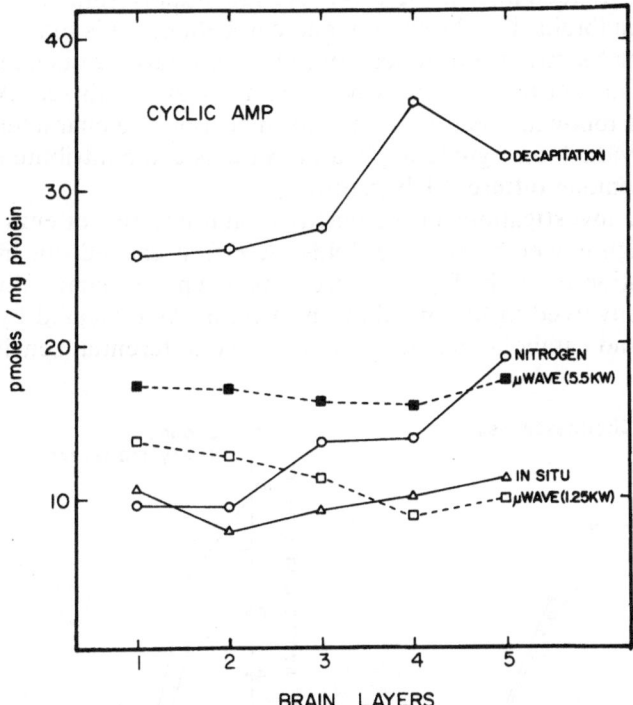

Fig. 6. A comparison of cAMP levels at various depths of the mouse brain following different methods of fixation. The cooling methods are represented by the solid lines, and the heating by the dashed lines. After the appropriate method of fixation, a dorsoventral plug of brain from the parietal region was removed in a cryostat ($-20°$) and divided into five wafers about 1.2 mm in thickness. The outermost portion of the brain was labeled layer 1, and the innermost layer 5. The symbols are values from the following methods of fixation: $\triangle$, *in situ* fixation with liquid nitrogen; $\bigcirc$, frozen intact in liquid nitrogen; $\ominus$, frozen after decapitation in liquid nitrogen; ■, microwave irradiation for 0.3 sec in a 5.5-kW oven; and □, microwave irradiation for 4 sec in a 1.25-kW oven. Each value is the mean of four to six determinations. (Adapted with permission from Lust *et al.*[8])

Table II
Endogenous Acetylcholine in Brain Regions of Mouse[a]

Brain region	Dislocation of the spine[41]	Microwave irradiation		
		Whole body[41] (1.3 kW, 7 sec)	Head[42] (6 kW, 0.30 sec)	Head[41] (5 kW, 0.25 sec)
Cortex	13.2 ± 0.74	18.4 ± 1.55	25.7 ± 3.4	24.8 ± 1.44
Hippocampus	15.8 ± 0.83	17.1 ± 1.77	31.0 ± 1.9	21.2 ± 2.32
Striatum	37.1 ± 2.06	40.4 ± 3.73	81.1 ± 1.9	76.8 ± 1.78
Brainstem	19.8 ± 1.07	20.8 ± 3.01	44.1 ± 1.9	29.1 ± 1.10
Midbrain	21.1 ± 1.24	23.2 ± 1.47	33.8 ± 2.2	29.0 ± 1.80
Cerebellum	3.7 ± 0.38	4.6 ± 0.97	16.6 ± 2.4	5.2 ± 1.05

[a] Values represent nmol/g ± S.E.M. Adapted from ref. 41.

gional brain concentration of ACh in the mouse using two microwave inactivation systems with similar power and exposure duration parameters. In spite of different assay techniques, the levels are remarkably similar except in portions of the hindbrain, i.e., brainstem and cerebellum. This may be attributed to factors such as strain differences but also may raise a question regarding the characteristics of the specific waveguide applicators utilized. As discussed more fully in a following section, nonuniform heating and characteristics of the design of individual waveguide applicator systems can contribute significantly to variability among different laboratories.

Extensive investigations of the inactivation properties of enzyme systems exposed to high-power microwave fields were first carried out in relation to the determination of ACh. Figure 7 presents a typical inactivation curve for the enzymes involved in the metabolism of ACh. As indicated by the figure, the anabolic and catabolic enzyme systems have differential sensitivity to mi-

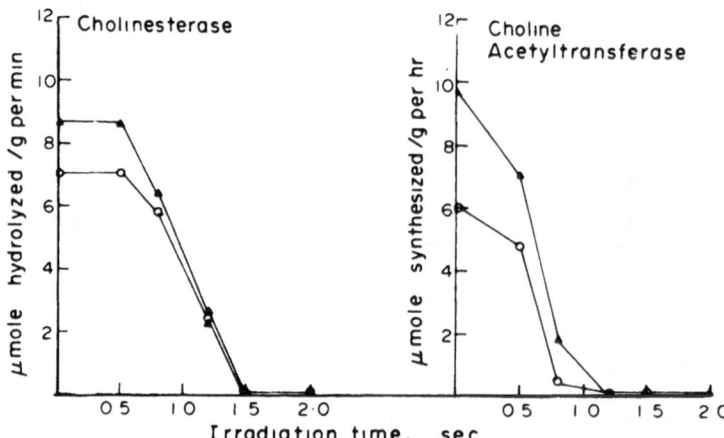

Fig. 7. Rate of inactivation of choline acetyltransferase and cholinesterase activity in brain regions of rats (150–200 g) exposed to microwave irradiation (1.25 kW, 2450 MHz) in a head-focused waveguide system. Brainstem (▲——▲), hippocampus (O——O). (Adapted with permission from Schmidt.[68])

crowave inactivation. In this case, the selection of 1.5 sec for this microwave system is a sufficient duration of exposure for determination of levels of ACh in the brain regions indicated. Stavinoha *et al.*[61] reported a detailed analysis of the regional heating characteristics of their microwave system (7.3 kW, 2540 MHz) in mouse brain. Utilizing both a rapid thermocouple and thermographs to assess the heating distribution following microwave irradiation, they were able to map levels of ACh and regional enzyme activities of both choline acetyltransferase (ChAT) and AChE as a function of relative temperature attained in various regions of the brain. These studies as well as other reports[32,41,63] demonstrated the importance of establishing adequate inactivation parameters for each metabolite in the brain regions of interest.

3.3.4. GABA and Other Amino Acids

Levels of GABA and some other amino acids are subject to postmortem change. In an early report, Lovell and Elliott demonstrated that GABA levels in tissue increased with time after decapitation and that levels of GABA in tissue frozen *in situ* were much lower than levels in similar but unfrozen tissue.[44] Shank and Aprison noted that levels of alanine as well as GABA increased post-mortem, whereas levels of aspartate, glutamate, glutamine, serine, and glycine were unchanged.[45]

Microwave irradiation has been successfully used by our laboratory and others to determine concentrations of GABA and other amino acids in rodent brain.[46–49] Rat whole-brain levels of GABA were equivalent in rats sacrificed by microwave irradiation and by decapitation into liquid nitrogen, whereas levels of GABA increased in brains sacrificed by decapitation and left at room temperature (20°C). Moreover, brains from animals previously sacrificed by microwave irradiation did not show an increase in GABA after exposure to 20°C for 30 min, indicating that microwave treatment stopped enzyme activity.[46] In the same study, the postmortem GABA rise in hypothalamus dissected from animals sacrificed by decapitation at 20°C was larger and more rapid than the rise seen in whole brain. This may be because of the high level of activity of the GABA synthetic enzyme, glutamate decarboxylase, in the hypothalamus.[50] Microwave irradiation has also been used to determine levels of GABA in discrete nuclei of hypothalamus and substantia nigra.[51] The GABA levels in the pars reticularis of the substantia nigra were fourfold higher in tissue dissected from rats sacrificed by decapitation than in tissue from rats sacrificed by microwave irradiation. Since freezing techniques require more than a minute to freeze deeper regions of rat brain, and since levels of GABA remain stable at room temperature in microwave-irradiated brains, allowing time for regional dissection, microwave irradiation is currently the best method available for measurement of *in vivo* levels of GABA.[46,51]

Microwave irradiation has also been used to determine levels of other amino acids in brain. Levels of glutamate in rat brain are similar following microwave irradiation or freezing, with little postmortem change in this amino acid.[47] In mouse brain, alanine, taurine, aspartic acid, threonine, serine, GABA, glutamine, lysine, histidine, arginine, glutamic acid, glycine, valine, leucine, tyrosine, and phenylalanine levels were determined and compared following sacrifice by decapitation into liquid nitrogen, decapitation at room temperature, and microwave irradiation.[52] There were marked postmortem elevations in GABA and alanine in the brains decapitated at room temperature. The other amino acid levels were unaffected by sacrifice technique.

Microwave enzyme inactivation appears to be the method of choice in studies involving the measurement of GABA and alanine. Although levels of other amino acids in whole brain do not appear to change rapidly after death, levels in specific brain regions may change. Since amino acids are heat stable, microwave irradiation would appear to be a sacrifice technique suitable for this class of metabolites.

3.3.5. Norepinephrine, Dopamine, and Serotonin

High-power microwave irradiation has been utilized in the determination of a number of other substances within the brain. These include the biogenic amines, norepinephrine (NE), dopamine (DA), and serotonin (5HT). Initial studies of levels of NE, DA, and 5HT in mouse or rat whole brain found comparable levels in animals sacrificed by microwave irradiation and by decapitation.[53,54] However, subsequent investigations revealed that DA levels in certain specific brain regions were different in tissue from microwave and decapitated animals.[54–56]

One such study reported that a portion (approximately 35%) of the total concentration of DA in mouse striatum was vulnerable to rapid postmortem degradation, as measured by the decrease in levels of DA and the increase in 3-methoxytyramine found in animals sacrificed by decapitation versus microwave.[57] Levels of DA in cerebellum, medulla–pons, hippocampus, diencephalon, and cortex were similar after decapitation or microwave in this study. Levels of NE in all regions were unaffected by sacrifice method. These authors also indicate preliminary findings of elevated levels of 5HT in a number of brain regions of the mouse following sacrifice by microwave compared with decapitation.

We reported that DA levels in the rat were markedly elevated in frontal and parietal cortex, remainder of cortex, hypothalamus, amygdala, and septal region after microwave irradiation.[54] These regions are located near the corpus striatum, nucleus accumbens, or olfactory tubercle, all of which have very high concentrations of DA. Diffusion from regions high in DA to adjacent regions low in DA might cause the observed elevations. On the other hand, NE regional levels measured in the same experiment were not different in microwaved or decapitated rats.

These reports suggest caution in interpreting studies of metabolite levels in whole brain which may mask significant alterations in regional distribution.[54,57]

3.3.6. Other Substrates

Histamine concentration has been determined in rat brain following microwave enzyme inactivation.[58] Histamine levels were significantly higher (threefold) in microwaved brain than in decapitated controls. Hough and Domino[58] concluded that this relative elevation of histamine content was secondary to the microwave treatment, possibly arising from metabolic formation from other than classical histidine pathways.

The use of microwave for determination of neuroactive peptides has been considered, particularly since both specific and nonspecific peptidases are distributed throughout the brain. Studies with mouse brain have shown similar distribution patterns of the opioid peptides (β-endorphin and enkephalins) in microwave-exposed animals as in decapitated animals. However, there was a small but consistently lower value from the decapitated brains, suggesting some postmortem degradation.[59] On the other hand, studies with substance P indicated that whole-brain levels were significantly elevated following microwave inactivation as compared to immersion into isopentane and decapitation.[60] These findings could not be replicated by merely heating the brain following decapitation. This study raises another important issue related to the interface between microwave-treated tissue and individual assay systems. Since radioimmunoassay is utilized by many laboratories for determination of peptides, it is possible that certain peptide fragments cleaved during the microwave heating process may result in additional immunoreactive substances. There are insufficient data to indicate a reason for the differences observed in substance P following microwave inactivation, but it does serve to emphasize the importance of documenting well the applicability of microwave irradiation for the metabolic systems under investigation.

3.4. Limitations

3.4.1. Nonuniform Heat Distribution

The precise distribution of heating and enzyme inactivation of the brain depends on the interaction between the pattern of microwave energy absorbed and the thermal diffusion properties of the tissue. The most efficient enzyme inactivation will occur if there is uniform microwave heating. The configuration of the field, which is responsible for the actual microwave heating pattern, is dependent on its frequency and the composition of the rat head, as well as the orientation of the head with respect to field polarity. The amount and precise pattern of microwave energy absorbed by a load (the rat's head) is dependent on the dielectric properties of the materials (hair, skin, muscle, fat, bone) composing that load and their characteristic distributions. Thus, the nonuniformity of dielectric loading of the rat head is sufficient to cause significant perturbations in the microwave field within the waveguide chamber, resulting in an inherent nonuniform distribution of microwave heating throughout the brain.

The enzyme inactivation pattern achieved, however, is dependent not only on the complex dielectric loading of the rat head but also on the precise orientation in relation to the polarization of the incident microwave field at the time of exposure. Studies using thermal probes[41] as well as more refined techniques of thermography[61] and enzyme histochemistry[23,62] have clearly confirmed this nonuniformity following microwave exposure (Fig. 8). Thus, the

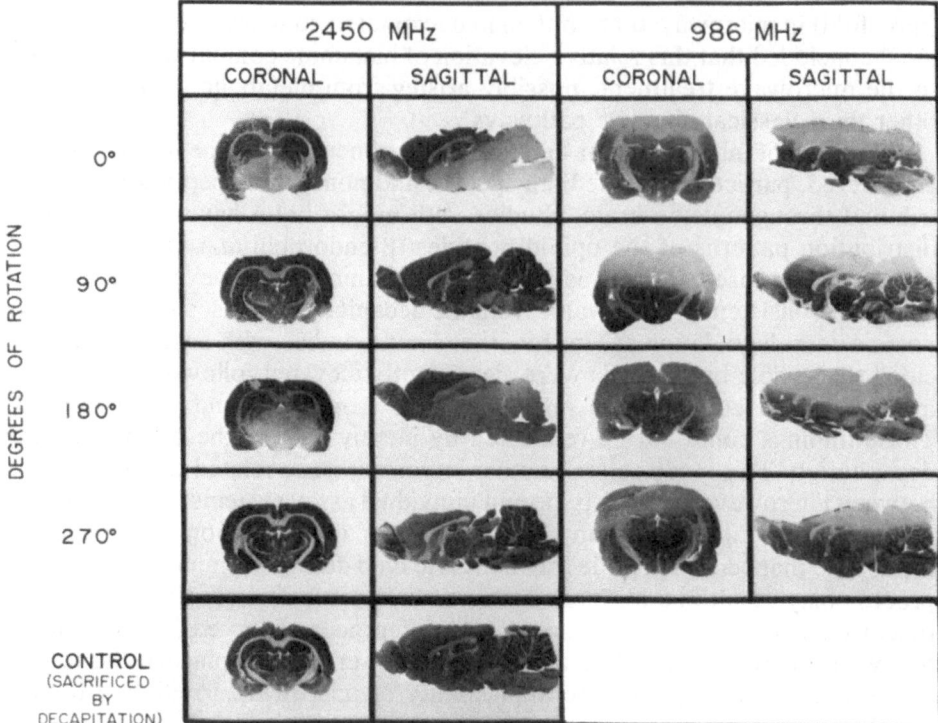

Fig. 8. Coronal and saggital sections of rat brain following exposure at 2450 MHz or 986 MHz at different orientations *vis à vis* the microwave field. Anesthetized rats (300 g) were inserted into a waveguide applicator at one of four angular positions (0°, 90°, 180°, or 270°). 0° and 180° describe the rat ear-to-ear axis parallel to the longitudinal axis of the waveguide; 90° and 270° describe ear-to-ear axis perpendicular to the longitudinal axis of the waveguide. Exposure durations for each frequency were designated to sacrifice the animal but leave residual enzyme activity in the brain (2450 MHz, 2.5 kW, 2.3 sec; 986 MHz, 1.75 kW, 6 sec). Following exposure, the brain was removed, frozen, and sectioned at 10 μm, at −2°C in a cryostat. The spatial pattern of heat distribution and enzyme inactivation in each section was characterized by cytochemical determination of succinic dehydrogenase activity. The resultant gradient of enzyme activity confirmed the nonuniformity of the microwave field generated within the brain. This gradient, however, was relatively independent of the degree of rotation *vis à vis* the microwave field at 986 MHz but was significantly angle dependent at 2450 MHz. (Adapted with permission from Meyerhoff *et al.*[23])

configuration of the heating and enzyme inactivation pattern within the brain will be a characteristic of the frequency of the microwave field and the waveguide applicator design, i.e., orientation of the animal's head relative to the incident microwave field. In the presence of this nonuniformity of microwave heating, it is essential that the duration of exposure be sufficient to inactivate the enzymes in the least heated regions. Moreover, in order to achieve reproducible regional inactivation once these parameters have been established, current microwave systems (2450 MHz) require consistent placement of the animal into the waveguide applicator within a plastic restraining tube. Therefore, it is necessary to determine microwave power and exposure parameters required to irreversibly inactivate the metabolic enzyme systems in multiple regions of the brain.[63]

3.4.2. Cellular/Subcellular Disruption

3.4.2a. Loss of Anatomic Integrity. Determination of the optimal duration of microwave exposure depends not only on the irreversibility of the inactivation of the enzyme systems involved and the stability of the substrates measured but also on maintenance of the structural integrity of the brain tissue.

Rapid deposition of microwave energy into the brain with resulting sudden vaporization of water content might be expected to cause disruption of the cytoarchitecture. Evidence from our laboratory indicates that at microwave exposure durations required for enzyme inactivation in the cyclic nucleotide system, there is significant disruption of myelin, resulting in major alteration of fiber bundles coursing through and delineating certain brain regions (Fig. 9).[54] Butcher and Butcher noted the inflated appearance of individual cell bodies and the presence of pyknotic nuclei in microwave-irradiated brain tissue.[62] Medina *et al.* have also examined brain tissue at a subcellular level and observed particular sensitivity of the mitochondria to microwave irradiation.[24] Synaptosomal preparations from microwave-irradiated brains demonstrate the presence of synaptosomes and synaptic vesicles but no mitochondria. Overexposure of brain to microwave energy results in vacuolization, shifting of brain regions, and destruction of cellular morphology. Although no significant effects of microwave irradiation on protein or water content of the brain have been observed within the parameters selected for enzyme inactivation,[32] prolonged exposure causing significant disruption of structural integrity may result in such alterations.

3.4.2b. Diffusion of Substrates. In the presence of nonuniform microwave heating, field concentrations produce "hot spots" which result in an inefficient rate of energy transfer by thermal conduction throughout the brain. Brain tissue in such "hot spot" areas can lose its anatomic features as a result of vacuolization and consequent cellular disruption. At such time, significant diffusion of substrates from one brain region to another might occur. Studies in our laboratory[54,55] using both 2450 MHz and 986 MHz microwave inactivation systems have presented evidence for diffusion of DA from regions high in DA content (corpus striatum, nucleus accumbens, olfactory tubercle) to secondary sites including frontal cortex and amygdala. As indicated in Section 3.3.5, these findings have also been confirmed in other laboratories.[56,57] Diffusion of other metabolites has not been reported. We have suggested that the appearance of a significant diffusion artifact for a given metabolite may be dependent on the concentration gradients existing between contiguous regions as well as on the size of the regions examined.[54] Thus, diffusion problems could arise in the measurement of any substrate when detailed regional dissection is contemplated following microwave inactivation.

3.4.3. Differential Rates of Enzyme Inactivation

The heat lability of different enzyme systems varies. The activity of mitochondrial ATPase from beef heart continues to be increased as temperatures

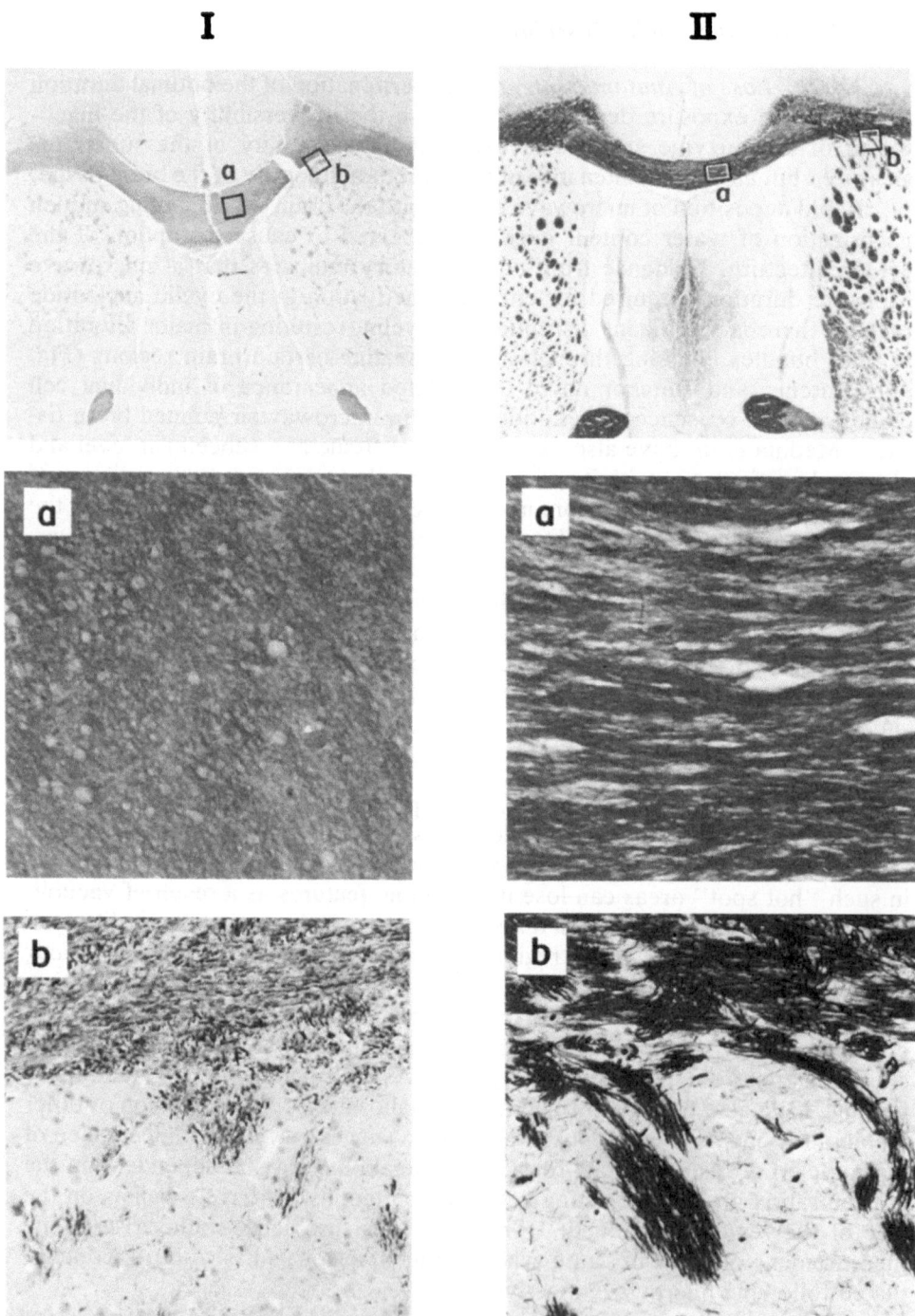

Fig. 9. Coronal sections of rat brain following (I) sacrifice by microwave irradiation at 2450 MHz (5 sec); (II) sacrifice by decapitation. Top sections were stained for myelin with luxol fast blue. Squares indicate areas enlarged to produce sections a and b, respectively. Sections Ia and IIa represent enlargement of the luxol fast blue stain of the corpus callosum. Sections Ib and IIb

are raised to 60–65°C, whereas creatine kinase from rabbit brain is inactivated at 48°C.[64,65] Studies using microwave inactivation, elevating brain temperatures to between 80 and 90°C, have demonstrated reduced activity of myokinase (6–12%) and catecholamine-O-methyltransferase (COMT) (18%), whereas the activities of enzyme such as ATPase, hexokinase, phosphatase, etc. were reduced to less than 0.5% of initial activity.[24,66] Studies of the relative sensitivity of anabolic/catabolic enzyme systems for substrates such as ACh, cAMP, cGMP, and DA to microwave inactivation have been reported. The anabolic enzymes AC, GC, and ChAT appear to be inactivated prior to their respective catabolic enzymes PDE or AChE.[67,68] This is also the case for DA, where tyrosine hydroxylase and DOPA decarboxylase activities are virtually eliminated at 81°C, although residual monoamine oxidase (MAO) (10%) and COMT (25%) activity remains.[57] In these studies, the investigators selected microwave exposure parameters necessary to achieve a temperature of 81°C (250 msec) as optimal for regional determination in mouse brain. Since a temperature of 90°C (300 msec) resulted in significant diffusion and tissue displacement, a compromise between adequate enzyme inactivation and maintenance of anatomic integrity was achieved. Our laboratory has similarly outlined criteria for selection of microwave exposure parameters for regional cAMP determination in rat brain.[63]

Slower microwave inactivation systems will require longer duration of exposure to reach critical inactivation temperatures for individual enzyme systems. As noted above, various enzymes may be particularly vulnerable to initial, brief, heat-induced increases in activity during the exposure period, resulting in potential artifactual changes in the substrates of interest. As the power of microwave systems increases, the time required to raise the temperature of the brain decreases significantly. As more energy is deposited in the brain over a shorter exposure time, permitting less opportunity for heat equilibration throughout the brain via thermal conduction, the problems of nonuniformity, generation of "hot spots," vacuolization, and loss of structural integrity become even more prominent. As efforts are made to accommodate for the nonuniformity of the field distribution by decreasing the exposure time and temperatures attained, there is a risk of residual low activity of some enzyme systems which can affect the stability of the substrate in a region of interest during periods of prolonged dissection at room temperature.

3.4.4. Other Problems

3.4.4a. Stress. In current microwave irradiation systems, the animal must be briefly immobilized in some type of plexiglass holder which is then inserted

---

represent enlargement of the interface of corona radiata and corpus striatum stained for axons by Bielschowski's method. Myelin shows a generalized decrease in staining in the microwave sections in all white matter. In the corpus callosum and corona radiata, individual myelinated bundles are much less distinct (Ia), and the tissue has a vacuolated and pale appearance. Axons appear extensively fragmented into argentophilic globules, with virtual disappearance of fine fibrils in the parenchyma of the corpus striatum (Ib). (Adapted with permission from Meyerhoff *et al.*[54])

into the waveguide applicator. This presacrifice procedure can significantly affect the very metabolites under consideration. Our laboratory has shown that cerebellar cGMP is very responsive to locomotor activity of the animal[39] and that cAMP in the pituitary is responsive to the stress of immobilization.[36] Others have observed elevations of cAMP in rat striatum following stress.[69] Techniques to habituate animals to the interface of the microwave procedure have also been developed.[34,69] In any case, these variables must be taken into consideration in examining any metabolic system utilizing microwave inactivation techniques.

3.4.4b. Microwave Inactivation System Variability. Parameters of the microwave power system must be considered in order to achieve reproducible enzyme inactivation. The net power absorbed by a given load is determined by the difference between the incident and reflected power. The incident power may vary from exposure to exposure with fluctuations in the line voltage, with the resolution capacity of the timing circuit, and with the age of the magnetron tube. The reflected power is dependent on the precise impedance match of the load to the source. The quality of this match depends on the frequency of the microwave output as well as upon the size, configuration, and dielectric characteristics of the load within the waveguide system.

In an attempt to control some of these variables, microwave systems should include the capability of in-line impedance matching to each animal with an appropriate tuner as well as continuous monitoring of the incident and reflected power with strip chart recorders.[70] A wide-frequency impedance match can be used to accommodate changes in the dielectric characteristics of the load during heating which would otherwise result in excessive amounts of reflected power. Electronic modifications have been designed to control precisely the leveling and duration of the output energy.[71] Moreover, periodic evaluations of the frequency spectrum of the power source and accuracy of the timing circuitry should be instituted.

3.5. Summary

Sacrifice by current microwave inactivation systems is very rapid, leaves the brain in a condition suitable for regional dissection, irreversibly inactivates enzymes, and appears to preserve *in vivo* levels of many heat-stable metabolites. Because of these advantages, microwave irradiation has become increasingly popular, and commercial systems are now available. However, the potential user should also be aware that current systems have certain limitations and disadvantages including: nonuniform heating throughout the brain resulting in differential rates of enzyme inactivation dependent on regional location; different heat sensitivities of various enzymes resulting in differential rates of enzyme inactivation and potential artifactual alterations in metabolite levels; possible diffusion of some metabolites within regions of the brain; requirement for immobilization of the animal prior to sacrifice; limitation of the size of the animal that can currently be microwaved; and the cost of adequate equipment.

The suitability of microwave inactivation techniques for a selected line of investigation must be considered carefully. Metabolic systems with heat-stable substrates and rapid postmortem turnover in the presence of an imbalance of anabolic/catabolic activity, resulting in an anoxic alteration in metabolite concentration, may be particularly suited to microwave methodology. However, as we have indicated, altered levels of substrate following microwave inactivation versus decapitation do not necessarily indicate that microwave sacrifice is more reliable for determination of that metabolite. Such differences in substrate concentration might also be attributable to limitations of microwave technology, i.e., to changes in structural integrity of the tissue with subsequent diffusion or to differential heat-related activation/inactivation of the enzyme systems involved. With these caveats in mind, microwave enzyme inactivation can be an effective and important technique for the neurochemist interested in the determination of *in vivo* levels of metabolites in the central nervous system.

4. FUTURE TECHNIQUES

Rapid enzyme inactivation to measure metabolites within the brain requires physical intervention having its own limitations. Both freezing and microwave inactivation techniques are useful for studies involving individual metabolites. The technology of freezing seems to have reached a plateau at present. Microwave inactivation systems may have more promise in terms of future development. There is a need for more uniform distribution of microwave energy throughout the brain in order to denature enzymes within milliseconds, yet maintain integrity of the cytoarchitecture of the tissue. For many metabolic systems, with improved uniformity of field distribution, microwave offers an ideal approach. The optimal method must be fast enough to raise temperature without significant heat activation of enzymes and yet maintain the uniform temperature rise for a period of time necessary to inactivate the enzymes. Some metabolic systems may have sufficiently different heat sensitivity of anabolic and catabolic enzymes that artifactual alterations in metabolite levels will naturally occur. As power is increased, even with more uniform distribution, the time required to raise the temperature to 80–90°C and maintain tissue integrity will become so brief that actual inactivation of enzymes will depend on thermal properties of the tissue to retain temperature for a sufficiently long period of time to result in *in situ* inactivation. Thus, compromises among exposure time required to raise brain temperature, the uniformity of field distribution, and critical inactivation temperatures for different enzyme systems must be foreseen.

Efforts to improve the uniformity of field distribution have been considered. The design of our present waveguide applicator for 2450 MHz has incorporated the use of internal shims to shape the field generated within the waveguide.[70] Based on our studies with field distribution using 986 MHz in the rat[23] (Fig. 8) and the theoretical concept that penetration of energy into a load in a microwave field is proportional to the wavelength, the possibility of designing waveguide systems capable of transmitting a complex of multiple frequencies has been considered.[23]

Another current limitation of microwave inactivation systems is the requirement for restraint prior to placement of the animal in the waveguide applicator. The prospect of exposing a free-moving animal in a microwave field capable of producing enzyme inactivation with a high degree of uniformity throughout brain regions would represent a significant technological advance. Additional theoretical knowledge of microwave field distribution and heating characteristics of complex biological systems within a waveguide should contribute significantly to the improved design of future microwave applicators for rapid *in vivo* inactivation of enzymes in the brain.[72]

Newer noninvasive techniques to examine functioning metabolic systems *in situ* hold the promise of eliminating the requirement for *in vivo* enzyme inactivation. Methods utilizing positron-emitting isotopes of compounds involved in cellular metabolism, e.g., [^{11}C]glucose and [^{18}F]fluorodeoxyglucose, in combination with positron-sensitive computerized axial tomography permit dynamic analysis of metabolic systems of the brain.[73,74] Nuclear magnetic resonance (NMR) of energy-related phosphate compounds, e.g., creatine phosphate and ATP, can also be carried out in living systems to monitor changes in metabolic activity.[75] Such techniques are presently limited by their resolution and, in the case of deoxyglucose, by assumptions regarding distribution and access of isotopes into physiologically relevant metabolic pools.

ACKNOWLEDGMENTS. The authors would like to express their appreciation to previous collaborators associated with the Department of Microwave Research, Walter Reed Army Institute of Research, for their input during the developmental stages of our microwave inactivation systems. We wish to thank Drs. John Holaday and David Lust for their critical review of the manuscript and Ginger McDowell for her fine typing support.

REFERENCES

1. Kerr, S. E., 1935, *J. Biol. Chem.* **110**:625–635.
2. Stone, W. E., 1938, *Biochem. J.* **32**:1908–1918.
3. Lowry, O. H., Passonneau, J. V., Hasselberger, F. X., and Schultz, D., 1964, *J. Biol. Chem.* **239**:18–30.
4. Jongkind, J. F., and Bruntink, R., 1970, *J. Neurochem.* **17**:1615–1617.
5. Ferrendelli, J. A., Gay, M. H., Sedgwick, W. G., and Chang, M. M., 1972, *J. Neurochem.* **19**:979–987.
6. Swaab, D. F., 1971, *J. Neurochem.* **18**:2085–2092.
7. Lust, W. D., Passonneau, J. V., and Veech, R. L., 1973, *Science* **181**:280–282.
8. Lust, W. D., Murakami, N., de Azerdo, F., and Passonneau, J. V., 1980, *Cerebral Metabolism and Neural Function* (J. V. Passonneau, R. A. Hawkins, W. D. Lust, and F. A. Welsh, eds.), Williams & Wilkins, Baltimore, pp 10–19.
9. Ehrlich, Y. H., Davis, L. G., and Brunngraber, E. G., 1978, *Brain Res. Bull.* **3**:251–256.
10. Takahashi, R., and Aprison, M. H., 1964, *J. Neurochem.* **11**:887–898.
11. Aprison, M. H., Kariya, T., Hingtgen, J. N., and Toru, M., 1968, *J. Neurochem.* **15**:1131–1139.
12. Richter, D., and Dawson, R. M. C., 1948, *Am. J. Physiol.* **154**:73–79.

13. Ponten, U., Ratcheson, R. A., Salford, L. G., and Siesjo, B. K., 1973, *J. Neurochem.* **21:**1127–1138.
14. Welsh, F. A., and Rieder, W., 1978, *J. Neurochem.* **31:**299–309.
15. Kant, G. J., Muller, T., Lenox, R. H., and Meyerhoff, J. L., 1980, *Biochem. Pharmacol.* **39:**1891–1896.
16. Skinner, J. E., Welch, K. M. A., Reed, J. C., and Nell, J. H., 1978, *J. Neurochem.* **30:**691–698.
17. Veech, R. L., Harris, R. L., Veloso, D., and Veech, E. H., 1973, *J. Neurochem.* **20:**183–188.
18. Quistorff, B., 1975, *Anal. Biochem.* **68:**102–118.
19. Quistorff, B., 1980, *Cerebral Metabolism and Neural Function* (J. V. Passonneau, R. A. Hawkins, W. D. Lust, and F. A. Welsh, eds.), Williams & Wilkins, Baltimore, pp. 42–52.
20. McCandless, D. W., and Rosberg, N. C., 1980, *Cerebral Metabolism and Neural Function* (J. V. Passonneau, R. A. Hawkins, W. D. Lust, and F. A. Welsh, eds.), Williams & Wilkins, Baltimore, pp. 53–55.
21. Stavinoha, W. B., Pepelko, B., and Smith, B., 1970, *Pharmacologist* **12:**257.
22. Lenox, R. H., Brown, P. V., and Meyerhoff, J. L., 1979, *Trends Neurosci.* **2:**106–109.
23. Meyerhoff, J. L., Gandhi, O. P., Jacobi, J. H., and Lenox, R. H., 1979, *IEEE Trans. Microwave Theor. Technol.* **27:**267–270.
24. Medina, M. A., Deam, A. P., and Stavinoha, W. B., 1980, *Cerebral Metabolism and Neural Function* (J. V. Passonneau, R. A. Hawkins, W. D. Lust, and F. A. Welsh, eds.), Williams & Wilkins, Baltimore, pp. 56–69.
25. Miller, A. L., and Shamban, A., 1977, *J. Neurochem.* **28:**1327–1334.
26. Guidotti, A., Cheney, D. L., Trabucchi, M., Doteuchi, M., and Wang, C., 1974, *Neuropharmacology* **13:**1115–1122.
27. Medina, M. A., Jones, D. J., Stavinoha, W. B., and Ross, D. H., 1975, *J. Neurochem.* **24:**223–227.
28. Medina, M. A., and Stavinoha, W. B., 1977, *Brain Res.* **132:**149–152.
29. Passonneau, J. V., Lust, W. D., and McCandless, D. W., 1979, *Techniques Life Sci.* [B] **212:**1–27.
30. Steiner, A. L., Ferrendelli, J. A., and Kipnis, D. M., 1972, *J. Biol. Chem.* **247:**1121–1124.
31. Schmidt, M. S., Schmidt, D. E., and Robison, G. A., 1971, *Science* **173:**1142–1143.
32. Jones, D. J., and Stavinoha, W. B., 1979, *Neuropharmacology of Cyclic Nucleotides* (G. Palmer, ed.), Urban and Schwarzenberg, Munich, pp. 253–281.
33. Uzunov, P., and Weiss, B., 1971, *Neuropharmacology* **10:**697–708.
34. Lenox, R. H., Kant, G. J., Sessions, G. R., Pennington, L. L., Mougey, E. H., and Meyerhoff, J. L., 1980, *Neuroendocrinology* **30:**300–308.
35. Lenox, R. H., Kant, G. J., and Meyerhoff, J. L., 1980, *Life Sci* **26:**2201–2209.
36. Meyerhoff, J. L., Kant, G. J., and Lenox, R. H., 1981, *Perspectives in Behavioral Medicine*, Volume II (R. B. Williams, ed.), Academic Press, New York (in press).
37. Lenox, R. H., Wray, H. L., Balcom, G. J., Hawkins, T. D., and Meyerhoff, J. L., 1979, *Eur. J. Pharmacol.* **55:**159–169.
38. Kant, G. J., Meyerhoff, J. L., and Lenox, R. H., 1980, *Biochem. Pharmacol.* **29:**369–373.
39. Meyerhoff, J. L., Lenox, R. H., Kant, G. J., Sessions, G. R., Mougey, E. H., and Pennington, L. L., 1979, *Life Sci.* **24:**1125–1130.
40. Weintraub, S. T., Modak, A. T., and Stavinoha, W. B., 1976, *Brain Res.* **105:**179–183.
41. Nordberg, A., and Sundwall, A., 1976, *Acta Physiol. Scand.* **98:**307–317.
42. Modak, A. T., Weintraub, S. T., McCoy, T. H., and Stavinoha, W. B., 1976, *J. Pharmacol. Exp. Ther.* **197:**245–252.
43. Stavinoha, W. B., Weintraub, S. T., and Modak, A. T., 1973, *J. Neurochem.* **20:**361–371.
44. Lovell, R. A., and Elliott, K. A. C., 1963, *J. Neurochem.* **10:**479–488.
45. Shank, R. P., and Aprison, M. H., 1971, *J. Neurobiol.* **2:**145–151.
46. Balcom, G. J., Lenox, R. H., and Meyerhoff, J. L., 1975, *J. Neurochem.* **24:**609–613.
47. Balcom, G. J., Lenox, R. H., and Meyerhoff, J. L., 1976, *J. Neurochem.* **26:**423–425.
48. Waniewski, R. A., and Suria, A., 1977, *Life Sci.* **21:**1129–1142.
49. Knieriem, K. M., Medina, M. A., and Stavinoha, W. B., 1977, *J. Neurochem.* **28:**885–886.
50. Albers, R. W., and Brady, R. O., 1959, *J. Biol. Chem.* **234:**926–928.
51. Tappaz, M. L., Brownstein, M. J., and Kopin, I. J., 1977, *Brain Res.* **125:**109–121.

52. Meyerhoff, J. L., Lenox, R. H., and Brown, N. D., 1977, *Neurosci. Abstr.* **2:**607.
53. Weintraub, S. T., Stavinoha, W. B., Pike, R. L., Morgan, W. W., Modak, A. T., Koslow, S. H., and Blank, L., 1976, *Life Sci.* **17:**1423–1428.
54. Meyerhoff, J. L., Kant, G. J., and Lenox, R. H., 1978, *Brain Res.* **152:**161–169.
55. Kant, G. J., Lenox, R. H., and Meyerhoff, J. L., 1979, *Neurochem. Res.* **4:**529–534.
56. Sharpless, N. S., and Brown, L. L., 1978, *Brain Res.* **140:**171–176.
57. Blank, C. L., Sasa, S., Isernhagen, R., Meyerson, L. R., Wassil, D., Wong, P., Modak, A. T., and Stavinoha, W. B., 1979, *J. Neurochem.* **33:**213–219.
58. Hough, L. B., and Domino, E. F., 1977, *J. Neurochem.* **29:**199–204.
59. Cheung, A. L., Stavinoha, W. B., and Goldstein, A., 1977, *Life Sci* **20:**1285–1290.
60. Kanazawa, I., and Jessell, T., 1976, *Brain Res.* **117:**362–367.
61. Stavinoha, W. B., Frazer, J., and Modak, A. T., 1978, *Cholinergic Mechanisms and Psychopharmacology* (D. J. Jenden, ed.), Plenum Press, New York, pp. 169–179.
62. Butcher, L. L., and Butcher, S. H., 1976, *Life Sci.* **19:**1079–1088.
63. Lenox, R. H., Meyerhoff, J. L., Gandhi, O. P., and Wray, H. L., 1977, *J. Cyclic Nucleotide Res.* **3:**367–379.
64. Dawson, D. M., Eppenberger, H. M., and Kaplan, N. U., 1967, *J. Biol. Chem.* **242:**210–217.
65. Andreoli, T. E., Lam, K. W., and Sanadi, D. R., 1965, *J. Biol. Chem.* **240:**2644–2653.
66. Nelson, S. R., 1973, *Radiat. Res.* **55:**152.
67. Jones, D. J., Medina, M. A., Ross, D. H., and Stavinoha, W. B., 1974, *Life Sci.* **14:**1577–1585.
68. Schmidt, D. E., 1976, *Neuropharmacology* **15:**77–84.
69. Corda, M. P., Biggio, G., and Gessa, G. L., 1980, *Brain Res.* **188:**287–290.
70. Lenox, R. H., Gandhi, O. P., Meyerhoff, J. L., and Grove, H. M., 1976, *IEEE Trans. Microwave Theor. Tech.* **24:**58–64.
71. Brown, P. V., Lenox, R. H., and Meyerhoff, J. L., 1978, *IEEE Trans. Biomed. Eng.* **25:**205–208.
72. Wang, J. J. H., 1978, U.S. Army Research Office, Final Technical Report, Project A-1943, Grant DAMD17-77-G-9422, Georgia Institute of Technology, Atlanta, Ga.
73. Raichle, M. E., Welch, M. J., Grubb, R. L., Jr., Higgins, C. S., Ter-Pogossian, M. M., and Larson, K. B., 1978, *Science* **199:**986–987.
74. Reivich, M., Kuhl, D., Wolf, A., Greenberg, J., Phelps, M., Ido, J., Cosella, N., Fowler, J., Hoffman, E., Alavi, A., Som, P., Sokoloff, L., 1979, *Circ. Res.* **44:**127–137.
75. Chance, B., Eleff, S., and Leigh, J. S., Jr., 1980, *Proc. Natl. Acad. Sci. U.S.A.* **77:**7430–7434.

Radioenzymatic Analyses

Stephen R. Philips

1. INTRODUCTION

Since the late 1960s, radioenzymatic assays have been developed for many of the neurotransmitters and, in some cases, for some of their precursors and metabolites. These techniques have largely replaced the less sensitive and less specific spectrofluorometric and bioassay procedures used previously and now provide one of the more commonly used means to measure picogram amounts of these substances in very small volumes of body fluids, in individual nuclei taken from specific brain regions, and, in some cases, in large individual cells of some simple animals. This chapter reviews the development of radioenzymatic assays for substances commonly believed to act as neurotransmitters in the CNS, for their associated precursors and metabolites, and for some of the compounds, notably the trace amines, which may also play a role in neurotransmission. In addition, applications of these techniques to transmitter determinations in very small quantities of biological specimens and, in some cases, in single cells are discussed.

2. CATECHOLAMINES, PRECURSORS, AND METABOLITES

2.1. Norepinephrine, Epinephrine, and Dopamine

Prior to the development of radioenzymatic assays, norepinephrine (NE), epinephrine (E), and dopamine (DA) were most commonly measured spectrofluorometrically by techniques that required relatively large samples of tissue and which, in spite of extensive extraction procedures, frequently lacked the specificity to eliminate interference from other tissue constituents with similar fluorescence spectra. Since the late 1960s, a variety of methods for the radioenzymatic determination of the catecholamines (CAs) have been published.

Stephen R. Philips • Psychiatric Research Division, University Hospital, Saskatoon, Saskatchewan S7N 0X0, Canada.

These, however, have all been modifications and refinements of two basic procedures which were adapted to specific analytical requirements.

Norepinephrine was first assayed radioenzymatically[1] by a technique that utilized the enzyme phenylethanolamine-N-methyltransferase (PNMT) to transfer a [^{14}C]methyl group from [^{14}C]methyl-S-adenosyl-L-methionine [^{14}C] SAM) to the amine nitrogen of NE[2] (Fig. 1). The labeled product, E, was separated by paper chromatography and measured by liquid scintillation spectrometry. The assay was capable of determining as little as 1–2 ng of NE in 10 mg of brain tissue and provided the means to establish the distribution of the amine, as well as its turnover rate, in various rat brain regions.[3]

Subsequent modifications to the assay were directed to measuring smaller quantities of NE. By incubating the enzyme directly with tissue homogenates rather than with tissue extracts and by using [^{3}H]SAM, which has a higher specific activity than [^{14}C]SAM, as the methyl donor, the assay sensitivity could be increased to 0.1 ng NE.[4] Known amounts of NE added to duplicate samples of the tissue homogenates served as internal standards. Norepinephrine was separated from the reaction mixture by paper chromatography. A similar technique has been used to measure the amine in median eminence and in pools of four to ten arcuate nuclei.[5] In other cases, perchloric acid (PCA) extracts of the samples have been incubated with PNMT and [^{3}H]SAM.[6] In those samples containing less than 10 pg NE/mg of tissue, NE was first concentrated by adsorption onto alumina and then eluted with a small volume of dilute HCl; an aliquot of the eluate was used for assay. As little as 25 pg of NE could be determined with a high degree of specificity in a variety of tissues and body fluids. The assay specificity arose from the preferential methylation of NE by PNMT and from the extensive series of purification steps to which the labeled product, E, was subjected. These included adsorption onto alumina, a subsequent elution into PCA, precipitation of excess [^{3}H]SAM by phosphotungstic acid, and, finally, extraction of E into diethylhexylphosphoric acid–toluene. This procedure or variations of it have been used to measure NE levels in brainstem and hypothalamic nuclei in the rat,[7] in human plasma,[8,9] and in CSF collected from the lateral ventricle of monkeys given intravenous amphetamine.[10]

Most of the radioenzymatic assays for NE and E and all of the published radioenzymatic methods for DA are based not on the transfer of a labeled methyl group from SAM to the nitrogen atom of the CAs by PNMT but rather

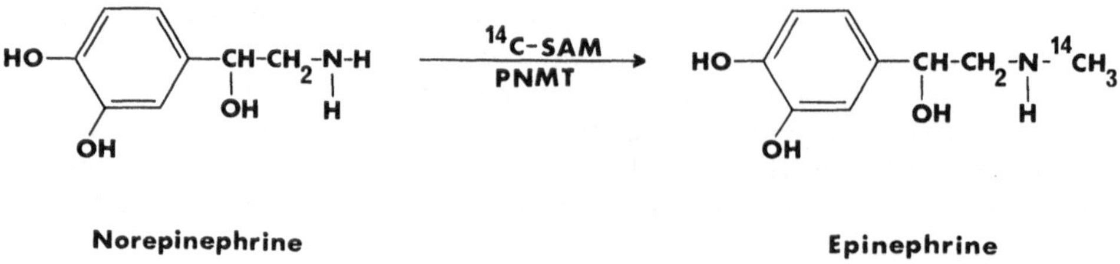

Norepinephrine   **Epinephrine**

Fig. 1. Enzymatic synthesis of epinephrine from norepinephrine. (Reproduced by permission of Academic Press, Inc.[201])

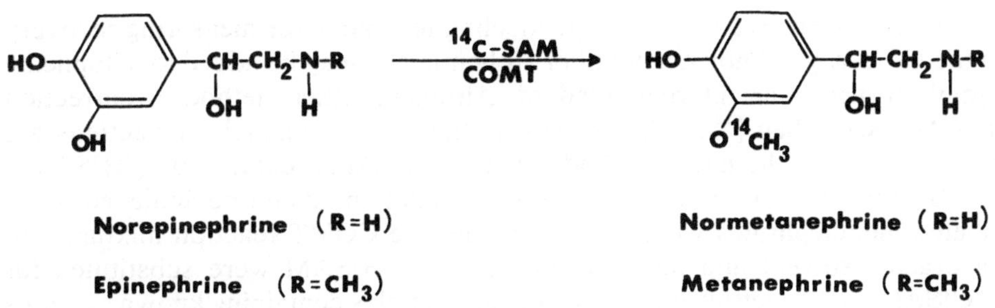

Fig. 2. Enzymatic synthesis of normetanephrine from norepinephrine and of metanephrine from epinephrine. (Reproduced by permission of Academic Press, Inc.[201])

on the transfer of a labeled methyl group from SAM to the 3-hydroxyl group of the CAs by the enzyme catechol-O-methyltransferase (COMT). The first published assay to use this principle to estimate NE in biological specimens employed [^{14}C]SAM as the methyl donor[11] (Fig. 2). Tracer quantities of [7-^{3}H] NE were added to each sample and carried through the entire procedure to estimate recovery. Catecholamines isolated from urine or plasma were incubated with the enzyme and [^{14}C]SAM. The reaction mixture was then made basic to destroy all unreacted [^{3}H]NE, and the newly formed normetanephrine (NMN), the O-methylated derivative of NE, was isolated on a column of cation exchange resin, oxidized to vanillin with NaIO$_4$ (Fig. 3), and purified by a series of extractions. Approximately 0.2 ng of CA could be determined.

The technique offered a comparatively sensitive means of measuring combined NE and E and, because of the tritiated internal standard, permitted corrections for recovery to be made. Furthermore, it provided a means of measuring other normally occurring catechols provided that appropriate chromatographic steps were incorporated. For example, NE and E could be determined simultaneously (Fig. 2) in a single plasma sample by isolating on a cation exchange column the labeled NMN and metanephrine (MN) formed in the enzyme reaction from NE and E, respectively. These substances were separated by thin-layer chromatography, oxidized to vanillin, and purified by extraction.[12] By using labeled tracers to allow a true recovery to be determined on each sample, as little as 0.25 ng of NE or E could be measured.

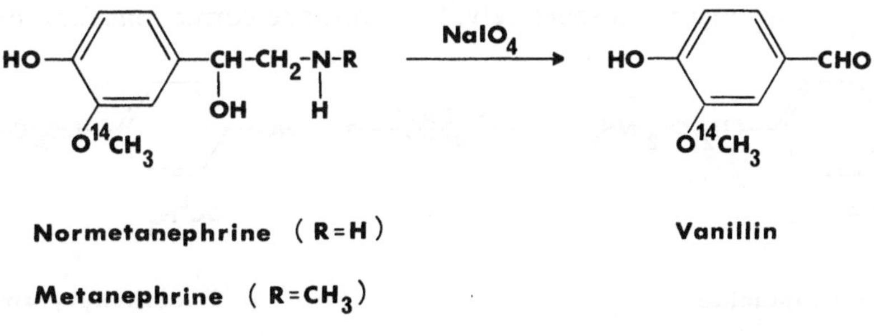

Fig. 3. Oxidation of normetanephrine and metanephrine by sodium metaperiodate. (Reproduced by permission of Academic Press, Inc.[201])

As an alternative to the double-labeling technique for measuring recovery in the CA assay, a known amount of authentic NE may be added to a duplicate sample to act as an internal standard. Although either method of correction may be used when [^{14}C]SAM serves as the methyl donor,[13] corrections are possible only by the latter method when [^{14}C]SAM is replaced by [^{3}H]SAM.

Significantly more sensitive CA determinations could be achieved if the labeled metanephrines were extracted from the COMT reaction mixture into an organic solvent and high-specific-activity [^{3}H]SAM were substituted for [^{14}C]SAM as the methyl donor.[14] Duplicate samples containing known amounts of added CAs were assayed to correct for intersample variations. Norepinephrine and E could be measured in less than 0.75 ml of plasma in amounts as low as 25–30 pg[14] with less than 0.5% contamination of either substance by the other.[15] Recently, sensitivities as low as 10–15 pg have been made possible by increasing the specific activity of the [^{3}H]SAM, by improving the extraction of tritiated methyl derivatives into the organic solvent, and by decreasing the molarity of total SAM in the enzyme reaction mixture from 10 μM to 1 μM.[16]

Although DA also occurs extensively in biological samples, it was not until 1973 that existing radioenzymatic assays for the CAs were adapted to permit DA determination.[17] Perchloric acid extracts of tissue homogenates and duplicate samples containing known amounts of NE and DA as internal standards were incubated with COMT and [^{3}H]SAM (Fig. 4). The O-methylated enzyme reaction products, isolated by extraction first into an organic solvent and then into dilute HCl, were reacted with NaIO$_4$ to oxidize the side chain at the β-hydroxyl group. [^{3}H]Vanillin, formed from NMN and MN, was separated from 3-methoxytyramine (3-MT), the methylated derivative of DA, purified, and counted. The 3-MT was extracted from the oxidizing medium into an organic solvent before counting. The CAs and DA could be determined in amounts as low as 100 and 200 pg, respectively, in the fetal rat brain. Although NE and E could not be distinguished separately, the assay was suitable for the specific application for which it was intended, since only trace amounts of E are present in the rodent brain.[18]

Modifications to this procedure, involving primarily a tenfold reduction in the reaction mixture volumes, have permitted assay sensitivities to be increased to 15 pg of NE and 45 pg of DA and have enabled these substances to be determined in isolated rat hypothalamic nuclei and nuclear subdivisions[19] as well as in various areas of the limbic system.[20] Norepinephrine and E still could not be distinguished separately, however, and corrections for cross con-

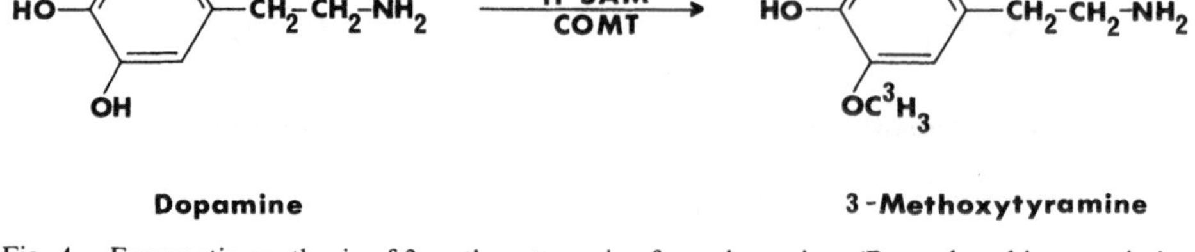

Dopamine　　　　　　　　　**3-Methoxytyramine**

Fig. 4.　Enzymatic synthesis of 3-methoxytyramine from dopamine. (Reproduced by permission of Academic Press, Inc.[201])

tamination were required for each assay, since each amine interfered to some extent in the determination of the others.

The original published assay for DA and the CAs or more recent modifications of that procedure have been widely used to measure these substances in a variety of tissues and body fluids. The subcellular distribution of endogenous NE in the rat hypothalamus[21] and the levels of both NE and DA in discrete hypothalamic regions[22] have been reported. By measuring DA levels in various brain regions after electrolytic lesions of the A8–A9–A10 region, the CNS projections of the nigral A8, A9, and A10 dopaminergic cell bodies have been determined in the rat brain.[23] Release of endogenous DA and NE has been measured from rat brain regions *in vitro*,[24,25] and endogenous NE release from brain synaptosomal fractions[26] and DA release from striatal synaptosomes[27] have been estimated. Finally, DA release has been measured *in vivo* from the caudate nucleus and nucleus accumbens septi of the gallamine-immobilized cat perfused by means of a push–pull cannula.[28]

An alternative and rather less complicated assay for DA, also published in 1973, was capable of measuring 1–3 ng of DA in the presence of up to 100 ng of NE.[29] Tissue CAs were extracted with PCA and methylated by reaction with [^{3}H]SAM in the presence of COMT. Tritiated 3-MT was isolated by paper chromatography and counted. If the reaction products were first extracted into an organic solvent and then back extracted into dilute HCl before chromatographic separation, the limit of detection was increased to 0.1–0.2 ng.[29] Dopamine and NE have been measured by this method in locust brain[30] and in regions of brains obtained post-mortem from human controls and schizophrenic patients.[31] A similar method has been used to determine NE, E, and DA simultaneously in various regions and nuclei in the rat brain.[32,33] In the latter cases, however, as little as 10 pg of the amines could be determined in tissue samples weighing less than 1 mg. Cross interferences among the three amines, although small, were constant, and hence, corrections were feasible.

Early attempts to measure the CAs and DA in plasma and other biological fluids had occasionally been unsuccessful because of the presence of endogenous inhibitors of COMT activity. As calcium was found to be the major inhibitor present in these samples, EGTA was often added to the plasma extract,[34] blood serum,[35,36] COMT reaction mixture,[37] or brain perfusate[38,39] to chelate calcium, and $MgCl_2$ was added to activate the enzyme. Alternatively, the CAs in tissue extracts or body fluids could be isolated on alumina and then subsequently eluted and assayed in order to permit their determination without interference from endogenous inhibitors or drugs.[40] Some of these procedures were unable to distinguish NE and E separately but were capable of determining as little as 25 pg of combined CAs in 300 μl of deproteinized serum.[35] Others,[34] although unable to obtain direct individual measurements of NE and E, succeeded in determining each amine individually by first O-methylating the combined CAs with COMT, then N-methylating NE separately with PNMT. Epinephrine was estimated by calculating the difference between the two procedures. The lower limit of detection for the amines using this technique corresponded to 20 pg of NE, 17 pg of E, and 127 pg of DA per milliliter of plasma.

Norepinephrine, E, and DA may be determined directly in a single sample of plasma or tissue by incorporating additional purification steps into the assay procedure. Thin-layer chromatography has frequently been used to separate the radioactive products extracted from the enzyme reaction mixture.[36,37] Corrections for recovery could be made by adding small amounts of [^{14}C]-labeled 3-MT, NMN, and MN before the extraction. It was possible to estimate 10–30 pg of each amine in as little as 50 μl of rat plasma.[37] The DA concentrations in rat pituitary stalk plasma[41] and extracts of median eminence[42] and the subcellular compartmentalization of endogenous DA in the rat anterior pituitary gland[43] have been determined by this procedure. Even more sensitive measurements could be obtained in plasma by assaying a duplicate sample to which known amounts of the CAs had been added as internal standards[36]; 1 pg of NE and E and 6 pg of DA could be detected. The high degree of sensitivity and low interference by one substance in the determination of the others observed in these assays was attributed largely to the thin-layer chromatographic separation of their [^{3}H]methyl derivatives. Although 3,4-dihydroxyphenylalanine (DOPA) cross reacted significantly in the DA assay, its interference was attributed to the presence of decarboxylase activity in the COMT preparation[37] and could be overcome by including a DOPA decarboxylase inhibitor in the enzyme reaction mixture.[36] This technique has recently been modified to enable the alkaloid salsolinol to be determined in plasma and neonatal rat tissue in amounts as low as 100 pg/ml of plasma and 500 pg/g of tissue.[44]

Sodium tetraphenylborate (NaTPB) has been found to enhance the extraction of O-methylated CAs from the enzyme reaction mixture into an organic solvent and to increase the pH range over which the extraction occurs. Its use has therefore become increasingly common in recent years.[39,40,45–47] By combining a more efficient extraction procedure with subsequent thin-layer chromatographic separation of the labeled methylated products and oxidation of NMN and MN to vanillin, some assays have been developed that are capable of measuring NE, E, and DA in tissue samples with a protein content of 100 μg or less and in plasma volumes of 20–100 μl.[45,46] They have been used to measure the *in vitro* release of NE and DA from the rat hypothalamus[48] and to establish the increased levels of these amines in primate CSF in response to amphetamine and phenylethylamine.[49] The levels of E in small brain regions,[50] of NE and DA in rat median eminence, medial basal hypothalmus, and suprachiasmatic–preoptic region,[51] and of NE, E, and DA in extracts of rabbit ear artery[52] have also been determined. In some cases, high-performance liquid chromatography has been used to separate the labeled enzyme reaction products; NE and E[53,54] and NE, E, and DA[55] have been measured in plasma, and NE and DA have been determined in rat pineal[56] by this procedure.

Changes in some of the enzyme reaction parameters have recently been shown to affect significantly the sensitivity of CA determinations.[57] Estimations were made on PCA extracts of tissue or deproteinized blood plasma. Bovine serum albumin was introduced into the enzyme reaction mixture to prevent COMT from binding to glass[58] and to increase enzyme activity and stability. Following the methylation reaction, many of the constituents of the mixture, including unreacted [³H]SAM, were precipitated by adding phosphotungstic acid in dilute HCl; the labeled reaction products were separated by thin-layer chromatography. Although the basic procedure permitted each of the CAs to be determined at the level of 15–20 pg, supplemental steps could be included to enable measurement of 1 pg of NE and E and 5 pg of DA. The assay was sufficiently sensitive to measure NE and E in 10 µl of rat plasma as well as CA turnover in small brain regions and the *in vitro* release of DA from striatal synaptosomes.[57]

Methylated enzyme reaction products have occasionally been derivatized to increase assay specificity. In some cases, these have been acetylated chemically before separation by paper[59] or thin-layer[60] chromatography. In others, acetylation of 3-MT has been carried out enzymatically by acetyl CoA in the presence of N-acetyltransferase[61] (Fig. 5). As little as 15 pg of DA could be determined. The assay has been used to measure DA in the ganglia and in individual neurons of several gastropod molluscs.[62]

Finally, the fluorescent dansyl derivatives of NMN, MN, and 3-MT have provided a very sensitive and specific means of determining NE, E, and DA, respectively.[63] Following methylation of the CAs and DA by COMT, the reaction products were danyslated, extracted into ethyl acetate, purified by two successive thin-layer chromatographic separations, and counted. The fact that the dansylated compounds could be separated chromatographically into very narrow bands imparted a high degree of specificity to the procedure. The technique permitted determination of 2–3 pg of NE, E, and DA and has enabled NE release from the hypothalamus and DA release from the caudate nucleus to be measured in artificial CSF perfused through a push–pull cannula implanted into these regions of the rat brain.[64]

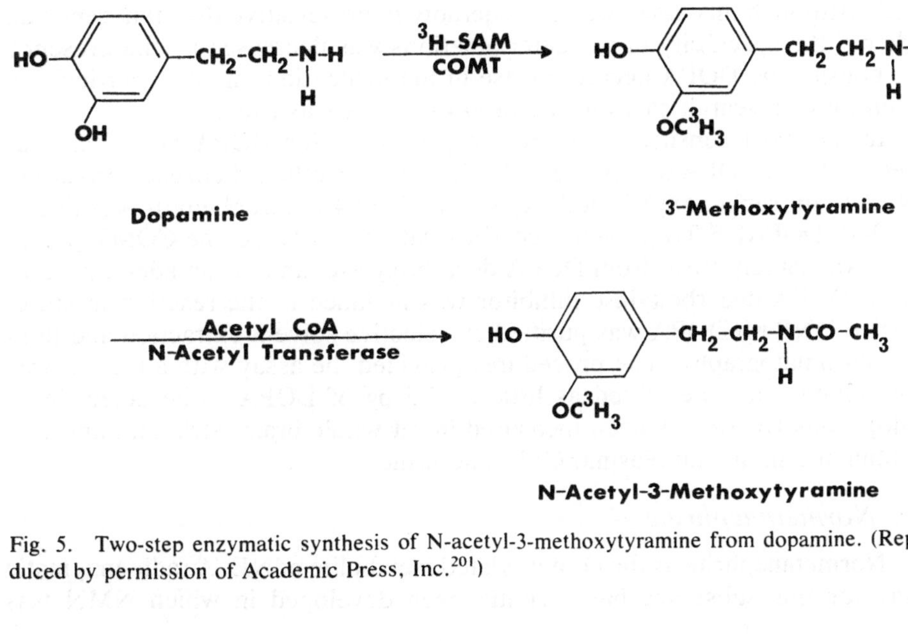

Fig. 5. Two-step enzymatic synthesis of N-acetyl-3-methoxytyramine from dopamine. (Reproduced by permission of Academic Press, Inc.[201])

2.2. DOPA

DOPA is the biosynthetic precursor of the CAs. It has only very recently been measured by radioenzymatic techniques. To date, all assays of this substance have been based on the transfer of a labeled methyl group from [³H] SAM to DOPA by the enzyme COMT. In the first reported procedure,[65] PCA extracts of rat brain regions were incubated with COMT according to conditions previously described for measuring the CAs.[19] The methylated product, 3-O-methyl-DOPA (3-O-MeDOPA), was isolated by a three-step process that included cation exchange, adsorption on activated charcoal, and anion exchange. The assay permitted DOPA to be determined in amounts as low as 50–200 pg. It has been used to measure the accumulation rate of DOPA in segments of substantia nigra in the presence of a DOPA decarboxylase inhibitor.[65]

Modifications of another technique originally developed for CA analysis[36] have recently enabled as little as 25 pg of DOPA to be measured in less than 200 µl of plasma or other fluid.[66] Plasma samples, as well as duplicates containing a known amount of DOPA as internal standard, were incubated with both DOPA decarboxylase and COMT. The two-enzyme system facilitated the decarboxylation of DOPA to DA and the subsequent methylation of DA to [³H]-3-MT; hence, DOPA and DA were measured together. Since 3-O-MeDOPA is also a substrate for the decarboxylase, the order in which the enzyme reactions occurred was of no consequence. A second pair of plasma samples was assayed to determine DA alone. DOPA decarboxylase was excluded from the reaction mixture, and a DOPA decarboxylase inhibitor was added to inhibit small amounts of decarboxylase present in the COMT preparation. Plasma DOPA could then be calculated from the difference between the two procedures. The assay has recently been extended to permit determination of CA and DOPA sulfates in plasma by including a sulfatase in the incubation mixture.[67] Although this assay was considerably more sensitive than that reported earlier,[65] the specificity of the determinations was decreased by the presence of a nonspecific DOPA decarboxylase in the incubation mixture, particularly in clinical or research cases in which α-MeDOPA was used.

In the most sensitive and specific procedure for DOPA yet published, labeled 3-O-MeDOPA was converted to its 2,4-dinitrofluorobenzene derivative, 3-[4-(2,4-dinitrophenoxy)-3-methoxyphenyl]-N-(2,4-dinitrophenyl)-L-alanine (DNP-MeDOPA).[68] To prevent decarboxylation of DOPA, the COMT preparation was largely freed from DOPA decarboxylase, and, as an added precaution, a DOPA decarboxylase inhibitor was included in the reaction mixture. Labeled DNP-MeDOPA was purified by selective solvent extraction and thin-layer chromatography. The procedures provided the assay with a high degree of specificity, and permitted as little as 0.5 pg of DOPA to be determined. Endogenous DOPA has been measured in rat whole brain, striatum, and cerebellum and in human plasma, CSF, and urine.[68]

2.3. Normetanephrine

Normetanephrine is the O-methylated metabolite of NE. A radioenzymatic assay for this substance has recently been developed in which NMN was converted to its N-methylated derivative, MN, by reaction with [^{3}H]SAM in the presence of PNMT.[69] Duplicate samples containing a known amount of NMN as internal standard were analyzed simultaneously. After the reaction had been completed, labeled MN was extracted into toluene:isoamyl alcohol (3:2, v/v), back extracted into dilute acetic acid, and isolated by thin-layer chromatography; it was completely separated from other potential PNMT substrates. The amount of product formed was proportional to the amount of NMN in the sample over the range 100 pg to 10 ng. The sensitivity of the method was such that NMN could be determined in 10 μl of urine.

The assay has been adapted to measure total (free and conjugated) NMN in plasma by subjecting the sample to acid hydrolysis,[70] then determining the amine as described for urine.[69] Approximately 20 pg of NMN could be detected, and the assay was linear from 25 pg to 10 ng. If, however, blood was first deproteinated, and NMN was extracted from the plasma onto a weak cation-exchange resin to minimize PNMT inhibition by the sample, the sensitivity could be increased to approximately 5 pg or 30 pg/ml of plasma.[71] Normetanephrine has also been determined with a sensitivity of 50 pg in 5 mg of brain tissue by incubating PCA extracts of the tissue with [^{3}H]SAM and PNMT.[72] Metanephrine was recovered as described for urine samples.[69] If required, the acid extract could be boiled to hydrolyze the conjugates to obtain a measure of total NMN in the brain.

2.4. Dihydroxymandelic Acid, Dihydroxyphenylacetic Acid, and Dihydroxyphenylglycol

3,4-Dihydroxymandelic acid (DOMA) and 3,4-dihydroxyphenylacetic acid (DOPAC) are the oxidative metabolites of NE and E and of DA, respectively. Both CA metabolites have been determined radioenzymatically. The DOMA was methylated by reaction with [^{14}C]SAM and COMT (Fig. 6), and the labeled product, [^{14}C]vanilmandelic acid (VMA), was oxidized to [^{14}C]vanillin by periodate.[73] Vanillin was purified by successive extractions into toluene and ammonium hydroxide, and its radioactivity was assessed by liquid scintillation counting. The amount of [^{14}C]vanillin formed was linearly related to the quantity of DOMA present in the sample over the range of 1–500 ng. Because the assay was relatively insensitive, it was first necessary to concentrate DOMA from relatively large samples by adsorption onto alumina. Approximately 1% of a 24-hr urine collection, 40 ml of blood, or 1 g of heart tissue was required for a single determination. Some interference by DOPAC was reported in determinations of DOMA.

More recently, DOMA and 3,4-dihydroxyphenylglycol (DOPEG), another major deaminated CA metabolite, have been measured in 50 μl of plasma or CSF and in small quantities of tissue following their methylation to VMA and 3-methoxy-4-hydroxyphenylglycol (MOPEG), respectively, by reaction with [^{3}H]SAM in the presence of COMT.[74] These substances were extracted into an organic solvent from the acidified reaction mixture, back extracted into dilute acetic acid, and separated by thin-layer chromatography. Norepinephrine and E could be determined simultaneously by isolating their respective meth-

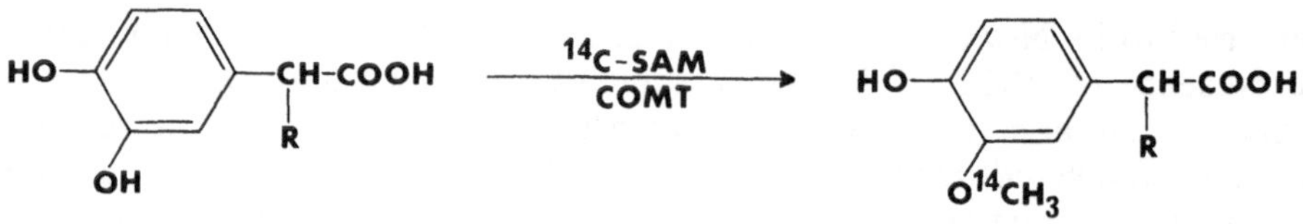

Fig. 6. Enzymatic synthesis of homovanillic acid (HVA) from dihydroxyphenylacetic acid (DOPAC) and of vanilmandelic acid (VMA) from dihydroxymandelic acid (DOMA). (Reproduced by permission of Academic Press, Inc.[201])

ylated derivatives, NMN and MN. The chromatographically separated compounds were then extracted into ammonium hydroxide, oxidized to vanillin by periodate, and counted. The assay had a sensitivity of 8 and 120 pg for DOPEG and DOMA, respectively, and 15 and 5 pg for NE and E.

The first published radioenzymatic assay for DOPAC in urine was similar to that reported for DOMA.[75] The DOPAC was concentrated from urine by adsorption to alumina and then methylated by reaction with [^{14}C]SAM in the presence of COMT (Fig. 6). The labeled product, [^{14}C]homovanillic acid (HVA), was extracted into ethyl acetate, isolated by paper chromatography, and counted. Because of the chromatographic step, HVA and VMA could be separated, and one did not interfere in the estimation of the other. Approximately 12 ng of DOPAC could be determined.

3,4-Dihydroxyphenylacetic acid is one of the principal metabolites of DA in tissue. By methylating DOPAC with high-specific-activity [^{3}H]SAM in the presence of COMT, amounts of DOPAC as low as 50 pg have been determined in PCA extracts of brain regions weighing 1–10 mg.[76] The reaction product was isolated from the enzyme reaction mixture on a column of Sephadex G-10 and extracted into ethyl acetate. The [^{3}H]VMA was also carried through the separation procedure but, if necessary, could be separated from [^{3}H]HVA by thin-layer chromatography.

[^{3}H]Homovanillic acid has also been separated from the reaction mixture by ion-exchange resin and solvent partition. If NE metabolites were present in sufficient quantities to interfere with the DOPAC determination, solvent partition could be replaced by thin-layer chromatography. Approximately 1 pmol (168 pg) or, if [^{3}H]HVA were separated twice by thin-layer chromatography, 0.5 pmol could be detected by this procedure. The assay has recently been used to measure DOPAC levels in the caudate nucleus and substantia nigra of the rat[77] and to examine some of the characteristics of DA uptake and DOPAC formation in dopaminergic terminals in the neurointermediate lobe of the rat and steer pituitary gland.[78]

Modifications of existing procedures for the CAs have enabled DOPAC and DA to be determined simultaneously in rat brain regions at levels of 150 pg and 250 pg, respectively.[79] In some cases, these substances have been measured in tissue and plasma samples at the level of 5–10 pg, whereas NE and E have been determined in amounts as low as 1 pg.[57] In other cases,[80,81] DOPAC and DOMA, along with DA, NE, and E, have been estimated in various rat brain regions with sensitivities in the 30- to 100-pg range. Simultaneous measurement of acid and amine levels have proved to be useful, since metabolic data on the CAs can be obtained by calculating the amine–metabolite ratios.

3. SEROTONIN

The first radioenzymatic assay for serotonin was published in 1973.[82] It was capable of measuring amounts of amine some 200 times smaller than those previously detectable with the most commonly used fluorometric procedures and three to four times smaller than that obtained by reacting the amine with [^{14}C]dansyl chloride and subsequently separating the product by thin-layer chromatography.[83] The assay was based on the conversion of serotonin to melatonin by a two-step enzymatic procedure: (1) N-acetylation of serotonin by acetyl CoA in the presence of rat liver N-acetyltransferase[84]; and (2) transfer of a tritiated methyl group from [^{3}H]SAM to the hydroxyl group of N-acetylserotonin by the pineal gland enzyme, hydroxyindole-O-methyltransferase (HIOMT)[85] (Fig. 7). The radioactive melatonin formed was extracted into toluene, reduced to dryness at 80°C, and counted. The amount of [^{3}H]melatonin measured was proportional to the amount of serotonin added to the reaction mixture over the range 50 pg to 20 ng. Although bufotenin, 5-hydroxytrypto-

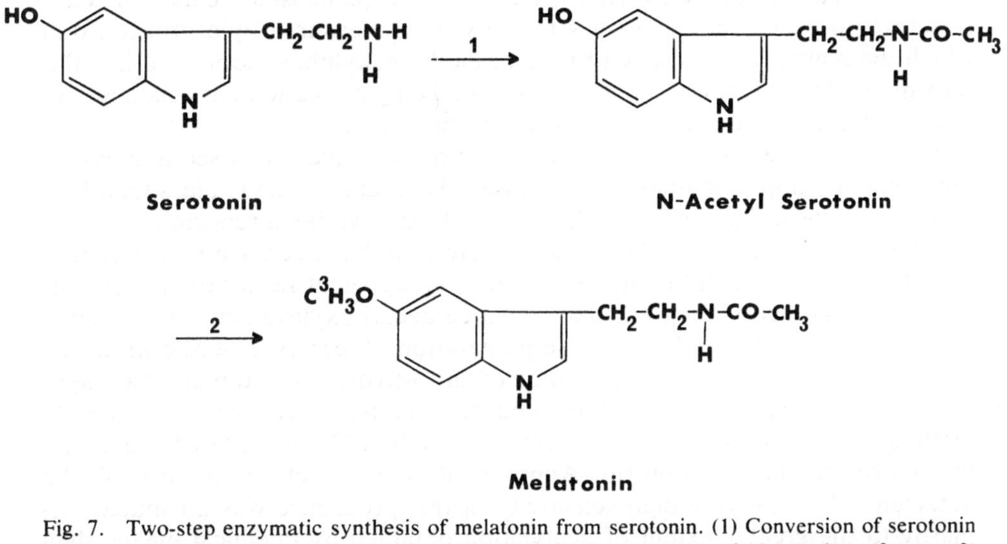

Fig. 7. Two-step enzymatic synthesis of melatonin from serotonin. (1) Conversion of serotonin to N-acetylserotonin by reaction with acetyl CoA in the presence of N-acetyltransferase. (2) Conversion of N-acetylserotonin to melatonin by reaction with [^{3}H]SAM in the presence of hydroxyindole-O-methyltransferase. (Reproduced by permission of Academic Press, Inc.[201])

phol, and N-acetylserotonin also yielded derivatives that could be extracted into toluene under the conditions of the assay, the first two compounds interfered significantly with serotonin determinations only when present in amounts ten times that of serotonin. Serotonin and N-acetylserotonin could both be determined in a sample by assaying duplicate aliquots of the sample in the presence and absence of acetyl CoA.

The use of a double enzymatic procedure provided the assay with significantly greater specificity than that of existing fluorometric procedures. It also allowed tissue blanks to be used, since acetyl CoA could be omitted from the incubation mixture to prevent the first enzyme step from occurring. The assay has been adapted to permit serotonin measurements to be made in very small amounts of tissue such as the nuclei in the rat hypothalamus and preoptic regions,[86,87] brainstem,[88] limbic system,[89] and raphe[90] as well as in large cortical regions.[91,92] The amine has also been analyzed in neuronal cell bodies of *Aplysia californica*.[93]

In one of the few significant modifications to the original radioenzymatic assay for serotonin, the amine was first separated from salts and indoles in biological fluids such as CSF by passing the sample at neutral pH through a column of Sephadex G-10.[94] Serotonin, which adsorbed to the gel, was then eluted with dilute acid, converted to [³H]melatonin, and extracted into toluene.[82] The labeled product was isolated by thin-layer chromatography. As little as 10–20 pg of serotonin could be measured in standard solutions. The technique has been used to determine the topographical distribution of serotonin in the cat spinal cord[95] and in brain cortical areas.[96] In addition, the *in vivo* release of serotonin has been measured in lateral ventricular perfusates collected from the anesthetized rat[97] and in superfusates collected from the ependymal surface of the caudate nucleus in "encéphale isolé" cats.[98] Release of serotonin from brain slices has recently been measured by a procedure in which the amine was extracted first into diethylhexylphosphoric acid in chloroform and then back into dilute HCl before being converted enzymatically to [³H]melatonin.[99] Procedures involving multiple-development thin-layer chromatographic separations of the radioactive product have been used to measure serotonin in intraparenchymal microvessels from rat brain and in less than 15% of a single pineal organ.[100] Sensitivities of 20 pg have been reported.

Finally, in at least one application, serotonin has been acetylated chemically by using acetic anhydride as the acetylating agent rather than enzymatically with the acetylating enzyme.[101] Since decarboxylase activity is known to be present in N-acetyltransferase preparations,[86] errors in serotonin determinations arising from decarboxylation of any 5-hydroxytryptophan that might be present in the sample were eliminated. N-Acetylserotonin was subsequently methylated by [³H]SAM in the presence of HIOMT, and labeled melatonin was extracted into chloroform. Approximately 35 pg of serotonin could be detected. The relatively high sensitivity of this procedure was attributed primarily to the greater extent of acetylation obtained by chemical means than is observed by enzymatic means and to the lower blank values obtained in the absence of decarboxylase activity in the medium.

4. ACETYLCHOLINE

Prior to the development of radioenzymatic assays, chemical or bioassay methods had most commonly been used to measure acetylcholine (ACh). The chemical procedures were, in general, neither specific nor sensitive enough to measure ACh in biological systems. Bioassay methods were more sensitive but became increasingly complicated as the need for specificity increased. Although several gas chromatographic procedures were available, sensitivity was limited, and some measured only the acyl component of the choline esters.

Recently developed radioenzymatic procedures have provided means of analysis that are both sensitive and specific. Two basically different techniques for ACh determination, as well as several modifications of each procedure, have been described. In the first, ACh was extracted from frozen brain tissue into 0.1 N PCA, separated electrophoretically from free choline (Ch) and other Ch-containing compounds, then hydrolyzed to Ch and reacetylated to [^{14}C] ACh by reaction with [^{14}C]acetyl CoA in the presence of choline acetyltransferase (ChAT)[102] (Fig. 8). Newly formed [^{14}C]ACh was separated electrophoretically from [^{14}C]acetyl CoA. The assay derived much of its specificity from the electrophoretic steps and was capable of measuring 10 ng of ACh.

Choline had been measured previously by radioenzymatic means.[103] Since Ch and ACh could be separated electrophoretically, both compounds could be determined simultaneously by isolating Ch from the electrophoretic strip and then reacting it with ChAT and [^{14}C]acetyl CoA.[104] This procedure has been used to measure both Ch and ACh in brain[104] and in spleen and iris.[105] A significant increase in sensitivity, to about 1.5 ng of each compound, could be obtained by extracting frozen brain tissue with 1 N formic acid:acetone (15:85, v/v).[106] Tissue lipids and lipid-soluble Ch-containing compounds were then extracted from the formic acid:acetone solution.[104] The degree of Ch overlap onto ACh could be determined by a double-isotope technique in which high-specific-activity tritiated Ch was used as a tracer.[107]

The alternative radioenzymatic assay for ACh involved its initial extraction

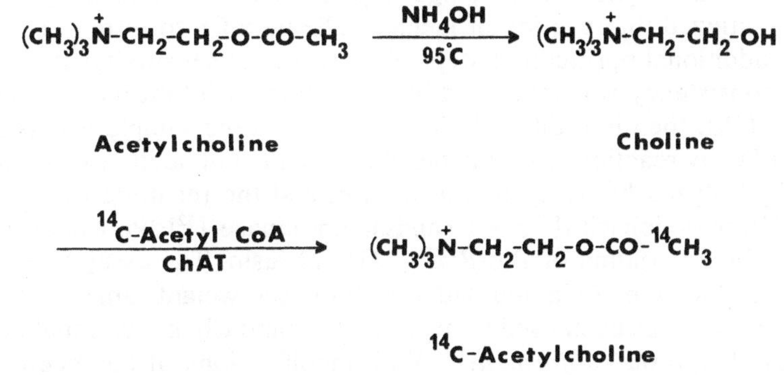

$$(CH_3)_3 \overset{+}{N}\text{-}CH_2\text{-}CH_2\text{-}O\text{-}CO\text{-}CH_3 \xrightarrow[95^\circ C]{NH_4OH} (CH_3)_3 \overset{+}{N}\text{-}CH_2\text{-}CH_2\text{-}OH$$

Acetylcholine   **Choline**

$$\xrightarrow[ChAT]{^{14}C\text{-}Acetyl\ CoA} (CH_3)_3 \overset{+}{N}\text{-}CH_2\text{-}CH_2\text{-}O\text{-}CO\text{-}^{14}CH_3$$

^{14}**C-Acetylcholine**

Fig. 8. Radioenzymatic determination of acetylcholine as [^{14}C]acetylcholine. (Reproduced by permission of Academic Press, Inc.[201])

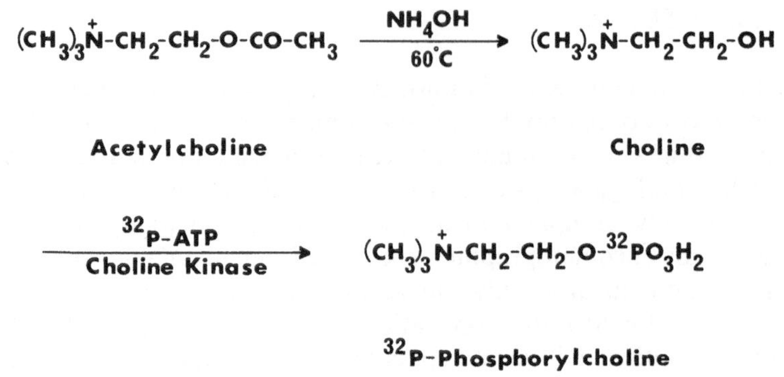

$$(CH_3)_3\overset{+}{N}\text{-}CH_2\text{-}CH_2\text{-}O\text{-}CO\text{-}CH_3 \xrightarrow[60°C]{NH_4OH} (CH_3)_3\overset{+}{N}\text{-}CH_2\text{-}CH_2\text{-}OH$$

Acetylcholine   **Choline**

$$\xrightarrow[\text{Choline Kinase}]{^{32}P\text{-ATP}} (CH_3)_3\overset{+}{N}\text{-}CH_2\text{-}CH_2\text{-}O\text{-}^{32}PO_3H_2$$

^{32}P-Phosphorylcholine

Fig. 9.   Radioenzymatic determination of acetylcholine as [^{32}P]phosphorylcholine. (Reproduced by permission of Academic Press, Inc.[201])

into formic acid: acetone and isolation by electrophoresis. The latter step imparted most of the specificity to the assay. Acetylcholine was then hydrolyzed to Ch and converted to [^{32}P]phosphorylcholine ([^{32}P]PPC) by reaction with [^{32}P]ATP in the presence of choline kinase[108,109] (Fig. 9). This technique took advantage of the relative stability of choline kinase compared to that of ChAT and of the lower cost of [^{32}P]ATP compared to [^{14}C]acetyl CoA. Acetylcholine could be measured in nanogram amounts in as little as 10 mg of tissue.

A third method for ACh determination has also been described[110] but, because of its limitations, has not been widely used. It is based on a single enzymatic reaction and the principle of isotope dilution. Although the assay was rapid and easy to perform, the sensitivity was low, and the enzyme kinetics on which the measurements depended were subject to modification by the presence of Ch, high concentrations of ions, or anticholinesterase drugs in the tissue extracts. Acetylcholine release from slices of electric organ of *Torpedo marmorata* have been measured by this procedure.[111]

Several adaptations and refinements of each of the two basic assays have been published. Most have avoided the use of columns or electrophoresis units by incorporating a liquid ion-exchange step into some stage of the procedure. Some have utilized the reaction of labeled ATP with Ch to form [^{32}P]PPC but have used additional purification steps to increase the sensitivity. In one case, the assay consistency was improved by extracting ACh first into 3-heptanone containing TPB, then into dilute HCl.[112] Choline in the sample was then converted to PPC by reaction with unlabeled ATP and choline kinase. Acetylcholine was hydrolyzed by acetylcholinesterase, and the resulting Ch converted to [^{32}P]PPC by reaction with [^{32}P]ATP and choline kinase. [^{32}P]Phosphorylcholine was purified on a column of Bio-Rad® AG1-X8 resin. The assay was capable of measuring 400 pg of ACh and had a useful tissue weight range of 0.2–8 mg of mouse brain. It has been used to measure ACh and Ch in individual neuronal cell bodies of *Aplysia californica*.[113] With modifications, it has been used to measure ACh in various tissues and subcellular fractions of rat brain cortex,[114,115] in subcellular fractions obtained by density gradient centrifugation,[116] and in the nervous system of the lobster.[117] Release of ACh has also been reported *in vitro* from superfused slices of rat cerebrum[118] and *in vivo* from the dorsal hippocampus and sensorimotor cortex during sensory stimulation and motor behavior.[119]

Other procedures have avoided use of the ^{32}P isotope with its short half-life and rapid decay characteristics. Both ACh and Ch[120,121] and, in some cases, acetyl CoA as well[122] have been determined simultaneously in small samples of brain tissue by phosphorylating Ch with unlabeled ATP, then hydrolyzing ACh and reacetylating the newly formed Ch with [^{14}C]acetyl CoA. The labeled product was separated from unreacted [^{14}C]acetyl CoA by extraction into butenenitrile containing TPB. Choline could be determined by omitting the initial phosphorylation of Ch and hydrolysis of ACh. Procedures similar to this and capable of detecting 10–25 pmol (1.6–4.0 ng) of either compound have been used to measure both ACh and Ch in nervous tissue.[123–125]

Significant increases in sensitivity and a greater convenience of measurement have recently been obtained by eliminating the double extraction of ACh from tissue homogenates into heptanone–TPB and then back into HCl, by increasing the specific activity of [^{32}P]ATP, and by decreasing the reagent blank.[126] As little as 6 pg of ACh could be assayed. Although the modified procedure could be used to measure ACh in *Aplysia* ganglion extract,[126] some constituent of mammalian nervous tissue produced anomalously high apparent ACh concentrations in mouse brain. It was therefore necessary to include the TPB extraction step when ACh measurements were made in those tissues. Nevertheless, ACh could be measured in as little as 8 μg of tissue; the assay has been used to establish ACh levels in discrete nuclei of the rat brain[127] and to measure its steady-state level in mouse brain in an *in vivo* turnover rate study.[128]

5. HISTAMINE

Histamine (HA) has been thought for some time to act as a transmitter in the CNS.[129] It was first measured radioenzymatically by a double-isotope technique in which a labeled methyl group was transferred from [^{14}C]SAM to tissue HA[130] (Fig. 10). Supernatants from homogenized tissue samples were incubated with [^{14}C]SAM in the presence of HA N-methyltransferase; the enzyme was a partially purified preparation obtained from guinea pig brain. Tracer amounts

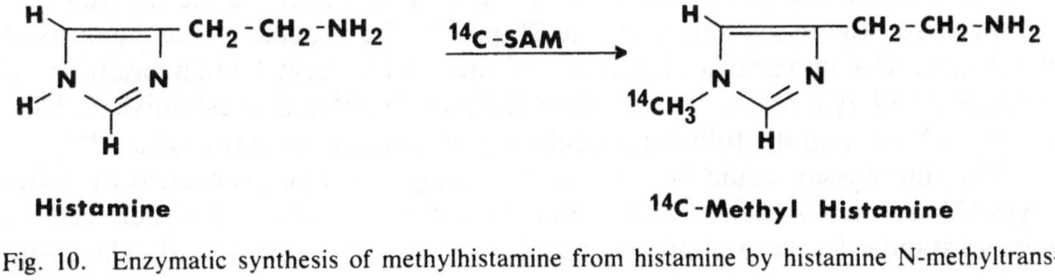

Histamine   **^{14}C-Methyl Histamine**

Fig. 10. Enzymatic synthesis of methylhistamine from histamine by histamine N-methyltransferase.

of [^{3}H]HA were added to correct for the rather low efficiency of the methylation reaction. Following incubation, the reaction was stopped with 1 N NaOH, and the solution was saturated with NaCl. Labeled methylhistamine (MeHA) was extracted into chloroform, reduced to dryness, and counted. Since [^{3}H]HA was diluted with endogenous HA, the ratio ^{14}C/^{3}H in [^{3}H,^{14}C]MeHA bore a linear relationship to the amount of unlabeled HA present and therefore permitted quantitative estimation of HA in tissue in amounts as low as 2 ng.

It soon became apparent that brain tissue extracts contained substances that interfered with the activity of the enzyme. If smaller volumes of more concentrated [^{14}C]SAM were used to methylate HA, the limit of sensitivity could be reduced to 0.2 ng, and HA could be measured in brain regions without preliminary concentration or purification of the homogenates.[131]

The subcellular localization of HA in rat brain regions[132] and the regional and subcellular distribution of HA in mouse brain[113] have been examined by this technique. In plasma, the effects of enzyme inhibitors were overcome by extracting HA from 10–12 ml plasma. The amine was then concentrated in a solution to which the enzymatic–isotopic procedure could be applied directly.[134,135] In some cases, the incubation conditions were altered to permit a more quantitative conversion of HA to [^{14}C]MeHA.[136] The extraction procedure was also modified to allow [^{14}C]MeHA to be recovered more reproducibly into chloroform and to minimize interference from other extractable [^{14}C]-containing material. These improvements increased the limit of detection to 100 pg of HA and enabled the amine to be measured directly in tissue homogenates, in 100 μl or less of plasma and serum, and in 20 μl of human urine. A similar procedure[137] has been used to determine HA levels in various regions of the rat brain.[138] Use of a single-isotope technique (see below) and purification of the labeled product by thin-layer chromatography have been reported to increase the assay sensitivity to 50 pg.[139]

Several single-isotope procedures have been developed to measure HA. [^{3}H]-S-Adenosyl-L-methionine was used as the methyl donor, and duplicate samples containing known amounts of authentic HA were used as internal standards. In one of the first published single-isotope assays, 100 pg of HA could be measured under optimum conditions.[140] The incubation volumes were relatively large (0.5 ml), and the [^{3}H]SAM was significantly diluted before use. The assay has been modified to study release of HA from sensitized human leukocytes.[141]

A double isotope microassay for HA in brain tissue in which the concentration of [^{14}C]SAM was increased and the total reaction volume was reduced to 30 μl has allowed more than 85% of the sample HA to be methylated and has increased the assay sensitivity to 20 pg.[142] The procedure has been used to estimate HA in various structures of the postmortem human brain,[143] to measure its allergic release from human leukocytes,[144] and to establish its half-life in rat brain regions following inhibition of histidine decarboxylase.[145]

The microassay could be adapted to a single-isotope procedure by using [^{3}H]SAM as methyl donor and duplicate samples containing added HA as internal standards. No preliminary purification of HA was required. The assay had a sensitivity of 10 pg and has been used to measure HA in various regions of the rat brain[142] and in individual hypothalamic nuclei and the median eminence of the rat.[146] Endogenous HA release from rat hypothalamic slices[147] and the presence of HA in some neurons isolated from the CNS of *Aplysia californica*[148] have also been demonstrated. With some minor modifications, the assay has been adapted to routine analysis of HA in 200 μl of blood, 1 ml of plasma, and 1–2 mg of biopsy tissues.[149] Recently, the allergic release of HA *in vitro* from whole human blood and rat peritoneal mast cells has been reported; HA was determined in only 20 μl of plasma.[150] The single-isotope procedure provided greater assay sensitivity and eliminated an earlier need to separate and concentrate leukocytes from whole blood.[144]

The single-isotope method has been adapted to measure drug-induced HA release from isolated basophils.[151] The high concentrations of codeine required to demonstrate HA release interfered with the methylation reaction and caused a concentration-dependent flattening of the HA standard curve obtained with single-isotope assays. Hence, $[^{14}C]HA$ was added to the reaction mixture as an internal standard to determine the recovery of HA in the presence of up to 70% enzyme inhibition. The use of this reverse double-isotope procedure permitted HA to be determined without external recovery measurements in the presence of various drug concentrations. Because of the relatively low specific activity of the $[^{14}C]HA$, however, significant quantities had to be added to a sample to be measurable. This had the effect of increasing the ^3H background and thus reducing the sensitivity of the assay compared to the single-isotope method. Thus, although the reverse double-isotope technique was satisfactory for measuring HA under conditions in which methylation inhibition was high or variable, it was not useful in determining very small amounts of HA.

Rat kidney, rather than guinea pig brain, has also been used as a source of histamine-N-methyltransferase.[152] The kidney enzyme, unlike the guinea pig preparation, was free of interfering enzyme activities and gave low blank values. In guinea pig brain preparations, a limiting factor in assay sensitivity was the presence of an interfering enzyme activity which led to the production of small amounts of chloroform-extractable labeled material. If the kidney enzyme were used, however, as little as 5 or 20 pg of HA could be detected, depending on whether the single- or double-isotope procedure was followed. Histamine could be measured in plasma and other body fluids in levels down to 0.2 ng/ml.

Microwave irradiation has recently become accepted as a method of sacrifice when estimating heat-stable substances undergoing rapid postmortem changes. When measured by the radioenzymatic technique, HA levels observed under these conditions have been reported to be much higher than those observed after decapitation.[137] Two possible explanations for this phenomenon have been advanced. In one, the labeled enzyme products were separated by thin-layer chromatography, and radioactivity was measured in sequential sections of the thin-layer plate. Although only one radioactive peak appeared when animals were killed by decapitation, two additional peaks were present if the animals were killed by microwave irradiation. These experiments indicated that chloroform-extractable artifacts were formed during irradiation and that these would lead to anomalously high values for HA unless they were separated from labeled MeHA before its estimation.[153]

It has also been suggested that the apparent elevation in brain HA following microwave irradiation may result, at least in part, from disruption of cellular integrity and the consequently increased extractability of tissue HA; a substantial increase in measured HA has been reported in tissues homogenized in larger than normal volumes of buffer or water, whether or not the animals were killed by microwave irradiation.[154]

Although the reason for the anomalously high HA levels in the brains of animals killed by microwave irradiation has not yet been established with certainty, it would appear that the artifacts may contribute to the measured amounts of "HA," and that labeled substances extracted from the enzyme reaction mixture should be subjected to chromatographic separation to isolate MeHA from other labeled material.

6. OCTOPAMINE AND PHENYLETHANOLAMINE

Octopamine (OA) was first identified in the posterior salivary gland of *Octopus vulgaris*[155,156] and has since been found in some mammalian tissues, in human urine, and in invertebrates. Although its role in the mammalian nervous system has not yet been clearly established, it may well function as a neurotransmitter in some invertebrates.[157]

Most early determinations of OA were made with relatively insensitive and tedious chromatographic separations. In 1969, however, a radioenzymatic method was introduced that permitted as little as 0.5 ng of OA to be measured in 0.5 ml of tissue homogenate.[158,159] In this assay, the β-hydroxylated amine was N-methylated by [^{14}C]SAM in the presence of PNMT (Fig. 11), and the labeled product, N-methyl-OA (synephrine), was extracted into toluene:isoamyl alcohol (3:2, v/v), taken to dryness to reduce the blank, and counted. Duplicate samples containing a known amount of OA as internal standard were also analyzed to allow correction for recovery. Phenylethanolamine (PEOHA) could be determined simultaneously in the same sample if the labeled enzyme reaction product, N-methylphenylethanolamine (N-MePEOHA), were extracted into a less polar solvent such as 3% isoamyl alcohol in toluene before [^{3}H]synephrine was extracted from the reaction mixture, and if the sample were not dried before counting.

This technique has been used to determine OA in sympathetically innervated rat tissues after monoamine oxidase (MAO) inhibition,[159] in various parts of the CNS in the rat,[160] in rat pineal and salivary gland,[161] in nervous tissue of snail,[162] lobster,[163,164] octopus[165] and other cephalopods,[166] and in the cockroach CNS.[167] Similar procedures have been used to measure OA in serum, plasma, and urine from patients with hepatic encephalopathy.[168–170] Phenylethanolamine has been measured in tissues of rats treated with phenylethylamine and an MAO inhibitor but could not be detected after an MAO inhibitor alone.[159]

The assay sensitivity for OA could be increased significantly by substituting high-specific-activity [³H]SAM for [¹⁴C]SAM as the methyl donor and by selectively extracting and evaporating the labeled product to separate it from the methyl donor and other radioactive contaminants.[171] Duplicate samples containing a known amount of OA were processed as internal standards. Phenylethanolamine could be measured in the same sample by analyzing a third portion of the tissue supernatant containing PEOHA as internal standard. The reaction mixture was extracted with heptane:isoamyl alcohol to remove [³H]MePEOHA and then with toluene:isoamyl alcohol to recover [³H]synephrine. The extracts were reduced to dryness at 40°C under reduced pressure, or at 80°C, respectively, and radioactivity in each substance was assessed by liquid scintillation counting. The assay was linear from 50 pg to 20 ng when authentic OA or PEOHA added to rat brain supernatants or Tris buffer was measured. Octopamine could be determined in rat brain regions weighing as little as 50 mg, in rat pineal glands, and in fetal rat brain.[171] It has also been reported in the earthworm CNS[172] and in the locust brain.[173] Phenylethanolamine has been measured by a similar procedure in amounts as low as 100 pg in the whole brain, in several brain regions, in peripheral tissues of the adult rat,[174] and in fetal rat brain.[175]

Although these assays have been claimed to be highly specific, the tissue values reported for PEOHA in rat brain[174] were higher than those obtained by more specific radioenzymatic procedures.[176] Phenylethanolamine was also detected radioenzymatically in the nervous system of the marine mollusc *Aplysia californica* and was suggested to play a role in synaptic transmission in that animal.[177] Subsequent studies, however, have been unable to confirm the presence of PEOHA in the *Aplysia* CNS.[178,179] Furthermore, OA has been shown to exist in ganglia and single neurons of *Aplysia*[178,179] in amounts that differ substantially from those originally reported.[180] It has become apparent that the originally published procedures for these amines, in which the [³H]-methylated products were isolated by simple organic extractions, lacked the specificity required to establish the presence or absence of very small amounts of these compounds such as those encountered in single cells or small numbers of cells.

Recent modifications of the OA assay have significantly increased its sensitivity and specificity. By using a longer incubation period and by extracting the reaction product first into toluene:isoamyl alcohol and then into dilute

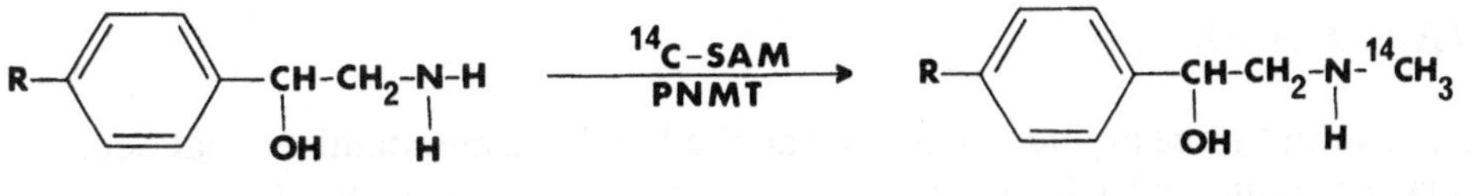

Octopamine (R=OH)

Phenylethanolamine (R=H)

N-Methyl Octopamine (R=OH)

N-Methyl Phenylethanolamine (R=H)

Fig. 11. Enzymatic synthesis of N-methyloctopamine from octopamine and of N-methylphenylethanolamine from phenylethanolamine. (Reproduced by permission of Academic Press. Inc.[201])

HCl, 20–60 pg of OA can be measured; its distribution in the rat CNS has been established by this procedure.[181] Octopamine has been measured in a similar way in the nervous systems of the locust and cockroach, with limits of detection of 10–15 pg[182]. In some cases, assay volumes have been reduced to allow very small amounts of OA and PEOHA to be determined.[178] If [^{3}H]synephrine were extracted into toluene : isoamyl alcohol, 0.2 pmol (30 pg) of OA could be measured even without subsequent extraction into acid; when ethyl acetate was substituted as the organic extraction solvent, the sensitivity was increased to about 0.05 pmol (8 pg).

Limitations in assay specificity imposed by simply extracting the enzyme reaction products from the incubation mixture have been largely overcome by separating the extracted products chromatographically, either before[183] or after[176] chemical derivatization. In the former case, both OA and PEOHA could be measured simultaneously in nervous tissue by extracting the labeled enzyme reaction products into ethyl acetate and then separating them by thin-layer chromatography. The developed chromatogram was sprayed with fluorescamine in acetone–triethanolamine, and the appropriate zones were located under UV light and removed for radioactive counting. Approximately 90 pg of OA and 100 pg of PEOHA could be determined.

Greater specificity and sensitivity were made possible by derivatizing the enzyme products before chromatographic separation. Following their extraction from the reaction mixture into ethyl acetate, the methylated products of PEOHA, *m*-OA, and *p*-OA were reacted with dansyl chloride reagent, extracted into benzene, and repeatedly separated by thin-layer chromatography.[176] As little as 10 pg of each amine could be detected. *p*-Octopamine contributed only traces of radioactivity to the *m*-OA and PEOHA zones. Although *m*-OA and PEOHA did contribute slightly to the radioactivity measured in the *p*-OA zone, the contamination did not affect *p*-OA determinations, since neither *m*-OA nor PEOHA was detectable in the normal rat brain. The assay provided a very specific and sensitive means of measuring PEOHA and OA and was for the first time able to distinguish between the positional isomers of OA. It has been used to measure PEOHA, *m*-OA, and *p*-OA levels in various rat brain regions in the presence and absence of an MAO inhibitor.[176] These substances have also been determined in the rat hypothalamus,[184] brainstem,[184] and denervated rat salivary glands.[185]

7. TRACE AMINES

The term "trace amines" has been applied to those monoamines that occur in nervous tissue and whose endogenous concentration is generally less than 100 ng/g of tissue.[186] Some of them, in particular β-phenylethylamine (PE), *m*-tyramine (*m*-TA), *p*-tyramine (*p*-TA), and tryptamine (T), have been proposed to exert their effects by acting as synaptic activators, either directly or indirectly, in the process of neurotransmission. Although the trace amines are more commonly measured by fluorometric, gas chromatographic, mass spectrometric, or immunoassay techniques, radioenzymatic assays have recently been developed for PE, TA, and T and for PEOHA and OA, the β-hydroxylated metabolites of PE and TA, respectively. The latter two substances were discussed earlier.

7.1. β-Phenylethylamine

β-Phenylethylamine has been determined in rat brain by a two-step enzymatic procedure in which the amine was oxidized to PEOHA by dopamine-β-hydroxylase (DBH) and then methylated to N-MePEOHA by [³H]SAM in the presence of PNMT[187] (Fig. 12). The radioactive product was extracted into toluene:isoamyl alcohol (97:3, v/v), taken to dryness to reduce the blank, and counted. Tissue blanks were used to correct for the N-methylation of endogenous tissue PEOHA, whereas internal standards assayed along with the tissue samples could be used to correct for losses encountered during the extraction steps. The double enzymatic procedure displayed marked specificity for PE and allowed as little as 100–200 pg of the amine to be determined. The assay has been used to measure PE in the rat brain, both endogenously and in the presence of a variety of drugs.[187] The amine levels reported were in excellent agreement with those obtained by mass spectrometric procedures.[188,189]

7.2. Tyramine

Tyramine has been determined in mammalian tissues by a procedure similar to that used to measure PE.[190] Tyramine was extracted from deproteinated

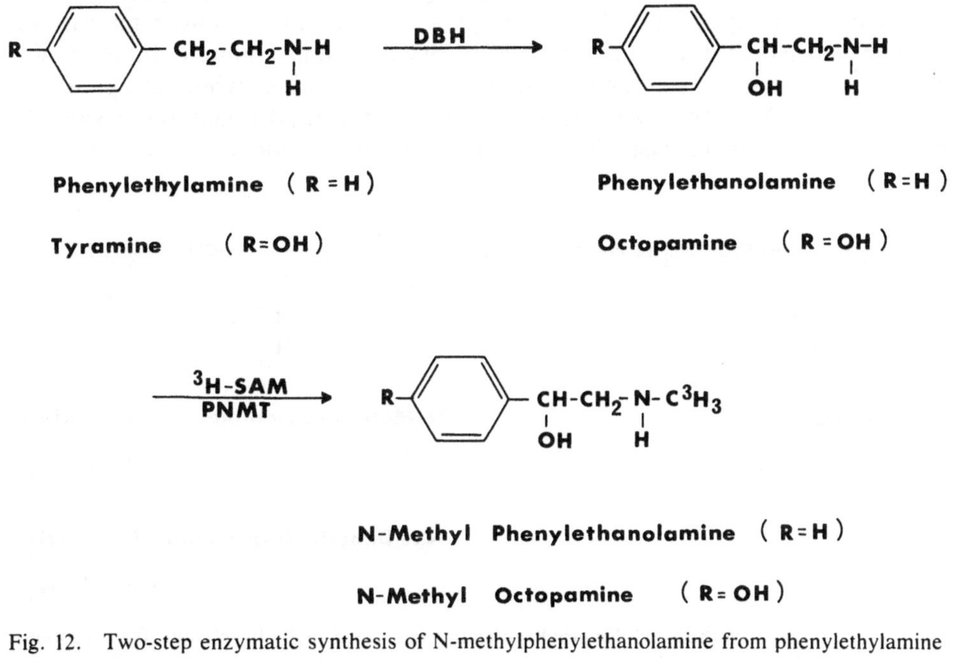

Fig. 12. Two-step enzymatic synthesis of N-methylphenylethanolamine from phenylethylamine and of N-methyloctopamine from tyramine.

tissue homogenates into methyl acetate and then back extracted into dilute HCl. It was oxidized to OA by incubation with catalase and DBH, then methylated to [^{3}H]synephrine by reaction with [^{3}H]SAM in the presence of PNMT (Fig. 12). [^{3}H]Synephrine was extracted into toluene: isoamyl alcohol (3:2, v/v), taken to dryness, and counted. The use of tissue blanks permitted corrections to be made for the small amounts of OA and PEOHA in the homogenates. Corrections for extraction losses were made by using internal standards. As little as 1 ng of TA could be measured in tissues, but the assay was unable to distinguish between the various TA isomers. Because of the double enzymatic reaction and solvent extractions, the assay displayed marked specificity for TA. Of the related amines tested for cross reactivity, only OA and PEOHA yielded significant amounts of extractable products. These substances could, however, produce high blank values and therefore large potential errors in TA measurements if present in large excess relative to TA. The amine has been measured in whole brain, in brain regions, and in several peripheral tissues of the rat.[190,191] Except for somewhat greater amounts found in the brain, the TA levels measured in these tissues were in agreement with those obtained mass spectrometrically.[192]

7.3. Tryptamine

Endogenous T has been measured radioenzymatically in mammalian tissues in amounts as low as 5 ng.[193] The amine was first extracted from deproteinated tissue homogenates into toluene and isoamyl alcohol. Following evaporation of the solvent, the extract was incubated with [^{14}C]SAM and partially purified rabbit lung N-methyltransferase (Fig. 13). The labeled enzyme reaction products, N-methyl-T and N,N-dimethyl-T, were extracted into a mixture of toluene and isoamyl alcohol, reduced to dryness, and counted. Interference by equivalent amounts of most normally occurring amines which are potential substrates for N-methyltransferase was less than 1%. Although the assay has been used to measure tissue T levels in the rat, both endogenously and in the

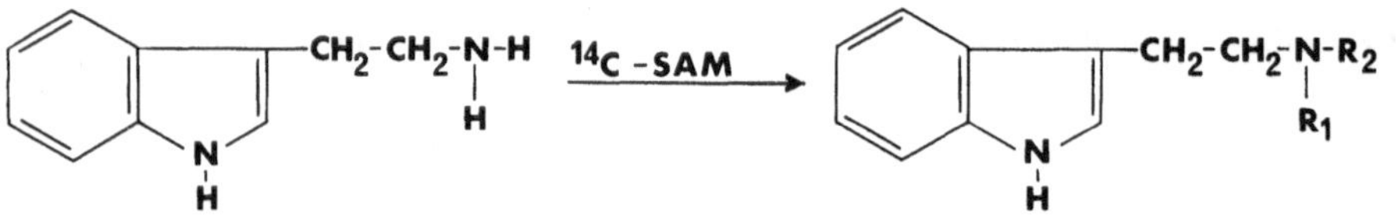

Tryptamine

N-Methyltryptamine (R_1 = $^{14}CH_3$)
(R_2 = H)

N,N-Dimethyltryptamine (R_1 = $^{14}CH_3$)
(R_2 = $^{14}CH_3$)

Fig. 13. Enzymatic synthesis of N-methyltryptamine and N,N-dimethyltryptamine from tryptamine by N-methyltransferase.

presence of a variety of drugs,[193,194] the values obtained were considerably greater than those reported by mass spectrometric[195] or gas chromato-graphic–mass spectrometric[196] procedures. It would appear, therefore, that the enzyme reaction products were not completely separated from all contaminants by a simple organic solvent extraction and that the radioenzymatic assay was less specific than mass spectrometric procedures.

8. TAURINE

Taurine has been reported to function as a neurotransmitter in the retina.[197] It arises directly as the decarboxylation product of cysteic acid and indirectly, via hypotaurine, as the decarboxylation product of cysteinesulfinic acid. Although most techniques for determining taurine are time consuming, lack sensitivity, or require expensive instrumentation, a radioenzymatic assay introduced some years ago provides a relatively rapid, sensitive, and highly specific method for measuring this substance in tissues.[198] It is based on the utilization of taurine in the synthesis of the cholic acid conjugate, taurocholic acid, and the dilution of the specific activity of labeled taurine by endogenous taurine present in PCA extracts of the tissue. Tissue extracts were first passed through ion-exchange resins to remove isethionic acid and amino acids, then incubated with labeled taurine and cholic acid in the presence of rat liver microsomal enzymes. Excess [^{3}H]cholic acid was removed from the aqueous phase by extraction into dichloromethane, and the enzyme reaction product, labeled taurocholic acid, was extracted into *n*-butanol and counted. Unlabeled taurine in amounts ranging from 2.5–200 nmol (300 ng–25 μg) was incubated simultaneously to provide a series of standards. Since the ratio of ^{3}H to ^{35}S in taurocholic acid was linearly related to the quantity of taurine in the sample, the amount of endogenous taurine in the tissue could be determined from the standard curve. The values for tissue taurine obtained by this method were in excellent agreement with those determined in the same tissue using an analytical amino acid analyzer.

9. GLUTAMINE

Glutamine (Gln) is a substrate for enzymatic reactions involved in the synthesis of purines, pyrimidines, pyridine nucleotide coenzymes, glucosamine, and some of the amino acids. It is usually measured by a procedure that involves ion-exchange chromatography and use of an amino acid analyzer or, alternatively, by collecting and determining the ammonia produced by the action of glutaminase. Recently, however, two radioenzymatic assays for Gln have been reported. In one of these, [^{14}C]bicarbonate was converted to [^{14}C] citrulline in the presence of carbamylphosphate synthetase, ornithine carbamyltransferase, and saturating concentrations of all substrates except Gln.[199] After the Gln-containing sample had been incubated, the enzyme reaction mixture was acidified and shaken to liberate excess $^{14}CO_2$. An aliquot of the mixture was then removed for radioactive counting. The formation of [^{14}C] citrulline was proportional to the amount of Gln in the sample over the range 4 to 350 pmol (600 pg–50 ng). Since ammonia could also serve as a nitrogen donor in the synthetic reaction, its complete removal from the sample was essential to the accurate estimation of Gln. The assay has been used to measure Gln in cultures of *Salmonella typhimurium*.[199]

In another radioenzymatic procedure for Gln, the amount of NH_4^+ present in the sample was measured before and after Gln hydrolysis.[200] The procedure was based on the reaction of α-[1-^{14}C]ketoglutarate with NH_4^+ to form [1-^{14}C] glutamate. Excess α-[1-^{14}C]ketoglutarate was subsequently decarboxylated by treatment with hydrogen peroxide. The NH_4^+ was the limiting factor in the reaction. After the reaction had been completed, the solution was heated to remove $^{14}CO_2$. The amount of radioactivity remaining in the mixture (as [1-^{14}C]glutamate) gave a direct measure of the amount of NH_4^+ originally present in the sample. [1-^{14}C]Glutamate formation was proportional to the amount of NH_4^+ present over the range 100 pmol to 10 nmol (15 ng–1.5 μg). Since the reaction catalyzed by L-glutamate dehydrogenase was reversible, the presence of $NADP^+$ and excess endogenous glutamate could cause erroneously high values for NH_4^+. For this reason, glucose-6-phosphate dehydrogenase was coupled to the assay to remove $NADP^+$. If Gln hydrolysis were coupled to the determination of NH_4^+, both NH_4^+ and Gln could be determined in plasma, urine, and bacterial cell extracts.

10. ADVANTAGES AND LIMITATIONS

Radioenzymatic methods of analysis provide the investigator with a relatively inexpensive means of measuring very small amounts of putative neurotransmitters and, in some cases, other biological compounds having similar structures in small volumes of body fluids, in milligram quantities of tissue, and in some large single cells and ganglia of primitive animals. In many cases, the assays are capable of measuring picogram quantities of these substances; hence, they compare favorably with mass spectrometric or gas chromatographic–mass spectrometric techniques which require much more expensive equipment.

Most assays are easy to perform and are relatively rapid; even in those procedures that require extensive product purification, 25–50 samples can be processed in one day. The required enzymes may all be prepared in the laboratory. The presence of other enzyme activities in these crude preparations may, however, interfere with the analysis of some substances unless inhibitors are included in the enzyme reaction mixture to block the interfering reaction.

Radioenzymatic assays are capable of providing very sensitive and specific measurements of a variety of substances provided that the enzyme reaction products are properly purified. Although it was originally claimed that a simple extraction of the reaction products into an organic solvent was sufficient to provide an assay with a high degree of specificity, it has since been shown that the procedure does not separate the compound of interest from all other interfering labeled compounds. Although volatile contaminants have occasionally been removed by evaporating the extract before liquid scintillation counting, nonvolatile contaminants remain with the labeled product. Therefore, subsequent ion-exchange or chromatographic separation or chemical derivatization followed by chromatographic separation are required to isolate the reaction product from other interfering substances. The purity of the product is particularly important for those substances that are normally present at very low levels in tissues or body fluids, since even minor contamination by other compounds normally present at much higher levels will result in substantial errors in measurement.

Finally, the amount of a substance determined radioenzymatically may be corrected for losses encountered throughout the procedure by assaying a duplicate sample to which a known amount of the substance has been added to act as an internal standard. Factors that influence the enzyme reaction in the sample will influence the reaction of the internal standard as well and thus allow for corrections to be made.

11. CONCLUSIONS

Radioenzymatic techniques for measuring neurotransmitters have been developed and applied to their determination in very small biological samples. Although the enzyme reaction itself often provides only limited specificity to the assay, a high degree of specificity can be obtained by subsequently extracting the reaction mixture with an organic solvent and isolating the enzyme reaction product by chromatography. The sensitivity of measurement permitted by these techniques, now at the picogram level in many cases, is providing a means to measure *in vivo* release of some neurotransmitters in the brain. Hence, by perfusing the tissue with fluid containing drugs or other transmitters, the effects of a variety of agents on release can be determined at specific sites within the brain. It should be possible to examine quantitatively neurotransmitter dynamics at the cellular level and perhaps eventually to examine in a more detailed and quantitative manner the neuronal pathways in the brain.

ACKNOWLEDGMENTS. I thank Mr. H. Miyashita for preparing the figures and Saskatchewan Health and the Medical Research Council of Canada for continuing financial support.

REFERENCES

1. Saelens, J. K., Schoen, M. S., and Kovacsics, G. B., 1967, *Biochem. Pharmacol.* **16**:1043–1049.
2. Axelrod, J., 1962, *J. Biol. Chem.* **237**:1657–1660.
3. Kovacsics, G. B., and Saelens, J. K., 1968, *Arch. Int. Pharmacodyn. Ther.* **174**:481–490.
4. Iversen, L. L., and Jarrott, B., 1970, *Biochem. Pharmacol.* **19**:1841–1843.
5. Cuello, A. C., Horn, A. S., Mackay, A. V. P., and Iversen, L. L., 1973, *Nature* **243**:465–467.
6. Henry, D. P., Starman, B. J., Johnson, D. G., and Williams, R. H., 1975, *Life Sci.* **16**:375–384.

7. Petty, M. A., and Reid, J. L., 1977, *Brain Res.* **136**:376–380.
8. Lake, C. R., Ziegler, M. G., and Kopin, I. J., 1976, *Life Sci.* **18**:1315–1325.
9. Falke, H. E., Punt, R., and Birkenhager, W. H., 1978, *Clin. Chim. Acta* **89**:111–117.
10. Ziegler, M. G., Lake, C. R., and Ebert, M. H., 1979, *Eur. J. Pharmacol.* **57**:127–133.
11. Engelman, K., Portnoy, B., and Lovenberg, W., 1968, *Am. J. Med. Sci.* **255**:259–268.
12. Engelman, K., and Portnoy, B., 1970, *Circ. Res.* **26**:53–57.
13. Nikodijevic, B., Daly, J., and Creveling, C. R., 1969, *Biochem. Pharmacol.* **18**:1577–1584.
14. Passon, P. G., and Peuler, J. D., 1973, *Anal. Biochem.* **51**:618–631.
15. Cryer, P. E., Santiago, J. V., and Shah, S., 1974, *J. Clin. Endocrinol. Metab.* **39**:1025–1029.
16. Hortnagl, H., Benedict, C. R., Grahame-Smith, D. G., and McGrath, B., 1977, *Br. J. Clin. Pharmacol.* **4**:553–558.
17. Coyle, J. T., and Henry, D., 1973, *J. Neurochem.* **21**:61–67.
18. Iversen, L. L., 1967, *The Uptake and Storage of Noradrenaline in Sympathetic Nerves*, Cambridge University Press, London, New York, p. 33.
19. Palkovits, M., Brownstein, M., Saavedra, J. M., and Axelrod, J., 1974, *Brain Res.* **77**:137–149.
20. Brownstein, M., Saavedra, J. M., and Palkovits, M., 1974, *Brain Res.* **79**:431–436.
21. Coyle, J. T., and Kuhar, M. J., 1974, *Brain Res.* **65**:475–487.
22. Cruce, J. A. F., Thoa, N. B., and Jacobowitz, D. M., 1978, *Pharmacol. Biochem. Behav.* **8**:287–289.
23. Kizer, J. S., Palkovits, M., and Brownstein, M. J., 1976, *Brain Res.* **108**:363–370.
24. Arnold, E. B., Molinoff, P. B., and Rutledge, C. O., 1977, *J. Pharmacol. Exp. Ther.* **202**:544–557.
25. Kant, G. J., and Meyerhoff, J. L., 1978, *Life Sci.* **23**:2111–2118.
26. Cotman, C. W., Haycock, J. W., and White, W. F., 1976, *J. Physiol. (Lond.)* **254**:475–505.
27. Patrick, R. L., and Barchas, J. D., 1976, *J. Pharmacol. Exp. Ther.* **197**:89–96.
28. Bartholini, G., 1976, *J. Pharm. Pharmacol.* **28**:429–433.
29. Cuello, A. C., Hiley, R., and Iversen, L. L., 1973, *J. Neurochem.* **21**:1337–1340.
30. Robertson, H. A., 1976, *Experientia* **32**:552–554.
31. Bird, E. D., Spokes, E. G. S., and Iversen, L. L., 1979, *Brain* **102**:347–360.
32. Van der Gugten, J., Palkovits, M., Wijnen, H. L. J. M., and Versteeg, D. H. G., 1976, *Brain Res.* **107**:171–175.
33. Versteeg, D. H. G., Van der Gugten, J., deJong, W., and Palkovits, M., 1976, *Brain Res.* **113**:563–574.
34. Weise, V. K., and Kopin, I. J., 1976, *Life Sci.* **19**:1673–1686.
35. De Champlain, J., Farley, L., Cousineau, D., and van Ameringen, M.-R., 1976, *Circ. Res.* **38**:109–114.
36. Peuler, J. D., and Johnson, G. A., 1977, *Life Sci.* **21**:625–636.
37. Ben-Jonathan, N., and Porter, J. C., 1976, *Endocrinology* **98**:1497–1507.
38. Reader, T. A., De Champlain, J., and Jasper, H., 1976, *Brain Res.* **111**:95–108.
39. Chiueh, C. C., and Kopin, I. J., 1978, *J. Neurochem.* **31**:561–564.
40. Gauchy, C., Tassin, J. P., Glowinski, J., and Cheramy, A., 1976, *J. Neurochem.* **26**:471–480.
41. Gudelsky, G. A., and Porter, J. C., 1979, *Life Sci.* **25**:1697–1702.
42. Pilotte, N. S., Gudelsky, G. A., and Porter, J. C., 1980, *Brain Res.* **193**:284–288.
43. Nansel, D. D., Gudelsky, G. A., and Porter, J. C., 1979, *Endocrinology* **105**:1073–1077.
44. Nesterick, C. A., and Rahwan, R. G., 1979, *J. Chromatogr.* **164**:205–216.
45. DaPrada, M., and Zurcher, G., 1976, *Life Sci.* **19**:1161–1174.
46. Saavedra, J. M., Kvetnansky, R., and Kopin, I. J., 1979, *Brain Res.* **160**:271–280.
47. Umezu, K., and Moore, K. E., 1979, *J. Pharmacol. Exp. Ther.* **208**:49–56.
48. Paul, S. M., Axelrod, J., Saavedra, J. M., and Skolnick, P., 1979, *Brain Res.* **178**:499–505.
49. Perlow, M. J., Chiueh, C. C., Lake, C. R., and Wyatt, R. J., 1980, *Brain Res.* **186**:469–473.
50. Scatton, B., Pelayo, F., Dubocovich, M. L., Langer, S. Z., and Bartholini, G., 1979, *Brain Res.* **176**:197–201.
51. Negro-Vilar, A., Ojeda, S. R., Advis, J. P., and McCann, S. M., 1979, *Endocrinology* **105**:86–91.
52. Head, R. J., and Berkowitz, B. A., 1979, *Blood Vessels* **16**:320–324.
53. Endert, E., 1979, *Clin. Chim. Acta* **96**:233–239.

54. Harapat, S., and Rubin, P., 1979, *J. Chromatogr.* **163:**77–80.
55. Klaniecki, T. S., Corder, C. N., McDonald, R. H., and Feldman, J. A., 1977, *J. Lab. Clin. Med.* **90:**604–612.
56. Fujiwara, M., Inagaki, C., Miwa, S., Takaori, S., Saeki, Y., and Nozaki, M., 1980, *Life Sci.* **26:**71–78.
57. Saller, C. F., and Zigmond, M. J., 1978, *Life Sci.* **23:**1117–1130.
58. Weinshilboum, R. M., and Raymond, F. A., 1976, *Biochem. Pharmacol.* **25:**573–579.
59. Fry, J. P., House, C. R., and Sharman, D. F., 1974, *Br. J. Pharmacol.* **51:**116–117P.
60. Martin, I. L., Baker, G. B., and Fleetwood-Walker, S. M., 1978, *Biochem. Pharmacol.* **27:**1519–1520.
61. McCaman, M. W., Ono, J. K., and McCaman, R. E., 1979, *J. Neurochem.* **32:**1111–1113.
62. McCaman, M. W., McCaman, R. E., and Stetzler, J., 1979, *Anal. Biochem.* **96:**175–180.
63. Philips, S. R., and Robson, A. M., (unpublished data).
64. Philips, S. R., Robson, A. M., and Boulton, A. A., 1982, *J. Neurochem.* **38:**1106–1110.
65. Hefti, F., and Lichtensteiger, W., 1976, *J. Neurochem.* **27:**647–649.
66. Johnson, G. A., Gren, J. M., and Kupiecki, R., 1978, *Clin. Chem.* **24:**1927–1930.
67. Johnson, G. A., Baker, C. A., and Smith, R. T., 1980, *Life Sci.* **26:**1591–1598.
68. Zurcher, G., and Da Prada, M., 1979, *J. Neurochem.* **33:**631–639.
69. Vlachakis, N. D., and DeQuattro, V., 1978, *Biochem. Med.* **20:**107–114.
70. Vlachakis, N. D., and Niarchos, A., 1979, *Clin. Chim. Acta* **99:**283–288.
71. Kobayashi, K., DeQuattro, V., Kolloch, R., and Miano, L., 1980, *Life Sci.* **26:**567–573.
72. Vlachakis, N. D., Alexander, N., and Maronde, R. F., 1980, *Life Sci.* **26:**97–102.
73. Sato, T., and DeQuattro, V., 1969, *J. Lab. Clin. Med.* **74:**672–681.
74. Vlachakis, N. D., Alexander, N., Velasquez, M. T., and Maronde, R. F., 1979, *Biochem. Med.* **22:**323–331.
75. Comoy, E., and Bohuon, C., 1972, *Clin. Chim. Acta* **36:**207–212.
76. Argiolas, A., Fadda, F., Stefanini, E., and Gessa, G. L., 1977, *J. Neurochem.* **29:**599–601.
77. Kebabian, J. W., Saavedra, J. M., and Axelrod, J., 1977, *J. Neurochem.* **28:**795–801.
78. Annunziato, L., and Weiner, R. I., 1980, *Neuroendocrinology* **31:**8–12.
79. Argiolas, A., and Fadda, F., 1978, *Experientia* **34:**739–740.
80. Fekete, M. I. K., Kanyicska, B., and Herman, J. P., 1978, *Life Sci.* **23:**1549–1556.
81. Fekete, M. I. K., Szentendrei, T., Herman, J. P., and Kanyicska, B., 1980, *Eur. J. Pharmacol.* **64:**231–238.
82. Saavedra, J. M., Brownstein, M., and Axelrod, J., 1973, *J. Pharmacol. Exp. Ther.* **186:**508–515.
83. Neuhoff, V., and Weise, M., 1970, *Arzneim. Forsch.* **20:**368–372.
84. Weissbach, H., Redfield, B. G., and Axelrod, J., 1961, *Biochim. Biophys. Acta* **54:**190–192.
85. Axelrod, J., and Weissbach, H., 1961, *J. Biol. Chem.* **236:**211–213.
86. Saavedra, J. M., Palkovits, M., Brownstein, M. J., and Axelrod, J., 1974, *Brain Res.* **77:**157–165.
87. Van de Kar, L., Levine, J., and Van Orden L. S. III, 1978, *Neuroendocrinology* **27:**186–192.
88. Palkovits, M., Brownstein, M., and Saavedra, J. M., 1974, *Brain Res.* **80:**237–249.
89. Saavedra, J. M., Brownstein, M., and Palkovits, M., 1974, *Brain Res.* **79:**437–441.
90. Massari, V. J., Tizabi, Y., Gottesfeld, Z., and Jacobowitz, D. M., 1978, *Neuroscience* **3:**339–344.
91. Palkovits, M., Saavedra, J. M., Jacobowitz, D. M., Kizer, J. S., Zaborszky, L., and Brownstein, M. J., 1977, *Brain Res.* **130:**121–134.
92. Reader, T. A., 1980, *Brain Res. Bull.* **5:**609–613.
93. Brownstein, M. J., Saavedra, J. M., Axelrod, J., Zeman, G. H., and Carpenter, D. O., 1974, *Proc. Natl. Acad. Sci. U.S.A.* **71:**4662–4665.
94. Boireau, A., Ternaux, J. P., Bourgoin, S., Hery, F., Glowinski, J., and Hamon, M., 1976, *J. Neurochem.* **26:**201–204.
95. Oliveras, J. L., Bourgoin, S., Hery, F., Besson, J. M., and Hamon, M., 1977, *Brain Res.* **138:**393–406.
96. Gaudin-Chazal, G., Daszuta, A., Faudon, M., and Ternaux, J. P., 1979, *Brain Res.* **160:**281–293.

97. Ternaux, J. P., Boireau, A., Bourgoin, S., Hamon, M., Hery, F., and Glowinski, J., 1976, *Brain Res.* **101**:533–548. ·

98. Ternaux, J. P., Hery, F., Hamon, M., Bourgoin, S., and Glowinski, J., 1977, *Brain Res.* **132**:575–579.

99. Elks, M. L., Youngblood, W. W., and Kizer, J. S., 1979, *Brain Res.* **172**:471–486.

100. Reinhard, J. F., Ozaki, Y., and Moskowitz, M. A., 1980, *Biochem. Pharmacol.* **29**:3020–3023.

101. Hammel, I., Naot, Y., Ben-David, E., and Ginsburg, H., 1978, *Anal. Biochem.* **90**:840–843.

102. Feigenson, M. E., and Saelens, J. K., 1969, *Biochem. Pharmacol.* **18**:1479–1486.

103. Smith, J. C., and Saelens, J. K., 1967, *Fed. Proc.* **26**:296.

104. Saelens, J. K., Allen, M. P., and Simke, J. P., 1970, *Arch. Int. Pharmacodyn. Ther.* **186**:279–286.

105. Consolo, S., Garattini, S., Ladinsky, H., and Thoenen, H., 1972, *J. Physiol. (Lond.)* **220**:639–646.

106. Toru, M., and Aprison, M. H., 1966, *J. Neurochem.* **13**:1533–1544.

107. Ladinsky, H., Consolo, S., and Bareggi, S. R., 1972, *Anal. Biochem.* **50**:460–466.

108. Reid, W. D., Haubrich, D. R., and Krishna, G., 1971, *Anal. Biochem.* **42**:390–397.

109. Haubrich, D. R., Reid, W. D., and Gillette, J. R., 1972, *Nature (New Biol.)* **238**:88–89.

110. Dunant, Y., and Hirt, L., 1976, *J. Neurochem.* **26**:657–659.

111. Dunant, Y., Eder, L., and Servetiadis-Hirt, L., 1980, *J. Physiol. (Lond.)* **298**:185–203.

112. Goldberg, A. M., and McCaman, R. E., 1973, *J. Neurochem.* **20**:1–8.

113. McCaman, R. E., Weinreich, D., and Borys, H., 1973, *J. Neurochem.* **21**:473–476.

114. Richter, J. A., and Shea, P. A., 1974, *J. Neurochem.* **23**:1225–1230.

115. Waldenlind, L., 1975, *Biochem. Pharmacol.* **24**:1339–1341.

116. Wonnacott, S., and Marchbanks, R. M., 1975, *Biochem. Soc. Trans.* **3**:102–106.

117. Hildebrand, J. G., Townsel, J. G., and Kravitz, E. A., 1974, *J. Neurochem.* **23**:951–963.

118. Richter, J. A., 1976, *J. Neurochem.* **26**:791–797.

119. Dudar, J. D., Whishaw, I. Q., and Szerb, J. C., 1979, *Neuropharmacology,* **18**:673–678.

120. Shea, P. A., and Aprison, M. H., 1973, *Fed. Proc.* **32**:430.

121. Shea, P. A., and Aprison, M. H., 1973, *Anal. Biochem.* **56**:165–177.

122. Shea, P. A., and Aprison, M. H., 1977, *J. Neurochem.* **28**:51–58.

123. Smith, J. E., Lane, J. D., Shea, P. A., McBride, W. J., and Aprison, M. H., 1975, *Anal. Biochem.* **64**:149–169.

124. Massarelli, R., Durkin, T., Niedergang, C., and Mandel, P., 1976, *Pharmacol. Res. Commun.* **8**:407–416.

125. Eckernas, S.-A., and Aquilonius, S.-M., 1977, *Acta Physiol. Scand.* **100**:446–451.

126. McCaman, R. E., and Stetzler, J., 1977, *J. Neurochem.* **28**:669–671.

127. Jacobowitz, D. M., and Goldberg, A. M., 1977, *Brain Res.* **122**:575–577.

128. Vocci, F. J., Karbowski, M. J., and Dewey, W. L., 1979, *J. Neurochem.* **32**:1417–1422.

129. Green, J. P., 1970, *Handbook of Neurochemistry*, Volume 4 (A. Lajtha, ed.), Plenum Press, New York, pp. 221–250.

130. Snyder, S. H., Baldessarini, R. J., and Axelrod, J., 1966, *J. Pharmacol. Exp. Ther.* **153**:544–549.

131. Taylor, K. M., and Snyder, S. H., 1971, *J. Pharmacol. Exp. Ther.* **179**:619–633.

132. Kuhar, M. J., Taylor, K. M., and Snyder, S. H., 1971, *J. Neurochem.* **18**:1515–1527.

133. Taylor, K. M., and Snyder, S. H., 1972, *J. Neurochem.* **19**:341–354.

134. Miller, R. L., McCord, C., Sanda, M., Bourne, H. R., and Melmon, K. L., 1970, *J. Pharmacol. Exp. Ther.* **175**:228–234.

135. Charles, T. J., Williams, S. J., Seaton, A., Bruce, C., and Taylor, W. H., 1979, *Clin. Sci.* **57**:39–45.

136. Beaven, M. A., Jacobsen, S., and Horakova, Z., 1972, *Clin. Chim. Acta* **37**:91–103.

137. Hough, L. B., and Domino, E. F., 1977, *J. Neurochem.* **29**:199–204.

138. Hough, L. B., and Domino, E. F., 1979, *J. Neurochem.* **32**:1865–1866.

139. Dent, C., Nilam, F., and Smith, I. R., 1979, *Agents Actions* **9**:34–35.

140. Kobayashi, Y., and Maudsley, D. V., 1972, *Anal. Biochem.* **46**:85–90.

141. Blumenthal, M. N., Roitman, B., Carlson, G., and Fish, L., 1974, *Biochem. Med.* **11**:312–317.

142. Taylor, K. M., and Snyder, S. H., 1972, *J. Neurochem.* **19**:1343–1358.

143. Lipinski, J. F., Schaumburg, H. H., and Baldessarini, R. J., 1973, *Brain Res.* **52**:403–408.
144. Levy, D. A., and Widra, M., 1973, *J. Lab. Clin. Med.* **81**:291–297.
145. Dismukes, K., and Snyder, S. H., 1974, *Brain Res.* **78**:467–481.
146. Brownstein, M. J., Saavedra, J. M., Palkovits, M., and Axelrod, J., 1974, *Brain Res.* **77**:151–156.
147. Taylor, K. M., and Snyder, S. H., 1973, *J. Neurochem.* **21**:1215–1223.
148. Weinreich, D., Weiner, C., and McCaman, R., 1975, *Brain Res.* **84**:341–345.
149. Subramanian, N., Mitznegg, P., and Domschke, W., 1978, *Acta Hepatogastroenterol. (Stuttg.)* **25**:456–458.
150. Taylor, K. M., Krilis, S., and Baldo, B. A., 1980, *Int. Arch. Allergy Appl. Immunol.* **61**:19–27.
151. Salberg, D. J., Hough, L. B., Kaplan, D. E., and Domino, E. F., 1977, *Life Sci.* **21**:1439–1446.
152. Shaff, R. E., and Beaven, M. A., 1979, *Anal. Biochem.* **94**:425–430.
153. Subramanian, N., Schinzel, W., Mitznegg, P., and Estler, C.-J., 1978, *Agents Actions* **8**:488–490.
154. Orr, E. L., and Eichelman, B., 1979, *J. Neurochem.* **33**:303–308.
155. Erspamer, V., 1948, *Acta Pharmacol. Toxicol. (Kbh.)* **4**:224–227.
156. Erspamer, V., 1952, *Nature* **169**:375–376.
157. Robertson, H. A., and Juorio, A. V., 1976, *Int. Rev. Neurobiol.* **19**:173–224.
158. Molinoff, P., and Axelrod, J., 1969, *Science* **164**:428–429.
159. Molinoff, P. B., Landsberg, L., and Axelrod, J., 1969, *J. Pharmacol. Exp. Ther.* **170**:253–261.
160. Molinoff, P. B., and Axelrod, J., 1972, *J. Neurochem.* **19**:157–163.
161. Jaim-Etcheverry, G., and Zieher, L. M., 1975, *J. Neurochem.* **25**:915–917.
162. Walker, R. J., Ramage, A. G., and Woodruff, G. N., 1972, *Experientia* **28**:1173–1174.
163. Barker, D. L., Molinoff, P. B., and Kravitz, E. A., 1972, *Nature (New Biol.)* **236**:61–63.
164. Wallace, B. G., Talamo, B. R., Evans, P. D., and Kravitz, E. A., 1974, *Brain Res.* **74**:349–355.
165. Juorio, A. V., and Molinoff, P. B., 1971, *Br. J. Pharmacol.* **43**:438–439P.
166. Juorio, A. V., and Molinoff, P. B., 1974, *J. Neurochem.* **22**:271–280.
167. Robertson, H. A., and Steele, J. E., 1974, *J. Physiol. (Lond.)* **237**:34–35P.
168. Lam, K. C., Tall, A. R., Goldstein, G. B., and Mistilis, S. P., 1973, *Scand. J. Gastroenterol.* **8**:465–472.
169. Rossi-Fanelli, F., Cangiano, C., Attili, A., Angelico, M., Cascino, A., Capocaccia, L., Strom, R., and Crifo, C., 1976, *Clin. Chim. Acta* **67**:255–261.
170. Kinniburgh, D. W., and Boyd, N. D., 1979, *Clin. Biochem.* **12**:27–32.
171. Saavedra, J. M., 1974, *Anal. Biochem.* **59**:628–633.
172. Robertson, H. A., 1975, *Experientia* **31**:1006–1007.
173. Robertson, H. A., 1976, *Experientia* **32**:552–554.
174. Saavedra, J. M., and Axelrod, J., 1973, *Proc. Natl. Acad. Sci. U.S.A.* **70**:769–772.
175. Saavedra, J. M., Coyle, J. T., and Axelrod, J., 1974, *J. Neurochem.* **23**:511–515.
176. Danielson, T. J., Boulton, A. A., and Robertson, H. A., 1977, *J. Neurochem.* **29**:1131–1135.
177. Saavedra, J. M., Ribas, J., Swann, J., and Carpenter, D. O., 1977, *Science* **195**:1004–1006.
178. McCaman, M. W., and McCaman, R. E., 1978, *Brain Res.* **141**:347–352.
179. Farnham, P. J., Novak, R. A., and McAdoo, D. J., 1978, *J. Neurochem.* **30**:1173–1176.
180. Saavedra, J. M., Brownstein, M. J., Carpenter, D. O., and Axelrod, J., 1974, *Science* **185**:364–365.
181. Buck, S. H., Murphy, R. C., and Molinoff, P. B., 1977, *Brain Res.* **122**:281–297.
182. Evans, P. D., 1978, *J. Neurochem.* **30**:1009–1013.
183. Harmar, A. J., and Horn, A. S., 1976, *J. Neurochem.* **26**:987–993.
184. David, J.-C., 1979, *Experientia* **35**:1483–1484.
185. Robertson, H. A., David, J.-C., and Danielson, T. J., 1977, *J. Neurochem.* **29**:1137–1139.
186. Boulton, A. A., 1976, *Trace Amines and the Brain* (E. Usdin, and M. Sandler, eds.), Marcel Dekker, New York, pp. 21–39.
187. Saavedra, J. M., 1974, *J. Neurochem.* **22**:211–216.
188. Durden, D. A., Philips, S. R., and Boulton, A. A., 1973, *Can. J. Biochem.* **51**:995–1002.
189. Willner, J., LeFevre, H. F., and Costa, E., 1974, *J. Neurochem.* **23**:857–859.
190. Tallman, J. F., Saavedra, J. M., and Axelrod, J., 1976, *J. Neurochem.* **27**:465–469.
191. Tallman, J. F., Saavedra, J. M., and Axelrod, J., 1976, *J. Pharmacol. Exp. Ther.* **199**:216–221.

192. Philips, S. R., Durden, D. A., and Boulton, A. A., 1974, *Can. J. Biochem.* **52:**366–373.
193. Saavedra, J. M., and Axelrod, J., 1972, *J. Pharmacol. Exp. Ther.* **182:**363–369.
194. Saavedra, J. M., and Axelrod, J., 1973, *J. Pharmacol. Exp. Ther.* **185:**523–529.
195. Philips, S. R., Durden, D. A., and Boulton, A. A., 1974, *Can. J. Biochem.* **52:**447–451.
196. Warsh, J. J., Coscina, D. V., Godse, D. D., and Chan, P. W., 1979, *J. Neurochem.* **32:**1191–1196.
197. Mandel, P., Pasantes-Morales, H., and Urban, P. F., 1976, *Transmitters in the Visual Process* (S. L. Bonting, ed.), Pergamon Press, New York, pp. 89–104.
198. Lombardini, J. B., 1975, *J. Pharmacol. Exp. Ther.* **193:**301–308.
199. Abdelal, A. T., and Ingraham, J. L., 1975, *Anal. Biochem.* **69:**652–654.
200. Kalb, V. F., Donohue, T. J., Corrigan, M. G., and Bernlohr, R. W., 1978, *Anal. Biochem.* **90:**47–57.
201. Philips, S. R., 1981, *Advances in Cellular Neurobiology* (S. Fedoroff and L. Hertz, eds.), Vol. 2, Academic Press, New York, pp. 355–391.

Two-Dimensional Polyacrylamide Gel Electrophoresis of Proteins

David L. Wilson

1. INTRODUCTION

O'Farrell[1] introduced the first high-resolution method for analyzing heterogenous populations of proteins. Earlier attempts had met with only limited success. O'Farrell's[1] two-dimensional polyacrylamide gel electrophoresis (2D-PAGE) technique separates proteins according to isoelectric point in one dimension and molecular weight in the second dimension. The independence of these separation criteria for proteins contributes to the high resolving power of the technique. After minor modifications, O'Farrell's[1] technique was first used to analyze nervous system proteins in my laboratory.[2-5]

Considerable advancement of a number of research areas was made possible by the one-dimensional polyacrylamide gels pioneered earlier.[6,7] The much higher resolution possible with 2D-PAGE offers the same magnitude of improvement. However, a number of researchers have continued to use only one-dimensional separations on research projects that would clearly benefit from the use of 2D-PAGE. The added complexity of 2D-PAGE may have discouraged some investigators, whereas others may have tried and failed to obtain adequate gels. I address this chapter especially to those who are using or who wish to develop the technique. With suggestions or through references to the literature, I hope to make 2D-PAGE an easier procedure. Its flexibility and its limitations are also explored.

2. BEFORE AND AFTER 2D-PAGE: LIMITATIONS

Following the development of polyacrylamide gel electrophoresis by Raymond and Weintraub,[8] Ornstein[6] and Davis[7] produced the first high-resolution

David L. Wilson • Department of Physiology and Biophysics, University of Miami, School of Medicine, Miami, Florida 33101.

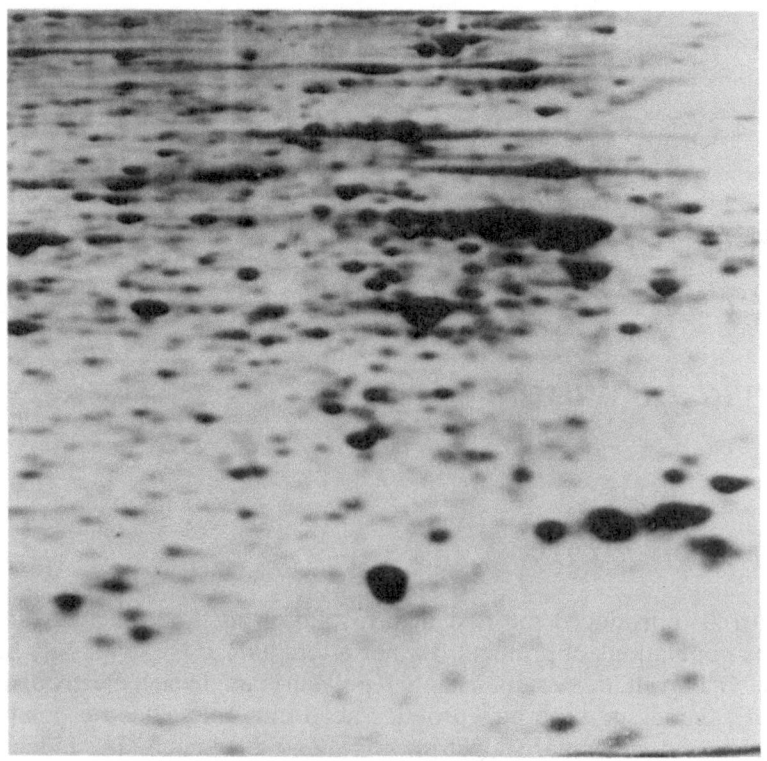

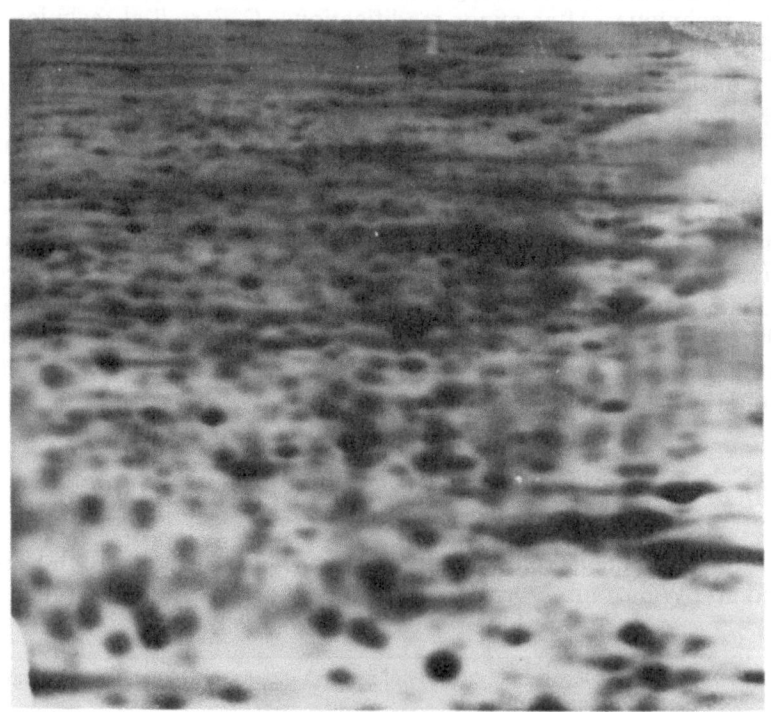

one-dimensional polyacrylamide gels. Both charge and size influenced protein migration on these gels. The addition of the ionic detergent sodium dodecyl sulfate (SDS) to the sample allowed most proteins to be separated according to molecular weight alone.[9,10] Some proteins, including glycoproteins, do not migrate strictly according to molecular weight.[11] If mercaptoethanol or an equivalent reducing agent is present, then disulfide bonds are split, and proteins migrate as individual polypeptides. Subsequently, the technique of isoelectric focusing was adapted to polyacrylamide gels. Thus, the one-dimensional polyacrylamide gel had evolved into a flexible system for separating proteins while measuring the important physical parameter of either size or charge. The gels were used to test purity, to do comparative studies, to detect changes in amounts or synthesis, and to analyze complex mixtures of proteins. The one-dimensional gel was made sensitive enough to allow for the study of single neurons.[12–14] Double-isotope experiments allowed the most critical comparison of control and experimental tissues on the same tube gel.[15] The development of slab gels and autoradiography on X-ray film also enhanced resolution in the comparison of samples.

However, the limitations of any one-dimensional analysis are troublesome. Even subcellular fractions or partially purified extracts usually contain more protein species than can be resolved on the gels. Consequently, for studies attempting to detect changes in proteins or differences in proteins, there was no way of confidently distinguishing altered *amounts* from altered species of proteins, and at best only the most abundant five to ten species of protein could be studied as individual proteins. Furthermore, sensitivity to changes in more minor protein species is radically impaired because these are masked by the more major, and perhaps more stable, proteins. Such tools as enzymology and immunology helped for those cases in which the identity of the protein was known in advance and when only one or a few proteins were of interest.

Although 2D-PAGE reduces many of the limitations of a one-dimensional analysis, it does not totally eliminate the problems. Coincidental comigration of proteins still is possible, although one protein species can often be observed to "crowd out" other proteins on the gels (see Section 3.4.1.). Even with the sensitivity-enhancing procedures to be discussed below, one is limited to studying the more abundant proteins in a complex sample. A typical cell may synthesize 10,000 different kinds of proteins. Under better conditions, the two-dimensional gels will allow one to detect only one-tenth of these. The wide range of concentrations of protein species in most samples will result in the most abundant species being very overloaded on the gels before the species in more minor abundance are at detectable levels. However, by subcellular fractionations, partial purifications, or adjustment of gel-running parameters

Fig. 1. Fluorograph of proteins from frog dorsal root ganglia following 2D-PAGE. Dorsal root ganglia from *Rana catesbeiana* were labeled with [^{3}H]leucine (top) or with [^{35}S]methionine (bottom). Subsequently, the ganglia were homogenized and subjected to 2D-PAGE. Experimental details were as described in Stone and Wilson.[47] In this case, Kodak Royal-X Pan film (4166) was exposed to the dried gels and developed with HC-110. The gel region shown in both cases extends from about 150,000 daltons (top) to 15,000 daltons (bottom) and from about pH 7 (left) to pH 4.5 (right).

(pH and molecular weight ranges), one can emphasize the analysis of certain classes of proteins in lesser abundance in a tissue.

Figure 1 shows an example of the complexity of protein species in nervous system tissue. The figure compares the newly synthesized proteins in frog dorsal root ganglia labeled with [3H] leucine (top) to those detected after [35S] methionine labeling (bottom). Many more spots are detected with methionine because of the higher specific activity of the label. All of the major proteins present after labeling with [3H]leucine are labeled by [35S]methionine, demonstrating that these proteins all have adequate numbers of methionine residues to be detected. The number of methionine-labeled proteins has begun to tax the resolving power of the gels in some regions. Longer exposures reveal even more protein species in some gel regions, but the film becomes overexposed, and detail is lost in more heavily labeled regions.

3. DETAILS OF THE 2D-PAGE TECHNIQUE

One key to successful gels is to read O'Farrell's[1] paper very thoroughly. Indeed, one would best benefit from what follows only after reading O'Farrell's paper,[1] especially as his terminology is used without further definition here.

In the few years since O'Farrell's paper appeared, a number of varients and improvements have been suggested by other researchers. Although the basic technique remains relatively unchanged, some of the options[2,16-18] are worth exploring before establishing a final procedure. Sources of some of the chemicals are important to insure good results, and O'Farrell[1] indicates these. From discussions with a number of researchers attempting to use 2D-PAGE, I conclude that many of the problems in running these gels come from mistakes in the preparation or mixing of the solutions (see Section 4).

3.1. Sample Preparation

O'Farrell used Nonidet® P-40 (NP-40) as a nonionic detergent to aid in solubilizing proteins during sample preparation; NP-40 is no longer made by Shell Chemical Co., but Triton X-100 apparently will substitute.[17] Ames and Nikaido[19] found that SDS was useful in solublizing membrane proteins for subsequent analysis by 2D-PAGE. Independently, we have shown that a similar solublization procedure, using SDS prior to addition of excess nonionic detergent, is generally useful for eukaryotic tissue extracts.[2] The SDS produces an artifactual clearing of material at a position in the acidic end of the isoelectric focusing gels but does not appear to otherwise disrupt the migration pattern of proteins. It is suspected that the SDS detaches from the proteins before or during the electrophoresis and may be isolated in micelles of the nonionic detergent.

For some samples, the addition of an inhibitor of proteolysis may be necessary. Even under the best of conditions, artifactual changes in proteins can occur with storage.[1,2] The normal procedure of heating samples in SDS should be avoided if possible, as artifacts can be formed.[2]

3.2. The First Dimension: pI

There are a number of different ranges of carrier ampholytes available. The ampholytes from different manufacturers are chemically distinct, and some may interact with sample proteins to disrupt or alter focusing. Once an acceptable ampholyte formulation is decided on, it is best to order an adequate supply of a single lot because, at least with some manufacturers, the pH gradients can vary from lot to lot. The ampholytes can be stored frozen in aliquots without appreciable deterioration for several months.

By using the O'Farrell[1] procedure and mixing ranges of carrier ampholytes, one can generate a wide variety of pH gradients in the first-dimension gel. The limit at the acidic end is pH 3.5 to 4. The limit at the basic end is about pH 8. Two published procedures allow more basic proteins to be analyzed. One uses a nonequilibrium pH gradient electrophoresis instead of isoelectric focusing.[20] For histones and perhaps other basic proteins, Willard et al.[21] propose a different solubilization procedure using dipalmitoyl-L-α-phosphatidylcholine and urea followed by some changes in the method for running the first-dimension gel. Single charge changes in a protein caused by mutation or posttranslational modification result in a detectable shift of isoelectric focusing position on 2D-PAGE.[1,22] Carbamylation of proteins, produced by heating in the presence of urea,[23] can be used to calibrate isoelectric focusing gels and estimate charge changes in proteins.[22,24]

A useful list of the isoelectric points of a few hundred proteins has been published.[25] Pharmacia® now has an isoelectric focusing calibration kit. It may be less expensive to buy the markers separately. There are limitations to the usefulness of such markers except as guides for a particular composition of isoelectric focusing gel. The isoelectric point for a protein depends on its neighboring environment and can be altered by denaturation, nonionic detergents, urea, ampholines, and other factors. The effects of such environmental factors vary from protein to protein.

Not only can the isoelectric point of a protein change, but the measurement of pH at different points in a gel also is difficult. Three procedures have been tried: (1) slicing the gel and placing each slice in water, then measuring the pH of the water after equilibration; (2) as in the first method but with 8 M urea; (3) using a microelectrode to measure pH directly from different points within the gel. The first two procedures have the problem of dilution effects on pH. The urea is added to more closely mimic the gel conditions, but other possible influences (polyacrylamide, NP-40, etc.) are ignored. The direct microelectrode measurement risks a bias on electrode readings caused by urea, NP-40, etc. Comparing the results of the three methods, we can get full pH unit differences in the pH measured at a given point in the gels. For this reason, we only cite general pH ranges for our published gels and use the identified tissue proteins as internal markers for pH.

3.3. The Second Dimension: Molecular Weight

Again, one has a number of options for emphasizing different molecular-weight ranges. A standard 10% acrylamide gel (9.46% acrylamide, 0.54% methylene-bis-acrylamide) is useful for proteins from 20,000 to 200,000 daltons. Proteins of high molecular weight can be better separated with a gel containing half the normal concentration of methylene-bis-acrylamide. This increases pore size in the second-dimension gel. The ultimate limit for high-molecular-weight proteins comes more from a failure to completely focus in the first dimension gel. This limit can be overcome somewhat with reduced acrylamide concentration in that gel, but the reduction may require addition of agarose to maintain stability. Cross-linking agents other than methylene-bis-acrylamide also have been reported to be effective.[26]

The fullest range of protein sizes on a single gel is probably best obtained using gradient gels, as O'Farrell[1] originally suggested. Low-molecular-weight proteins and peptides can present special problems in terms of solubility in weaker acid fixers and in terms of acrylamide concentration adjustments for adequate resolution. One unusual case we have found concerns substance P, a small peptide of known amino acid composition and sequence. We estimate its pI to be 11.5 using the method of Edsall and Wyman.[27] Even in the presence of SDS, substance P migrates toward a negative electrode, requiring electrode reversal to detect on gels (E. Hershberger and D. L. Wilson, unpublished data).

Molecular weight (electrophoresis) calibration kits are available from a number of manufacturers. These consist of a set of proteins and polypeptides of known molecular weight. Be careful to avoid gel filtration kits that contain proteins with more than one polypeptide chain. It may be less expensive to buy the proteins separately. As indicated above, not all proteins migrate in SDS gels strictly according to molecular weight. The addition of urea to the second dimension can markedly alter the migration of some proteins in SDS gels.

3.4. The Third Dimension: Abundance

The methods for studying amounts of proteins and radioactivity on the gels following 2D-PAGE are rapidly increasing. There are general staining procedures for most proteins and specific staining procedures for particular proteins. Radiolabeling of proteins can be done during their synthesis, after synthesis, or with labeled binding agents. These specific binding agents range from receptor ligands to antibodies. Many of these continue to show affinity even to the denatured proteins resulting from 2D-PAGE. Unfortunately, most enzymatic activities are lost because of the denaturation of the proteins. Automated methods of radioactivity quantitation have been developed for 2D-PAGE, and further analysis of the separated proteins by such methods as proteolytic digestion has been described. The presently available options are discussed below.

3.4.1. Staining Procedures

Most proteins are easily fixed in the gels during staining (25% trichloroacetic acid, 0.1% Coomassie brilliant blue R; Serva blue R apparently substitutes for Coomassie). Small peptides may present special fixation problems, but a range of acids will fix most proteins in the gels.

Coomassie blue is an inexpensive and very sensitive dye for protein detection on the gels. Unfortunately, it is no longer being made by the original company. A variety of substitutes have different properties. Independent of source, some proteins take up more stain per unit weight than do others. Thus, intensity of stain cannot be used to determine absolute protein amounts or stoichiometry unless a calibration has been done for each protein.

A more sensitive stain for most proteins has been developed.[28,29] This silver staining technique is 100 times as sensitive as Coomassie blue. A simpler and less expensive version of the silver staining procedure has recently been described.[30] We have found silver staining to be very useful for detecting minor proteins, although it is not as straightforward as the Coomassie procedure for routine work. Spots can appear to be in different relative abundance when the results of the two staining procedures are compared.

Proteins in more major abundance in a tissue or cell extract can produce stained spots that occupy 10 mm^2 or more of gel surface area. It might be thought that these could engulf more minor species of protein migrating nearby. However, we have repeatedly observed more minor proteins migrating at the very edge of the major protein spot. These can readily be seen in autoradiographs if the more minor species are labeled and the major proteins are not (see Stone *et al.*[31] for an example). The background labeling in the center of the unlabeled major spots actually can be less than in surrounding regions of the gels. This brings into question the normal procedures for subtracting average surrounding background when quantitatively analyzing autoradiographs.

3.4.2. Gel Drying

After destaining in acetic acid, and if necessary embedding a fluorescent compound (see Section 3.4.3 below), it has become routine to dry the gels onto a backing such as Whatman No. 3 filter paper. Gels can be photographed after they are dry (and this is best for comigration studies comparing label pattern with staining pattern because of possible unevenness and minor changes in gel size during drying), but the quality of the photographic image is usually somewhat better when the wet gel is used. Dried gels are much more easily stored than wet ones and give much higher sensitivity for autoradiography.

We have found the BioRad® drier (Model 224), which uses heating and links to a laboratory vacuum, to be effective. There are a number of similar items on the market. A much less expensive drier can be made with Bel Art's (Pequannock, NJ) Number 1255 linear porous plastic, with an appropriate backing to allow a vacuum hook-up. We have found such an apparatus adequate for drying gels of less than 15% acrylamide without cracking. Once dried, the gels can be stored for months without problems. We store our dried gels between plastic sheets in a binder to prevent them from curling or being damaged. A few drops of water on a dried gel can destroy it. Mike Hall and I have successfully reexposed [^{14}C]-labeled gels to X-ray film after several years of storage. G. W. Perry and C. Leonardi (personal communication) have even taken protein spots (actin) from our stained, dried gels after several years and obtained a normal proteolytic digest by the Cleveland[32] method.

3.4.3. Autoradiography and Fluorography

For intense radiation such as ^{32}P, drying of the gels is not strictly necessary, but I will presume in what follows that dried gels with radioactively labeled proteins are to be used to expose film. After fixation or after staining and destaining, dried gels can be placed next to X-ray film for exposure. For autoradiography, where the decay particle itself exposes the film, we have found Kodak no-screen X-ray film to be superior for ^{35}S and ^{14}C. When counts of these isotopes on the gels are low, or when ^{3}H is to be used, fluorography is necessary. Bonner, Lasky, and Mills[33,34] have published a procedure for embedding PPO into the gels by use of dimethyl sulfoxide (DMSO). They suggest preflashing the X-ray film to obtain linear responses for low levels of tritium. Two other methods of fluorographic enhancement are now available, but neither is as effective as the DMSO–PPO procedure in our hands. Dino Rulli, in my laboratory, has compared Enhance® (New England Nuclear) and sodium salicylate[35] with DMSO–PPO. Enhance® is easier to use and avoids contact with DMSO solutions, but for best results, gels should not be dried at higher temperatures, probably because a more volatile phosphor is used. In our hands, after gels are dried at 50°C or higher, the DMSO–PPO treatment gives greater sensitivity to ^{35}S (^{14}C) than Enhance® or sodium salicylate. Sodium salicylate[35] offers a much less costly option than either DMSO–PPO or Enhance®. Thorough drying of the gel is critical to reduce nonuniform response to radioactivity across a gel. With any of the fluorographic procedures, exposure is at -70°C.

The new Kodak XAR X-ray film is superior to Kodak no-screen X-ray film for ^{3}H and ^{35}S (^{14}C) when DMSO–PPO treatment is used. With either Enhance® or salicylate, Dino Rulli (personal communication) finds that the XAR remains superior for ^{3}H, but not for ^{35}S. For high-energy decays from isotopes such as ^{32}P and ^{125}I, intensifying screens greatly improve detection of spots on the gels.[36]

With one-dimensional gels, double-label methods (using the same amino acid labeled with one isotope for experimental cells or tissue and a second isotope for control cells or tissue) allowed mixing of samples during homogenization and a resulting higher sensitivity to change than is possible when samples are compared on different gels.[15,37,38] Of course, one can use double-labeled gels with 2D-PAGE as well. Three methods for making the analysis of double-labeled gels easier have been described.[39–41] None of these have the full range or sensitivity of direct counting of cut-out spots, and the number of counts in samples must be adequate to allow adjustments in amount of sample loaded. Otherwise, for example, we have found that spots with low levels of ^{14}C can be mistaken for ^{3}H-containing spots.

3.4.4. Quantitation of Radioactively Labeled Proteins after 2D-PAGE

We have demonstrated that the recovery of counts in labeled proteins after 2D-PAGE is independent of the amount of total or individual proteins loaded onto the gels over considerable ranges of protein amounts.[31] More than 60 μg of individual proteins or over 300 μg of nervous tissue extract can be loaded. Furthermore, such gels show adequate reproducibility to allow quantitative analysis to be done.

The simplest and most direct way to determine the amount of radioactivity in individual spots on the gels is to cut out and count the gel pieces.[2,4,31] Wet or dried gels can be used. A scalpel is used to cut out stained spots accurately, or a photograph of the autoradiograph can be used to locate labeled but unstained spots. The negative is placed in an enlarger, and its image is projected onto the dried gel. With proper alignment, very precise cutting of particular labeled spots is possible, as can be demonstrated by repeating the autoradiography of the gel after spot cutting.[49] Normalization techniques for this simple method have been described.[5,31]

Much more sophisticated, and more automated, methods are now available for gel analysis. These involve densitometric scanning of autoradiographs from the gels followed by computer analysis and comparisons.[16,42–44] Corrections must be made for nonlinearity in film exposure, for different densities and kinds of label, as well as for nonuniformity in the gels and background labeling. The background correction may present problems, as indicated in Section 3.4.1. Unfortunately, expense and complexity place these sophisticated methods out of reach for many laboratories. We have tried three different half-way methods using scanning devices without the sophisticated computer back-up, but these have been no less effort than the simple cut-out-and-count method and can introduce scatter and error unless care is taken in calibration and analysis.

There is no question that the automated analysis methods are useful when a study of a very large number of gels and protein spots is anticipated. However, in one study, G. W. Perry and I have used the simple punch-and-count method for 80 protein spots on each of 32 gels and could have cut out even more spots.

3.5. Comigration Analysis

Other important uses of 2D-PAGE include the identification of known proteins in complex tissue samples, the analysis of purity of a protein sample, and the comparison of protein species in different samples, tissues, or subcellular fractions. Comigration analysis[2] allows a very accurate comparison of different samples to be made. One sample, "A," is mixed with a second sample, "B," and proteins in the mixture are separated by 2D-PAGE. The stain (dried gel) or label (film) pattern is compared with the pattern obtained with sample A alone and sample B alone on 2D-PAGE. If a particular spot is being analyzed, it is best to adjust the amount of A and B to give about equal concentrations or counts in that spot on the gels. Comigration of the spot(s) from A and B can then be tested on the mixed gel. One useful property of the gels when overloaded for a particular protein is our finding that nearby proteins are not necessarily absorbed into the large spot but can be "pushed" to the periphery (see Section 3.4.1.). Of course, it is possible for two different proteins to have the same pI and molecular weight. For this reason, comigration analysis alone is not a conclusive test of identity. Further analysis of the proteins separated by 2D-PAGE is necessary to more completely establish identity, as indicated below.

3.6. Continued Analysis of Proteins following 2D-PAGE

A number of procedures are now appearing for the continued analysis of proteins following 2D-PAGE. In addition to the above procedures for staining and for the detection of labeled proteins, some notable new procedures include a method for peptide mapping from spots on the gels, antibody-linked fluorescent probes for detecting and localizing specific proteins, and the use of tagged ligands for detecting receptors and other proteins. It is possible to recover enzyme function after SDS-polyacrylamide gel electrophoresis.[45]

The Cleveland method[32] of peptide mapping by limited proteolysis can be applied to spots after 2D-PAGE.[46] This allows confirmation that comigrating proteins actually are the same. It also allows for the detection of similarities in proteins that have been modified and consequently migrate at a different gel position. Precursor–product relationships as well as posttranslational modifications can be demonstrated in this way. Antibodies and specific ligands can be spread onto the gels or bathed with the gels. After rinsing, binding to specific proteins can be detected by previous radiolabeling or fluorescent labeling of the antibody or ligand. Ligand binding frequently occurs despite the denatured state of the proteins.

4. TROUBLESHOOTING 2D-PAGE

A partial listing of problems and possible causes of problems associated with 2D-PAGE follows. Once a laboratory has achieved good, reproducible gels, a record should be kept of all new solutions and chemicals used. This will allow errors in solution preparation to be corrected easily, without having to remake all solutions.

4.1. No Spots on Gel after Staining

1. Reversed polarity of electrodes for first or second dimension electrophoresis.
2. Current not passing through gel.
3. Inadequate protein loaded. A reasonable amount of protein for observing staining patterns of complex samples is 50 μg. The best resolution is obtained with less,[1] but more can be loaded to study minor species or samples with low specific radioactivity.[31]
4. Inadequate protein solubilization. Try heating in 1% SDS before adding other components, but watch for artifactual spots.[2]
5. Excess nucleic acid present and causing blocking. Use DNase and RNase.[1]

    6. A bad lot or source of Coomassie stain is being used. As a check, try another stain.

4.2. Staining Pattern Has Streaks in Isoelectric Focusing Direction (Horizontal Streaks)

    1. Impurities in water or in chemicals. Contaminants in second dimension running buffer chemicals or agarose (or from fingers!) will produce lines across the entire slab gel beyond the ends of the first-dimension gel.
    2. Inadequate current through first-dimension gel.
    3. Acrylamide concentration too high.
    4. Clogging at top of first-dimension gel.
    5. See 4 and 5 in Section 4.1.
    6. Room temperature low enough to allow urea to precipitate in place in gel.

4.3. Staining Pattern Has Streaks in Molecular Weight Direction (Vertical Streaks)

    1. Inadequate equilibration time.
    2. Contaminants in sample. Especially with tissue containing considerable myelin (e.g., nerve segments), we observe some reproducible background streaks in both staining and labeling patterns in particular pH ranges. We have been unable to eliminate such streaks by extraction of the sample with lipid solvents.
    3. Dirt on second-dimension gel plates.
    4. Low SDS concentration in running buffer.

4.4. Proteins at Basic End of Gel Show Some Streaking and Variability from Gel to Gel

    1. Unfortunately, this is a normal condition. If proteins of interest are in this region of the gel, try one of the alternative procedures for focusing basic proteins, as described above.

4.5. Isoelectric Focusing Pattern (pH Gradient) Is Shifted or Distorted

    1. Different batches of ampholines can cause such variation.
    2. Upper or lower running buffer for first dimension is not correctly made.
    3. Incorrectly made gel solutions or impurities in chemicals.

4.6. Molecular Weight Distribution Has Shifted

1. Acrylamide or bisacrylamide concentration changed.
2. Running buffer for second dimension gel improperly made (can also influence voltage and current readings as well as gel running time).
3. Improper lower gel buffer.

4.7. X-ray Film Has Spurious Marks, Lines, or Background Exposure

1. Film near another source of radioactivity during storage or exposure (^{32}P or ^{125}I in freezer during exposure of fluorographs presents a problem).
2. Improper handling during film development.
3. Old film or aged film-processing solutions.
4. Incomplete gel drying (sticky gel; acetic acid still present).

4.8. Uneven Tracking Dye Front during Second Dimension Run

1. Uneven current across gel. Check for bubbles at bottom of slab gel. Temperature gradients also can produce uneven currents.

4.9. Distorted Spots

1. Improper "stacking" between lower and upper gel in second dimension.

5. TWO-DIMENSIONAL PAGE AS A TOOL FOR NEUROSCIENCE

Two-dimensional PAGE has obvious advantages for use in the study of many biological phenomena. Rather than needing to know in advance which specific proteins are important in an event or process, one can screen a large number of proteins simultaneously. Those that are altered can then be selected for more detailed study and identification.

Studies in cellular and molecular biology now frequently make use of 2D-PAGE. Neurobiology has also felt the impact of the technique. Two-dimensional PAGE offers a superb test for purity of a protein and a means of monitoring purification procedures. Two-dimensional PAGE also has use in comparative studies of proteins from different tissues and from different organisms.[47] Studies on the regulation of neuronal gene expression can use 2D-PAGE to detect changes in the relative rates of synthesis of particular proteins.[5,48] An early use of 2D-PAGE was to detect and identify proteins and peptides in nervous tissue.[2] As more purified neuronal proteins and peptides become available, this use for 2D-PAGE should allow for a rapid biochemical mapping of these proteins and peptides in different areas of the brain. For studies of modification of proteins associated with neuronal functions, 2D-PAGE offers high sensitivity to charge changes. Thus, phosphorylation, glycosylation, and any other modification inducing even a single charge change will produce a detectable shift in the position of the polypeptide after 2D-PAGE. Finally, the potential for clinical uses of 2D-PAGE, especially in diagnosis with CSF or blood, appears high.

As an example of how one can use these gels to demonstrate and approach problems of mechanism, George Stone and I were able to show that there was a profound selection of proteins for rapid transport down axons. Only a small subset of the proteins synthesized in the cell body are found among the rapidly transported species. We also were able to approach the question of the mechanism of rapid transport and to rule out a number of theories with our demonstration that neither actin nor tubulin were among the abundant, newly synthesized proteins being rapidly transported.

George Stone and I in collaboration with Robert Schmidt have been able to show that a number of cytoskeletal "linker" proteins also are not appreciably transported in newly synthesized form. These include α-actinin (the actin-linking protein that forms the Z line in skeletal muscle), the myosin light chains, and the more abundant MAPs (microtubule-associated proteins).

6. CONCLUSIONS

Two-dimensional PAGE is one of the best techniques available today for the analysis of proteins. It is a highly sensitive method with great resolving power. I hope that the suggestions in this chapter will make the procedure easier to use on a wider variety of studies.

ACKNOWLEDGMENTS. The research described in this chapter was supported by NIH Grants NS14328 and NS12393. I thank Gary Perry and Robert Rubin for their suggestions.

REFERENCES

1. O'Farrell, P. H., 1975, *J. Biol. Chem.* **250**:4007–4021.
2. Wilson, D. L., Hall, M. E., Stone, G. C., and Rubin, R. W., 1977, *Anal. Biochem.* **83**:33–44.
3. Stone, G. C., Wilson, D. L., and Hall, M. E., 1978, *Brain Res.* **144**:287–302.
4. Wilson, D. L., Hall, M. E., and Stone, G. C., 1978, *Gerontology* **24**:426–433.
5. Hall, M. E., Wilson, D. L., and Stone, G. C., 1978, *J. Neurobiol.* **9**:353–366.
6. Ornstein, L., 1964, *Ann. N.Y. Acad. Sci.* **121**:321–349.
7. Davis, B. J., 1964, *Ann. N.Y. Acad. Sci.* **121**:404–427.
8. Raymond, S., and Weintraub, L. S., 1959, *Science* **130**:711.
9. Shapiro, A. L., Vinuela, E., and Maizel, J. V., Jr., 1967, *Biochem. Biophys. Res. Commun.* **28**:815–820.
10. Weber, K., and Osborn, M., 1969, *J. Biol. Chem.* **244**:4406–4412.
11. Tung, J. S., Knight, C. A., 1972, *Anal. Biochem.* **48**:153–163.
12. Hyden, H., Bjurstam, K., and McEwen, B., 1966, *Anal. Biochem.* **17**:1–15.

13. Wilson, D. L., 1971, *J. Gen. Physiol.* **57**:26–40.
14. Gainer, H., 1971, *Anal. Biochem.* **44**:589–605.
15. Wilson, D. L., and Berry, R. W., 1972, *J. Neurobiol.* **3**:369–379.
16. Garrels, J., 1979, *J. Biol. Chem.* **254**:7961–7977.
17. Anderson, N. G., and Anderson, N. L., 1978, *Anal. Biochem.* **85**:331–340.
18. Anderson, N. L., and Anderson, N. G., 1978, *Anal. Biochem.* **85**:341–354.
19. Ames, G. F.-L., and Nikaido, K., 1976, *Biochemistry* **15**:616–623.
20. O'Farrell, P. Z., Roodman, H. M., and O'Farrell, P. H., 1977, *Cell* **12**:1133–1142.
21. Willard, K. E., Giometti, C. S., Anderson, N. L., O'Connor, T. E., and Anderson, N. G., 1979, *Anal. Biochem.* **100**:289–298.
22. Steinberg, R. A., O'Farrell, P. H., Friedrich, U., and Coffino, P., 1977, *Cell* **10**:381–391.
23. Bobb, D., and Hofstee, B. H. J., 1971, *Anal. Biochem.* **40**:209–217.
24. Anderson, N. L., and Hickman, B. J., 1979, *Anal. Biochem.* **93**:312–320.
25. Malamud, D., and Drysdall, J. W., 1978, *Anal. Biochem.* **86**:620–647.
26. O'Connell, P. B. H., and Brady, C. J., 1976, *Anal. Biochem.* **76**:63–73.
27. Edsall, J. T., and Wyman, J., 1958, *Biophysical Chemistry*, Academic Press, New York, pp. 507–510.
28. Switzer, R. C., Merril, C. R., and Shifrin, S., 1979, *Anal. Biochem.* **98**:231–237.
29. Merril, C. R., Switzer, R. C., and Van Keuren, M. L., 1979, *Proc. Natl. Acad. Sci. U.S.A.* **76**:4335–4339.
30. Merril, C. R., Goldman, D., Sedman, S. A., and Ebert, M. H., 1981, *Science* **211**:1437–1438.
31. Stone, G. C., Wilson, D. L., and Perry, G. W., 1980, *Electrophoresis '79* (B. J. Radola, ed.), W. de Gruyter, Berlin, pp. 361–382.
32. Cleveland, D. W., Fischer, S. G., Kierschner, M. W., and Lalmmli, U. K., 1977, *J. Biol. Chem.* **252**:1102–1106.
33. Bonner, W. M., and Laskey, R. A., 1974, *Eur. J. Biochem.* **46**:83–88.
34. Laskey, R. A., and Mills, A. D., 1975, *Eur. J. Biochem.* **56**:335–s341.
35. Chamberlain, J. P., 1979, *Anal. Biochem.* **98**:132–135.
36. Swanstrom, R., and Shank, P. R., 1978, *Anal. Biochem.* **86**:184–192.
37. Wilson, D. L., 1974, *J. Neurochem.* **22**:465–467.
38. Strumwasser, F., and Wilson, D. L., *J. Gen. Physiol.* **67**:691–702.
39. Kronenberg, L. H., 1979, *Anal. Biochem.* **93**:189–195.
40. McConkey, E. H., 1979, *Anal. Biochem.* **96**:39–44.
41. Choo, K. H., Cotton, R. G. H., and Danks, D. M., 1980, *Anal. Biochem.* **103**:33–38.
42. Lutin, W. A., Kyle, C. F., and Freeman, J. A., 1978, *Electrophoresis '78* (N. Catsimpoolas, ed.), Elsever/North-Holland, Amsterdam, pp. 93–106.
43. Bossinger, J., Miller, M. J., Vo, K. P., Geiduschek, E. P., and Xuong, N. H., 1979, *J. Biol. Chem.* **254**:7986–7998.
44. Taylor, J., Anderson, N. L., Coulter, B. P., Scandora, A. E., and Anderson, N. G., 1980, *Electrophoresis '79* (B. J. Radola, ed.), W. de Gruyter, Berlin, pp. 329–339.
45. Hager, D. A., and Burgess, R. R., 1980, *Anal. Biochem.* **109**:76–86.
46. Gard, D. L., Bell, P. B., and Lazarides, E., 1979, *Proc. Natl. Acad. Sci. U.S.A.* **76**:3894–3898.
47. Stone, G. C., and Wilson, D. L., 1979, *J. Neurobiol.* **10**:1–12.
48. Hall, M. E., and Wilson, D. L., 1979, *Brain Res.* **168**:414–418.
49. Perry, G. W., and Wilson, D. L., 1981, *J. Neurochem.* **37**:1203–1217.

The Identification of Subcellular Fractions of the Central Nervous System

Suella W. Henn and Fritz A. Henn

1. INTRODUCTION

The understanding of cellular function should ultimately come from a comprehension of molecular events and their mode of integration into a working cell. A principle step in that integration takes place through the structuring of biochemical reactions at the subcellular level. This structure involves the components common to all cells, such as the nucleus, endoplasmic reticulum, mitochondria, lysosomes, and plasma membranes as well as such specialized features as the synaptic endings found on meurons in the CNS. A valid way of trying to understand the workings of a cell involves studying an isolated organelle. This requires the development of methods of purification and criteria for determining the degree of purity. This chapter is concerned with the assessment of advantages and limitations of subcellular purification techniques and the identification and evaluation of the resultant fractions.

2. SYNAPTOSOMES

2.1. Review of Preparative Methods

Synaptosomes were first prepared by a sucrose density gradient method developed by Gray and Whittaker[1] and De Robertis and co-workers.[2] Basically, the procedure involves a series of differential centrifugations of the brain homogenate in 0.32 M sucrose (see Fig. 1). The crude mitochondrial pellet, P_2, is then further resolved on sucrose or isotonic Ficoll–sucrose[3-5] discontinuous gradients. If the P_2 pellet is washed twice and then applied to a gradient of

Suella W. Henn and Fritz A. Henn • University of Iowa, Department of Psychiatry, College of Medicine, Iowa City, Iowa 52242. Present address: Long Island Research Institute, State University of New York, Stony Brook, New York 11794.

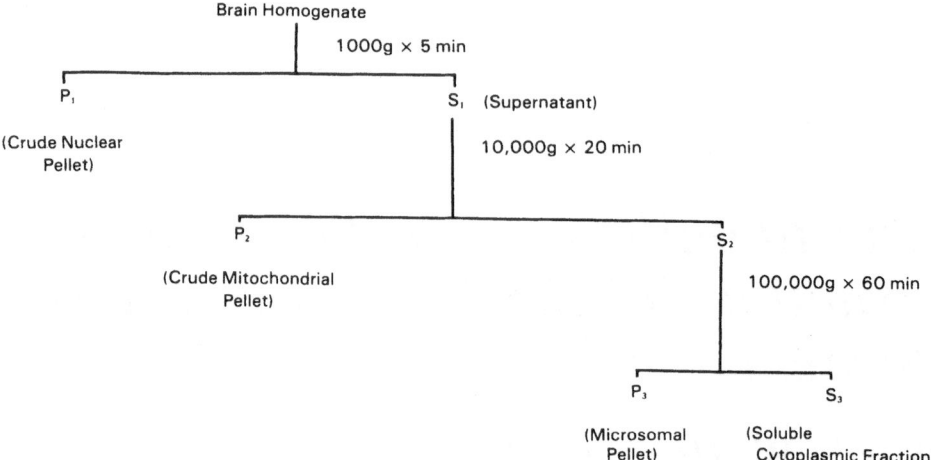

Fig. 1. Differential centrifugation scheme for fractionation of a brain homogenate. Pellets are usually washed and recentrifuged for analytical work. They can then be applied to density gradients for further centrifugation and purification. See text for references detailing specific procedures.

three or more steps, a lower yield with greater homogeneity of particles is obtained than by a procedure without washings of the P_2 and only a two-step gradient.[6] A similar procedure has been developed for the preparation of synaptosomes from chick brain.[7] The yield of synaptosomes was highest in the fraction layering between 13 and 20% Ficoll–sucrose, and there were a large number of empty profiles and mitochondria in the fraction interfacing between 8 and 13% Ficoll–sucrose.[7,8]

An improved and more rapid method for the preparation of synaptosomes of high purity using low centrifugal forces has been reported.[9] The P_2 pellet is suspended in 0.3 M sucrose and layered over 0.8 M sucrose and centrifuged at 9000 g for 25 min. Particles dispersed in the 0.8 M sucrose solution were gradually diluted to 0.4 M sucrose and were seen to be 89% synaptosomes by counting of electron micrographs. It has been demonstrated that glial cells can form vesicles that maintain characteristic uptake processes and sediment with synaptosomes when prepared by the above methods.[10] This would indicate the need for great care to be taken before assigning any particular metabolic or functional role to nerve ending particles without even greater attention being paid to the purity of the preparations.

More recent improvements in the preparations of synaptosomes include the development of a rapid (less than 2 hr) procedure which uses flotation of the crude mitochondrial pellet suspended in 12% Ficoll–sucrose into an upper layer of 7.5% Ficoll–sucrose and a top layer of 0.32 M sucrose.[11] Synaptosomes banded at the 7.5–12% interface were metabolically active and were contaminated with about 4% mitochondria as judged by electron microscopic and enzyme studies. However, actual counts of the number of vesicle-containing synaptosomes are not reported. Large-scale preparations of synaptosomes have also been developed using a zonal rotor technique.[12]

Further purification of the synaptosome fraction collected at the 7.5–14%

Ficoll–sucrose interface has been shown following centrifugation on discontinuous sucrose gradients.[13] This allows the preparation of synaptosomes essentially free of myelin membrane fragments as indicated by the specific activity of the 2',3'-cyclic nucleotide 3'-phosphohydrolase, an enzyme found in the myelin-forming oligodendrocytes.

2.2. Criteria for Purity

The problem of detecting contaminants in any subcellular fraction is a difficult one to solve. Among criteria used to assess purity are morphological determinations with electron microscopy, enzyme profiles, receptor enrichment, and, most recently, immunologic identification. The general features discussed in the following sections hold for all subcellular fractions. Several criteria should always be used if at all possible, and the analysis should include an evaluation of the degree of contamination with other subcellular fractions. In other words, one should not follow only the enrichment of a single putative marker for the fraction being sought. Several markers should be followed and be shown to separate from putative markers for other possible contaminating fractions. These caveats are especially important for fractions as complex as synaptosomally derived ones.

2.2.1. Electron Microscopy

Electron microscopy is frequently used to illustrate the particle composition of each of the differential and density gradient subfractions in the synaptosomal preparations reported in the literature. It is difficult to compare electron microscopic results because of different ways of selecting the field, evaluating the particles, and calculating the percentage figures. In most cases, only estimates of purity can be made without more detailed serial section work. With Ficoll gradients, enrichments in synaptosomes of 60–70% have been reported.[4,8,14]

Particles are identified as synaptosomes by virtue of the fact that the membrane-limited rounded structures contain synaptic vesicles and one or more mitochondria.[4,14,15] The fractions also contain free membranes, membrane vesicles, and mitochondria. Occasionally, myelin fragments are also seen. With sucrose gradients, synaptosomal enrichments of 20–60% have been reported.[15–18] Hajós, in his low-centrifugal-force sucrose gradient procedure,[9] reports an enrichment of about 90% in synaptosomes as judged by electron microscopy.

A morphological assessment of synaptosomal fractions obtained by Ficoll gradients has been reported by Joo and Karnushina.[8] They counted all particles with an area of 0.1 μm^2 or greater for a 100-μm^2 area from 10 sections prepared from pellets of material sedimenting at the 4–8%, 8–13%, and 13–20% Ficoll interfaces. Their criteria for synaptosomes were the presence of four or more synaptic vesicles.

Unfortunately, electron microscopy is often used nonquantitatively as supportive evidence. Selective fields or random sections are not sufficient to quantitate contaminants. Neither are estimates of the amount of a specific contaminant such as "free mitochondria" sufficient by themselves to define the purity of a synaptosomal preparation. Rather, positive evidence on the proportion of total membrane in a sample that comes from synaptic contact regions is necessary.

2.2.2. Enzymology

A stronger case for the relative degree of purity of the synaptosomes is obtained when electron microscopy is combined with assays for a variety of subcellular marker enzymes. Table I presents various enzymes that have been used to define certain subcellular fractions. These can also be used as negative markers to estimate contaminants. However, a number of cautions must be borne in mind when evaluating such data.

Synaptosomes from any region of the CNS will be heterogeneous with respect to transmitters and the enzymes involved in their synthesis.

Contamination with various subcellular constituents for which there is no specific marker available is possible.

Table I
Markers Used to Identify Subcellular Fractions

Subcellular fraction	Marker	Reference
Synaptosome	specific acetylcholinesterase	12
	glutamic acid decarboxylase	21
	choline acetyltransferase	100
Synaptosomal plasma membranes	Na^+–K^+-ATPase	101
	acetylcholinesterase	13
	5′-nucleotidase	14
Postsynaptic junctional densities	tubulin antisera	59, 60
Nuclei	DNA	64, 65, 67
	DNA polymerase	62
	RNA polymerase	64, 65
	adenylate kinase	62
Mitochondria	cytochrome c oxidase	74
	monoamine oxidase	75
	succinic dehydrogenase	74
	rotenone-sensitive NADH: cytochrome c reductase	75
	rotenone-insensitive NADH: cytochrome c reductase	76
Microsome	NADPH: Cytochrome c reductase	76
	glucose-6-phosphatase	80
	antimycin A-insensitive cytochrome c reductase	81
	glutamine synthetase	90
Lysosomes	Acid phosphatase	94
	β-N-acetylglucosaminidase	94
Plasma membranes	Na^+–K^+ ATPase	101
	5′-nucleotidase	14
	acetylcholinesterase	13

Leakage of soluble enzymes and adsorption to other particles may occur. A particular enzyme may be low in a fraction in which it is expected to be present in relatively high specific activity, or it may be present in a fraction in which no particles or very few from which it arises are present simply because of redistribution during isolation procedures.

Whittaker and his colleagues[1,19] initially used bound acetylcholine as a marker for the P_2 fraction and succinic dehydrogenase (SDH) as a marker for mitochondria enclosed in the synaptosomes. The use of specific acetylcholinesterase is one way of defining neuronal membrane as opposed to nonneuronal contaminants. Leskawa and co-workers[12] compared the activity of specific acetylcholinesterase with nonspecific acetylcholinesterase. The former is thought to be located on the outside of the nerve terminal, whereas the latter is considered to be associated with nonneuronal (glial) elements.

In general, no enzyme can be considered to be a positive marker for all areas of synaptic contact. One problem with acetylcholinesterase is that it is not present on cells using other neurotransmitters such as amino acids or catecholamines. In specific cases, this problem can be overcome by using specially chosen areas such as *Torpedo* end plates for starting materials. Similar problems exist for enzymes specific to other transmitter systems. However, several are very useful as positive markers. Glutamic acid decarboxylase (GAD) is the final enzyme involved in the synthesis of GABA (γ-aminobutyric acid). This is a good marker of synaptic activity because of the ubiquitous nature of GABAergic transmission in the CNS. It has been shown by immunocytochemical methodology[21] to be localized in synaptic regions. Similarly, choline acetyltransferase is a marker for cholinergic synapses, and dopamine β-hydroxylase is a useful marker for noradrenergic nerve endings.

2.2.3. Transmitter Systems

As mentioned above, one severe limitation of work done with CNS synaptosomes is that nerve endings representing a variety of neurotransmitters are involved. Whittaker has attempted to overcome this difficulty by utilizing a simpler system, the electromotor system of *Torpedo*.[22] This system provides large quanitities of readily accessible cholinergic neurons containing abundant amounts of acetylcholine, choline acetyltransferase (ChAT), and acetylcholinesterase.

Balazs and his co-workers[23] prepared glomerulus particles from rat cerebellum internal granular layer. This synaptic complex, enclosed in a glial sheath, consists of axodendritic synapses between mossy fiber terminals and granule cell dendrites. The granule cells are excited by the mossy fiber terminals for which there is no generally acepted biochemical marker. It may be cholinergic.[24,25] The granule cells are inhibited by the Golgi cells. Glutamic acid decarboxylase activity was enriched in the glomerulus particles with a relative specific activity of 2.5 compared with that of ChAT.

Wilkin and co-workers[26] report an enrichment of "cholinergic" synapses from rat cerebella as indicated by choline acetyltransferase in fractions subjected to centrifugation in discontinuous and continuous sucrose gradients.

When compared with the other transmitter enzyme assayed, glutamate decarboxylase (GAD), the GAD:ChAT relative specific activity ratio was 0.2 for the "cholinergic" synaptosomes. The peak of the ChAT activity was in a lighter part of the gradient (about 1.2 M) compared with that of the GAD activity (about 1.3 M). Studies such as these indicate some promise for isolating synapses utilizing different transmitters.

2.2.4. Receptors

With the development of specific binding assays for neurotransmitter receptors, an apparently ideal positive marker for synaptic membranes became available. Receptors are tightly bound membrane proteins which do not diffuse or migrate during isolation procedures. Although there appear to be both postsynaptic and presynaptic receptors, initially it was felt that these receptors would be confined to the synaptic region. Subsequent studies indicate that this may not be true for all receptors. Henn and co-workers[27] first suggested that dopamine receptors might be found on glia. This has been confirmed in retina using various lesion techniques.[28] This plus the finding of β-receptors on gliomas[29] suggests that catecholamine receptors may not be accurate markers of synaptic membranes. Similar types of studies have suggested a partial glial localization for other neurotransmitter receptors.

However, one binding ligand may be relatively specific for neuronal membranes and widespread enough in the CNS to be of general value as a positive marker. That is kainic acid. This ligand binds glutamatelike receptor sites but clearly defines a site different from glutamate.[30] Evidence from binding studies suggests that kainic acid binding is localized to synaptic regions in hippocampus.[31] Binding studies that we have carried out on isolated glial fractions and primary astrocytic cultures revealed the absence of these receptors, tending to confirm that kainic acid binding may prove to be a reliable marker of synaptic membranes.

Another useful and general neuronal marker that is restricted to the cell surface is the receptor for tetanus toxin. The localization of this marker has been described by Mirsky and colleagues.[32] The tetanus toxin receptor has been used to identify neuronal elements in differentiation studies and should have potential for use in studies of bulk-isolated fractions. Cholera enterotoxin has also been used as a neuronal marker. Although not uniformly present in neurons, it has been used in a novel isolation procedure[33] using magnetic microspheres.

2.2.5. Immunology

The most specific and precise markers to any cell or cellular component ought to be antibodies to specific membrane components. Until recently, the problem involved the need for pure antigens to produce pure antibodies. However, with the advent of monoclonal antibody techniques, more work has begun on CNS specific antibodies. Even prior to these newer techniques, several laboratories defined antibodies with some specificity for various CNS structures. The most interesting antigens described thus far on synaptic membranes have been described by Bock and her colleagues.[34] Jorgensen and Bock[35] have described a series of antibodies to synaptic plasma membrane proteins. The synaptic membrane proteins are called D_1, D_2, and D_3, and all are enriched in synaptic membrane preparations about 3.5-fold over their level in crude brain homogenates. These antigens could not be identified in cultures of primary astroglial cells,[36] and they appear in all CNS regions. They appear to be good neuronal markers. Another protein studied by this group is specific to the nervous system and is found on vesicle membranes. This synaptic vesicle membrane marker is called synaptin.[37] Another antiserum that appears to be specific for the neuronal surface antigen call adhesion molecule has been reported by Rutishauser and co-workers.[38]

Very recently, monoclonal techniques have yielded antibody that appears uniformly on the CNS neuronal membrane surface but is not present on other known cell types. This antigen, called A4, completely shares the binding pattern of tetanus toxin receptors in its neuronal specificity.[39] It is not present in the peripheral nervous system and appears as early as the tenth embryonic day in rat. It should prove to be a useful immunologic marker for future studies. Thus, it appears that a number of antigens are becoming defined, and ever greater numbers of cell-specific antibodies should become available through the use of monoclonal techniques.

3. SYNAPTOSOMAL PLASMA MEMBRANES

Synaptosomal plasma membranes are composed of three components as detected in ultrastructural studies. These include the presynaptic membrane, the postsynaptic membranes, and the synaptic junction, a thickening between the pre- and postsynaptic membranes. Many attempts have been made to isolate the synaptosomal plasma membrane (SPM) so that its structure and function may be studied at the biochemical level.[5,6,13,14,40–42] All of these workers have stressed the point that it is necessary to start with as pure a synaptosomal fraction as possible before attempting to isolate the SPM fraction. Otherwise, there are too many membrane fragments from a variety of sources contaminating the preparation.

Procedures are open to many slight modifications in centrifugation forces, types of media, and composition of density gradients, all of which can affect yields and degrees of contamination by mitochondria, microsomes, lysosomes and their membranes, glial membrane fragments, and myelin. Basically, the purified synaptosomal fraction is subjected to hypoosmotic conditions, and the resulting membrane fragments are then centrifuged in a sucrose or Ficoll–sucrose gradient. Using enzymatic markers as criteria for degree of contamination of the SPM, purity estimates of 75–90% have been obtained. Major contaminants have been shown to be mitochondria, microsomes, and lysosomes. Mitochondrial contaminants are assayed via cytochrome oxidase, monoamine oxidase, and succinic dehydrogenase. Microsomal contamination is followed by NADPH:cytochrome c reductase, although this enzyme activity does not parallel the RNA content, indicating that the enzyme may be on the synaptic plasma membrane as well as on endoplasmic reticulum.[41] Lysosomes are assayed by acid phosphatase and β-N-acetylglucosaminidase. It is possible that small amounts of these enzymes are integral to the neuronal plasma membrane and function to help maintain the integrity of the membrane.[14] Membranes arising from myelin or glia are usually assayed by examining 2',3'-cyclic nucleotide 3'-phosphohydrolase activity.[5,13] This enzyme is localized on oligodendroglia but is not helpful in determining astrocytic contamination.

It is difficult to decide on a specific marker for SPM; (Na^+1-K^+1) ATPase enrichment is often followed to indicate synaptic membrane purification.[13,14,40,42,43] It is generally considered the most specific marker for the synaptosomal membrane.[45,46] A 5- to 13-fold enrichment of the enzyme has been reported.[13,14,42] However, high activities of this enzyme are also found in astrocytic plasma membranes.[45,46] Other plasma membrane marker enzymes that have been used as markers include acetylcholinesterase and 5'-nucleotidase.[13,14,40] However, histochemical analysis has shown acetylcholinesterase to be associated with axonal and dendritic membranes,[47,48] endoplasmic reticulum, small neuronal elements in neuropile,[49] and to some degree with presynaptic endings.[50] 5'-Nucleotidase activity frequently parallels that of acetylcholinesterase, and, thus, it too may be present in axonal, dendritic, or glial membranes in higher concentrations than in synaptic membranes. Histochemical analysis has demonstrated the activity in glial cells.[51] Mena et al.[13] also found high levels of NADPH:cytochrome c reductase activity copurifying with SPM and suggest that it may be a microsomal contaminant copurifying with the synaptosomal plasma membranes as has been proposed,[41] or it may be an intrinsic component of the synaptosome.[52]

Polyacrylamide gel electrophoresis has also been used to compare the polypeptide compositions of possible contaminating membranes and of SPMs.[13,40,41] In the presence of sodium dodecyl sulfate, mitochondrial and myelin membranes had distinct compositions when compared to each other or to synaptic plasma membranes and microsomal membranes. Synaptic plasma membranes and microsomal membranes show a nearly identical polypeptide composition even though electron microscopy shows that fractions containing these components are morphologically distinct.[13] It has been suggested that synaptic membrane specialization resides with a minor polypeptide component[13] or perhaps in a polysaccharide component. One of the distinctive SPM polypeptide components migrates with isolated tubulin,[13] cross reacts with antisera to tubulin,[53] and has a polypeptide map similar to that of tubulin.[54]

4. POSTSYNAPTIC JUNCTIONAL DENSITIES

Synaptosomal membranes can be digested with sodium deoxycholate, and intact postsynaptic junctional densities released. When the component proteins are investigated with sodium dodecyl sulfate polyacrylamide gel electrophoresis, a major component of these structures comigrates with tubulin, which is the subunit protein of microtubules.[55,56] Independent studies[57,58] confirm that the major protein of the synaptic junction densities has a molecular weight close to that of tubulin. Immunuhistochemical staining with an antiserum raised against electrophoretically purified tubulin from microtubules shows tubulin antigen in microtubules and in most postsynaptic junctional densities.[59,60] There is no tubulin antigen apparent in the synaptic cleft or in the presynaptic axon terminal. Also, all postsynaptic junctions do not stain with the antisera, which may indicate that tubulin is not common to all postsynaptic junctional densities. Because of their distinct morphological appearance and resistance to detergents, unlike other membrane fractions, these junctional complexes may be prepared in relatively pure form.

5. NUCLEI

The analysis of nuclei from central nervous system tissue is difficult because of the heterogeneity of the cell types involved. In order to get homogeneous nuclei, it is necessary either first to purify the individual cell type or to fractionate further the crude nuclear pellet. Methods have been devised to isolate cell nuclei free of contamination from endothelial cells and extranuclear material.[61] Common isolation procedures use either aqueous or nonaqueous solvents for homogenization, depending on the desired results. Aqueous methods allow leakage of nuclear components into the media, whereas nonaqueous ones damage the nuclear membrane. A system containing hexylene glycol and piperazine-N,N'-bis(2-ethane sulfonic acid) and $CaCl_2$ as the isolation medium has been developed[62] and appears to combine some advantages of both types of media. Nuclei prepared with this system conform to criteria for purity of nuclei that are generally agreed on. These include morphological integrity and lack of cytoplasmic contaminants. The morphology of nuclei as detected by light and electron microscopy and their chemical composition as indicated by RNA/DNA, protein/DNA, acid-soluble protein/DNA, and phospholipid/DNA ratios are similar to those obtained for nuclei obtained from other tissues. Fractions of nuclei show an appropriate partition of nuclear and cytoplasmic enzymatic activities.

Nonhistone chromosomal proteins have been the subject of much study because they are thought to be involved in the control of gene expression in which the DNA is released for transcription from the DNA–histone complex. For example, Fujitani and Holoubek[63] have analyzed the composition of nonhistone chromosomal proteins from nuclei isolated from rat cerebral cortex, cerebellum, and total residual brain, since RNA synthesis has been shown to be highest in brain cortex and lowest in cerebellum.[64] The ratio of nonhistone chromosomal protein to DNA was 0.52 in cortex, 0.18 in cerebellum, and 0.38 in the rest of the brain. These ratios are proportional to the genomic activities in these nuclei. However, there are no qualitative differences in the electrophoretic patterns of the proteins from these different sections of brain. It is likely that the specific regulatory proteins represent only a very small proportion of the total nonhistone chromosomal proteins and so are not detectable by polyacrylamide gel electrophoresis.

Unfractionated nuclei from various brain regions are of limited use since neurons, astroglia, and oligodendroglial cells undoubtedly perform quite distinct functions and would be expressing these through various transcriptions of their nuclear DNAs. Austoker *et al.*[65] have developed a procedure for separating rat brain nuclei into five classes using zonal discontinuous sucrose density gradient centrifugation and have characterized the resulting fractions. The morphological criteria used to evaluate the fractions are summarized by Rappoport and co-workers.[66] Neuronal nuclei have a single, dense, centrally located, spherical nucleolus and pale chromatin. Astrocytic nuclei have large pale oval or round nuclei with two, three, or more paracentral nucleoli. Nuclei of oligodendrocytes are smaller, round, and dense with tightly packed chromatin and peripheral nucleoli. Microglial nuclei are very difficult to distinguish from oligodendroglial nuclei.

Fractionated nuclei show some interesting differences. They all possess the same amount of DNA.[65,67] Neuronal nuclei have the most protein, about 30% more than astrocytes and more than twice the amount of oligodendroglia.[65,67] The RNA/DNA ratio is also highest in neuronal nuclei and lowest in oligodendroglial nuclei. RNA polymerase activity has been found to be highest in astrocytic nuclei sedimenting closest to the neuronal nuclei.[65] DNA polymerase activity in fractionated brain nuclei has also been investigated.[68] The highest activity was found in neuronal nuclei, with astroglial less, and oligodendroglial even less. *In vivo* experiments on infant rats examining the incorporation of [^{3}H]thymidine into nuclear DNA showed the highest incorporation into oligodendroglial nuclei.

Nonhistone chromosomal proteins from fractionated brain nuclei have been examined by several groups,[67,69] and a limited specificity of these proteins has been demonstrated. Neuronal chromatin characterized with polyacrylamide gel electrophoresis shows a high proportion of high-molecular-weight nonhistone proteins as compared with oligodendroglial nuclei in which these appear to be absent.[69] In other studies[67] utilizing sodium dodecyl sulfate/polyacrylamide gel electrophoresis, electrofocusing electrophoresis, and two-dimensional electrophoresis to analyze nonhistone chromosomal proteins, one polypeptide band with molecular weight of 10,000 and pI of 8.5 was found to be present in the neuronal nuclear fraction and missing from all other nuclear fractions of adult rats. Interestingly, this polypeptide was present in all classes of nuclei of infant (10-day-old) rats. Further characterization of this protein remains to be done as well as a determination of its function. If regulatory proteins are indeed present among nonhistone chromosomal proteins, it might be expected that they are present in very small quantities and will not be detected by these methods. The fractionation of brain cell nuclei into different cell types will make it possible to make progress from the morphological identification of classes of nuclei to neurochemical criteria for identification.

6. MITOCHONDRIA

Brain mitochondria are extremely heterogeneous with respect to their enzyme components. This is the case whether the mitochondria are isolated from neuron- and glia-enriched fractions of brain[70] or from sucrose density gradients of crude mitochondrial preparations from brain tissue.[71-73] Lai and co-workers[72] have developed methods for separating "free" mitochondria from synaptosomal mitochondria. These fractions are relatively pure, metabolically active, and well coupled.

Marker enzymes used to identify mitochondria and to distinguish them from other subcellular fractions include cytochrome *c* oxidase and succinate dehydrogenase for the inner mitochondrial membrane[74]; rotenone-sensitive NADH:cytochrome *c* reductase for the inner membrane[75]; rotenone-insensitive NADH:cytochrome *c* reductase for the outer membrane[76]; and monoamine oxidase for the outer membrane.[75] Even a particular enzyme can be heterogeneous, as has been shown to be the case with monoamine oxidase.[77,78] Crude brain mitochondrial fractions centrifuged on Ficoll–sucrose density gradients show a heterogeneous profile with respect to substrate specificity and pH optima for monoamine oxidase.

7. MICROSOMES

Microsomes are those components of a tissue homogenate that can be sedimented by ultracentrifugation of the mitochondrial supernatant. In this fraction can be found both smooth and rough endoplasmic reticulum, polysomes, and Golgi membranes.[79] Among relatively specific marker enzymes used to identify microsomes are glucose-6-phosphatase,[80] antimycin A-insensitive cytochrome *c* reductase,[81] and NADPH:cytochrome *c* reductase.[76] Microsomes with similar acitivities can also be prepared by low-speed centrifugation in the presence of Ca^{2+}.[82-84]

Other activities demonstrated to be present in the microsomal fractions from brain tissue are ceramide galactosyltransferase, cerebroside sulfotransferase on Golgi membranes,[85] and tubulin synthetase.[86] Glutamine synthetase has been localized on the membranous system of the microsomes[87-89] and has been determined to be localized to astrocytes by ultrastructural immunocytochemistry[90]; thus, it can serve as a marker of astrocytic microsomes.

8. LYSOSOMES

Usual methods for isolating lysosomes combine procedures of differential and density gradient centrifugation.[91] However, the preparation of pure lysosomes from brain tissue is particularly difficult because of the variety of cell types, the inclusion of lysosomes with axons and nerve endings, the cosedimentation with lysosomes of large numbers of mitochondria, particularly from bovine brain,[92] and the association of lysosomal hydrolases with endoplasmic reticulum and Golgi apparatus. A new technique has been developed in which the sedimentation rate of mitochondria has been increased by including succinate or other salts that can be taken up by mitochondria and presumably increase the density by causing swelling.[93] This "chemical field" technique has facilitated the purification of bovine brain lysosomes.

Numerous hydrolytic enzymes are present in lysosomes and can be used as marker enzymes. These include acid phosphatase, acid glycerophosphatase, acid deoxyribonuclease, acid ribonclease, cathepsin, β-glucuronidase, β-galactosidase, N-acetyl-β-D-glucosaminidase, aryl sulfatase, sialidase, and cerebroside galactosidase.[94] Lysosomal particles in an isotonic sucrose suspension can be stained *in vitro* with the basic fluorochrome acridine orange. *In vivo* staining of lysosomes is also possible with acridine orange.

As was mentioned in the section on synaptosomal plasma membranes, acid phosphatase activity is present on synaptic plasma membranes. This may reflect the small numbers of lysosomes found at nerve endings and a slight contamination of the SPM with lysosomal membrane fragments.

9. PLASMA MEMBRANES

The isolation and characterization of plasma membrane from mammalian CNS is plagued by the heterogeneity of cell types. Even the plasma membranes of individual cells have been shown to be nonuniform, with many cell types displaying functional polarity. De Pierre and Karnovsky[95] review most helpfully the problems and productive approaches to take for a study of plasma membranes. Ideally, one would first obtain a suspension of individual cells which could then be separated into the different cell types on the basis of plasma membrane markers. These would include enzymes in which the active site faces the external medium, e.g., acetylcholinesterase, and binding sites for neurotransmitters and surface antigens. The technology is now becoming available with the fluorescence-activated cell sorter[96] to use fluorescence-labeled antibodies to tag immunologically and functionally distinct cells and to separate them from a cellular suspension.

Much of the work done to date with CNS plasma membranes has been with synaptic plasma membranes (*q.v.*) Only a few techniques have been described for the preparation of plasma membranes from separated cells as opposed to solid nervous tissue. These include purification of plasma membranes from neuronal- and glial-enriched fractions,[46,96] cultured neuroblastoma,[97] dissociated immature neurons,[98] and myelinated axon-enriched fractions which are then demyelinated.[99]

The problem in using a single marker, not only for plasma membranes but for any subcellular fraction, is that the use of a single marker can lead to a circular process. Defining a fraction by means of a marker leads to the isolation of a fraction with that marker but not necessarily the fraction one desires. For example, the (Na^+l-K^+l) ATPase is utilized as a plasma membrane marker in the CNS but might be found on a variety of membranes from CNS. Thus, without multiple markers and immunochemical localization of the marker errors in fraction isolation will result. Other markers that can be used for CNS plasma membranes include 5′-nucleotidase, acetylcholinesterase, adenylate cyclase, and more recently receptors, one of which, kainic acid might be specific for neuronal membranes. In preparing plasma membranes, it is important to assay for contaminating membranes from organelles such as mitochondria or microsomes which can rupture in the isolation procedure and then copurify with the plasma membrane.

10. CONCLUSION

The complexity of the central nervous system is reflected in the heterogeneity of cell types and the great diversity and anatomic organization among the various cell types. The identification of subcellular organelles would be expected to reflect, at least in part, the cellular origins of the organelle and the functions that particular cell was expressing *in vivo*. Although a wide variety of cell markers are currently in use as a means of identifying subcellular organelles, one might expect further refinements as the cellular sources of the organelles are determined.

We are presently limited by the inherent difficulties of separating into cell types as interwined and heterogeneous a system as the mammalian central nervous system. However, separation techniques utilizing differential and density gradient centrifugation, lesioning studies, sieving of carefully dissected material, and, potentially, the fluorescence-activated cell sorter are facilitating the preparation of cell fractions enriched in particular cell types. That, coupled with enzymatic markers, morphological and immunologic criteria, and neurotransmitter receptors, allows a clearer determination of the cell type being studied. Tissue culture studies of isolated cell types, especially primary cultures, are of increasing value as is, in some cases, the use of specific mutants to allow analysis of the functions of a given cell type.

REFERENCES

1. Gray, E. G., and Whittaker, V. P., 1962, *J. Anat.* **96**:79–87.
2. DeRobertis, E., Pellegrino de Iraldi, A., Rodriquez de Lores Arnaiz, G., and Salganicoff, L., 1962, *J. Neurochem.* **9**:23–35.
3. Abdel-Latif, A. A., 1966, *Biochim. Biophys. Acta* **121**:403–406.
4. Autilio, L. A., Appel, S. H., Pettis, P., and Gambetti, P. L., 1968, *Biochemistry* **7**:2615–2622.
5. Morgan, I. G., Wolfe, L. S., Mandel, P., and Gombos, G., 1971, *Biochim. Biophys. Acta* **241**:737–751.
6. Cotman, C. W., 1974, *Methods Enzymol.* **31**:445–452.
7. Oestreicher, A. B., and van Leeuwen, C., 1975, *J. Neurochem.* **24**:251–259.
8. Joo, F., and Karnushina, I., 1975, *J. Neurochem.* **24**:839–840.
9. Hajós, F., 1975, *Brain Res.* **93**:485–489.
10. Henn, F. A., Anderson, D. J., and Rustad, D. G., 1976, *Brain Res.* **101**:341–344.
11. Booth, R. F. G., and Clark, J. B., 1978, *Biochem. J.* **176**:365–370.
12. Leskawa, K. C., Yohe, H. C., Matsumoto, M., and Rosenberg, A., 1979, *Neurochem. Res.* **4**:483–504.
13. Mena, E. E., Hoeser, C. A., and Moore, B. W., 1980, *Brain Res.* **188**:207–231.
14. Cotman, C. W., and Matthews, D. A., 1971, *Biochim. Biophys. Acta* **249**:380–394.
15. Babitch, J. A., 1973, *Brain Res.* **49**:135–150.
16. Bretz, U., Baggiolini, M., Hauser, R., and C., Hodel, 1974, *J. Cell Biol.* **61**:466–480.
17. Ross, L. K., Andreoli, V. M., and Marchbanks, R. M., 1971, *Brain Res.* **25**:103–119.
18. Whittaker, V. P., 1968, *Biochem. J.* **106**:412–417.

19. Whittaker, V. P., Michaelson, J. A., and Kirkland, R. J. A., 1964, *Biochem. J.* **90**:293–303.
20. Whittaker, V. P. and Barker, L. A., 1972, *Methods of Neurochemistry*, Volume 2 (R. Fried, ed.), Marcel Dekker, New York, pp. 1–52.
21. Roberts, E., 1978, *Interactions between Putative Neurotransmitters in the Brain*, (S. Garattini, J. F. Pujol, and R. Samanin, eds.), Raven Press, New York, pp. 89–107.
22. Whittaker, V. P., 1976, *Prog. Brain Res.* **45**:45–64.
23. Balazs, R., Hajos, F., Johnson, A. L., Riejneirse, G. L. A. Tapia, R., and Wilkin, G. P., 1975, *Brain Res.* **86**:17–30.
24. Crawford, J. M., Curtis, D. R., Voorhoeve, P. E., and Wilson V. J., 1966, *J. Physiol. (Lond.)* **186**:139–165.
25. McCance, I., and Phillis, J. W., 1968, *Int. J. Neuropharmacol.* **7**:447–462.
26. Wilkin, G. P., Reijneirse, G. L. A., Johnson, A. L., and Balazs, R., 1979, *Brain Res.* **164**:153–163.
27. Henn, F. A., Anderson, D. J., and Sellstrom, A., 1977, *Nature* **266**:637–638.
28. Memo, M., Riccardi, F., Trabucchi, M., and Spano, P., 1981, *Adv. Biochem. Psychopharmacol.* **26**:41–51.
29. Gilman, A. G., and Nirenberg, M., 1971, *Proc. Natl. Acad. Sci. U.S.A.* **68**:2165–2168.
30. London, E. D., and Coyle, J. T., 1979, *Eur. J. Pharmacol.* **56**:287–290.
31. Foster, A. C., Mena, E. E., Monaghan, D. T., and Cotman, C. W., 1981, *Nature* **289**:73–75.
32. Mirsky, R., Wendon, L. M. B., Black, P., Stolkin, C., and Bray, D., 1978, *Brain Res.* **148**:251–259.
33. Kronick, P. L. Campbell, G. LeM., and Joseph, K., 1978, *Science* **200**:1074–1076.
34. Bock, E., 1978, *J. Neurochem.* **30**:7–14.
35. Jorgensen, O. S., and Bock, E., 1974, *J. Neurochem.* **23**:879–880.
36. Bock, E., Jorgensen, O., Dittmann, L., and Eng, L., 1975, *J. Neurochem.* **25**:867–870.
37. Bock, E., Jorgensen, O. S., and Morris, S. J., 1974, *J. Neurochem.* **22**:1013–1017.
38. Rutishauser, U., Gall, W. E., and Edelman, G. M., 1978, *J. Cell Biol.* **79**:382–393.
39. Cohen, J., and Selvendran, S. Y., 1981, *Nature* **291**:421–423.
40. Levitan, I. B., Mushynski, W. E., and Ramirez, G., 1972, *J. Biol. Chem.* **247**:5376–5381.
41. Gurd, J. W., Jones, L. R., Mahler, H. R., and Moore, W. J., 1974, *J. Neurochem.* **22**:281–290.
42. Babitch, J. A., Breithaupt, T. B., Chiu, T.-C., Garadi, R., and Helseth, D. L., 1976, *Biochim. Biophys. Acta* **433**:75–89.
43. Hosie, R. J. A., 1965, *Biochem. J.* **96**:404–412.
44. Jones, D. H., and Matus, A. I., 1974, *Biochim. Biophys. Acta* **356**:276–287.
45. Henn, F. A., Haljamae, H., and Hamberger, A., 1972, *Brain Res.* **43**:437–443.
46. Henn, F. A., and Hamberger, A., 1976, *Neurochem. Res.* **1**:261–273.
47. Shute, C. C. D., and Lewis, P. R., 1966, *Z. Zellforsch.* **69**:334–343.
48. Novikoff, A. B., 1967, *The Neuron* (H. Hydén, ed.), Elsevier, Amsterdam, pp. 255–318.
49. Kokko, A., Mautner, H. G., and Barnett, R. J., 1969, *J. Histochem. Cytochem.* **17**:625–640.
50. Levin, S. J., and Bodansky, O., 1966, *J. Biol. Chem.* **241**:51–56.
51. Torack, R. M., and Barrnett, R. J., 1964, *J. Neuropathol. Exp. Neurol.* **23**:46–59.
52. Miller, E. K., and Dawson, R. M., 1972, *Biochem. J.* **126**:805–821.
53. Yen, S. H., Liem, R. K. H., Kelly, P. T., Cotman, C. W., and Shelanski, M. L., 1977, *Brain Res.* **132**:172–175.
54. Feit, H., Kelly, P., and Cotman, C. W., 1977, *Proc. Natl. Acad. Sci. U.S.A.* **74**:1047–1051.
55. Walters, B. B., and Matus, A. I., 1974, *J. Anat.* **119**:415.
56. Walters, B. B., and Matus, A. I., 1975, *Biochem. Soc. Trans.* **3**:109–112.
57. Banker, G., Churchill, L., and Cotman, C. W., 1974, *J. Cell Biol.* **63**:456–465.
58. Cotman, C. W., Banker, G., Churchill, L., and Taylor, D., 1974, *J. Cell Biol.* **63**:441–455.
59. Matus, A. I., Walters, B. B., and Mughal, S., 1975, *J. Neurocytol.* **4**:733–744.
60. Westrum, L. E., and Gray, E. G., 1976, *Brain Res.* **105**:547–556.
61. Siakotos, A. N., 1974, *Methods Enzymol.* **31**:452–457.
62. Wray, W., Conn, P. M., and Wray, V. P., 1977, *Methods Cell Biol.* **16**:69–86.
63. Fujitani, H., and Holoubek, V., 1974, *J. Neurochem.* **23**:1215–1224.
64. McEwen, B. S., Plapinger, L., Wallach, G., and Magnus, C., 1972, *J. Neurochem.* **19**:1159–1170.

65. Austoker, J., Cox, D., and Mathias, A. P., 1972, *Biochem. J.* **129:**1139–1155.
66. Rappoport, D. A., Maxcy, P., Jr., and Daginawala, H. F., 1969, *Handbook of Neurochemistry*, Volume 2 (A. Lajtha, ed.), Plenum Press, New York, pp. 241–254.
67. Tsitilou, S. G., Cox., D., Mathias, A. P., and Ridge, D., 1979, *Biochem. J.* **177:**331–346.
68. Stambolova, M. A., Cox, D., and Mathias, A. P., 1973, *Biochem. J.* **136:**685–695.
69. Tashiro, T., Mizobe, F., and Kurokawa, M., 1974, *FEBS Lett.* **38:**121–124.
70. Hamberger, A., Blomstrand, C., and Lehninger, A. L., 1970, *J. Cell Biol.* **45:**221–234.
71. Van den Berg, C. J., 1973, *Metabolic Compartmentation in the Brain* (R. Balazs and J. E. Cremer, eds.), Macmillan, London, pp. 137–166.
72. Lai, J. C. K., Walsh, J. M., Dennis, S. C., and Clark, J. B., 1975, *Metabolic Compartmentation and Neurotransmission*, (S. Berl, D. D. Clarke, and D. Schneider, eds.), Plenum Press, New York, pp. 487–496.
73. Lai, J. C. K., and Clark, J. B., 1976, *Biochem. J.* **154:**423–432.
74. Sottocasa, G. L., Kuylenstierna, B., Ernster, L., and Bergstrand, A., 1967, *Methods Enzymol.* **10:**448–463.
75. Schnaitman, C., Erwin, V. G., and Greenawalt, J. W., 1967, *J. Cell Biol.* **32:**719–735.
76. Sottocasa, G. L., Kuylenstierna, B., Ernster, L., and Bergstrand, A., 1967, *J. Cell Biol.* **32:**415–438.
77. Youdim, M. B. H., 1976, *J. Neural Transm.* **38:**15–29.
78. Achee, F. M., Togulga, G., and Gabay, S., 1974, *J. Neurochem.* **22:**651–661.
79. Reid, E., and Williamson, R., 1974, *Methods Enzymol.* **31:**713–733.
80. Nordlie, R. C., and Arion, W. J., 1966, *Methods Enzymol.* **9:**619–625.
81. Crane, F. L., 1957, *Plant Physiol.* **32:**619–625.
82. De Marchena, O., Herndon, R. M., and Guarnieri, M., 1974, *Brain Res.* **80:**497–502.
83. Kamath, S. A., and Narayan, K. A., 1972, *Anal. Biochem.* **48:**53–61.
84. Kamath, S. A., and Rubin, E., 1972, *Biochem. Biophys. Res. Commun.* **49:**52–59.
85. Siegrist, H. P., Burkart, T., Wiesmann, U. N., Herschkowitz, N. N., and Spycher, M. A., 1979, *J. Neurochem.* **33:**497–504.
86. Jorgensen, A. O., and Heywood, S. M., 1974, *Proc. Natl. Acad. Sci. U.S.A.* **71:**4278–4282.
87. Salganicoff, L., and De Robertis, E., 1965, *J. Neurochem.* **12:**287–309.
88. Sellinger, O. Z., and de Balbian Verster, F., 1962, *J. Biol. Chem.* **237:**2836–2844.
89. Sellinger, O. Z., de Balbian Verster, F., Sullivan, R. J., and Lamar, C., Jr., 1966, *J. Neurochem.* **13:**501–513.
90. Norenberg, M. D., and Martinez-Hernandez, A., 1979, *Brain Res.* **161:**303–310.
91. Koenig, H., 1974, *Methods Enzymol.* **31:**457–477.
92. Overdijk, B., Hooghwinkel, G. J. M., and Lisman, J. J. W., 1978, *Enzymes of Lipid Metabolism* (S. Gatt, L. Freysz, and P. Mandel, eds.), Plenum Press, New York, pp. 601–610.
93. Lisman, J. J. W., De Haan, J., and Overdijk, B., 1979, *Biochem. J.* **178:**79–87.
94. Koenig, H. 1969, *Handbook of Neurochemistry*, Volume 2 (A. Lajtha, ed.), Plenum Press, New York, pp. 255–301.
95. De Pierre, J. W., and Karnovsky, M. L., 1973, *J. Cell Biol.* **56:**275–303.
96. Henn, F. A., 1980, *Advances in Cellular Neurobiology*, Volume I (S. Fedoroff and L. Hertz, eds.), Academic Press, New York, ppa. 373–403.
97. Glick, M. C., Kimhi, Y., and Littauer, U. Z., 1973, *Proc. Natl. Acad. Sci. U.S.A.* **70:**1370–1372.
98. Hemminki, K., 1975, *Methods Cell Biol.* **9:**247–257.
99. DeVries, G. H., Matthieu, J.-M., Beny, M., Chicheportiche, R., Lazdunski, M., and Dolivo, M., 1978, *Brain Res.* **147:**339–352.
100. Fonnum, F., 1969, *Biochem. J.* **115:**465–472.
101. Cotman, C. W., Herschman, H., and Taylor, D., 1971, *J. Neurobiol.* **2:**169–180.

Cell Isolation

Jørgen Clausen

1. INTRODUCTION

Since the initial discoveries of the cell structure by Schwann,[1] the study of the cellular structure of central and peripheral nervous systems (CNS and PNS, respectively) became possible after the development of photographic techniques and of the process of precipitation of silver salts in nervous structures.[2] By this technique, details of nerve fibers and shapes of neurons were revealed. On this basis, the neuron doctrine was formulated, i.e., that the neuron is a genetic, structural, and functional unit. The neuron was also demonstrated to be a trophic unit in which regeneration always was initiated from the axonal part still in contact with the original cell body. Although it is obvious that the function of the whole nervous system is more than the sum of the elements of which it is composed, neurochemists have obtained valuable information by studying separated and enriched homogeneous cellular populations of CNS and PNS.

The neurochemist who wants to work with isolated cellular populations of CNS or PNS must have elementary knowledge of the anatomy and histology of CNS and PNS. Furthermore, he must be aware that in addition to the vulnerability of the cells to be isolated, two other factors determine the possibilities for isolation of brain cells: (1) the integration of cellular contacts between different types of brain cells and (2) their cellular junctions, i.e., desmosomes (maculae and zonula adherens), gap junctions,[3-6] and tight junctions, influence the possibilities for isolation of brain cells.

The desmosomes are the matrix material that binds the cells together. The gap junction is a 2- to 4-nm gap between cells containing specialized proteins (connexones) which coordinate the electrical and metabolic activities between cells. Finally, the tight junctions are areas between cells where the intercellular space is eliminated. These areas contain the so-called transmembrane proteins and connexones.[7,8]

Jørgen Clausen • Neurochemical Institute, Copenhagen, Denmark and Institute of Biology and Chemistry, University of Roskilde, DK 4000 Roskilde, Denmark.

2. CELLULAR COMPOSITION OF CNS AND PNS

The CNS is defined anatomically as the nervous tissue inside meninges. The brain is covered by pia, and nutritients may be transferred to the brain either through the capillary system or through the spinal fluid compartment. The spinal fluid is formed by the plexus choroideus.

The functional unit of the nervous system, the neuron, represents a wide variety of nerve cells. The central part of the neuron with its nucleus is called the perikaryon. Cytoplasmic processes are given off from the perikaryon and may be classified as axons and dendrites. The axons may be naked or enclosed in a myelin sheath, i.e., the remnants of the plasma membrane and cytoplasm of oligodendroglia (in CNS) or Schwann cells in the peripheral nervous system.

Different types of neurons exist, i.e., those with long axons (Golgi type I neurons) and those with short axons (Golgi type II neurons). Several dendrites may arise from a single neuron. Different neurons are encountered. Thus, the primary motor neurons of the spinal cord are multipolar cells with rather long branching dendrites and an axon which forms the axis cylinder of a motor nerve fiber terminating in the motor ending in a muscle. These long axons acquire a myelin sheath. On the other hand, the pyramidal cells of the cerebral cortex are multipolar and have the shape implied by their name. The apical part of the cell, oriented towards the surface of the cortex, is a thick branching process. From the base, shorter branching processes are given off, and special collaterals occur. The third basal process is a long slender axon equipped with collaterals. This axon continues to a distant part of the CNS.

Another example of neuronal diversity is the special neuronal interrelationship that exists in the cerebellar cortex. One basket cell may form several synapses with different Purkinje cells. In the cerebellar cortex, the prevalent neuronal type is the excitatory neuron, the granule cell, comprising 70–80% of the total cell population. Inhibitory neurons are the Purkinje, Golgi, basket, and satellite cells.

The connective tissue of the nervous system is the glial system of the CNS and Schwann cells of the PNS.[9] The glial cells include ependyma, astrocytes, oligodendroglia, microglia, and satellite cells of the CNS. The ependymal cells form a single layer of columnar epithelial cells lining the ventricles of the brain and the central canal of the spinal cord. These cells are interconnected by gap junctions, causing free permeability of macromolecules from the ventricular spinal fluid to the brain tissue and vice versa. In certain areas including the eminentia mediana and recessus pinealis, tight junctions occur instead of gap junctions, probably causing a more selective permeability in these areas. Finally, in the eminentia mediana the so-called tanycytes occur, i.e., ependymal cells that contain a basal dendrite penetrating the brain tissue and ending in a terminal body in close contact with a fenestrated capillary.

The astrocytes may be of two types, protoplasmic or fibrous actrocytes. The former are characterized by their numerous freely branching protoplasmic processes; they are also called the Mossey cells. The fibrous astrocytes, which are found chiefly in the white matter, have long unbranched fibers. Both types are attached to the blood vessels by one or more processes that terminate in perivascular feet. Some of their other fiber branches may terminate at the perikarya of some of the nerve cells.

The oligodendroglia are smaller than the astrocytes. In the prenatal and neonatal period, they form the myelin membrane and, even in the adult brain, they are connected to the myelin sheath. They also function as satellites to nerve cells. The microglia cells of mesodermal origin are found in both white and gray matter. They are small and may be either multipolar or bipolar. They function as the scavenger cells of the CNS.

Of the three types of capillaries—the discontinuous, the fenestrated, and the continuous type—only the fenestrated and continuous types occur in CNS. The fenestrated capillaries thus occur in the areas of CNS where no blood–brain barrier is present (epiphysis cerebri, plexus choroideus ventriculi tertii, subfornical organ, organum vasculosum laminae terminalis, neurohypophysis, and area postrema). The capillaries are characterized by endothelial cells interconnected by junctions (probably tight junctions) between which are fenestra (defects in areas of about 50 nm). These areas are covered by only a thin diaphragm.[10]

The continuous capillaries contain a complete layer of endothelial cells. In the CNS, these capillaries contain several mitochondria (10% of the volume of the endothelial cells) as do epithelial cells possessing transport capacities. The endothelial cells are connected by special junctions. Below the endothelial cells is a 30–40 nm thick basal lamina. The perivascular space seems relatively narrow or is lacking. In the basal lamina, phagocytosing pericytes are layered. The astrocytes form foot processes on the capillaries in such a way that the basal laminae of the two cell types are fused. Gap but not tight junctions are found between the foot processes. Tracer experiments with horseradish peroxidase[11] seem to support the view that the endothelial cells and not the foot processes are the blood–brain barrier.

The density of capillaries in the CNS represents a parameter of regional difference in metabolic activity[12] but seems not to be related to the density of neurons.[13]

On the surface of the brain is the pia mater. Here, the fibrous astrocytes extend their foot processes to the surface, forming a continuous layer on the brain surface. The foot processes form gap junctions, thus allowing free diffusion of macromolecules. Above this is a basal layer, some collagen and flattened mesothelial cells which do not constitute a continuum.

Outside the pia is the arachnoidea mater, consisting of several layers of mesothelial cells connected with tight junctions and forming a barrier that allows the spinal fluid to pass only via special canals and by means of vesicles in the mesothelial cells.

The plexus choroideus is localized in the ventricles and consists of many villi containing (1) cubical epithelium cells towards the ventricle, (2) a continuous layer of flattened mesothelial cells, and (3) fenestered capillaries. Towards the ventricular system, the cubical cells are connected by tight junctions, thereby forming the blood–cerebrospinal fluid barrier.

In the PNS, the perikaryon may be present in the periphery (e.g., the sympathetic ganglia or the dorsal root ganglia) or in the spinal cord (e.g., the perikarya of the motor neurons). The fibers may be myelinated (e.g., ventral root nerves) or unmyelinated (some of the dorsal roots). The PNS nerves may terminate in synapses at muscle end plates or arise from sensory apparatuses. In the PNS, the blood–brain barrier also exists, since the endothelial cells of vasa vasorum of the peripheral nerves are connected in tight junctions. In the PNS, the Schwann cells, analogous to the oligodendroglia of the CNS, form the myelin sheath. In myelinated peripheral nerves, the Schwann cells, like the oligodendroglia cells in the CNS, form a myelin sheath by curling around the axon in the prenatal or neonatal period. The Schwann cells also enclose several unmyelinated axons, but here the multilayered myelin structure is not formed.

3. GENERAL PRINCIPLES FOR CELL ISOLATION

3.1. Advantages and Pitfalls of Using Enriched Brain Cell Fractions

From the introduction to the present chapter, it is obvious that the fiber network of neurons and glia cells is so interdigitated that complete isolation of intact cells, including intact fibers (dendrites and axons) from CNS and PNS, seems impossible. Also, the variable density of tight and gap junctions may be a factor determining the quality and cellular yield of certain areas of CNS.

The nervous structure seems to be looser in structure the younger and more immature the brain is. Therefore, the cellular yield and morphological integrity of the cells depend on the maturity of the brain tissue processed.

Like the slice technique of brain tissue, the isolation of individual cell types from CNS and PNS permits the elimination of the blood–brain and blood–spinal fluid barriers, thus permitting direct studies of the brain cells. In contrast to the tissue slice technique, cell isolation procedures also permit studies *in vitro* of different cell types, although these are often seriously damaged. In studies of pathologically changed nervous tissues or even autopsy specimens, the neurochemist must remember that the cellular composition in such conditions is changed.

Guanieri et al.[15] have found that pretreatment of brain tissue with enzymes, e.g., trypsin and collagenase, inactivates specific receptor functions. Thus, both enzymes diminished opiate receptor binding of naloxone by 55 and 70%, respectively. Furthermore, collagenase diminished the muscarinic cholinergic binding by up to 95%.

In pathologically changed brain tissues, the cellular composition may be altered. Thus, in neuroimmunologic conditions, e.g., multiple sclerosis, allergic encephalomyelitis, lymphocytes accumulate perivascularly and in border zones of demyelinated areas; microglia also accumulates. Furthermore, autopsy specimens may contain seriously damaged nerve cells.

Under normal conditions, with an optimal supply of oxygen, brain cells are dependent on the oxidation of glucose; glucose $+ 6\ O_2 + 38\ ADP \rightarrow 6\ CO_2 + 6\ H_2O + 38\ ATP$. In ischemia, e.g., during brain cell isolation, the lack of oxygen will cause the cells to utilize anaerobic glycolysis: glucose +

2 ADP → 2 lactate + 2 ATP. Thus, since only 2 ATP are formed by the latter process, the rate of anaerobic glycolysis should be increased 19 times to generate 38 ATP. Practically, it can only be enhanced five- to sevenfold. Thus, energy must be available from other sources, i.e., by utilization of other energy-rich reserves such as phosphocreatine (PCR). The oxygen supply in an aerobic working brain is only 0.11 μmol/g and thus covers only 1–2 sec of the aerobic glycolysis. The $\sim$ P $\approx$ P Cr + ATP + ADP + 2 glucose + 2.9 $\times$ glycogen is 20 μmol/g. Since the consumption of $\sim$P is about 30 μmol/g per min, there is only 40 sec normal energy supply in a normal brain.

As a consequence of ischemia, and extracellular $[K^+_e]$ increases from 3 mM to 30 mM.[16] Since the increase in $[K^+]$ is slower in young animals than in adult animals, young animals will be better sources for isolation of brain cells. Isolated brain cells seem more resistant to ischemia than cells *in situ*, probably because of a more efficient oxygen supply to the free cells than to the cells in the ischemic brain. Although brain cells thus seem sensitive to even few minutes' ischemia, isolated brain cells (neurons and glia cells) seem able to accumulate potassium and amino acids to an extent that may indicate that, in part, the membrane function of free brain cells is intact although the intra–extracellular ratio is lower than that in the intact tissue.[17,18]

3.2. Factors Determining the Possibilities for Separation and Isolation of Brain Cells

Isolation of brain cells was first done by hand dissection.[19] Perikarya and surrounding neuropils could be isolated only from selected areas of the brain. By means of microdissection, only a few brain cells could be isolated for biochemical studies. Therefore, methods were developed to transfer brain tissue to a suspension. McIlwain[20] demonstrated that incubation of brain slices or chopped material for 10 min at 37°C in a buffered salt medium containing papain made it possible to obtain a suspension of brain cells partially retaining their processes. Then, in 1956, Korey *et al.*[21] attempted to isolate glial cells mechanically from a buffered sucrose solution by gradient centrifugation.

The separation and isolation of brain cells must thus be based on methods that enable the dissociation of the cellular network at gap and tight junctions as well as at the basal laminae.

Three approaches are available for dissociation of brain cells:

1. Preincubation of brain slices or chopped tissue in a buffered medium containing either a proteolytic enzyme or an ion-complexing agent.[22,23]
2. Direct mechanical treatment by passing fresh brain tissue through nylon or stainless steel meshes.[24–28]
3. Preincubation of brain slices or chopped material in buffered solutions followed by mechanical disruption.[18]

Initially McIlwain[20] demonstrated that incubation of sliced or chopped brain material for 10 min at 37°C in a buffered salt solution containing papain was effective in liberation of individual brain cells. Similarly, Norton and Poduslo[23] made use of preincubation with 1% trypsin. Alternatively, Rappaport and Howze[22] used tetraphenyl boron (TPB), a potassium-complexing agent. The beneficial effect of the use of proteolytic enzymes or ion-complexing agents seems doubtful, however, probably because of uneven penetration to target areas, i.e., to the junctions, and because of irreversible effects on the cellular surfaces.

This view was supported by the systematic study by Hamberger *et al.*[18] who preincubated tissue for 30 min at 37°C in flasks gassed with 100% oxygen containing a medium: Tris HCl 35 mM (pH 7.6), NaCl 120 mM, Na-phosphate buffer 5mM (pH 7.6), glucose 20 mM, $MgCl_2$ 2.5 mM, ADP 2.5 mM, and Ficoll (Pharmacia) 2%. After incubation for 30–120 min, the brain cells were further liberated by mild mechanical disruption by passing the tissue material 10–15 times through 100-μm nylon mesh attached to a plastic syringe. Hamberger *et al.*[18] demonstrated that the buffered salt solution gave larger mean cellular size, better retained cell processes, and more intact nuclei than a 0.32 M sucrose solution. It was also demonstrated that addition of enzymes to the incubation buffer (DNase, trypsin, neuraminidase, hyaluronidase) did not have any beneficial effect. This is, however, a point on which authors do not agree. Liberated DNA may cause liberated cells to adhere; therefore, some authors recommend use of DNase. Some authors, e.g., Wilkin *et al.*,[29] make use of sequential treatment with trypsin, DNase, and soya bean trypsin inhibitor. Others favor the addition of tetraphenylboron (TPB) which, however, may have direct toxic effects.[30,31]

3.3. Principles for Separation of Dissociated Brain Cells

3.3.1. Elementary Background

Ultracentrifugation in step gradients or continuous gradients of sucrose and/or Ficoll has been used to separate neurons from glial cells.[32] The major obstacle to the isolation of these two cellular types is the admixture of blood capillaries and ependymal cells. Hamberger *et al.*[18] solved this problem by step-gradient centrifugation in 10%, 12.5%, 15%, 20%, and 30% Ficoll and a mixture of 20% Ficoll and 25% sucrose for 120 min at 55,000 g. In this gradient, membranes, myelin, axons, and undisrupted tissue went to the top. The glia were enriched in 12.5% Ficoll. The unseparated neuron–glia complexes as well as the blood capillaries were located in the zone corresponding to 15–30% Ficoll. The enriched neuronal fraction was at 30% Ficoll, and free cell nuclei went to the bottom (Fig. 1).

3.3.2. Isolation of Neurons and Glia from Whole Brains or Cerebellum

Disrupted brain cells from the cortex may be used as a source for isolation of neurons and astrocytes. Two different gradient materials, i.e., sucrose and/ or Ficoll, may be used. A Ficoll gradient may be used to enrich neurons and glia (astrocytes) in zones corresponding to 30% and 12.5% Ficoll, respectively[33,34] (see above) (Figs. 2, 3).

This separation principle has been further elaborated by Farooq and Norton[35] and Nagata *et al.*[34] Farooq and Norton[35] disrupted the brain material by

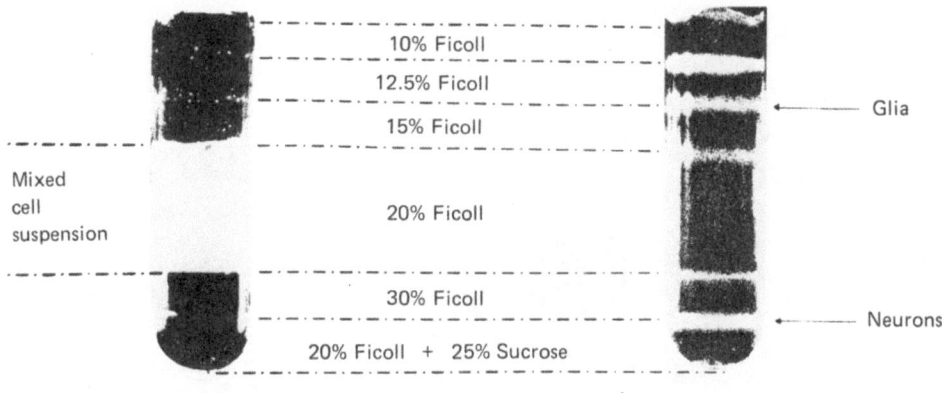

Fig. 1. Ficoll–sucrose density gradient used for the separation of neuronal and glial cells by the method of Blomstrand and Hamberger. The mixed cell suspension is applied in 20% Ficoll in the central part of the gradient (left). The fractions obtained after centrifugation for 120 min at 55,000 *g* (right) are, from the top, membranes, myelin, axons, undisrupted tissue, smaller membrane, glial-enriched fraction, unseparated neuronal and glial cells, blood capillaries, neuronal and glial cells, neuron-enriched fraction, and free cell nuclei. From Hamberger *et al.*[18]

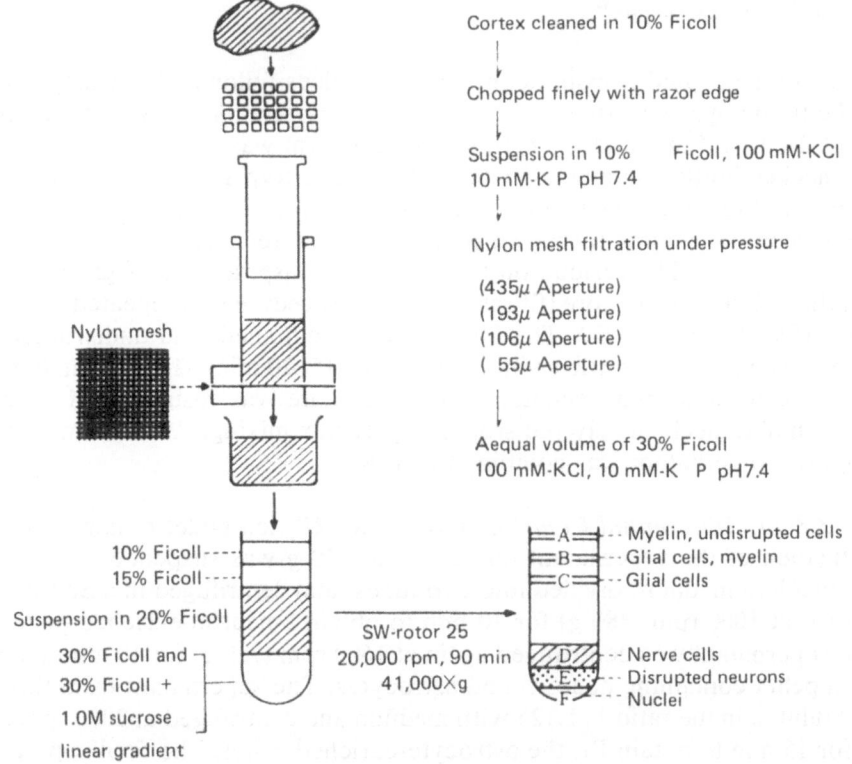

Fig. 2. A schematic diagram of the procedures for separating cells from adult cerebral cortex from rats. From Nagata *et al.*[34]

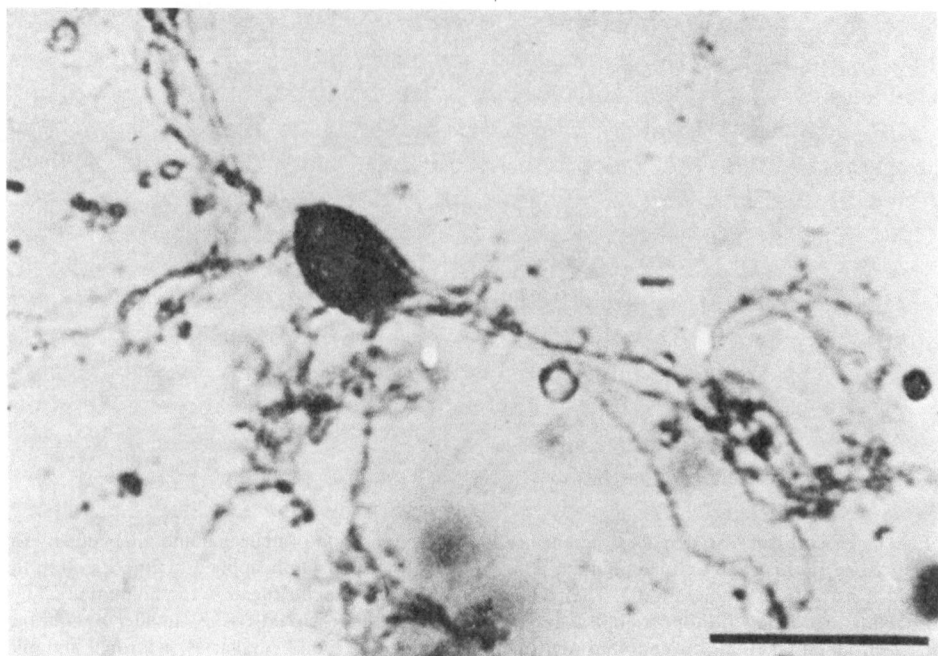

Fig. 3.   A glial cell with finely arborized processes, stained with acidic thionine. The bar is equal to 5 μm. From Nagata *et al.*[34]

means of acetylated trypsin acting on brain slices followed by disaggregation of the tissue by successive aspiration steps. A nozzle having a diameter at the tip of 2.2 to 2.4 mm and a length of 3.4 to 3.5 cm was connected to a 250-ml two-necked bottle containing 30 ml medium. The trypsinized brain slices were aspirated through the nozzle into the bottle with a slight vacuum. The coarse suspension was poured through a 40-mesh nylon screen having openings of 420 μm (filtrate F_1). The residue on the screen was aspirated through the nozzle and then filtered once more (filtrate F_2). The procedure was repeated two more times (filtrates F_3 and F_4). The filtrates were combined, and undisrupted material was allowed to settle spontaneously for 15–20 min. The supernatant (S_1) of free cells was then decanted. The precipitate was resuspended in 20 ml medium and mechanically redisrupted by vortex mixing. After 15 min of settling, the supernatant was combined with S_1.

3.3.2a. Differential Centrifugation. The cell-rich pellet obtained by centrifugation of the supernatant (cf above) at 720 *g* was suspended in 64 ml of 7% Ficöll in medium, divided into two tubes, and centrifuged in a Sorvall HG-4 rotor at 1000 rpm (280 *g*) for 10 min to obtain a neuron-enriched pellet, P_1. The supernatant was centrifuged again at 1600 rpm (720 *g*) for 10 min to obtain P_2, a pellet containing neurons and astrocytes. The supernatant from this step was diluted in the ratio 1:1.125 with medium and centrifuged at 2000 rpm (1120 *g*) for 15 min to obtain P_3, the astrocyte-enriched pellet. The final supernatant was discarded.

3.3.2b. Density Gradient Centrifugation. Each of the pellets, P_1, P_2, and P_3, was suspended gently in 38 ml of 7% Ficoll in medium, and each suspension was divided and layered onto two discontinuous gradients. These were made up in 39-ml tubes of the Spinco SW-27 rotor and consisted of (from the bottom up) 5 ml each of 32% Ficoll, 28% Ficoll, 22% Ficoll, and 10% Ficoll, all made up in medium. These tubes were contrifuged at 8000 rpm (8500 g) for 5 min. The layers at each interface and the pellets were removed with a Pasteur pipet. The composition of the fractions is as described in Table I.

From cerebellum, Yanaghiara and Hamberger[36] were able by means of the Ficoll gradient technique to enrich Purkinje and granule cells. Thus, by means of a discontinuous Ficoll gradient (from the bottom: 1.5 M sucrose, 40, 30, 23, 15, 12, and 10% Ficoll) and by centrifugation at 81,000 g (4°C, 100 min), the Purkinje cells were localized at 23% Ficoll, and the granule cells at 40% Ficoll. The granule cells but not the Purkinje cells could be further purified by a subsequent centrifugation in a Ficoll continuous density gradient (20–35%). A similar method has been used by Hazama and Uchimura.[37]

The second gradient type utilizes sucrose as a density medium. Iqbal and Tellez-Nagel[38] demonstrated that mechanically processed chopped brain tissue could be separated into neuron- and glia-enriched fractions in a discontinuous sucrose gradient (50, 45, 40, and 35% sucrose at 4500 g for 10 min). Here, the

Table I
Particle Composition of Fraction from Discontinuous Ficoll Gradients[a]

Fraction[b]	Composition
P_1/10	Broken-off processes and small debris
P_1/22[c]	Astrocytes, clumps of undisrupted tissue, a few neurons
P_1/28	Small neurons, nuclei, oligodendroglia, capillary fragments, erythrocytes
P_1/32[c]	Mostly neurons with extensive processes; some capillaries, cell fragments, and nuclei
P_1/P	Purified neuronal perikarya with little contamination
P_2/10	Broken-off processes
P_2/22[c]	Astrocytes, neurons, and clumps of undisrupted tissue
P_2/28	Small neurons, oligodendroglia, erythrocytes, free nuclei, capillary fragments
P_2/32[c]	Similar to P_1/32, mostly neurons with processes, some capillaries, occasional astrocytes and oligodendrocytes
P_2/p[c]	Similar to P_1/p, purified neuronal perikarya with little contamination
P_3/10	Broken-off processes
P_3/22[c]	Astrocytes are the major cell type; some free processes, capillary fragments, debris; very few neurons
P_3/28	Few cells of any kind
P_3/32	Few cells of any kind
P_3/p	Few cells of any kind

[a] From Farooq and Norton.[35]

[b] The fractions are designated by a term indicating from which pellet they are derived and on which layer they occur. For example, P_1/32 is the interface layering out on 32% Ficoll from the gradient of the P_1 pellet, whereas P_2/p is the pellet fraction from the gradient centrifugation of P_2.

[c] Fractions selected for more extensive analysis.

glia cells were enriched in the 35–40% interface (contaminated with large neurons and few capillaries). The neurons were localized in the upper (40–45%) sucrose and in the lower (45–50%) sucrose interface fraction (90–95% pure neurons). The glia fraction was further processed on a second sucrose gradient: 40% and 35% sucrose. Here, the glia cells were at the 35–45% interface. The method of Iqbal and Tellez-Nagel gave the highest cellular yields on young rat brains as compared to adult rat brains, and the method could also be used for isolation of brain cells from human autopsy brains. A similar method for the separation and isolation of neurons, astrocytes, and oligodendrocytes was elaborated by Chao and Rumsby.[39]

3.3.3. Oligodendroglia Cells

In 1967, Fewster et al.[40] demonstrated that oligodendroglia could be obtained from mechanically disrupted white matter in 20% Ficoll on a mixed gradient of 1.5 ml 30% Ficoll and 0.8 ml 1.35 M sucrose (lower layer). The oligodendroglia were enriched in the interface between 30% Ficoll and 1.35% M sucrose (Fig. 4). Studies by Fewster et al.[41] revealed that the oligodendroglia could also be isolated on a pure Ficoll discontinuous gradient consisting of 4 ml 43% and 47 ml 30% Ficoll and 10 ml of disrupted brain suspension in 20% Ficoll. Here, the oligodendroglia were enriched at the 30–45% Ficoll interface. The cellular yield was twice as high in suspensions from fetal brains as in those from adult brains.

By means of the sucrose gradient method of Iqbal and Tellez-Nagel,[38]

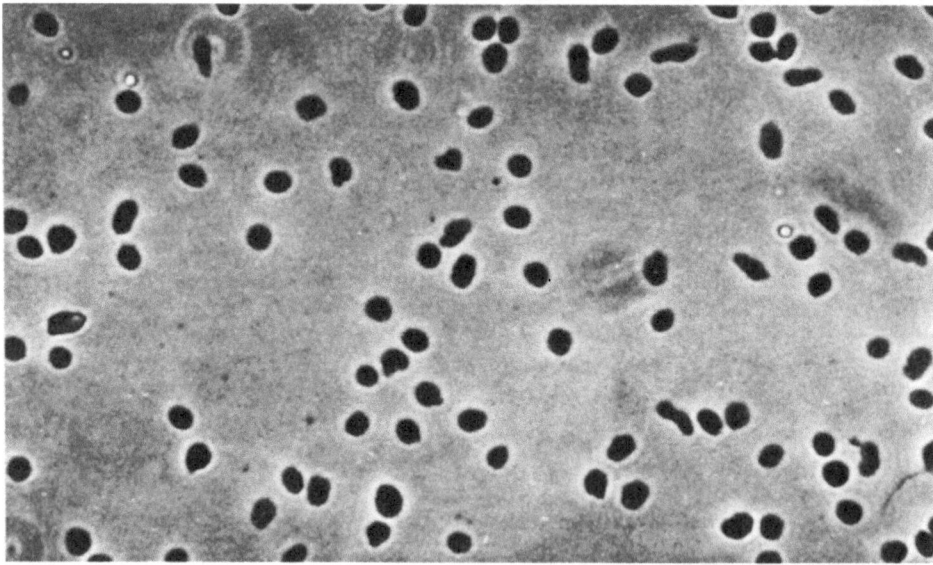

Fig. 4.   Oligodendrocytes from fetal brain. Oligodendrocytes prepared from bovine centrum ovale. Photographed in phase contrast from interface layer removed after differential centrifugation. Magnification, ×330. From Fewster et al.[40]

Freysz *et al.*[42] isolated the oligodendroglia at the 40–45% sucrose interface.[43] Similarly, Poduslo[45] and Poduslo and Norton[44] isolated oligodendroglia in a discontinuous sucrose gradient (1.55, 1.40, 0.9 M sucrose) at the 1.55–1.4 M interface. The method has been further elaborated by pretreatment of the brain sample with trypsin and DNase.[46]

3.3.4. Isolation of Brain Microvessels

As mentioned above, the vascular system of the brain is of the fenestrated or the continuous type. The microvessels are defined as those having a diameter less than 300 μm. Since many pharmacologically active agents as well as nutrients and infectious agents may cross the microvascular walls of the CNS, interest in these entities has been increasing in recent years. Techniques for isolation of CNS microvessels have relied on the ability of brain microvessels to resist homogenization and withstand forces used for separating glia and neuronal cells (see above).[47–54] The principle for isolation of microvessels in these methods is as described by Brendel *et al.*[50] Mechanically disrupted brain tissue is passed through successively smaller nylon screens. The filtrate retained on the 75-μm nylon mesh screen was the microvessels. The microvessels thus isolated seem to have intact basement membranes but defects in the structural integrity of the endothelial cells.[55]

Recently, Williams *et al.*[56] described the isolation of viable endothelial cells with minimal contamination by pericytes, neurons, and glia. Brain tissue (2 mm^3) suspended in HEPES buffer was treated with collagenase (30 min, 37°) and then mechanically disrupted and centrifuged at 1000 *g* for 20 min. The floating cake was isolated, resuspended, recentrifuged, and once more treated with collagenase. After resuspension in HEPES buffer containing 10 g/liter bovine serum albumin, a crude microvascular fraction was pelleted at 100 *g* (5 min). The microvessels were freed of erythrocytes, pericytes, and nuclei by passing through a column filled with 0.45-mm diameter glass beads. The column exclusively retained the microvessels.

White[57,58] isolated microvessels from brain slices preincubated with Krebs–Ringer–phosphate buffer (pH 7.4, 37°) for 25 min. After washing in 0.32 M sucrose, the slices were homogenized in 2.0 M sucrose. The material that floated on the top of 2.0 M sucrose was suspended in 0.32 M sucrose containing 1 mM MgCl$_2$ and centrifuged at 600 *g* for 5 min. The sediment was resuspended in 10 ml 45% (w/w) sucrose and layered on 4 ml of 2.0 M sucrose containing 1 mM MgCl$_2$. The mixture was overlaid with 0.32 M sucrose and centrifuged at 64,000 *g* for 2 hr. The capillaries were now located in the 45%/2.0 M sucrose. This method permitted tracer experiments on protein synthesis in the microvessels, since radioactive amino acids were added to the preincubation mixture (Fig. 5).

3.3.5. Isolation of Schwann Cells

The isolation of this type of supporting cell of the PNS represents a special problem since the Schwann cell is specifically stimulated by the associated

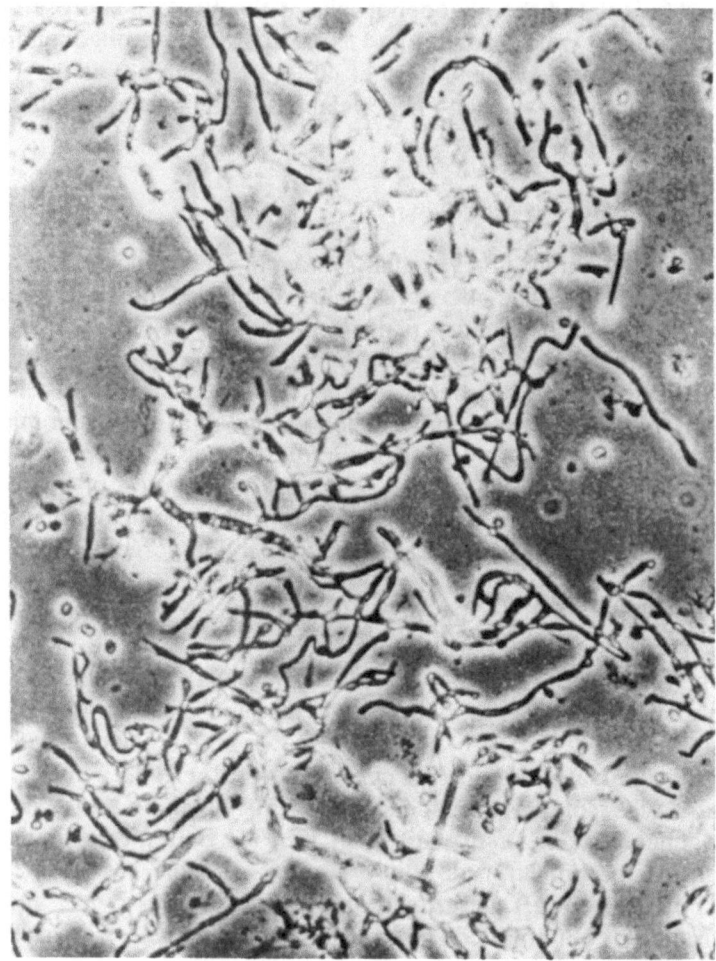

Fig. 5. The appearance of the microvascular fraction in the light microscope. Microvascular fraction was resuspended in 0.32 M sucrose. Phase contrast at a magnification of ×125. From White.[58]

axon.[59] Spencer et al.[59] demonstrated the possibility of isolating neuronally nonstimulated Schwann cells by nerve transection. This initiates a proliferation of Schwann cells parallel to the disappearance of the axon and the myelin. The Schwann cells may be harvested from the endoneuronal tissue of the nerve stumps 5–12 weeks after nerve transection. As reported by Wood,[60] Schwann cells may also be isolated by culturing dorsal root ganglia, because in certain media, Schwann cell proliferation is stimulated, unlike that of other cell types.

3.3.6. Isolation of Subsets of Brain Cells

Traditional techniques involving density gradient centrifugation of mechanically and/or enzymatically disintegrated nervous tissue have made it possible to isolate fractions enriched in neurons and glial cells. However, these techniques are not able to distinguish and separate subtypes with specific surface receptors such as certain gangliosides, specific intracellular enzymes (e.g., esterases), or certain membrane-bound proteins (e.g., antibody-coated cells or glycoprotein surface receptors). Cell sorting based on these specificities are now available. Three major techniques may be used: (1) fluorescence-activated cell analysis, (2) the magnetic microsphere technique, and (3) affinity chromatography.

3.3.6a. Fluorescence-Activated Cell Analysis. Hulett *et al.*[61] developed the fluorescence-activated cell analysis principle. Fluorescent-positive cells are separated from fluorescent-negative cells.

Cells are separated on the basis of whether they have the esterase that hydrolyzes fluorescein diacetate (FDA). Fluorescein diacetate can enter the cells, and if the esterase is present, the substrate is split to free fluorescein which cannot pass the cell membrane. The esterase-positive cells thus become fluorescent (fluorochromasia). After incubation with FDA, the esterase-positive cells are separated from the esterase-negative cells by passing the cell suspension through a small glass nozzle where a liquid stream is formed. The stream is darkfield illuminated by an exciting beam of blue light from a mercury arc. The yellow–green fluorescence emitted from the esterase-positive cells is focused by a microscope onto a photomultiplier tube through a yellow barrier filter. When fluorescent cells activate the photomultiplier, a charging voltage pulse is applied to the stream which now breaks into droplets because of low-power ultrasonic vibration. The cells that are esterase positive are now separated from the main stream of cells by deflection by an electric field between a pair of statically charged deflection plates.

3.3.6b. The Magnetic Microsphere Technique. Separation based on the presence or absence of certain membrane receptors may also be utilized in the magnetic microsphere technique of Kronick *et al.*[62] Here, magnetic hydrogel microspheres are made from allylamine, isothiocyanate, hydroxymethacrylate, methacrylylic acid admixed with ammonium persulfate, and powdered magnetite. During stirring, microspheres are formed of which 99% are magnetic and are separable by a divergent 5000-G magnet. The microspheres are reacted with a spacer group of diaminoheptane and thereafter with cholera toxin which is a receptor for ganglioside G_{M1}. The labeled microsphere beads are incubated with a brain cell suspension. The ganglioside-positive cells attached to the microspheres may be separated from free microspheres by centrifugation in fetal calf serum. The separation of the microsphere-bound cells from the unbound cells is performed by passing them by gravity flow through a 1-mm polyvinyl chloride tube wound six times around the circumference of the pole pieces of an 8-in electromagnet. The divergence of the field across these edges was 10^3G/cm from 10^4G.

3.3.6c. Affinity Chromatography. Affinity chromatography with immobilized receptor, i.e., specific antibodies, may also be used. Although simple

in principle, this technique suffers from false absorption phenomena,[63] but it
has been used for separation of neurons from chick spinal cord.[64]

3.3.6d. Practical Performance of Subset Separation. By combination of
one of the abovementioned techniques with methods already mentioned for
enrichment of neurons and glia cells, Campbell and colleagues[63,65] have separated cells from cerebellum into subsets.

The basic principles are as mentioned above, but in addition, use is made
of the fact that antibodies formed in experimental animals or by the hybridoma
technique are able to differentiate subsets of brain cells. Schachner, *et al.*[66]
Poduslo *et al.,*[67] and Campbell and colleagues[63,65] separated cells from trypsin-
and DNase-treated cerebellum in a discontinuous gradient of bovine serum
albumin (10, 15, and 31%). Three fractions were isolated: A (interface to 10%
albumin), B (10–15% interface), and C (15–31% interface).

Indirect immunofluorescence studies for the presence of the glial fibrillary
acidic protein and the S-100 protein revealed fraction B to be enriched in glia.
Fluorescence activation analysis of cells from fractions B and C coupled with
a fluorescence-labeled anti-corpus-callosum antibody revealed 68–72% of the
cells in fractions A and B but only 6.1–10.1% of fraction C to be antibody
positive. Campbell's studies[63,65] may thus be promising with regard to further
subtyping of brain cells which now may be enriched by combination of gradient
centrifugation and physical and immunochemical methods.

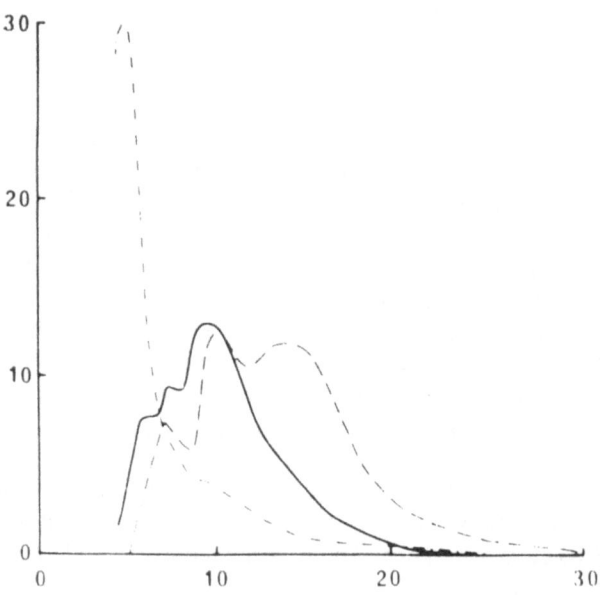

Fig. 6.   Size distribution of mixed cell suspension (——), neuronal fraction (– – – –), and glial
fraction (–·–·–·–). Abscissa, diameter (μm); ordinate, % population. From Hamberger.[18]

4. CHARACTERIZATION OF SEPARATED AND ISOLATED BRAIN CELLS

4.1. Morphology

The morphological criterial mentioned in the Introduction (Section 1) should be used. However, since isolated brain cells may change size because of osmotic shock and chemical treatment, the environmental factors (e.g., composition of medium) should be strictly defined. In the medium used by Hamberger,[18] the rabbit neuronal fraction displayed a size distribution with peak values corresponding to 10–20 μm, whereas the size distribution of glia cells revealed a peak at 7 μm. Thus, neurons are larger cells than glia cells (Fig. 6). Neurons have a large nucleus, a single nucleolus, and abundant cytoplasm. The astrocytes have much smaller perikarya, a prominent nucleus, little visible cytoplasm, but branched processes. Oligodendroglia are the smallest cell type, similar in size to erythrocytes; it is difficult to distinguish these cells from nuclei, since they only have a small rim of cytoplasm. They may have a small process (Fig. 4).

4.2. Biochemical Criteria

4.2.1. Use of Enzymes

One of the few enzymes predominantly localized to glia may be carbonic anhydrase (Table II).[34,68] Enzymatic activities specific for brain endothelial cells are alkaline phosphatase (E.C. 3.1.3.1), γ-glutamyltranspeptidase (E.C. 2.3.2.1),[69,70] and 2′,3′-cyclic nucleotide 3′-phosphohydrolase (Table III). The activity of 5′-nucleotidase (E.C. 3.1.3.5) which is high in neuronal cells is minimal in CNS microvessels. Other enzymatic activities such as lysosomal enzymes occur in both neurons and glia.[42]

Table II
*Carbonic Anhydrase Activity in Separated
Neuronal and Glial Fractions*

Experiment	C (Glial)	D (Neuronal)
1	4.51×10^{-5}	2.12×10^{5}
2	10.20×10^{-5}	1.47×10^{5}
3	16.35×10^{-5}	0.78×10^{5}
4	6.12×10^{-5}	1.17×10^{5}
5	4.92×10^{5}	2.35×10^{5}
Average	8.42×10^{5}	1.51×10^{-5}

[a] The enzyme activity was determined as the rate relative to that of diluted blood giving the same reaction velocity at pH 8.0 on the same protein basis. From Nagata *et al.*[34]

4.2.2. Proteins

The so-called α-albumin is related to the glial fibrillary acid protein. It is a marker of astroglia.[71,72] Gheuens *et al.*[73] have used a "sandwich" radioimmunoassay for evaluation of separation of astrocytes from neurons. The S-100 protein is also associated with glia.[33]

4.2.3. Lipids

Apart from galactolipids which are enriched in oligodendroglia,[74] all other lipids seem to occur in both glia and neurons.

5. APPLICATION OF PROCEDURES FOR ISOLATION OF BRAIN CELLS

In addition to what has already been mentioned, isolated brain cells may be used for several purposes including research on metabolic and structural properties of enriched (i.e., nearly homogeneous) brain cell populations. Furthermore, the absence of a blood–brain barrier may permit toxicological and pharmacological studies. Although the present chapter is not comprehensive, Table IV exemplifies these applications.

Important discoveries made with the cell isolation technique include the proof that myelin basic protein occurs in oligodendroglia[41,75] and that oligodendroglia preferentially incorporate linoleic acid over linolenic acid.[75] Other examples are the finding that herpesvirus types 1 and 2 have higher affinity for glia than for nerve cell perikarya[77] and that isolated neurons show more depressed protein synthesis than glia after methyl mercury intoxication.[78]

6. CONCLUSION

Methods are now available for isolation of separate fractions enriched in neurons, Purkinje cells, glia (antrocytes), oligodendroglia, and Schwann cells.

Table III
2′,3′-Cyclic Nucleotide 3′-Phosphohydrolase Activity in Separated Neuronal and Glial Fractions (Unit/mg Protein)[a]

Fraction	Cerebral cortex	White matter	
C (Glial)	3.04	C upper phase[b]	2.96
D (Neuronal)	0.18	C pellet[b]	1.36
			0.12

[a] From Nagata *et al.*[34]
[b] Fraction C, obtained from the cerebral white matter suspension, was further separated by layering the sample on 0.8 M sucrose and centrifuging at 50,000 *g* for 15 min.

Table IV
Survey of Some Applications of Isolated Nervous Cells for Basic and Applied
Research

Cell type	Object and/or result of study	Authors
Neurons and glia	Methyl mercury intoxication: neuronal protein synthesis more affected than that of glia cell	Syversen[78]
Neurons (Purkinje cells)	Herpesvirus affinity higher for glia cells than for neurons	Vahlne et al.[77]
Neurons from locus		
Oligodendroglia	Synthesis of basic protein	Fewster et al.[41] McDermott et al.[75]
	Biosynthesis of long-chain fatty acid preferentially from linolenic acid rather than linolenic acid	Fewster et al.[76]
	Developmental changes	Cohen et al.[79]
	Actions of transmitters	Holden et al.[80]
Schwann cells	Interaction with axons	Hall[81] Bunge et al.[82]

The glia cells are now available with nearly intact cellular processes in contrast to neurons where breakage of neuronal processes occurs. The cell separation techniques utilize mechanical and/or enzymatic disintegration of nervous tissue followed by density and/or gradient centrifugation. Further refinement of the techniques involves use of physical and immunochemical methods.

REFERENCES

1. Schwann, T., 1939, *Mikroscopische Untersuchungen*, p. 174, cited in Greenfield, J. G., Blackwood, W., Meyer, A., McMenemey, W. H., and Norman, R. M., 1960, *Neuropathology*, Edward Arnold, London.
2. Cajal, S. R. y., 1897, *Rev. Trimest. Micrograph. Madrid*, 2:29.
3. Caspar, D. L. D., Goodenough, D. A., Makowski, L., and Phillips, W. C., 1977, *J. Cell Biol.* 74:605–628.
4. Goodenough, D. A., Paul, D. L., and Culbert, K. E., 1978, *Birth Defects* 14:83–97.
5. Loewenstein, W. R., Kanno, Y., and Socolar, S. J., 1978, *Fed. Proc.* 37:2645–2650.
6. Unwin, P. N. T., and Zampighi, G., 1980, *Nature* 283:545–549.
7. McNutt, N. S., and Weinstein, R. S., 1973, *Prog. Biophys. Mol. Biol.* 26:47–101.
8. Staehelin, L. A., 1974, *Int. Rev. Cytol.* 39:191–283.
9. Baumann, N., and Hauw, J. J., 1979, *Pathol. Biol.* 27:169–177.
10. Møller, M., van Deurs, B., and Westergaard, E., 1978, *Cell Tissue Res.* 195:1–15.
11. Reese, T. S., and Karnovsky, M. J., 1967, *J. Cell Biol.* 34:207–217.
12. Friede, R. L., Flemming, L. M., and Knoller, M., 1963, *J. Neurochem.* 10:263–277.
13. Dunning, H. S., and Wolff, H. G., 1936, *Trans. Am. Neurol. Assoc.* 62:150–154.
14. Dunning, H. S., and Wolff, H. G., 1937, *J. Comp. Neurol.* 67:433–450.
15. Guarnieri, M., Krell, L. S., McKhann, G. M., Pasternak, G. W., and Yamamura, H. I., 1975, *Brain Res.* 93:337–342.
16. Hansen, A. J., 1977, *Acta Physiol. Scand.* 99:412–420.
17. Bocci, V., 1966, *Nature* 212:826–827.
18. Hamberger, A., Eriksson, O., and Norrby, K., 1971, *Exp. Cell Res.* 67:380–388.
19. Hyden, H., 1959, *Nature* 184:433.
20. McIlwain, H., 1954, *Proc. Univ. Otago Med. School.* 32:17.

21. Korey, S. R., Orchen, M., and Brtoz, M., 1958, *J. Neuropathol. Exp. Neurol.* **17**:561.
22. Rappaport, C., and Howze, G. B., 1966, *Proc. Soc. Exp. Biol.* **121**:1010.
23. Norton, W. T., and Poduslo, S. E., 1970, *Science* **167**:1144.
24. Azcurra, J. M., Lodin, Z., and Sellenger, O. Z., 1969, *Abstr. Second Meeting Int. Soc. Neurochem.*, p. 76.
25. Flangas, A. L., and Bowman, R. E., 1968, *Science* **161**:1025.
26. Rose, S. P. R., 1967, *Biochem. J.* **23**:102.
27. Satake, M., and Abe, S., 1966, *J. Biochem. (Tokyo)* **57**:72.
28. Sellinger, S. P. R., Azcurra, J. M., Johnson, D. E., Ohlson, W. G., and Lodin, Z., 1971, *Nature* **230**:253.
29. Wilkin, G. P., Balazs, R., Wilson, J. E., Cohen, J., and Dutton, G. R., 1976, *Brain Res.* **115**:181–199.
30. Friedman, T., and Epstein, C. S., 1967, *Biochim. Biophys. Acta* **138**:622.
31. Utsumi, K., and Packer, L., 1907, *Arch. Biochem.* **122**:509.
32. Poduslo, E. and Norton, W. T. 1975, *Methods Enzymol.* **35**:561–579.
33. Nagata, Y., Mikoshiba, K., and Tsukada, Y., 1974, *J. Neurochem.* **22**:493–503.
34. Nagata, Y., Mikoshiba, K., and Tsukada, Y., 1976, *Asian Med. J.* **19**:13–43.
35. Farooq, M., and Norton, W. T., 1978, *J. Neurochem.* **31**:887–894.
36. Yanagihara, T., and Hamberger, A., 1973, *Brain Res.* **59**:445–448.
37. Hazama, H., and Uchimura, H., 1974, *Exp. Cell Res.* **87**:412–415.
38. Iqbal, K., and Tellez-Nagel, I., 1972, *Brain Res.* **45**:296–301.
39. Chao, S. W., and Rumsby, M. G., 1977, *Brain Res.* **124**:347–351.
40. Fewster, M. E., Scheibel, A. B., and Mead, J. F., 1967, *Brain Res.* **6**:401–408.
41. Fewster, M. E., Einstein, E. R., Csejtey, J., and Blackstone, S. C., 1974, *Neurobiology* **4**:388–401.
42. Freysz, L., Farooqui, A. A., Adamczewska-Goncerzewicz, Z., and Mandel, P., 1979, *J. Lipid Res.* **20**:503–507.
43. Iqbal, K., Grundke-Iqbal, I., and Wisniewski, H. M., 1977, *J. Neurochem.* **28**:707–716.
44. Poduslo, S. E., and Norton, W. T., 1972, *J. Neurochem.* **19**:727–736.
45. Poduslo, S. E., 1975, *J. Neurochem.* **24**:647–654.
46. Snyder, D. S., Raine, C. S., Farooq, M., and Norton, W. T., 1980, *J. Neurochem.* **34**:1614–1621.
47. Landers, J. W., Chason, J. L., Gonzalez, J. E., and Palutke, W., 1962, *Lab. Invest.* **11**:1253–1259.
48. Siakotos, A. N., and Fleischer, S., 1969, *Lipids* **4**:234–239.
49. Joo, F., and Karnushina, I., 1973, *Cytobiology* **8**:41–48.
50. Brendel, K., Meezan, E., and Carlson, E. C., 1974, *Science* **185**:953–955.
51. Goldstein, G. W., Wolinsky, J. S., Csejtey, J., and Diamond, I., 1975, *J. Neurochem.* **25**:715–717.
52. Mrsulja, B. B., Mrsulja, B. J., Fujimoto, T., Klatzo, I., and Spatz, M., 1976, *Brain Res.* **110**:361–365.
53. Sessa, G., Orlowski, M., and Green, J. P., 1976, *J. Neurobiol.* **7**:51–61.
54. Selivonchick, D. P., and Roots, B. I., 1977, *Lipids* **12**:165–169.
55. DeBault, L. E., Kahn, L. E., Frommes, S. P., and Cancilla, P. A., 1979, *In Vitro* **15**:473–478.
56. Williams, S. K., Gillis, J. F., Matthews, M. A., Wagner, R. C., and Bitensky, M. W., 1980, *J. Neurochem.* **35**:374–381.
57. White, F. P., 1979, *Neuroscience* **5**:173–178.
58. White, F. P., 1980, *J. Neurochem.* **35**:88–94.
59. Spencer, P. S., Weinberg, H. J., Krygier-Brévart, V., and Zabrenetzky, V., 1979, *Brain Res.* **165**:119–126.
60. Wood, P. M., 1976, *Brain Res.* **115**:361–375.
61. Hulett, H. R., Bonner, W. A., Barrett, J., and Herzenberg, L. A., 1969, *Science* **166**:747–749.
62. Kronick, P. L., Campbell, G. L., and Joseph, K., 1978, *Science* **200**:1074–1076.
63. Campbell, G. L., 1979, *Fed. Proc.* **38**:2386–2390.
64. Dvorak, D. J., Gipps, E., and Kidson, C., 1978, *Nature* **271**:564–566.
65. Campbell, G. L., Schachner, M., and Sharrow, S. O., 1977, *Brain Res.* **127**:69–86.

66. Schachner, M., Wortham, K. A., Ruberg, M. Z., Dorfman, S., and Campbell, G. L., 1977, *Brain Res* **127**:87–97.
67. Poduslo, S. E., McFarland, H. F., and McKhann, G. M., 1977, *Science* **197**:270–272.
68. Giacobini, E., 1961, *Science* **134**:1524–1525.
69. Orlowski, M., Sessa, G., and Green, J. P., 1974, *Science* **184**:66–68.
70. Djuricic, B. M., and Mrsulja, B. B., 1977, *Brain Res.* **138**:561–564.
71. Bignami, A., and Dahl, D., 1977, *J. Histochem. Cytochem.* **25**:466–469.
72. Ludwin, S. K., Kosek, J. C., and Eng, L. F., 1976, *J. Comp. Neurol.* **165**:197–207.
73. Gheuens, J., Noppe, M., Karcher, D., and Lowenthal, A., 1980, *Neurochem. Res.* **5**:757–768.
74. Norton, W. T., Abe, T., Poduslo, S. E., and DeVriest, G. H., 1975, *J. Neurosci. Res.* **1**:57–75.
75. McDermott, J. R., Iqbal, K., and Wisniewski, H. M., 1977, *J. Neurochem.* **28**:1081–1088.
76. Fewster, M. E., Ihrig, T., and Mead, J. F., 1975, *J. Neurochem.* **25**:207–213.
77. Vahlne, A., Svennerholm, B., Sandberg, M., Hamberger, A., and Lycke, E., 1980, *Infect. Immun.* **28**:675–680.
78. Syversen, T. L. M., 1977, *Neuropathol. Appl. Neurobiol.* **3**:225–236.
79. Cohen, J., Balazs, R., Hajos, F., Currie, D. N., and Dutton, G. R., 1978, *Brain Res.* **148**:313–331.
80. Holden, J. S., Suter, C., and Usherwood, P. N. R., 1978, *J. Physiol. (Lond.)* **276**:4P–5P.
81. Hall, S. M., 1978, *Neuropathol. Appl. Neurobiol.* **4**:165–176.
82. Bunge, M. B., Williams, A. K., Wood, P. M., Uitto, J., and Jeffrey, J. J., 1980, *J. Cell Biol.* **84**:184–202.

Principles of Compartmentation

N. M. van Gelder

1. INTRODUCTION

Compartmentation in biochemical terms refers to the separation of metabolic processes that together constitute a metabolic cycle. Examples of such cycles come readily to mind and include protein synthesis, the urea cycle, and the metabolism of glutamic acid. Typically, all cycles require a number of precisely sequenced metabolic steps which are not incorporated into a single anatomic entity. Therefore, any substance not immediately broken down by only one enzyme to CO_2, H_2O, or another excretion product may be considered to exhibit a compartmentalized metabolism. Each enzyme constitutes a distinct catalytic surface and hence a unique structure.[1,2] In the course of its metabolism, a substance may be passed on many times from one specialized affinity site to another while its molecular structure is being rearranged, added to, or broken down. Thus, in addition to involving more than one anatomic element, compartmentation must also always include some type(s) of transfer or transport system(s). This is especially true when biochemical compartmentation involves more than one intracellular membrane-enclosed organelle or different cells of the same organ.

Several reasons may be invoked to explain the need for compartmentation. At least one of these appears to be strictly chemical. Most naturally occurring substances exhibit stable chemical bonding. A sequence of catalytic surfaces with specialized recognition sites is often required to render a molecule sufficiently chemically unstable to become susceptible to nucleophilic or electrophilic modification.[3] The glycolytic steps leading to the eventual oxidation of glucose clearly fall into this category. The metabolism of glucose also exemplifies a second reason for compartmentation. Once cleavage of stable chemical bonds has occurred, the energy liberated must be incorporated into chemical storage forms that are readily available to tissues. This entails controlled electron transfer, and such processes require that the compounds implicated in this phenomenon remain closely aligned relative to one another. It is especially this

N. M. van Gelder • Centre de Recherche en Sciences Neurologiques, Département de Physiologie, Faculté de Médecine, Université de Montréal, Montreal, Québec H3C 3J7, Canada.

sequence of metabolic steps which requires compartmentation. Although the glycolytic portion of glucose metabolism involves a series of cytoplasmic enzymes that do not appear to be closely aligned, the energy-generating sequence of that metabolism needs to be compartmentalized and subcompartmentalized in mitochondria.

The ability to separate a precursor from its product by accumulating the two at different locations within the same cell or within different cells presents another advantage of compartmentation. Large quantities of precursor can be stockpiled in this manner, thereby insuring a steady but controlled rate of supply of the needed product independent of the immediate "nutritional" status of the tissue. Fats, glycogen, and many free amino acid pools all represent precursors present far in excess of those immediately required to provide energy or substrates for anabolic reactions. For substances exhibiting direct biological action once liberated, i.e., chemical messengers, the separation of the "storage" site from a "liberation" site becomes essential. This arrangement insures an adequate reserve while at the same time allowing the quantities liberated to be precisely controlled by modulation of the strength/duration parameters of the appropriate stimulus. Not surprisingly, therefore, nervous tissue more than any other type of tissue is highly specialized with respect to both anatomic and metabolic compartmentation.

This chapter does not intend to review any particular type of compartmentalized metabolism, since other sections specifically discuss many of these in detail. Rather, an attempt has been made here to present a general overview of the principles of compartmentation. The references cited evidently had to be somewhat arbitrarily selected; those listed were found helpful in the initiation of a literature search for information on several areas of particular interest.

2. ANATOMIC COMPARTMENTATION

The cytoarchitecture of the central nervous system, unlike that of most other organs, is composed of a large number of different cell types.[4] Aside from being a major separate compartment in the body, each cellular component of the CNS constitutes an anatomic compartment incorporating a series of specialized biochemical functions. Many of these functions, especially the more subtle ones, have not as yet been elucidated. Nevertheless, it is now generally accepted that major specialized functions can be assigned to broad anatomic subdivisions of the CNS and, hence, that certain biochemical events may be considered markers for these subdivisions.[5]

2.1. Blood–Brain Barrier Systems

Microcapillaries[6,10] and closely apposing (micro)glial elements together with the choroid plexus and the villi of the arachnoid membrane form this "compartment" which is primarily involved in selectively permitting access by substances to CNS parenchyma[7,8] and preventing the accumulation of toxic wastes in nervous tissue and cerebrospinal fluid.[8,9] Biochemically, this compartment is typified by a host of often asymmetric and very selective transport/exchange mechanisms.[11]

2.2. Oligodendrocytes

These glia are the principal cells responsible for the synthesis, turnover, and maintenance of myelin. To date, the two most specific biochemical markers for oligodendrocytes and their product, myelin, appear to be certain gangliosides (GM_4)[14] and basic myelin protein.[13] Whether or not the distribution of these constituents of nervous tissue is exclusively confined to oligodendrocytes or to myelin itself is not completely established. Yet, the specialized function of oligodendrocytic elements in the CNS[18] demands that a number of enzymes or proteins related to myelin metabolism be especially heavily concentrated in this structure.[15,16] That the localization of such enzymes has so far not proven to be exclusive is perhaps related to the fact that myelin is essentially a condensed representation of the normal cell membrane. Specific biochemical markers may possibly be found only at the specialized interfaces of the axolemma and the glial membrane.[12,17] Immunogenic attack occurs most readily at these sites.

2.3. Astrocytes

Together with certain microglia, these cells are found in close apposition to neurons. Immunohistochemistry,[21,23] various tissue preparations,[24] and direct biochemical studies[19,22] all provide strong evidence that the presence of glutamine synthetase (E.C. 6.3.1.2) serves as a biochemical marker for astrocytes.[20,24,25] One of the principal functions for these cells appears to be to metabolically support neuronal elements.[26–37] In addition to glucose, which serves as an almost exclusive energy substrate for neurons, the astrocytes are able to utilize a number of amino acids (e.g., leucine) and ketoacids (e.g., succinate) as energy sources. A second, equally important role is one of protecting neuronal surfaces from prolonged exposure to substances capable of modifying neuronal excitability. The astrocytic membrane is thus rich in transport systems that accumulate chemical messengers for subsequent rapid metabolic transformation.

Every type of glial cell found in the central gray of the CNS has been shown to contain high levels of carbonic anhydrase (E.C. 4.2.1.1) in contrast to neurons in which this enzyme seems to be practically absent. These types of (satellite) cells thus must play a very important role in maintaining the interstitial pH.[31,35]

2.4. Neurons

By virtue of their great diversity in shape, size, axodendritic extensions, and location, the neurons show the greatest diversity in anatomic compartmentation.[4] Unfortunately, this type of compartmentation does not appear to correspond closely to biochemical specialization which correlates more specifically with function. This realization has been brought into focus by the recent rapid advances in cytochemistry, immunohistochemistry, and autoradiography of tissue preparations representing cytoarchitectural complexities ranging from purified enzymes to intact brain regions. Many of these studies have revealed a large number of metabolic compartments in the CNS, encompassing intraneuronal as well as intercellular involvement. Although they no doubt represent a great oversimplification, neuronal compartments may be classified into three broad categories according to their metabolic function.

2.4.1. Central Support

As defined by their anatomic configuration, the perikarya of neurons are often situated quite distant from the endings of axodendritic extensions. All such extensions are provided with metabolites/precursors originating in the soma and distributed by a number of intraaxonal and intradendritic(?) transport systems.[41] Compounds requiring the entire genetic directive apparatus[48] involving the DNA–RNA–Golgi complex (promodulators, peptides, receptor subunits, enzymes) are especially apt to be transported to the appropriate sites by intraaxonal/-dendritic flow. At these sites, the "precursors" then undergo the final (bio)chemical modifications for eventual secretion, liberation on impulse generation, or incorporation into and formation of structural proteins in membranes.[52] This type of compartmentation in essence appears to exist as a consequence of the fact that peptide sequencing requires an elaborate structural guiding system[42,50] composed of the nucleus, nucleolus, Golgi apparatus–endoplasmic reticulum,[44,47] and associated ribosomal arrangements.[37]

The structural skeleton supporting these biochemical events occupies most of the perikaryal space together with the mitochondria. Protein turnover[43] must therefore be especially active in neuronal perikarya.[45,50] Compartmentation involving amino acid transport in association with free amino acid protein bound pools should be a characteristic feature of such neuronal structures.[46] Hormones, in general, influence this phenomenon on several levels: intercellular communication, differentiation, and transport.[38–40,49,51] Much of this type of compartmentalized metabolism is summarized in Fig. 1.

2.4.2. Local Support

The mitochondria, being the principle but perhaps not the only site where oxidation of glucose provides the large ATP requirements of neurons, are the type of metabolic compartment ideally suited to provide energy at any location in the widely dispersed network of cytoplasmic prolongations. Streamlined and smooth-surfaced externally, these organelles displace rapidly towards sites in neuronal extensions where the energy requirements are high: synaptic contacts. Whether this displacement is entirely a consequence of continually shifting intraaxonal streaming or whether active participation of the mitochondrion itself (chemotaxis?) occurs as well is not known. Chemotaxis, if it occurs, is most likely accomplished by enzymes on the external mitochondrial membrane

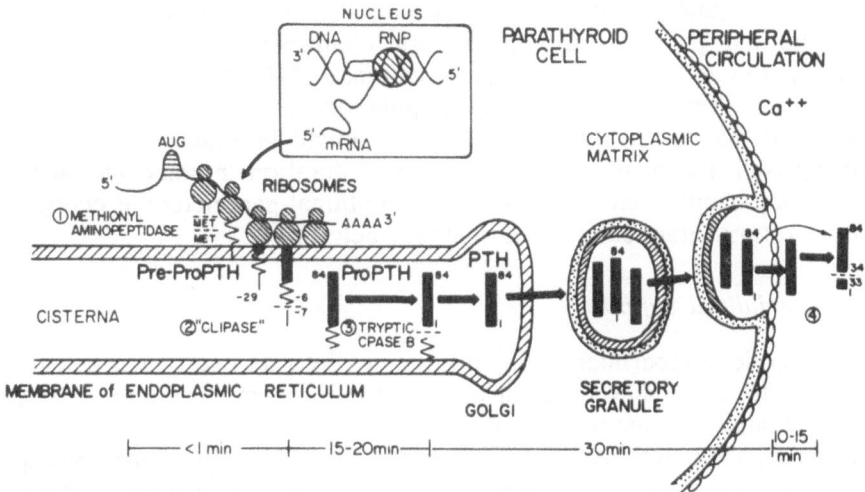

Fig. 1. Scheme depicting the proposed intracellular pathway of the biosynthesis of parathyroid hormone. Preproparathyroid hormone (preproPTH), the initial product of synthesis on the ribosomes, is converted into proparathyroid hormone (proPTH) by removal of (1) the NH_2-terminal methionyl residues and (2) the NH_2-terminal sequence (-29 through -7) of 23 amino acids during or within seconds after synthesis, respectively. The conversion of preproPTH probably occurs during transport of the polypeptide into the cisterna of the rough endoplasmic reticulum. By 20 min after synthesis, proPTH reaches the Golgi region and is converted into PTH by (3) removal of the NH_2-terminal hexapeptide. The PTH is stored in the secretory granule until released into the circulation in response to a fall in the blood concentration of calcium. The time needed for the events is given below the scheme. (From Potts *et al.*[46] with permission of authors and Academic Press, Inc.)

being attracted through the sol/gel cytoplasm to their substrates entering across the cell membranes. Besides providing energy on site, these organelles are major regulators of the intraneuronal acid–base balance (e.g., α-ketoglutarate, NH_3), redistribution of nitrogen among amino acids (transamination), submolecular group transfer, and, directly or indirectly, phosphorylation–dephosphorylation of proteins in relation to synaptic transmission.[53–57]

There is, moreover, some suggestion that the mitochondrial population within nervous tissue may not be homogeneous with respect to their metabolic properties. Costa-Tiozzo and Gal[52] showed, for example, that synaptosomal preparations of relatively high homogeneity exhibited different transport properties for L-tryptophan and norepinephrine depending on whether or not mitochondria excluded from synaptosomes were added to their preparations. Dennis and Clark[54] demonstrated similar differences between nonsynaptic and synaptic mitochondria with respect to glutamate synthesis. What is not established at this time is whether the nonsynaptic mitochondria may be further subdivided on the basis of their presence in neurons, glia, or blood vessels or whether the neuronal cell bodies contain mitochondria having different properties from those present in their synaptic terminals. These problems may be especially important when dealing with the energy reserves of the nervous system. It is quite conceivable that damage to one selective population of mitochondria might not be reflected by large changes in the overall metabolism of nervous tissue (total ATP, phosphocreatine, cAMP, etc.). Yet the selective nonreversible impairment of a small population of mitochondria could have pronounced effects on brain function. This might be one explanation why, following ischemia or hypoxia of even short duration, permanent functional damage may occur even though most conventional parameters of energy metabolism return (practically) to normal (see below).[111,115]

2.4.3. Specialized Support

Metabolic compartmentation truly comes into its own when considered in relation to the unique function of neurons: the communication of physiological signals from one nerve cell to another to produce a meaningful interpretation of, and reaction to, the environment. The complex of anatomic structures represented by the presynaptic terminal, the postsynaptic (receptor) region, and the closely apposing surrounding glial extensions can in certain aspects be considered as one metabolic unit which is compartmentalized anatomically by membranes and metabolically by a selective distribution of receptors,[77] enzymes,[65] and transport[59,62,64] and storage sites[58,65-67] at and in membranes and within organelles enclosed by these membranes.[58-68] This metabolic unit is of course of special interest to investigators studying the metabolism of synaptic transmission,[58,59,61,64,75] receptor biochemistry[71-73] including structure,[69,79] configuration,[70,74] and binding,[76,78] uptake, release, and inactivation of chemical messengers,[60,62] precursor–chemical messenger relationships, and biochemical changes in the membrane related to permeability alterations.[61]

The relevance of studies on synaptosomes arises from the fact that this cellular unit, representing the nerve terminal, can provide information on metabolic events directly connected to the physiological events: the uptake, release, and metabolism of neurotransmitter substances. The interesting development that certain drugs *in vivo* may act selectively on synaptosomes and that, for example, an increase of synaptosomal GABA may be correlated with an anticonvulsant action[64] illustrates one of the uses of studying synaptosomes. The effects of nerve terminal depolarization on transmitter metabolism can also be studied with the aid of synaptosomes. Bradford and co-workers[58,61] were able in this manner to demonstrate that synaptic metabolic events were triggered by the depolarizing process *per se*.

Levi and Raiteri,[66] in a series of papers, demonstrated that many studies using tracer quantities of neurotransmitter substances were not necessarily dealing with net uptake of such substances in nerve terminals but rather demonstrated (an enhanced) exchange of molecules across the membrane with altering physiological conditions. The distinction between exchange and net flux of a compound across membranes is essential in understanding the metabolic changes directly implicated in synaptic transmission. Ease of release may regulate the probability of a response occurring but does not necessarily determine the amplitude of the response or the period during which repetitive stimulation will elicit a response. The latter would be much more dependent on intrasynaptosomal events such as the number of biological packages available for release, precursor supply, synthesizing enzyme activity, etc. The "same" enzyme present in different cell types may be very differently regulated metabolically. Such was recently shown to be the case for glutaminase (E.C. 3.5.1.2) which, in the nerve terminal, is much more sensitive to ammonia and glutamate inhibition than is its form in cultures of glia.[65]

One of the major difficulties associated with the study of such phenomena is that the anatomic complexity of the synaptic unit makes it difficult to study individual biochemical events. Furthermore, when this is accomplished, the true regulation of the compartmentalized metabolism has been destroyed. Hence, the better the anatomic or biochemical isolation, the less the likelihood that the findings are immediately applicable to events occurring *in vivo*. It must be borne in mind, therefore, that the many methods used to study this metabolic unit in particular all have certain disadvantages. Needless to say, this does not negate the many findings, but it does require particular judgment regarding the interpretation of such findings, depending as it does on both the tissue preparations used and the particular methodology employed.

3. BIOCHEMICAL COMPARTMENTATION

The fact that metabolic compartmentation can represent a biochemical cycle that often does not correspond to recognized functional anatomic compartments has to some extent created a certain confusion. In this respect, the availability of radiochemicals of high specific activity has contributed as much to a certain dichotomy in results as it has helped in elucidating such cycles. Because, by definition, biochemical compartmentation includes the transfer or transport of substances, a clear distinction must be made between methods aimed at determining the flow of relatively few representative molecules through a sequence of metabolic steps and those aimed at determining the delay (i.e., accumulation) in this sequential flow-through process. Each such delay represents potentially the formation of a reservoir (pool) the size of which is determined by (at least) three biochemical steps: (1) the inward or supply transport rate (Fig. 2, 1); (2) the rates of intracompartmental enzymic reaction(s) if molecular alteration does occur (Fig. 2, 3); and (3) the rate of outward transport or passive/directed liberation to the next sequence of events (Fig. 2, 5). These major steps are in turn often influenced by a number of selective biochemical events. Studies aimed at investigating the various metabolic pools of a substance and their changes with physiological conditions may therefore, as a first approximation, divide the many metabolic events into two classes, each of which then represents a single step: one responsible for supplying the pool, and the other responsible for emptying the pool.[86,133]

In order to designate a substance as undergoing biochemical compartmentalized metabolism, the following minimum criteria need to apply. (1) Its total tissue (organ) content must be shown to represent at least two fractions distinguishable by differences in their metabolic fate, irrespective of whether this occurs as a consequence of anatomic barriers.[80] (2) Transfer of individual molecules from one pool to another must not be direct but rather must entail either a delay and/or a precursor–product relationship; irrespective of the nature of the separation, the metabolic connection between the two pools must exhibit a rate limit.[81,83]

3.1. The Separation of Metabolic Pools

As recounted by Ratner,[85] the availability of the stable isotopes deuterium and ^{15}N allowed biochemists for the first time to follow the biological fate of individual atoms in a molecule. Perhaps the first proof of compartmentation was obtained by Schoenheimer and his collaborators when they showed that the nitrogen of an amino acid administered to animals received a different fate and was distributed more heterogeneously than was the carbon skeleton of that same amino acid. This clearly indicated that certain portions of a molecule were selectively redistributed among other substances. The fact that the nitrogen and carbon skeleton portions of an amino acid molecule were not metabolized as a unit indicated that selective removal and separation of the α-nitrogen atom from the remaining molecule had occurred. These two molecular subunits were thus divided between different metabolic pools.

Subsequently, it became apparent that not only individual atoms of a molecule might be metabolized separately but also that the tracer molecules, when injected, did not necessarily mix homogeneously with other molecules of the same substance derived from endogenous metabolism (reviewed by Berl[81]). This latter observation in particular gave rise to the notion that separate metabolic pools of the same substance existed which, although biochemically related, were nevertheless differentially affected under specific physiological conditions.[82,84,86]

Simply stated (Fig. 2), when a labeled precursor of substance A is introduced into a homogeneous metabolic system (e.g., a set of purified enzymes), the formation of labeled A* and of a labeled metabolite derived from A* will exhibit defined rates (Fig. 2, 1,3,5). Moreover, the specific activity of that metabolite can never exceed, and usually is less, than that of compound A* or, as the case may be, of the precursor from which A* originated. Also, in such a homogeneous system, direct introduction of labeled compound A (TRACER a*) instead of its precursor (Fig. 2,2) should result in a similar relationship between substance A and its metabolite. One reason why, especially in a tissue, the specific activity of the metabolite is usually less than that of A* and of its precursor is that the label (isotope a*) is used to mark a few representative molecules of the total population (specific activity = cpm/mmol). If (unequal) endogenous pools of A (Fig. 2, 3) and its metabolite are present, then obviously a* will be diluted by endogenous A [i.e., A(a*), Fig. 2] as will the metabolite* on mixing with its unlabeled endogenous pool. To some extent, the dilution factor can be calculated by determining the quantity (pool size) of endogenous metabolites (see below, however).

Compartmentalized metabolism became indicated when on introduction of a* (Fig. 2, 6), the resulting metabolite pool (i.e. Fig. 2, 5) exhibited a higher

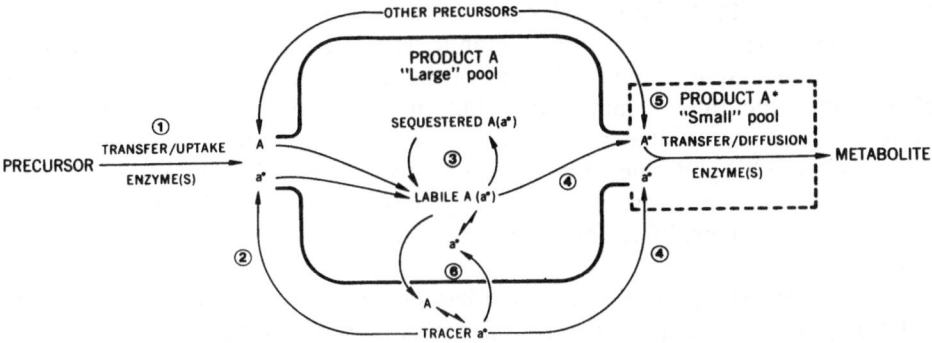

Fig. 2. Scheme of the compartmentalized metabolism of a substance A and the fate of tracer molecules a* when introduced into the system. The compartments may represent actual separate anatomic structures or may indicate biochemical separation by means of a series of transport and enzymic mechanisms. These may be studied individually or can be considered to represent in combination two major metabolic steps, one regulating access of A to a pool, the other egress of the molecules and their metabolism. For example, in terms of the compartmentalized metabolism of glutamic acid, read: PRECURSOR = glucose; PRODUCT A = glutamic acid; METABOLITE = glutamine; TRACER a* = isotope-labeled glutamic acid; LABILE A(a) = so-called releasable glutamic acid; SEQUESTERED A(a) = protein incorporated; OTHER PRECURSORS = metabolites of the tricarboxylic acid cycle (ketoacids).

specific activity than appeared possible if the tracer molecules of a* had first mixed with its endogenous pool (i.e., Fig. 2,3). This thus suggested that the labeled tracer molecules did not mix with the endogenous pool A or, if they did, with only a small fraction thereof (Fig. 2, 4). Although there is some indication that the introduction of isotopes into a molecule can alter its metabolism, the observed quite large discrepancy in the anticipated specific activities of the substance A pool and of its metabolites(s) clearly could not be accounted for by such an isotope effect. The alternative explanation, therefore, was that tracer a* gave rise directly to the metabolite without first entering the endogenous pool of A (Fig. 2, 4).[69,82,84]

One of the clearest examples of a compartmentalized metabolism is represented by Fig. 3 which also illustrates how that phenomenon may be used to discern when this type of metabolism becomes established in the increasing complexity of CNS cytoarchitecture. It has been known for years that the injection of labeled glucose (PRECURSOR, Fig. 2) results in isotopic labeling of glutamic acid (PRODUCT A) and its metabolites (glutamine, aspartate, and GABA).[80,81] When glutamic acid is directly introduced into the brain of a kitten (TRACER a*), the highest specific activity appears in glutamic acid followed by glutamine and then aspartic acid (Fig. 2, 1). However, with maturation of the CNS, which to a large extent represents glial proliferation and synaptogenesis, the glutamine pool (METABOLITE, Fig. 2) becomes increasingly more radioactively labeled relative to glutamic acid, whereas the specific activity of aspartic acid never surpasses that of either glutamine or glutamic acid.

One explanation for this observation in terms of Fig. 2 is that on maturation of the cortex, the glutamic acid tracer, rather than entering the metabolic pools at steps 1, 2, or 6, starts to enter at steps 4 or 5. Hence, the tracer molecules no longer mix with the large, unlabeled glutamic acid pool A but starts to enter directly into the "small" pool of glutamic acid that immediately gives rise to glutamine (and aspartic acid). Studies of this type thus not only reveal details of a compartmentalized metabolism but, in addition, may also indicate the preferential anatomic sites in which a substance is (further) metabolized. In this case, the "small" glutamic acid pool seems to be situated preferentially in glial and/or synaptic elements, since the appearance of the radioactive anomaly coincides with the proliferation of just these structural elements in the maturing CNS.

The methodology outlined above has proven to be a powerful technique in helping to indicate the existence of metabolic pools. It furthermore permits the estimation of the turnover rates of these pools if the isotope redistribution among the various metabolites is followed as a function of time. Such studies do require certain precautions, however.

The total endogenous amount of a substance A to be diluted with isotope-labeled molecules has to be known. This is required to determine the maximum specific activity theoretically possible and to permit comparison of the specific activity of the resulting metabolite(s) relative to that activity.

Since tissue content or metabolic turnover of a substance may vary with differing physiological/environmental conditions, those conditions must be kept constant throughout the duration of the experiment (e.g., monitored anaesthesia). If no precautions are taken, the specific activities of the different pools may vary unpredictably because of fluctuations in rates of synthesis or metabolism of precursor (step 1), of the original isotope pool (step 3), or of the pools of metabolites (step 5). Maintaining constant conditions is therefore especially important if estimations of relative turnover rates are desired.[93,94,101]

The type of isotope label must be considered. A "uniformly" labeled substance indicates molecules in which the isotopes are distributed randomly among a chain of the species of atoms designated as the label; some molecules may contain more than one isotope. Any type of biochemical degradation of such a molecule may lead to a wide dispersion of the isotope among a number

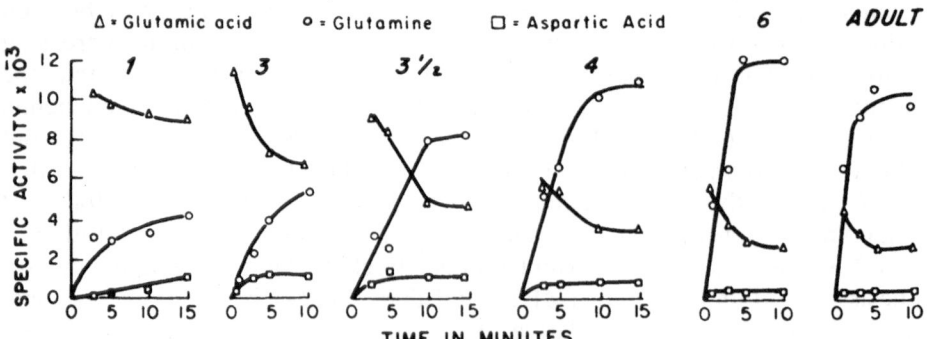

Fig. 3. Specific activity (counts/min per mol) of glutamic acid, glutamine, and aspartic acid in brain cortex following application of L-[U-^{14}C]glutamic acid (20 μl, 1.8 μCi; 120 to 205 μCi/mol) to the surface of the brains of kittens. Age in weeks indicated by the figures above the curves. (Data from Berl[173] with permission of author and Am. Soc. of Biol. Chemists.)

of metabolites. Estimation of turnover rates may therefore be exaggerated. Isotope labeling of a molecule at a specified position allows a better estimation of turnover rates catalyzed by defined enzymic mechanisms; e.g., $[1\text{-}^{14}C]$glutamate on α-decarboxylation yields $^{14}CO_2$ + unlabeled GABA; $[U\text{-}^{14}C]$glutamate yields $^{14}CO_2$ + labeled glutamine, GABA, aspartate, and glutamate (among others).

The specific activities of compounds to be used are limited. High specific activities accelerate radiodecomposition in a labeled compound and may jeopardize the chemical purity of the tracer material.

3.2. The Separation of Endogenous Pools

The procedure referred to above, although useful to indicate temporal discontinuity in the metabolism of a substance, does not indicate which *in vivo* conditions give rise to such discontinuity.[88,90,97,98] Furthermore, that fraction of the endogenous content in which the radioactivity is concentrated is always, by definition, designated as the "small" pool.[87,95] Implicit in this concept is the understanding that the "large" pool is less active metabolically. However, when experiments are performed *in vivo* or, in general, in structurally intact tissue preparations, it becomes apparent that with the choice of the appropriate labeled precursor, the so-called "small" metabolic pool may in fact turn out to be a conceptual artifact. This "pool" usually represents in reality that fraction of a reservoir that is *in the process* of being transferred to a site at which certain metabolic events transform it into a product (Fig. 2, 5). When anatomic organization is taken into consideration, such studies thus indicate that a substance contained at one storage site flows (in)to another anatomic site where it forms a metabolite pool. When a tracer substance is introduced, it may either first enter the storage site, thereby causing dilution of the specific activity (Fig. 2, 2,5), or it may directly access the site of further metabolism (Fig. 2, 4). In that case, the specific activity of the metabolite will be greater than that of the reservoir pool.

3.2.1. Endogenous Product Accumulation

Once formed from an endogenous precursor, a substance may accumulate at its site of formation (Fig. 2, "large" pool A). This process, being enzymic in nature, suggests that accumulation will be limited by gradual product inhibition. Moreover, if such a substance causes strong intracellular pH shifts, the presence or absence of certain neutralizing mechanisms should represent another factor determining the degree of accumulation. Finally, the ability of a site to sequester a substance (Fig. 2, 3) may also be influenced by membrane permeability and ionic transmembrane gradients (Na^+/K^+ balance, Ca^{2+}/Mg^{2+}, etc.) which in turn are strongly related to the physiological condition (state of excitation) of a cell.[61,68,89]

Because of all of these variables, the size of the endogenous pool in a specified environmental/physiological state represents a dynamic equilibrium.[99] If appropriate detection techniques are available, such as chromatography, radioimmune methods, fluorescence, etc., steady-state pools can be quantified directly. On the other hand, radioisotopes can only be contemplated for this purpose if *steady-state* labeling of the pool under specified conditions is achievable. Single ("bolus") injections of tracer material representing either precursor or product compound will yield little meaningful data in this case, since the diverse rate-limiting factors determining the eventual pool size would introduce too much variability in the data over the short life-span of the bolus (Figs. 2, 3,6). The latter technique can only accurately reflect initial rate exogenous uptake, since only total isotope content needs to be determined in that case. Even then, it should be noted that because of the process of exchange (Fig. 2, 6), the exogenous tracer pool is continuously being altered, thus preventing application of Michaelis–Menten kinetics to that process.[94,100] The only manner in which this problem can be resolved is by studying the influx (uptake) and efflux processes separately, as was accomplished recently for GABA.[94]

3.2.2. Exogenous Uptake

This phenomenon may be defined as the formation of a metabolic pool by uptake of a compound, present exogenously, into the anatomic compartment in which it becomes concentrated (Fig. 2, 6). It is best measured by determining the rate of exogenous disappearance of that compound combined with the proportional rate of its accumulation within the tissue. The discrepancy between the two processes should represent its metabolism or incorporation into another pool (which then would increase). Exogenous uptake also depends on a number of other events.[91,92,96]

3.2.2a. Diffusion or Rate of Access to Accumulation Sites. A plot of velocity versus substrate is linear through the origin and (almost) independent of metabolic conditions (edema or hyperosmolarity does create an effect).

3.2.2b. Adsorption. This is a "nonspecific" process which will, however, be strongly influenced by physicochemical parameters of the tissue preparation and of the compound (ionization forces, isomers, etc.). It occurs not quite independently of temperature but may be influenced by metabolic conditions (membrane structure) as well as by the external environment (pH, ions, competitors, etc.). A plot of V versus S usually extrapolates through or near the origin (substract diffusion), exhibits saturation, and shows competitive or noncompetitive kinetics.

3.2.2c. Receptor Binding. This is a special type of adsorption since it signifies a process of adsorption entailing the involvement in the membrane of an exact tertiary protein structure which is complementary to the structure of the compound to be bound. These specialized affinity regions may represent sites implicated in producing a physiological response (ionic permeability) or sites representing the first step in a sequence of events leading to inward transport and accumulation, i.e., formation of a metabolic reservoir. Both types of sites, by definition, are present on the external side of a membrane (organelle), are influenced by the chemical composition of the external milieu, and, in the case of transport, by the ability to sequester. Since specific binding is a biological process, the receptors or transport sites operate at 37°C in the presence of the usual external ionic environment. A plot of V versus S must satisfy true Michaelis–Menten kinetics when diffusion and nonspecific binding have been taken into consideration and no longer contribute to the measurements obtained. The required conditions for application Michaelis–Menten kinetics are (1) unlimited substrate of constant concentration (specific activity), (2) a plot of V versus S yielding a rectangular hyperbola, and (3) minimal "product" accumulation, i.e., initial rate conditions. Failure to satisfy any one of these criteria renders application of (unmodified) Michaelis–Menten rate kinetics to the data invalid.

The size (content, concentration) of the metabolic pool and its rate of formation reflect the sum total of the rate of diffusion into the extracellular space, specific transporter binding, inward transmembranal movement and accompanying sequestration, and outward movement represented by exchange diffusion at near-maximum pool size or the transfer rate to the next pool.

Because many of the processes outlined above entail very small concentrations, quantitative measurements are often limited by the sensitivity of the detection methods. Since radioactivity represents a very sensitive method of detection, it has been used extensively in such studies. However, when dealing with substances of high specific activity (10^5–10^6 cpm/μmol is not uncommon), the detection of even 10^3 counts in tissue preparations may represent only a fraction of a microliter of medium contamination or a fraction of a percent of the initial quantity of substance used at the onset of incubation. Many studies intent on determining binding or those concerned with autoradiography (localization), and uptake versus exchange transport phenomena, may therefore report on quantitatively negligible alterations.

When an endogenous metabolic pool is in dynamic equilibrium, the rate of replenishment (Fig. 2, 1) is matched by the rate of transfer to the next compartment (Fig. 2, 5). During these steady-state conditions, the endogenous pool should not increase when a precursor or the compound itself is quantitatively added exogenously or as a tracer material. Although inward transport into the pool will occur, this will be compensated for by an acceleration of the outflow from that pool, provided steady-state conditions are maintained. The accelerated outflow may represent exchange diffusion–i.e., inward and outward transport rates are equal (Fig. 2, 6)—or a stimulated transfer (metabolism) to the next compartment (Fig. 2, 5). Moreover, during maximum pool saturation, the pool will not change quantitatively, although a trace molecule may indicate rapid turnover of pool content. Thus, entrance of label under these circumstances may indicate uptake (Fig. 2, 2,6), enhanced turnover (Fig. 2, 3,4), or direct entrance into an adjacent metabolite compartment (Fig. 2, 4,5). In addition, because of exchange transport, the exogenous specific activity of the tracer material may be continuously changing during incubation (Fig. 2,

6). If, furthermore, the metabolite compartment also releases a radioactive metabolite by the exchange process, it is evident that the potential for error or misinterpretation of the data is large (see Section 2.2.2).

For these reasons, changes in pool size(s) often require, as a minimum, the chemical quantification of such pools rather than reliance on tracer techniques alone. Although the latter may appear more convenient and elegant because of the ease of detection, the results from "older" studies which in part or totally relied on chemical determinations should not be ignored nor disallowed when mechanisms of compartmentation are formulated. As one such example, one may cite the case of GABA which in recent years has been shown to be actively transported into subfractions of brain homogenates.

This phenomenon is largely detectable only when labeled GABA of extremely high specific activity is used for purposes of autoradiography or the study of isolated tissue subfractions (synaptosomes, etc.). Yet, as early as 1958 and in subsequent series of investigations,[89] the work of Elliott and collaborators demonstrated that quantitative uptake and enlargement of the endogenous cerebral GABA pool seem only to occur when the tissue is structurally intact and incubated under conditions closely approximating physiological conditions. Later studies using tracer techniques demonstrated rapid turnover of exogenous GABA and "selective" uptake, depending on the degree of structural integrity of the tissue (cubes versus slices) and by possibly two different Michaelis–Menten saturation constants. Although these findings revealed the rapid flow of individually labeled molecules from a steady-state pool and the specific structures in which this may occur, they would have missed the fact that nervous tissue can only store appreciable quantities of a substance when the cytoarchitecture is intact (retention capacity). This parameter, when changed, thus provides a good measure of tissue damage. In general, most studies seem to indicate that amino acids are concentrated quantitatively in intact nervous tissue, thereby causing an enlargement of an endogenous (free) pool, with only a small fraction of that exogenously derived amino acid pool being metabolized further.

4. TISSUE COMPARTMENTATION

In view of the abovementioned complexities in studying compartmentation, a combination of techniques and tissue preparations is usually required to elucidate a series of metabolic compartments involving several cytoarchitectural elements.

4.1. Intact Animal

Tracing a compound through its entire metabolic cycle within the CNS must first take into consideration the route of administration.[103] Intracarotid introduction partially or entirely bypasses the hepatic portal system and introduces high concentrations into the cerebral circulation.[102,106,111] Unless continuous, this method of administration permits the CNS only a limited time to capture the substance from the circulation.[112] Entrance into the CNS depends on cerebral circulatory conditions[107,110,111] and blood–brain barrier systems[108]; the latter can now be accurately estimated. It is usually expressed as a percentage of the rate of entrance of a substance freely permeable through the barrier systems, usually deuterated water.[10] There is great need for these types of studies, since they permit a correlation of physiological (EEG[103]) or behavioral parameters[106] of brain function with information on intact blood–brain barrier systems[108] and *in vivo* cerebral circulation.[107,110,111] Especially in the domain of brain energy requirements, this information appears essential if the newer methods using positron emission tomography[97,112] are to receive the proper interpretation.

Intraperitoneal and subcutaneous injections provide alternative methods to introduce a substance into the circulation.[105] In this case, however, maximum concentrations are always lower than can be achieved with the same dose on direct infusion. On the other hand, steady blood concentrations can be maintained for longer periods, and this method may be considered to represent a compromise between bolus administration and creation of steady-state conditions. Except when permanent implantation of cannulas is envisaged, those are most easily obtained by oral administration.[115] Steady-state conditions are usually reached with a dosage schedule equivalent to six biological half-lives, and the maximum concentration achievable approximates 1.5 times the number of doses given per half-time. Oral administration must take into consideration digestive absorption processes and the hepatic circulation. Much higher doses are therefore required to reach desired blood concentrations, and the oral route increases the chance for formation of metabolites exerting their own biological action, or the induction of enzymes.[114] The method may thus be expensive and experimentally may introduce complicating factors.

Especially when the blood–brain barrier systems do not permit access of a circulating substance, such a compound may be introduced intraventricularly into the cerebrospinal fluid.[104] This seems, however, to represent an unnatural route of entrance into the CNS.[7] This can be illustrated by means of a simple test: hypoglycemic coma is reversed in seconds by intravenous glucose; intraventricular administration is quite ineffective under these circumstances. A parenchyma-to-CSF transport gradient (in general), CSF flow, and CSF–blood exchange mechanisms insure that most substances are rapidly prevented from (re)entering nervous tissue via the CSF.

Direct application to exposed brain surfaces (superfusion of cortex, push–pull cannula, etc.)[113] may be considered as another alternative to bypass the blood–brain barrier systems. Although prolonged direct exposure of nervous tissue can, and usually does, create anatomic and biochemical damage, at the same time, limited diffusion into the surrounding tissue will occasionally permit some semblance of ''natural'' distribution in a circumscribed region. The technique does require precise definition of the experimental conditions, preferably in terms of anatomic parameters. Great care is required that the pH and the molarity of the solutions, as well as the volume, are precisely regulated.

One most effective manner to study a number of aspects of compartmentalized metabolism *in vivo* is to use developing animals.[124] The blood–brain barrier systems[108,120] are less established, and because of smaller body size, effective concentrations are easier to reach with less material.[26] Furthermore, defined stages of CNS maturation over a period ranging from prenatal to neonatal life often make it possible to correlate the appearance of a biochemical event[117,122] with the development of a specific anatomic structure (mitochondria, nerve terminals, neuroglia, etc.). Combined with the use of certain tissue preparations, the gradual establishment of the compartmentalized metabolism for a substance in the CNS[116] can be joined to developing physiological function and the anatomic substratum.[118,119,121,125]

Most techniques adopted to elucidate metabolic compartmentation *in vivo* require the use of labeled tracer substances. The label is either in the form of an isotope (e.g., ^{14}C, ^{3}H), a substitute isotope (e.g., ^{18}F for H), or a chemical analogue that at a precise stage of the biochemical cycle represents a dead-end inhibitor of compartmentalized metabolism (e.g., fluorocitrate, deoxyglucose). Such methods have revealed a wealth of information about turnover rates and the number of metabolic compartments implicated as well as delineating, in conjunction with developmental studies and autoradiography, certain nervous structures associated with such metabolism. The techniques have been less successful in providing an understanding of the mechanisms involved in regulating the steady-state metabolic pools and the changes that occur in association with alterations in the physiological state of the CNS. *In vivo* studies directed towards these aims have recently been attempted by basing the interpretation of results from steady-state metabolic pool determinations on the existence of hypothetical compartments.[123]

4.2. Tissue Preparations

The benefits of using nervous tissue directly, without having first to bypass chemical and physical barriers, are clear. The extracellular and, to a limited extent, even the intracellular biochemical environment can be precisely controlled. A judicious combination of tissue preparations varying in anatomical complexity allows the separate study of individual biochemical compartments, their localization in specific anatomic elements, and, eventually, a combination of all of these findings to present an integration of the biochemical compartment as a function of physiological states and nervous tissue cytoarchitecture. One other great advantage of such preparations lies in the fact that altered physiological or anatomic parameters, induced directly or produced first *in vivo*, may also be reflected in the environmental medium to which the viable tissue preparations are exposed.

4.2.1. Brain Slices

Popular with neurochemists because slices appeared to accurately reflect metabolic events *in vivo*,[126,129,133] this type of preparation was once severely criticized by neurophysiologists as representing a "nonphysiological" preparation. However, when it became established that neurons in slices could respond to physiological stimulation and other methods of controlled depolarization, this tissue preparation was rapidly accepted in electrophysiology.[127,128] Somewhat ironically, it now seems necessary to warn that brain slices do not mimic intact nervous tissue in several ways. First, they represent severely deafferented tissue and as such may more closely resemble undercut brain regions. Moreover, when slices are continually perfused as is especially the custom in physiological investigations, the interstitial environment may be very different from that actually encountered in an (enclosed) intact brain region. Influx and efflux of substances to and from interstitial spaces may be less natural. Little is as yet known how the combined metabolic processes of neurons and satellite cells, in association with those of microcapillaries and interstitial fluid flow, regulate the immediate chemical environment of neurons *in vivo*.[130,131] The introduction of substances either in the medium or iontophoretically, regulation of pH in a bicarbonate- as opposed to phosphate-buffered medium, effective concentration, and many other factors of this nature have in most instances been selected by trial and error.[132] They therefore may not accurately reflect *in vivo* conditions.

In relation to compartmentation, especially with regard to uptake, binding, and turnover studies, at least one compartment in addition to those present *in vivo* is introduced artificially. This compartment is formed by the cut surface(s) of a slice. In this region, substances from the medium may have abnormal access to intracellular membranes and create diffusion and uptake artifacts.[128] Also, variations in the thickness of slices, which is difficult to control from one experiment to another, will alter gaseous and particulate penetration parameters. In this respect, 400-μm slices in combination with a 95% O_2/5% CO_2 atmosphere (37°C) represents the best compromise in that an approximately 200-μm ''center'' sandwiched between two 100-μm ''damaged'' layers provides physiologically and biochemically patent tissue. For these many reasons, one important criterion for proper utilization of tissue slices is the reproducibility of results from slice to slice and from experiment to experiment.

When such reservations are taken into consideration, brain slices can be used for many purposes because of the large flexibility in modifying experimental conditions and the wide range in choices of methods. Antibodies can penetrate, and, thus, extracellular immunohistochemistry is applicable. Isotope studies in connection with autoradiography, receptor binding, cell-selective uptake and sequestration, and collection of liberated substances combined with differing states of physiological or chemical excitation are all phenomena amenable to study in this tissue preparation.

One specialized brain region behaves in several ways as both an *in vivo* preparation and as a brain slice without many of the usual disadvantages of either one. The retina can be used *in vitro* as a physiologically viable preparation,[139,142] is always of practically the same thickness, and does not expose artificial compartments to an incubation medium. Moreover, many of the individual anatomic structures can be isolated with less damage than often occurs when slices of other CNS regions are used.[137,140] The various systems of compartmentalized metabolism in the retina seem to reflect the specialized physiological function of a CNS region as well as certain species- (genetic-) directed variations.[136,141,143] The former is important to determine alterations in the compartmentalized metabolism of a substance in terms of the physiological function of a brain region, whereas the latter emphasizes that the metabolism of a substance may differ in subtle ways between species even though its biological role phylogenetically remains the same.[134,135,138] Finally, retinal excitation is possible by using the natural stimulus, i.e., light, and strength, duration, area stimulated, and wavelengths can be experimentally controlled very precisely. Stimulation of tissue slices usually requires either electrodes or depolarizing chemicals, neither of which may resemble natural *in vivo* activating conditions since they usually cause massive synchronized depolarization of the entire tissue.[140]

4.2.2. Tissue Cultures

This type of experimental preparation resembles less closely the natural anatomic complexities of nervous tissue. Usually representing a predominance of one particular cell type, it lacks some of the biochemical and functional characteristics of an integrated nervous system.[148] In this respect, primary tissue cultures obtained from outgrowths of nervous tissue in early stages of development may biochemically be more normal than cell lines cultivated from tumors.[144,151] A specific cell culture, on the other hand, is most useful to investigate whether certain events incorporated in the compartmentalized metabolism of a substance (transport, storage, enzymes) are particularly concentrated in certain cell types of the nervous tissue.[149,150,152] They lend themselves to all techniques available for *in vitro* studies including those examining their physiological and pharmacological properties.[146,147] Cell cultures are also used extensively to investigate cinematographically synaptogenesis and myelinization, two dynamic processes which are especially difficult to investigate during *in vivo* development.[145] Combined with immunohistochemistry, autoradiography, and sensitive enzyme methods, these preparations have demonstrated their utility many times over.

Cultures are an especially important adjunct to studies involving subfractionated tissue preparations (see Section 4.2.3), since in the latter instance, homogeneity by experimental anatomic criteria may not always correspond to actual anatomic purity. For example, membrane fractions as they become more pure tend to fuse to form vesicles irrespective of whether they derive from glial or neuronal membranes; similarly, mitochondrial fractions from nervous tissue can be derived from many cell types including those in blood vessel. Whether or not they exhibit differing metabolic properties in addition to those they have in common is difficult to verify. Cultures of neuronal or specific glial origin are able to resolve a number of these problems.[153]

4.2.3. Cell Subfractions

These include many different preparations but may be defined as representing purified preparations of cellular organelles: synaptic vesicles, mitochondria, presynaptic and postsynaptic membrane fractions, nuclei and nu-cleoli, enzymes purified to single protein species, etc. Analysis of biochemical properties of the fractions as they become more homogeneous during stepwise purification provides an opportunity to uncover differences in metabolic reg-ulatory mechanisms operating in isolation as opposed to their properties when present in more complex anatomical arrangements. Several drawbacks must be considered when biochemical studies are contemplated using any tissue subfraction.

It may take several hours after death before the fractions are isolated. Selective alterations or deletions of certain metabolic properties, the nature of which is not always known, can occur during this period. During long isolation procedures, some thought should be given to possible bacterial contamination, especially when tissue is obtained post-mortem or from slaughtered animals. How seriously such contamination may interfere with the results is really not known at this time. It probably will very much depend on the type of bio-chemical events that are under study.

The quantities of purified fractions obtainable are often quite small and may require prodigious effort and time on the part of the investigator. The work as outlined, for example, by Schally[164] in isolating TRH certainly ex-emplifies many of the problems associated with the purification of certain tissue constituents. Because of the small quantities, the verification of purity, which is essential if such fractions are to be useful, may be difficult. It requires a high degree of sophistication in terms of equipment as well as direct chemical and immunologic techniques. To derive maximum profit from the latter tech-nique, it is necessary first to purify the antibody to homogeneity.[175]

With respect to heterogeneous structures containing more than one type of organelle, criteria of purity entail both biochemical and anatomic verifica-tion. For example, criteria for purity of synaptosomal fractions theoretically would require electron microscopic examination of the isolated fractions as well as the determination of enzyme activity associated specifically with certain presynaptic, postsynaptic, or glial membranes. In reality, this procedure is impractical, since the preparation of fractions for electron microscopy and subsequent examination requires several days, and determinations of enzymic markers may well use up a large part of the isolated fractions in addition to delaying their use for the actual studies intended. Hence, purification criteria are usually satisfied by performing the analysis on the first few fractions isolated, once the techniques have been standardized, with the implicit assumption that these techniques will then always yield fractions of similar quality. Such an assumption is not necessarily valid, since even small variations in density gradients, initial endogenous ion concentrations (Ca^{2+}/Mg^{2+} ratios), and other conditions may drastically alter the contamination of the fractions at the density boundaries. In addition, few biochemistry laboratories have routine access to electron microscopes. For many such reasons, to define tissue fractions in terms of precisely described experimental manipulation combined with good reproducibility of the results may empirically be as good a criterion as trying to define purity of tissue fractions by electron microscopy and enzyme markers if these are not performed routinely throughout the entire duration (months?) of the series of experiments carried out.

Some fractions are obviously more easily isolated than others, but especially with respect to the isolation of the same organelles from different types of cells or different intracellular locations (e.g., nerve terminals versus glial or perikaryal mitochondria), certain limitations of the methodology must be accepted in order to obtain consistently a homogeneous tissue fraction. Here, an eclectic standard based on reproducibility of techniques combined with a certain caution in the interpretation of the data has proven in many instances to yield very valuable information. On the other hand, the increasing sophistication of methods available to localize proteins and other molecules is gradually diminishing the existing dichotomy between biochemical and anatomic compartmentation.[154-173] This appears essential if that compartmentation is to be understood in terms of the physiological function of the CNS.[173]

In summary, the present chapter has attempted to outline the principles of compartmentation with respect to both the biochemical and cytoarchitectural organization of an organ, in particular, the CNS. A wide spectrum of methods and experimental strategies are available to study this phenomenon. The list of references represents an effort to provide access to most of the information required to decide on the proper experimental approach within the limits of instrumentation and specialization available in a particular laboratory. It is clear that any one of the methods to be used has certain drawbacks, and, thus, that discrepancies between data from different laboratories may be anticipated. Rather than a weakness, this often represents an advantage, since the different approaches to the same problem have often yielded far more information than the individual articles seemed to suggest.

REFERENCES

1. Cornforth, J. W., 1976, *Science* **193**:121–125.
2. Cram, D. J., and Cram, J. M., 1974, *Science* **183**:803–809.
3. Walsh, C. (ed.), 1979, *Enzymatic Reaction Mechanisms*, W. H. Freeman, San Francisco, pp. 891–916.
4. Glees, P., 1973, *Metabolic Compartmentation in the Brain* (R. Balazs, and J. E. Cremer, eds.), Macmillan, London, pp. 209–231.
5. Bock, E., 1978,. *J. Neurochem.* **30**:7–14.
6. Kolber, A. R., Bagnell, C. R., Krigman, M. R., Hayward J., and Morell, P., 1979, *J. Neurochem.* **33**:419–432.
7. Lajtha, A., and Ford, D. H. (eds.), 1968, *Brain Barrier Systems (Prog. Brain Res.* Volume 29), Elsevier, Amsterdam.
8. Lund-Andersen, H., 1979, *Physiol. Rev.* **59**:305–352.
9. McGale, E. H. F., Pye, I. F., Stonier, C., Hutchinson, E. C., and Aber, G. M., 1977, *J. Neurochem.* **29**:291–297.
10. Oldendorf, W. H., and Braun, L. D., 1976, *Brain Res.* **113**:219–224.
11. Pardridge, W. M. and Oldendorf, W. H., 1977, *J. Neurochem.* **28**:5–12.
12. Aguayo, A. J., Charron, L., and Bray, G. M., 1976, *J. Neurocytol.* **5**:565–573.
13. Benjamins, J. A., and Morell, P., 1978, *Neurochem. Res.* **3**:137–174.
14. Cochran, F. B., Yu, R. K., Ando, S., and Ledeen, R. W., 1981, *J. Neurochem.* **36**:696–702.
15. Eylar, E. H., Jackson, J. J., and Kniskern, P. J., 1979, *Neurochem. Res.* **4**:249–258.

16. Hashim, G., Sharpe, R. D., Carvalho, E. F., and Stevens, L. E., 1976, *J. Immunol.* **116**:126–130.
17. Rasminsky, M., 1978, *Physiology and Pathobiology of Axons* (S. G. Waxman, ed.), Raven Press, New York, pp. 361–376.
18. Weinberg, H. J., and Spencer, P. S., 1976, *Brain Res.* **113**:363–378.
19. Berl, S., Lajtha, A., and Waelsch, H., 1961, *J. Neurochem.* **7**:186–187.
20. Henn, F. A., Goldstein, M. N., and Hamberger, A., 1974, *Nature* **249**:663–664.
21. Martinez-Hernandez, A., Bell, K. P., and Norenberg, M. D., 1977, *Science* **195**:1356–1358.
22. Minchin, M. C. W., 1977, *Exp. Brain Res.* **29**:515–526.
23. Norenberg, M. D., and Martinez-Hernandez, A., 1979, *Brain Res.* **161**:303–310.
24. Schousboe, A., Hertz, L., Svenneby, G., and Kvamme, E., 1979, *J. Neurochem.* **32**:943–950.
25. Walum, E., 1979, *Biochem. Biophys. Res. Commun.* **88**:1271–1274.
26. DeVivo, D. C., Leckie, M. P., and Agrawal, H. C., 1975, *J. Neurochem.* **25**:161–170.
27. Drummond, R. J., and Phillips, A. T., 1977, *J. Neurochem.* **29**:101–108.
28. Tursky, T., Ruscak, M., Lassanova, M., and Ruscakova, D., 1979, *J. Neurochem.* **33**:1209–1215.
29. Tursky, T., Ruscak, M., and Lassanova, M., 1979, *Physiol. Bohemoslov.* **23**:43–49.
30. Weyne, J., van Leuven, F., Demeester, G., and Leusen, I., 1977, *J. Neurochem.* **29**:469–476.
31. Giacobini, E., 1962, *J. Neurochem.* **9**:169–177.
32. Kornblath, J. A., 1980, *Can. J. Biochem.* **58**:840–850.
33. Mandel, P., Roussel, G., Delaunoy, J.-P., and Nussbaum, J.-L., 1978, *Dynamic Properties of Glia Cells* (E. Schoffeniels, G. Franck, L. Hertz, and D. B. Tower, eds.), Academic Press, New York, pp. 619–630.
34. Parthe, V., 1981, *J. Neurosci. Res.* **6**:119–131.
35. Schousboe, A., Nissen, C., Bock, E., Sapirstein, V. S., Juurlink, B. H. J., and Hertz, L., 1980, *Tissue Culture in Neurobiology* (A. Vernadakis and E. Giacobini, eds.), Raven Press, New York, pp. 397–409.
36. Davis, B. D., 1971, *Nature* **231**:153–157.
37. Fulks, R. M., Li, J. B., and Goldberg, A. L., 1975, *J. Biol. Chem.* **250**:290–298.
38. Goldfine, I. D., Jones, A. L., Hradek, G. T., Wong, K. Y., and Mooney, J. S., 1978, *Science* **202**:760–762.
39. Gullis, R. J., and Rowe, C. E., 1975, *Biochem. J.* **148**:197–208.
40. Holtzman, E., 1977, *Neuroscience* **2**:327–355.
41. Kurokawa, M., Tashiro, T., Mizobe, F., and Setani, N., 1973, *FEBS Lett.* **34**:165–168.
42. Lajtha, A., Dunlop, D., Patlak, C., and Toth, J., 1979, *Biochim. Biophys. Acta* **561**:491–501.
43. Mains, R. E., and Eipper, B. A., 1980, *Ann. N.Y. Acad. Sci.* **343**:94–108.
44. O'Malley, B. W., and Means, A. R., 1974, *Science* **183**:610–619.
45. Parks, J. M., Ames, A., and Nesbett, F. B., 1976, *J. Neurochem.* **27**:987–997.
46. Potts, J. T., Kronenberg, H. M., Habener, J. F., and Rich, A., 1980, *Ann. N.Y. Acad. Sci.* **343**:38–54.
47. Sauvageau, M. A., 1975, *Nature* **258**:208–214.
48. Sutherland, E. W., 1972, *Science* **177**:401–408.
49. Takahashi, Y., Araki, K., and Suzuki, Y., 1971, *Brain Res.* **32**:179–188.
50. Tomkins, G. M., 1975, *Science* **189**:760–763.
51. Wickner, W. 1980, *Science* **210**:861–868.
52. Costa-Tiozzo, R., and Gal, E. M., 1980, *Neurochem. Res.* **5**:23–36.
53. Dennis, S. C., Lai, J. C. K., and Clark, J. B., 1977, *Biochem. J.* **164**:727–736.
54. Dennis, S. G. C., and Clark, J. B., 1978, *J. Neurochem.* **31**:673–680.
55. Simpson, D. P., and Adam, W., 1975, *J. Biol. Chem.* **250**:8148–8158.
56. Von Korff, R. W., and Kerpel-Fronius, S., 1975, *J. Neurochem.* **25**:767–778.
57. Blaustein, M. P., and Goldring, J. M., 1975, *J. Physiol. (Lond.)* **247**:589–615.
58. Bradford, H. F., Bennett, G. W., and Thomas, A. J., 1973, *J. Neurochem.* **21**:495–505.
59. Chaplin, E. R., Golberg, A. L., and Diamond, I., 1976, *J. Neurochem.* **26**:701–707.
60. Coyne, L. M., 1979, *J. Theor. Biol.* **79**:455–472.
61. De Belleroche, J. S., and Bradford, H. F., 1977, *J. Neurochem.* **29**:335–343.
62. Fishman, P. H., and Brady, R. O., 1976, *Science* **194**:906–915.

63. Hammerstad, J. P., Cawthorn, M. L., and Lytle, C. R., 1979, *J. Neurochem.* **32:**195–202.
64. Iadorola, M. J., and Gale, K., 1979, *Eur. J. Pharmacol.* **59:**125–129.
65. Kvamme, E., 1979, *GABA—Biochemistry and CNS Functions* (P. Mandel and F. V. deFeudis, eds.), Plenum Press, New York, pp. 111–138.
66. Levi, G., and Raiteri, M., 1978, *Proc. Natl. Acad. Sci. U.S.A.* **75:**2981–2985.
67. Sieghart, W., and Karobath, M., 1974, *J. Neurochem.* **23:**911–915.
68. Zisapel, N., and Zurgil, N., 1979, *Brain Res.* **178:**297–310.
69. Anderson, C. M., Zucker, F. H., and Steilz, T. A., 1979, *Science* **204:**375–380.
70. Buu, N. T., Puil, E., and van Gelder, N. M., 1976, *Gen. Pharmacol.* **7:**5–14.
71. Enna, S. J., and Synder, S. H., 1975, *Brain Res.* **100:**81–97.
72. Friedberg, F., 1977, *Horizons Biochem. Biophys.* **4:**63–90.
73. Glowinski, J., and Iversen, L. L., 1966, *J. Neurochem.* **13:**655–669.
74. Krogsgaard-Larsen, P., Johnston, G. A. R., Curtis, D. R., Game, A., and McCulloch, R. M., 1975, *J. Neurochem.* **25:**803–809.
75. Meldrum, B., Pedley, T., Horton, R., Anlezark, G., and Franks, A., 1980, *Brain Res. Bull.* **5**(Suppl. 2):685–690.
76. Napias, C., Bergman, M. O., Van Ness, P. C., Greenlee, D. V., and Olson, R. W., 1980, *Life Sci.* **27:**1001–1011.
77. Nistsi, A., and Constanti, A., 1979, *Prog. Neurobiol.* **13:**117–235.
78. Reisine, T. D., Wastek, G. J., Speth, R. C., Bird, E. D., and Yamamura, H. I., 1979, *Brain Res* **165:**183–187.
79. Toffano, G., Guidotti, A., and Costa, E., 1978, *Proc. Natl. Acad. Sci. U.S.A.* **75:**4024–4028.
80. Balazs, R., and Cremer, J. E. (eds.), 1973, *Metabolic Compartmentation in the Brain: Models of Metabolic Compartmentation*, Macmillan, London, pp. 121–186.
81. Berl, S., 1973, *Metabolic Compartmentation in the Brain* (R. Balazs and J. E. Cremer, eds.), Macmillan, London, pp. 3–17.
82. Lajtha, A., Berl, S., and Waelsch, H., 1959, *J. Neurochem.* **3:**322–332.
83. Palade, G., 1975, *Science* **189:**347–358.
84. Patel, A. J., Johnson, A. L., and Balazs, R., 1974, *J. Neurochem.* **23:**1271–1279.
85. Ratner, S., 1979, *Ann. N.Y. Acad. Sci.* **325:**189–206.
86. van Gelder, N. M., 1978, *Can. J. Physiol. Pharmacol.* **56:**362–374.
87. Benjamin, A. M., and Quastel, J. H., 1976, *J. Neurochem.* **26:**431–441.
88. DiRocco, R. J., and Hall, W. G., 1980, *J. Neurosci. Res.* **6:**13–19.
89. Elliott, K. A. C., and van Gelder, N. M., 1958, *J. Neurochem.* **3:**28–40.
90. Hawkins, R. A., and Miller, A. L., 1978, *Neuroscience* **3:**251–258.
91. Lähdesmäki, P., Karppinen, A., Saarni, H., and Winter, R., 1977, *Brain Res.* **38:**295–308.
92. Levi, G., Gallo, V., Ciotti, T., and Raiteri, M., 1979, *J. Neurochem.* **33:**1043–1053.
93. Levi, G., Gallo, V., and Raiteri, M., 1980, *Neurochem. Res.* **5:**281–295.
94. Moscowitz, J. A., and Cutler, R. W. P., 1980, *J. Neurochem.* **35:**1394–1399.
95. Nadler, J. V., White, W. F., Vaca, K. W., Redburn, D. A., and Cotman, C. W., 1977, *J. Neurochem.* **29:**279–290.
96. Oja, S. S., and Kontro, P., 1980, *J. Neurochem.* **36:**1303–1308.
97. Sokoloff, L., Reivich, M., Kennedy, C., Des Rosiers, M. H., Patlak, C. S., Pettigrew, K. D., Sakurada, O., and Shinohara, M., 1977, *J. Neurochem.* **28:**897–916.
98. Stump, D. A., and Williams, R., 1980, *Brain Lang.* **9:**35–46.
99. van Gelder, N. M., 1982, *Taurine in Nutrition and Neurology* (H. Pasantes-Morales and R. Huxtable, eds.), Plenum Press, New York, pp. 239–256.
100. Wheeler D. D., and Hollingsworth, R. G., 1979, *J. Neurosci. Res.* **4:**265–289.
101. Wheeler, D. D., 1980, *Pharmacology* **21:**141–152.
102. Cremer, J. E., Heath, D. F., Teal, H. M., Woods, M. S., and Cavanagh, J. B., 1975, *Neuropathol. Appl. Neurobiol.* **3:**293–311.
103. Dirks, B., Hanke, J., Krieglstein, J., Stock, R., and Wickop, G., 1980, *J. Neurochem.* **35:**311–317.
104. Gainer, H., Barker, J. L., and Wollberg, Z., 1975, *J. Neurochem.* **25:**177–179.
105. Gaitonde, M. K., Wharton, J., and Holt, E., 1977, *J. Neurochem.* **29:**127–133.
106. Gorell, J. M., Dolkhart, P. H., and Ferrendelli, J. A., 1976, *J. Neurochem.* **27:**1043–1049.

107. Gregoire, N. M., Gjedde, A., Plum, F., and Duffy, T. E., 1978, *J. Neurochem.* **30:**63–69.
108. Lajtha, A., and Sershen, H., 1980, *The Blood–Retinal Barriers* (J. G. Cunha-Vaz, ed.), Plenum Press, New York, pp. 119–132.
109. Möhler, H., Patel, A. J., and Balazs, R., 1974, *J. Neurochem.* **23:**1281–1289.
110. Mrsulja, B. B., Lust, W. D., Mrsulja, B. J., Passonau, J. V., and Klatzo, I., 1976, *J. Neurochem.* **26:**1099–1103.
111. Norberg, K., and Siesjö, B. K., 1976, *J. Neurochem.* **26:**345–352.
112. Phelps, M. E., Kuhl, D. E., and Mazziota, J. C., 1981, *Science* **211:**1445–1448.
113. Roberts, F., Pearce, M., and Taberner, P. V., 1980, *Brain Res.* **201:**431–435.
114. Tyce, G. M., Ogg, J., and Owen, C. A., 1981, *J. Neurochem.* **36:**640–650.
115. Woznicki, D. T., and Walker, J. B., 1980, *J. Neurochem.* **34:**1247–1253.
116. Berl, S., and Purpura, D. P., 1966, *J. Neurochem.* **13:**293–304.
117. Borg, J., Ramaharobandro, N., Mark, J., and Mandel, P., 1980, *J. Neurochem.* **34:**1113–1122.
118. Himwich, W. (ed.), 1973, *Biochemistry of the Developing Brain*, Volume 1, Marcel Dekker, New York.
119. Hommes, F. A., and Van den Berg, C. J., (eds.), 1975, *Normal and Pathological Development of Energy Metabolism*, Academic Press, New York.
120. Lajtha, A., Dunlop, D., Patlak, C., and Toth, J., 1979, *Biochim. Biophys. Acta* **561:**491–501.
121. Levitsky, D. A. (ed.), 1979, *Malnutrition, Environment, and Behavior*, Cornell University Press, Ithaca.
122. Nehlig, A., Moncotel, D., and Lehr, P. R., 1980, *Life Sci.* **27:**483–489.
123. van Gelder, N. M., 1981, *Amino Acid Transmitters* (P. Mandel and F. V. DeFeudis, eds.), Raven Press, New York, pp. 115–125.
124. Vanucci, S. J., and Vanucci, R. C., 1980, *J. Neurochem.* **34:**1100–1105.
125. Winick, M., 1975, *Adv. Neurol.* **13:**193–246.
126. Benjamin, A. M., Verjee, Z. H., and Quastel, J. H., 1980, *J. Neurochem.* **35:**78–87.
127. Dingledine, R., Dodd, J., and Kelly, J. S., 1980, *J. Neurosci. Methods* **2:**323–362.
128. Lynch, G., 1980, *Annu. Rev. Neurosci.* **3:**1–22.
129. McIlwain, H., and Bachelard, H. S. (eds.), 1971, *Biochemistry and the Central Nervous System*, Churchill Livingstone, Edinburgh, pp. 61–97.
130. Neal, M. J., and Bowery, N. G., 1979, *Brain Res.* **167:**337–343.
131. Orrego, F., Miranda, R., and Saldate, C., 1976, *Neuroscience* **1:**325–332.
132. Teyler, T. J., 1980, *Brain Res. Bull.* **5:**391–403.
133. Weil-Malherbe, H., and Gordon, J., 1971, *J. Neurochem.* **18:**1659–1672.
134. Ames, A., and Parks, J. M., 1976, *J. Neurochem.* **27:**1017–1025.
135. Cohen, A. I., McDaniel, M., and Orr, H., 1973, *Invest. Opthalmol.* **12:**686–693.
136. Ehinger, B., 1972, *Brain Res.* **46:**297–311.
137. Lam, D. M. K., Su, Y. Y. T., Chin, C. A., Brandon, C., Wu, J.-Y., Marc, R. E., and Lasater, E. M., 1980, *Brain Res. Bull.* **5**(Suppl. 2):137–140.
138. Pasantes-Morales, H., Salceda, R., and Lopez-Colomé, A. M., 1980, *J. Neurochem.* **35:**172–177.
139. Santamaria, L., Drujan, B. D., Svaetichin, G., and Negishi, K., 1971, *Vision Res.* **11:**877–887.
140. Svaetichin, G., Negishi, K., Fatehchand, R., Drujan, B. D., and Selvin de Testa, A., 1965, *Prog. Brain Res.* **15:**243–266.
141. Starr, M. S., 1974, *J. Neurochem.* **23:**337–344.
142. van Gelder, N. M., and Drujan, B. D., 1978, *Brain Res.* **159:**137–148.
143. Voaden, M. J., Lake, N., Marshal, J., and Morjaria, B., 1978, *J. Neurochem.* **31:**1069–1076.
144. Baetge, E. E., Bulloch, K., and Stallcup, W. B., 1979, *Brain Res.* **167:**210–214.
145. Banker, G. A., 1980, *Science* **209:**809–810.
146. Henn, F. A., 1976, *J. Neurosci. Res.* **2:**271–282.
147. Barker, J. L., and Ransom, B. R., 1978, *J. Physiol. (Lond.)* **280:**331–354.
148. Hertz, L., and Schousboe, A., 1980, *Brain Res. Bull.* **5**(Suppl. 2):389–395.
149. Hösli, E., and Hösli L., 1978, *Neuroscience* **5:**145–152.
150. Kvamme, E., and Olson, B. E., 1979, *FEBS Lett.* **107:**33–36.
151. Nicklas, W. J., and Browning, E. T., 1978, *J. Neurochem.* **30:**955–963.
152. Schon, F., and Kelly, J. S., 1974, *Brain Res.* **66:**275–288.

153. Sellström, A., Sjöberg, L.-B., and Hamberger, A., 1975, *J. Neurochem.* **25**:393–398.
154. Bubenik, G. A., Brown, G. M., and Grota, L. J., 1976, *J. Histochem. Cytochem.* **24**:1173–1177.
155. Castejon, O. J., and Castejon, H. V., 1972, *Acta Histochem. (Jena)* **43**:153–163.
156. Descarries, L., and Droz, B., 1970, *J. Cell Biol.* **44**:385–399.
157. Fonnum, F., 1970, *J. Neurochem.* **17**:1029–1037.
158. Gutnick, M. J., and Prince, D. A., 1981, *Science* **211**:67–70.
159. Hösli, E., and Hösli, L., 1976, *Exp. Brain Res.* **26**:319–324.
160. Majcen, Z., and Brzin, M., 1971, *Histochemie* **25**:217–224.
161. Negishi, K., Hayashi, T., Nakamura, T., and Drujan, B. D., 1979, *Neurochem. Res.* **4**:473–482.
162. Privat, A., 1976, *J. Microsc. Biol. Cell* **27**:253–256.
163. Rojik, I., and Fehér, O., 1980, *Exp. Brain Res.* **39**:321–326.
164. Schally, A. V., 1978, *Science* **202**:18–28.
165. Stretton, A. O. W., and Kravitz, E. A., 1968, *Science* **162**:132–134.
166. Storm-Mathisen, J., and Iversen, L. L., 1979, *Neuroscience* **4**:1237–1253.
167. Stuart, A. E., Hudspeth, A. J., and Hall, Z. W., 1974, *Cell Tissue Res.* **153**:55–61.
168. van Gelder, N. M., 1965, *J. Neurochem.* **12**:231–237.
169. van Gelder, N. M., and Drujan, B. D., 1980, *Brain Res.* **200**:443–455.
170. Wu, J.-Y., 1976, *GABA in Nervous System Function* (E. Roberts, T. N. Chase, and D. B. Tower, eds.), Raven Press, New York, pp. 7–55.
171. Yalow, R. S., 1978, *Science* **200**:1236–1245.
172. Young, J. A. C., Brown, D. A., Kelly, J. S., and Schow, F., 1973, *Brain Res.* **63**:479–486.
173. Berl, S., 1965, *J. Biol. Chem.* **240**:2047–2054.

10

Diagnosis of Hereditary Neurological Metabolic Diseases

A. Lowenthal, N. Chamoles, K. Adriaenssens, and R. Humbel

1. INTRODUCTION

Metabolic studies in man can reveal interesting and useful information. Clinical manifestations that appear at first to be a single metabolic disease often turn out to be genetic variants attributed to mutations, either in different loci or in a single locus (polyallelism).

Most inborn errors of metabolism are autosomal recessive disorders; some recessives and some rare dominant disorders are X-linked.

More than 2000 single-gene disorders, of the estimated 50,000 different human genes, have been reported, but most are rare. It is estimated that about 1% of all live-born infants will present symptoms of such disorders some time in life and that roughly 4% of the population are carriers for one of these mutant genes (in about 200 of these mutant genes, the precise enzyme defect is known). It has been demonstrated that 6 to 9% of the population of children's hospitals are affected by single-gene diseases.

The detection of a metabolic disease always begins with clinical observations. Demonstration of the rise and/or decline of metabolites in body fluids or tissues (biochemical anomalies) by screening procedures and/or specific techniques leads to the discovery of enzyme-deficiency-specific enzyme assays.

The following list of clinical observations may indicate that a particular patient is suffering from a metabolic disorder:

1. A positive family history of a specific metabolic disease.
2. A family history of a clinical syndrome and/or unexplained death in one or more siblings or close relatives.

A. Lowenthal • Laboratory of Neurochemistry, Born Bunge Foundation, Universitaire Instelling Antwerpen, Antwerp, Belgium.   *N. Chamoles* • Laboratorio de Neuroquimica, Clinica del Sol, Buenos Aires, Argentina.   *K. Adriaenssens* • Provinciaal Instituut voor Hygiëne, Antwerp, Belgium.   *R. Humbel* • Centre Hospitalier de Luxembourg, Luxembourg.

3. A positive screening test in a newborn.
4. Peculiar appearance of the patient (e.g., in mucopolysaccharidoses).
5. Growth impairment (skull, height, weight) and malformation of the genitals.
6. Neurological findings such as developmental (or mental) retardation, progressive neurological deterioration, seizures, somnalence, coma, ataxia, tumors, hypo- or hypertonia, etc.
7. Anomalies of the hair, the skin, the sense organs, the bones, and the joints.
8. Organomegaly (liver, spleen, tongue).
9. Abnormal odor or color of the urine.
10. Unexplained clinical features: anemia, jaundice, cyanosis, edema, acid–base, water, and electrolyte disequilibrium; gastrointestinal symptoms; renal disease; liver disease; recurrent bleeding; recurrent infections.

The finding that severe clinical symptoms, such as mental subnormality, can be prevented by early treatment (phenylalanine-poor diet in phenylketonuria) led to the development of neonatal mass screening procedures directed towards detection of metabolic disorders in the neonatal period, before the appearance of clinical symptoms.

The development of microassays for specific enzymes has led to the establishment of enzyme deficiency tests that can be carried out on readily available cells (erythrocytes, leucocytes, fibroblast cultures, cultured amniotic fluid cells); this has made it possible to establish with greater or lesser certainty the differential diagnosis among homozygotes, heterozygotes, and normals, which in turn permits prenatal diagnosis of metabolic diseases in high-risk pregnancies and more precise genetic counseling in high-risk families.

Table I
Screening Techniques

Test	Finding
Color of urine	
Odor of urine	
Reducing test (urine)	Reducing sugars
Ferrichloride test (Phenistix®) (urine)	Phenylpyruvic acid
Cyanide nitroprusside test (urine)	Cystine, homocystine
Dinitrophenylhydrazine test (urine)	Ketones, keto acids
Alcian blue test (urine)	Mucopolysaccharides
Chromatography (urine or blood)	Amino acids, sugars, indoles, imidazoles, guanidines, purines, pyrimidines, mucopolysaccharides, oligosaccharides
Bacterial inhibition assays	
Guthrie tests (blood)	Galactose, amino acids
Bacterial assay (Paigen) (blood)	Galactose
Beutler test (blood)	Galactose
Radioimmunoassay (blood)	Enzymes
Electrophoresis (blood)	Enzymes and isoenzymes
Enzyme multiple auxotroph assay (blood)	Mass screening, urea cycle diseases[1]

Table II
Urine Color

Color and cause	Suspected metabolic disorder
Rosy to red	
Porphyrins	Hereditary porphyrias
Hemoglobins	Hereditary hemolytic anemias
Myoglobins	Rhabdomyolysis
	Glycogenosis type V and type VII
Erythrocytes (trouble)	Renal stones in cystin/lysinuria, oxalosis, xanthinuria, gout; hereditary disorders of hemostasis
Urate (precipitate)	Gout, Lesch–Nyhan syndrome
Brown to black	
Alkapton	Alkaptonuria
Urobilin, methemoglobin, hemosiderin, dipyrroles (degradation products of hemoglobin)	Hereditary hemolytic anemias
p-Hydroxyphenylpyruvic acid	Tyrosinosis
Blue	
Indigo	Blue diaper syndrome
Orange to green	
Bilirubin to biliverdin	Hereditary hyperbilirubinemias

In this chapter, we deal only with genetic diseases showing neurological symptoms, associated or not with disorders of viscera such as liver, spleen, kidneys, etc., where a hereditary anomaly is identified. We discuss the associations between clinical symptoms and biochemical anomalies to reach a better classification of neurological degenerative diseases and metabolic diseases. We conclude with directions for therapy.

2. METHODOLOGY

The methods used to study these diseases can be divided into two groups: screening techniques and techniques for the identification of metabolic disorders.

2.1. Screening

Screening is performed with thin-layer chromatography or gas chromatography by the standard methods shown in Table I. Tables II and III display the conclusions that can be drawn from the color or the odor of the urine. A review of problems related to screening has been published by Milunsky.[2]

2.2. Identification of the Disease

Identification of the disease is reached by quantitative biochemical methods, anatomopathological methods often associated with biochemical methods, or methods from other disciplines.

Table III
Urine Odor

Odor	Metabolite(s)	Suspected metabolic disorder
Musty	Phenylacetic acid	Phenylketonuria
Maple syrup	Keto acids of branched amino acids	Maple syrup urine disease
Malt	α-Hydoxybutyric acid	Methionine malabsorption, oasthouse disease
Rotten cabbage	Methionine	Tyrosinosis
Rotten eggs	H_2S	Cystin/lysinuria, homocystinuria
Rotten fish	Trimethylamine-N-oxide	Trimethylaminuria
Sweaty feet	Isovaleric acid	Isovaleric aciduria
Cat's urine	?	β-Methylcrotonyl aciduria
Acetone	Ketones (acetone, butanone, pentanone, hexanone, acetone)	Propionic aciduria Methylmalonic aciduria Glycogenoses, diabetes mellitus

2.2.1. Quantitative Biochemistry

Quantitative biochemistry can be carried out by determinations of amino acids, carbohydrates, glycogen, mucopolysaccharides, lipids, fatty acids, proteins, and lipoproteins. In addition, the metabolites of all of these substances can be determined. Determinations are carried out not only on serum, plasma, urine, or cerebrospinal fluid but also on blood cells (erythrocytes, leukocytes), on skin fibroblasts after culture, and sometimes on amniotic fluid, tears, or hair roots.[3,4] Determinations can be made after loading with different substrates or metabolites.

2.2.2. Anatomopathological Examinations

Anatomopathological examinations, which remain indispensible for all neurological examinations, can be performed by rectal, appendicular, cerebral, or neuromuscular biopsy, and also by conjunctiva or skin biopsy. Similar examinations can also be made post-mortem.

These morphological examinations are carried out with the help of optical microscopy or electron microscopy and histochemistry.[5] The latter will rest on the quantitative biochemical methods mentioned above.

2.2.3. Other Methods

Other methods are ophthalmological, such as those for hyperornithinemia (gyrate atrophy)[6] and for Tay-Sachs disease (cherry red spot), or radiological (adrenal calcification in Wolman's disease).[7]

3. PHYSIOPATHOLOGICAL CONSIDERATIONS

One has to take into account that samples are very often examined at a late or final stage of a disease. Discovery of lipidosis is the result of postmortem anatomopathological observations in cells of the nervous system which, when stained with dyes specific for lipids, showed inclusions; lipid content of these inclusions was then identified and extracted. Only later was it discovered that such anomalies were accompanied by others: α-albumin or glial fibrillary acidic proteins and GM_2 ganglioside in the brain of Tay–Sachs[8]; mucopolysaccharides in the urine and tissues of GM_1 lipidosis.

Similar problems have to be taken into account for the other diseases. What are the toxic elements in hyperargininemia[9]: excess of arginine, increased ammonia levels, the synthesized monoguanidine compounds, or the lysine cystinuria? Why, in urea cycle diseases, does guanidinosuccinic acid urinary excretion not increase as that of other monoguanidine compounds? Phenylketonuria cannot be understood without considering hyperphenylalaninemia at its inception. Neither can one accept that because a diet low in phenylalanine provides a treatment for phenylketonuria, the disease is caused by phenylalanine intoxication: it is now known that in man and animals phenylalanine intoxication does not produce the symptomatology seen in phenylketonuria. The same holds true for Refsum disease, where a diet low in phytanic acid slows down the development but does not cure the neurological lesions which are the consequence of a block in phytanic acid metabolism. In the same family, serious metabolic anomalies can affect one member although others remain unaffected (α-aminoadipic aciduria).[10] Carriers of biochemical anomalies need not present any clinical symptoms (histidinemia).[11]

These examples all show that a concept based only on the final stage of a disease has to be abandoned and that two facts must be borne in mind: (1) the anomalies in human genetic metabolic disease, even those caused by only one metabolic block, have far more complex implications, and (2) the study of the metabolic anomaly must include thorough biochemical and concurrent examinations.

These remarks can be further amplified by considerations of neurological genetic metabolic diseases.

There are multiple forms of the same disease (metachromatic leukodystrophy, Niemann–Pick disease, etc.). Some of such multiple forms are explained by multiple enzyme forms (or isoenzymes); in other cases, the explanation is not so simple. Leigh disease and type C of Niemann–Pick disease[12] can be explained by the presence of an inhibiting factor or the absence of an activating factor of the enzyme.

Garrod's hypothesis (one gene, one enzyme) may not be confirmed in some diseases: in mucolipidoses II and III, many enzyme activities are increased, and in mucosulfatidoses, the sulfatase deficiency is not specific for one substrate.

4. CLASSIFICATION OF HUMAN METABOLIC DISEASES IN NEUROLOGY

4.1. The Aminoacidopathies

The diagnosis of aminoacidopathy (Table IV) is usually established by thin-layer chromatography or automatic column chromatography or, after dansylation, by identification of metabolites and by enzymatic measurements.

Table IV
Aminoacidopathies

Aminoacidopathy	Clinical findings	Biochemistry
Hyper-β-alaninemia	Somnolence, seizures	↗β-Alanine (B, U, CSF)ᵃ; ↙β-alanine-α-ketoglutarate transaminase?
Alanine with diabetes mellitus	Mental retardation, microcephaly, dwarfism	↗Alanine, ↗glucose (B, U)
α-Aminoadipicaciduria	Mental retardation	α-Aminoadipic acid (U)
Hyperammonemia	Protein intolerance, vomiting, ataxia	↗NH₃, ↗glutamine, ↗alanine (B, U)
Type 1	Irritability, lethargy, coma, mental retardation	↙Carbamyl phosphate synthetase (liver, leukocytes)
Type 2	Same as type 1	↗NH₃, ↗glutamine, ↗alanine (B, U), ↗pyrimidines, ↙ornithine carbamyltransferase (liver, jejunum mucosa)
Hyperargininemia	Oligofrenia, epilepsy	↗Arginine (B), ↗NH₃ (B), monoguanidines (U), ↙arginase (erythrocytes, leukocytes, fibroblasts, liver)
Argininosuccinic aciduria	Mental retardation, hair abnormality (trichorrhexis nodosa)	↗Argininosuccinic acid (B, U, CSF), ↙Argininosuccinase (erythrocytes, liver, fibroblasts)
Blue diaper syndrome	Mental retardation, constipation, nephrocalcinosis, blue discoloration of urine	↗Indican (U) → indigo blue, intestinal tryptophan absorption defect
Citrullinemia		↗Citrulline (B, U, CSF), ↙argininosuccinate synthetase (fibroblast)
Dibasic aminoaciduria type 2	Mental retardation, diarrhea, vomiting	↗Cystine, lysine, ornithine, arginine (U), NH₃ (B), renal arginine, cystine, lysine, ornithine transport defect
Hyperglycinemia	Mental retardation, seizures, spacticity	↗Glycine (B, U, CSF), ↙glycine synthetase, increased CSF proteins
Hartnup disease	Mental retardation, ataxia, psychiatric changes, pellagralike rash	Generalized aminoaciduria, renal and intestinal transport defect, monoaminomonocarboxylic acids
Histidinemia	Mental retardation(?), speech defect	↗histidine (B, U, CSF), ↙histidase (skin, liver)
Homocystinuria	Mental retardation, lens luxation, marfanoid stature	↗homocystine, ↗methionine (B, U, CSF), ↙cystathionine synthetase (liver, fibroblast)
Hydroxylysinuria	Mental retardation, seizures	↗Bound hydroxylysine (B, U)
Hydroxyprolinemia	Mental retardation, hematuria	↗Hydroxyproline (B, U), ↙hydroxyproline oxidase (liver)
Hyperlysinemia (persistent)	Mental retardation, ectopia lentis	↗Lysine (B, U), ↙lysine ketoglutarate reductase
Hyperlysinemia	Episodic vomiting, spasticity, seizures, coma, mental retardation	↗Lysine, ↗NH₃ (B, U), ↙l-lysine NAD reductase (liver)

(continued)

Table IV. (Continued)

Aminoacidopathy	Clinical findings	Biochemistry
Maple syrup urine disease	Mental retardation, seizures, lethargy, coma, characteristic odor of urine	↗Leucine, valine, isoleucine (B, U) and keto acids (B, U), branched-chain keto-acid decarboxylase
Hyperornithinemia	Mental retardation, gyrate atrophy	↗Ornithine (B, U, CSF), ↙ornithine oxoacid aminotransferase (liver)
Phenylketonuria	Mental retardation, seizures, fair hair, blue eyes	↗Phenylalanine (B, U, CSF), ↙tyrosine, ↙phenylalanine hydroxylase (liver)
Hyperpipecolemia	Mental retardation, hepatomegaly	↗Pipecolate (B)
Hyperprolinemia Type 1	Mental retardation, deafness, hematuria	↗Proline (B, U), ↙Δ^1-pyrroline-5-carboxylate reductase (liver)
Type 2	Mental retardation, seizures	↗Proline (B, U), ↙Δ^1-pyrroline-5-carboxylate dehydrogenase
Saccharopinuria	Mental retardation	↗Saccharopine (B, U), ↙saccharopine dehydrogenase (liver, fibroblast)
Tyrosinemia	Mental retardation, cornea dystrophy, hyperkeratosis skin	↗Tyrosine (B, U), tyrosyl derivatives (U), ↙cytosol tyrosine transaminase (liver)
Valinemia	Mental retardation, vomiting	↗Valine (B, U), ↙valine aminotransferase (leukocytes)

[a] Abbreviations: B, blood; U, urine; CSF, cerebrospinal fluid.

These diseases are principally characterized by biochemical anomalies and fundamentally give rise to nonspecific clinical symptoms (oligophrenia, epileptic seizures, behavior disorders, etc.).

They can also affect organs other than the nervous system (Table IV). There are few or no anatomopathological lesions. For the physiopathology, the following points must be brought out. (1) A number of diseases appear on the same metabolic pathway: diseases of the aromatic amino acids (phenylketonuria, tyrosinemia, alcaptonuria), of the sulfur amino acids (homocystinuria, cystathionuria, methioninemia), of the urea cycle (hyperammonemia, OTC deficiency, citrullinemia, argininosuccinic aciduria, hyperargininemia, hyperornithinemia), and of the lysine–saccharopine pathway (lysinemia, sacchar-

Table V
Carbohydrate Disorders

Disorder	Clinical findings	Biochemistry
Fructosemia	Fructose intolerance	↙Fructose-1,6-diphosphatase, ↗fructose
Galactosemia	Mental retardation, cataract, hepatomegaly	↗Galactose (B, U), ↙Galactose-1-phosphate uridyltransferase (liver, erythrocytes, fibroblast, amniotic cells)
Glycogenose	Hypotonia	↗Glycogen (liver, muscle)
Type II	Cardiomegaly	↙Acid maltase

opinuria). (2) Some anomalies are common to several diseases: primary cystinuria and secondary cystinuria in hyperargininemia, hyperphenylalaninemia, shared by phenylketonuria and tyrosinemia, and hyperammonemia of all urea cycle diseases. (3) Some anomalies are secondary: in the plasma of hyperphenylalaninemia, the secondary anomaly is a decrease of the content of amino acids other than phenylalanine. Secondary anomalies include phenylketonuria in pheylalaninemia and the excretion of large quantities of monoguanidines[9] in the urine of patients affected by hyperargininemia. Such anomalies could play an important role in the pathology of the diseases.

4.2. Carbohydrate Disorders

Fructosemia and galactosemia are examples of carbohydrate disorders; another is the glycogenose which produces rather characteristic clinical and anatomopathological symptoms. The identification can be made through thin-layer chromatography and enzyme measurements (Table V).

Table VI
Mucopolysaccharidoses

Mucopolysaccharidosis	Clinical symptoms	Biochemistry
I_H	Hurler (oliogophrenia)	↙ α-L-Iduronidase (leucocytes), ↗ heparan and dermatan sulfate (urine)
I_S (Scheie)	Normal intelligence	↙ α-L-Iduronidase (leucocytes), ↗ heparan and dermatan sulfate (urine)
II	Hunter (normal intelligence)	↙ α-L-Iduronidase, ↗ heparan in greater amounts and dermatan sulfate (urine)
III_A	San Filippo	N-Acetyl-α-D-glucosaminidase normal or increased (serum), ↙ heparan sulfate sulfamidase (serum), ↗ heparan sulfate in urine
III_B	San Filippo	↙ N-Acetyl-α-D-glucosaminidase (serum), ↙ heparan sulfate sulfamidase (serum), ↗ heparan sulfate in urine
IV Morquio (different forms)	Normal intelligence, athetosis	↗ Keratan sulfate (urine)
V Scheie (or I_S)	Normal intelligence	↙ α-L-Iduronidase, ↗ heparan and dermatan sulfate (urine)
VI Marotteaux–Lamy	Normal intelligence	↙ Arylsulfatase B, ↗ dermatan sulfate (urine)
VII	Similar to Hurler syndrome	↙ Glycuronidase (serum), ↗ chondroitin sulfate (urine)
VIII(?)		↙ β-D-Glycuronidase, glycosaminoglycans normal (urine)

Table VII
Mucolipidosis (MLD)

Disorder	Biochemical findings
Mucolipidosis I (sialidosis)	Acid hydrolase (serum) normal, foam cells in bone marrow, abnormal excretion of oligosaccharides (urine), glycosaminoglycans normal (urine)
Mucolipidosis II	↗α-Mannosidase
Mucolipidosis III	↗All serum hydrolase (except acid phosphatase), same enzymes are decreased in the cells, urinary excretion normal
Mucosulfatidosis	↙Multiple sulfatase (arylsulfatase) (serum and leukocytes), ↗glycosaminoglycans and sulfatides stored (urinary excretion of chondroitin and heparan sulfate and sulfates)
GM₁ gangliosidosis	↙β-D-Galactosidase (serum and leukocytes), ↗keratan sulfate and galacto-oligosaccharides (urine), storage of galactose-containing compounds (GM₁ gangliosides, keratan sulfate, and galacto-oligosaccharides)
Fucosidosis	↙α-L-Fucosidase (serum), ↗keratan sulfate in urine

4.3. Diseases of Glycoconjugation (Mucopolysaccharidoses, Mucolipidoses, and Oligosaccharidoses)

Coarse face, skeletal anomalies accompanied by dwarfism, oligophrenia, corneal clouding, and hepatosplenomegaly are the main clinical features of these lysosomal diseases.

Seven or eight different mucopolysaccharidoses are usually mentioned (Table VI). Urinary excretion of glycosaminoglycans is increased, and glycosaminoglycans are accumulated in tissues in these diseases.

In mucolipidoses, glycosaminoglycans, sphingolipids, and glycolipids are accumulated without abnormal urinary excretion, except in GM₁ gangliosidosis and mucosulfatidosis (Table VII). In this group, GM₁ gangliosidosis is a combined gangliosidosis and mucopolysaccharidosis.[13]

In the third group, oligosaccharides rich in mannose or fucose or aspartyl groups[14] are stored (Table VIII). Diagnosis rest on three things[15]:

1. Detection of vacuolated lymphocytes among bone marrow cells. In mucopolysaccharidoses, one often finds granules in the vacuoles.

Table VIII
Oligosaccharidoses

Disorder	Biochemical findings
Mannosidosis	↙α-D-Mannosidase, isozyme at pH 3.5 only (serum), mannose-rich oligosaccharides
Fucosidosis	↙α-L-Fucosidase, ↗fucose-rich oligosaccharides, glycolipids, and keratan sulfate (urine)
Aspartyl glycosaminuria	↙Aspartylglycosamine aminohydrolase, ↙aspartylglycosamine and aspartyl oligosaccharides

2. Characterization[16] and determination[17,18] of glycosaminoglycans, oligosaccharides,[19] and glycolipids. These occur mainly in urine.

3. Determinations of glycosidic hydrolases and lysosomal enzymes (α-galactosidase, β-galactosidase, α-mannosidase, α-fucosidase, hexosaminidases, β-glycuronidase, arylsulfatase A and B,[20] N-acetyl α-D-glucosaminidase, α-L-iduronidase) in leukocytes, with suitable substrates and at suitable pH.

The increase in mucolipidoses II and III of serum lysosomal enzymes except acid phosphatase and the multiple sulfatase deficiency[21] of mucosulfatidoses, both give pictures different from that of classical metabolic diseases (one gene, one enzyme).

Recently articles were published concerning the problems discussed here. We shall mention two of them. E. H. Kolodny et al.[22] mention three cases in which parents of patients affected with metachromatic leukodystrophy in whom arylsulfatase A activity is decreased, suffer from a neurological disease. The authors consider the neurological disease to be multiple sclerosis. We wish to draw attention to the fact that the given clinical information is not convincing. In our opinion, to accept the diagnosis of multiple sclerosis, an indisputable clinical observation is required as well as an electrophoretic analysis in agar gel or agarose revealing the immunological reaction which always accompanies this disease.

D. M. Swallow et al.[23] demonstrate that the electrophoretic mobilities of different glycoprotein enzymes are modified in sialidoses and mucolipidoses. They explain this phenomenon either by the fact that the sialic acid was not normally removed from the enzymes or by a combination inter-medium of these enzymes with sialic traces.

Table IX
Disorders of Lipid and Fatty Acid Metabolism

Disorder	Symptoms	Histological findings	Biochemical findings
Tay–Sachs (GM$_2$ gangliosidosis)	Severe dementia with epilepsy, cherry-red spot	Membranous bodies	GM$_2$ ganglioside increased in tissues; hexosominidase A decreased (serum, fibroblast, amniotic fluid, leukocytes, hair)[24]
GM$_1$ gangliosidosis	Dementia with amaurosis	Membranous bodies	GM$_1$ gangliosides increased in tissue, mucopolysaccharides in urine; decrease of β-galactosidase
Gaucher	Ataxia or neurologically normal	Inclusions in central nervous system, spleen, liver, bones	Increased glucocerebrosides; decrease of β-glucosidase
Niemann–Pick (different forms)	Dementia and epilepsy, splenomegaly and hepatomegaly	Inclusions in central nervous system, spleen, liver, bone marrow	Increase of sphyngomyelins; decrease of sphyngomyelinase (leukocytes)

(continued)

Table IX. (Continued)

Disorder	Symptoms	Histological findings	Biochemical findings
Krabbe	Diffuse sclerosis	Absence of myelin, globoid cells	Normal myelin; decrease of β-galactosidase and galactocerebroside (leukocytes, fibroblasts, cells of amniotic fluid)
Fabry[25]	Angiokeratom cornea verticillata, polyneuropathy,[26] cardiovascular lesions		Increase of ceramide di- and trihexosides (quotient 6/1 in men and 1/1 in women); abnormal mucopolysaccharides[27]; decrease of galactosidase in fibroblasts, skin, and amniotic fluid cells
Metachromatic leukodystrophy (different forms)	Diffuse sclerosis with dementia and epilepsy, spastic diplegia[28]	Demyelination and increase of metachromatic substances	Increased sulfatides in nervous system and urine[29]; decrease of arylsulfatase A in leucocytes, fibroblasts, and cells of amniotic fluid[30]
Refsum	Heredoataxia and polyneuritis, albuminocytological dissociation	Heredoataxia and increase of tissue lipids	Increased phytanic acid and failure of oxidation of phytanic acid
Ceroid lipofuchsinosis	Dementia and epilepsy	Inclusions	?
Wolman[31]			Increase of cholesterol esters and triglycerides
Absence of β-lipoproteins	Heredoataxia and polyneuropathy[32]	Heredoataxia	Absence of β-lipoproteins and acanthocytosis
Schilder	Diffuse sclerosis and adrenal insufficiency	Diffuse sclerosis	Increase of long-chain fatty acids
Adrenomyeloneur opathy	Spastic paraplegia with adrenal insufficiency		Increase of long-chain fatty acids

4.4. Disorders Due to Lipids and Fatty Acids

These diseases (Table IX) present very typical clinical, biochemical, and anatomopathological pictures. Diagnosis can be established by using thin-layer chromatography and gas chromatography, enzyme and isoenzyme measurements (Tay–Sachs disease, metachromatic leukodystrophy) of urine and blood specimens, and especially by histological and histochemical techniques. A distinction must be made between diseases that are solely neurological such as Tay–Sachs disease and those that also affect the viscera such as Niemann–Pick disease. A difference should also be made between diseases solely due to lipids and those solely due to fatty acids.

4.5. *Organic Acidurias: Errors of Metabolism that Cause Acute Life-Threatening Illness in the first Weeks of Life*

Organic acidurias are characterized by an abnormal excretion of nonamino carboxylic acids. We include here disorders with metabolic acidosis caused by an acidic metabolite specifically related to the metabolic lesion (Table X). We associate with them, for therapeutic reasons, another group that can present a similar dangerous situation: the primary hyperammonemias (Table XI). These patients are usually born in good health, near term, and of normal size because before birth the placenta provides effective hemodialysis and prevents the accumulation of toxic products. Acute illness appears after an interval that is shorter or longer as the metabolic block is more or less lethal. The physiological immaturity of some enzyme systems and the normal hypercatabolism of the newborn contribute to precipitation of the illness.

The appearance of symptoms can be related to the introduction or increase of protein intake, to a stress situation, or to infections that increase the endogenous hypercatabolism. The major clinical symptoms of this group of disorders are not specific, and they overlap those of frequent and common illness of the neonatal period: anoxia, sepsis, transient hyperglycemia, tetany, pyloric stenosis, intracranial bleeding, and atypical respiratory distress syndrome.

Most of these patients began their illness with the refusal of food, violent vomiting, and irritability, followed by seizures (focal or generalized, tonic–clonic or myoclonic, of a changing pattern from hour to hour), severe hypotonia and/or crisis of hypertonia, respiratory troubles with periodic rhythm or apneic spells, and lethargy leading to coma and death if an appropriate therapy is not applied. The primary biochemical abnormalities are metabolic acidosis or res-

Table X
Organic Acidurias

Propionic acidemia
Methylmalonic aciduria
3-Methylcrotonylglycine–3-hydroxyisovaleric aciduria
Isovaleric acidemia
2-Methylacetoacetic–2-methyl-3-hydroxybutyric aciduria
Pyroglutamic acidura
Pyruvate carboxylase deficiency
Pyruvate dehydrogenase deficiency
Congenital lactic acidosis
Succinyl CoA–3-ketoacid CoA transferase deficiency
2-Ketoadipic aciduria
Butyric and *n*-hexanoic acidemia
Oasthouse urine disease
Maple syrup urine disease
D-Glyceric acidemia type I
D-Glyceric acidemia type II
Glutaric aciduria type I[33]
Glutaric aciduria type II
3-Hydroxy-3-methylglutaric aciduria

Table XI
Urea Cycle Diseases

Carbamylphosphate synthetase deficiency
Ornithine transcarbamylase deficiency
Citrullinemia, acute neonatal form
Citrullinemia type II
Argininosuccinic acidemia, neonatal form
Argininemia

piratory alkalosis, hyperammonemia, hypoglycemia, and ketosis, alone or in combinations.

The presence of abnormal smell must be carefully investigated. An appropriate therapy will cause the child to improve. The damage to the nervous system is related to the recurrences and the duration of the crisis.

In other cases, there is normal growth with development of various neurological deficits: ataxia, mild or severe mental retardation, extrapyramidal signs, seizures, etc. The records of these patients also show periods of refusal of food with crises of irritability, vomiting, and lethargy.

Both groups of cases frequently have a history of multiple unexplained perinatal deaths in the patients' siblings.

4.5.1. Initial Tests

1. Blood: pH, electrolytes, sugar, Ca^{2+}, Mg^{2+}, urea, transaminase, hemocultures, ammonium.
2. Urine: pH reducing sugars, ketones, abnormal odors, sodium and potassium balance.
3. Cerebrospinal fluid: routine and bacteriological examination.

4.5.2. Specialized Investigations

1. Urinary orotic acid determination.
2. Organic acids screened by gas chromatography or liquid partition chromatography adapted for organic acid analysis. The quantitation of this compound and the definite identification requires a combination of gas chromatography and mass spectrophotometry.
3. Autopsies should be performed immediately after death, and samples of liver, kidney, muscle, and brain should be stored frozen.

4.5.3. Prenatal Diagnosis

Although a direct chemical test on the amniotic fluid or sometimes on the mother's urine can detect the presence of some abnormal organic acids, the enzymatic assay on cultured amniotic fluid cells is still essential for the prenatal diagnosis of these disorders.

4.6. Diseases with Metal or Trace Element Accumulation

Among these, we cite Wilson's disease, caused by a generalized accumulation of copper, and Hallervorden–Spatz disease, caused by an accumulation of iron in the basal and mescencephalon ganglia. Trace metal measurements are carried out after low-temperature ashing and atomic absorption.

4.7. Diseases That Are Difficult to Classify

We have grouped in Table XII diseases that are difficult to classify.

Table XII
Diseases That Are Difficult to Classify

Disease	Clinical findings	Biochemistry
Acrodermatitis enteropatica	Irritability, tremor, ataxia, dermatitis, paronychia, alopecia	↙Zinc (B), intestinal transport defect for zinc
Adenosine deaminase deficiency	Hypotonia, irritability, nystagmus, mental retardation, combined immunodeficiency	↗Deoxy-ATP, -ADP (erythrocytes), ↙adenosine deaminase (erythrocytes)
Aldolase A deficiency	Mental retardation, hepatomegaly, anemia	Fructose biphosphate aldolase (erythrocytes)
Ataxia–telangiectasia	Ataxia, telangiectasia, infection	↙IgA (B), DNA excision repair enzyme
Hypercalcemia (idiopathic)	Mental retardation, aortic stenosis	↗Ca^{2+} (B)
Cockayne syndrome	Microcephaly, mental retardation, photosensitive dermatitis, dwarfism	Slow DNA synthesis
Crigler–Najjar syndrome	Kern icterus	↗Bilirubin (conjugated) (B), ↙UDP-glucoronosyltransferase
Formiminotransferase deficiency	Mental retardation	↗Folate (B), figlu-test + (U), ↙formiminotransferase
Lesch–Nyhan syndrome	Mental retardation, choreoathetosis	↗Uric acid (B), ↙hypoxanthine–guanine phosphoribosyl transferase (erythrocytes, fibroblast, amniotic cells)
Hypomagnesemia (primary)	Tetany	↙Mg^{++} (B), intestinal transport defect for Mg
Leigh disease	Brainstem syndrome	↗Lactate, pyruvate, ketoglutarate, ↗alanine, ↙pyruvate carboxylase, inhibition of thiamine triphosphate activity by urinary inhibitor
Menkes syndrome	Mental retardation, kinky hair (pili-torti)	↙Cu (B), intestinal transport defect
Methemoglobinemia	Mental retardation, cyanosis	↗Methemoglobin, ↙NADH-linked methemoglobin reductase (diaphorase) (erythrocytes)

Table XII. (Continued)

Disease	Clinical findings	Biochemistry
Porphyria (acute intermittent)	Neuralgia, paralysis, abdominal pain, seizures	↗Porphobilinogen (B, U), ↗ALA (U), ↙uriporphyrinogen I synthetase (erythrocytes)
Pseudouridinuria	Mental retardation	↗Pseudouridine (U)
Pyridoxine dependency with seizures	Seizures, mental retardation	↗Xanthurenic acid after tryptophan (U), ↙glutamate decarboxylase (kidney)
Thyroid dyshormonogenesis (5 types)	Mental retardation, goiter, hypothyroidism	Disturbances of thyroid metabolism (↙T_4, ↗TSH) (B), several enzymes
Triosephosphate isomerase deficiency	Mental retardation, hemolytic anemia	↗Dihydroxyacetone phosphate (erythrocytes), ↙Triosephosphate isomerase (erythrocytes)
Xanthurenic aciduria	Mental retardation	↗Xanthurenic acid (U), ↙kynureninase (liver)
Xeroderma pigmentosum of Sanctis and Cacchione	Microcephaly, mental retardation, choreoathetosis, spasticity, deafness, xeroderma, skin cancer (photosensitivity), eye lesions	DNA excision repair enzyme

5. THERAPEUTIC CONCLUSIONS

5.1. Diet Deprived of Accumulated Substances

It is appropriate to present diets deprived of the accumulated substances (phenylalanine in phenylketonuria, galactose in galactosemia, fructose in fructosemia, leucine, isoleucine, and valine in leucinosis, proteins in hyperammonemia and hyperlysinemia).

In other diseases such as hyperornithinemia with gyrate atrophy, diets with low doses of arginine should be prescribed.[6]

In lysinuria resulting from the lack of intestinal absorption of arginine and lysine, a diet enriched in citrulline has been recommended.[34]

5.2. Vitamins

Vitamin B_6 is useful in homocystinuria,[35] as are biotin in propionicaciduria and vitamin B_{12} in methylmalonicaciduria.

5.3. Detoxication Therapies

Penicillamine dissolves copper in Wilson's disease and helps to improve patient condition.

5.4. Treatment of Organic Acidurias

We wish to add here some additional indications concerning treatment of organic acidurias as new possibilities have been discovered in this field[36]:

1. Treatment of the acute situation. Plasmapheresis and/or peritoneal dialysis should be applied promptly to prevent biochemical induction of neurological deterioration.
2. Correction of acidosis or alkalosis and fluid balance.
3. Restriction of nitrogen intake to the absolute minimum for anabolic needs.
4. Correction of the hypoglycemia; provision of sufficient calories to prevent tissue catabolism.
5. Administration of multiple vitamins (500 times the minimum requirement) in the hope that a vitamin-dependent defect is present.

5.5. Substitution Therapy by Enzymes

Erythrocyte loading, liposomes, etc. have been discussed elsewhere.[37]

5.6. Prevention

Prevention implies genetic diagnosis in members of the family and methods for prenatal detection, e.g., amniotic fluid puncture for biochemical measurement or fibroblast cultures and enzyme measurements.[38]

REFERENCES

1. Naylor, E. W., in press.
2. Milunsky, A., 1981, *Am. J. Med.* **70**:7–8.
3. Hösli, P., Schneck, L., Amsterdam, D., and Volk, B. W., 1977, *Lancet* **1**:285–287.
4. Antoon, J., Janssens, M., Plakke, T., Trijbels, F. J. M., Sengers, R. C. A., and Monnens, L. A. M., 1981, *Clin. Chim. Acta* **113**:213–216.
5. Martin, J. J., Ceuterick, C., and Libert, J., 1979, *Prosp. Pediatr.* **36**:449–467.
6. Sipha, I., Rapola, J., Simell, O., and Vannas, A., 1981, *N. Engl. J. Med.* **304**:867–871.
7. Schaubn, J., Janka, G. E., Christomanou, H., and Sandhoff, K., 1980, *Eur. J. Pediatr.* **135**:45–54.
8. Karcher, D., Lowenthal, A., and Zeman, W., 1972, *Adv. Exp. Med. Biol.* **19**:151–162.
9. Terheggen, H. G., Lowenthal, A., Lavinha, F., and Colombo, J. P., 1975, *Arch. Dis. Child.* **50**:57–62.
10. Lormans, S., and Lowenthal, A., 1974, *Clin. Chim. Acta* **57**:97–101.
11. Van Sande, M., De Raedt, R., and Lowenthal, A., 1969, *Tijdschr. Geneeskd.* **20**:1028–1031.
12. Christomanou, H., 1980, *Hoppe Seylers Z. Physiol. Chem.* **361**:1489–1502.
13. Suzuki, K., 1968, *Science* **159**:1471–1472.
14. Humbel, R., Marchal, C., and Fall, M., 1974, *J. Pediatr.* **84**:456.
15. Humbel, R., 1975, *Helv. Paediatr. Acta* **30**:191–200.
16. Humbel, R., and Chamoles, N. A., 1972, *Clin. Chim. Acta* **40**:290–293.
17. Humbel, R., 1971, *Glycosaminoglycanes Urinaires dans les Mucopolysaccharidoses*, Thesis, Poitiers.
18. Humbel, R., 1974, *Clin. Chim. Acta* **52**:173–177.

19. Humbel, R., and Collart, M., 1975, *Clin. Chim. Acta* **60**:143–145.
20. Humbel, R., 1982, *Clin. Chim. Acta* (in press).
21. Eto, Y., Numaguchi, S., Tahara, T., and Rennert, O. M., 1980, *Eur. J. Pediatr.* **135**:85–90.
22. Kolodny, E. H., Srinivasan, S., Raghavan, S. S., Lott, I. T., and Sergay, S. M. 1981, *Neurology (New York)* **31**:88–98.
23. Swallow, D. M., O'Brien, J. S., Hoogeveen, A. T., and Buck, D. W., 1981, *Ann. Hum. Genet.* **45**:29–37.
24. O'Brien, J. S., 1973, *Fed. Proc.* **32**:191–199.
25. Brady, R. O., Uhlendorf, B. W., and Jacobson, C. B., 1971, *Science* **172**:174–175.
26. Kocen, R. S., and Thomas, P. K., 1970, *Arch. Neurol.* **22**:81–88.
27. Matalon, R., Dorfman, A., Dawson, G., and Sweeley, C. C., 1969, *Science* **164**:1522–1523.
28. Austin, J. H., 1957, *Neurology (Minneap.)* **7**:415–426.
29. Stumpf, D., and Austin, J., 1971, *Arch. Neurol.* **24**:117–124.
30. Leroy, J. G., Van Elsen, A. F., Martin, J. J., Dumon, J. E., Hulet, A. E., Okada, S., and Navarro, C., 1973, *N. Engl. J. Med.* **288**:1365–1369.
31. Lake, B. D., and Patrick, A. D., 1970, *J. Pediatr.* **76**:262–266.
32. Miller, R. G., Davis, C. J. F., Illingworth, D. R., and Bradley, W., 1980, *Neurology (Minneap.)* **30**:1286–1291.
33. Leibel, R. L., Shih, V. E., Goodman, S. I., Bauman, M. L., McCabe, E. R. B., Zwerdling, R. G., Bergman, I., and Costello, C., 1980, *Neurology (Minneap.)* **30**:1163–1168.
34. Rajantie, J., Simell, O., Rapola, J., and Perheentupa, J., 1980, *J. Pediatr.* **97**:927–932.
35. Gröbe, H., 1980, *Eur. J. Pediatr.* **135**:199–204.
36. Batshaw, M. L., and Brusilow, S. W., 1980, *J. Pediatr.* **97**:893–900.
37. Adriaenssens, K., Karcher, D., Lowenthal, A., and Terheggen, H. G., 1976, *Clin. Chem.* **22**:323–326.
38. Martin, J. J., Leroy, J. G., Ceuterick, C., Libert, J., Dodinval, P., and Martin, L., 1981, *Acta Neuropathol. (Berl.)* **53**:87–91.

Human Brain Postmortem Studies of Neurotransmitters and Related Markers

E. D. Bird and L. L. Iversen

1. INTRODUCTION AND HISTORICAL ASPECTS

During the last two decades many new techniques have been developed to measure neurotransmitters and neurotransmitter-related enzymes and receptor sites in the central nervous system. The application of such neurochemical techniques, together with histochemical staining methods, has greatly expanded our knowledge of transmitter-specific pathways in the animal brain. The possibility that neurotransmitter abnormalities underlie a variety of neuropsychiatric disorders in man has also prompted similar studies in normal and pathological human brain.

One of the earliest and most important examples of this genre was the discovery of the dopamine abnormality in Parkinson's disease. Although Blaschko in 1939 had stressed the biological importance of dopamine (DA) as a precursor of norepinephrine,[1] it was not until Carlsson and Waldeck utilized a sensitive spectrophotofluorimetric method for measuring catecholamines that DA was found to be localized in the basal ganglia.[2,3] These methods were rapidly applied by Ehringer and Hornykiewicz[4] to the examination of postmortem brain tissue from patients dying with Parkinson's disease, and the deficiency of dopamine that was found in the basal ganglia of the Parkinson's disease brain led to the development of the use of L-DOPA to correct this deficiency in patients.

Huntington's disease (H.D.), a dominantly inherited disorder characterized by degeneration of cells in the basal ganglia, is considered to be the clinical converse of Parkinson's disease: choreiform uncontrollable movements occur

E. D. Bird • McLean Hospital, Harvard Medical School, Belmont, Massachusetts 02178. *L. L. Iversen* • Neurochemical Pharmacology Unit, Medical Research Council Centre, Medical School, Cambridge CB2 2QH, England.

early in the disorder rather than tremor and rigidity. No specific biochemical abnormalities in peripheral tissues have been discovered in this disorder. However, examination of the H.D. postmortem brain has revealed a loss of GABAergic neurons from the basal ganglia.[5,6]

Three separate laboratories in Great Britain have studied postmortem brains of patients dying with Alzheimer's disease and have found a marked decrease in the activity of choline acetyltransferase, the biosynthetic enzyme for acetylcholine, in the cerebral cortex.[7-9] This has led to a number of therapeutic trials with cholinelike agents with variable responses.

Schizophrenia, a nondegenerative disorder of the brain, is currently undergoing active neurochemical investigation. Interest has focused on the catecholamine systems in the postmortem brain, and some of the results obtained are discussed below.[10]

Attention is also beginning to be directed to other disorders whose symptoms can be modified by neuropharmacological agents, e.g., dystonia, Gilles de la Tourette Syndrome, and tardive dyskinesia. However, not enough postmortem brain samples have been available in these disorders so far for neurochemical studies, so they are not discussed in this chapter.

2. SELECTION, COLLECTION, HANDLING, AND STORAGE

Postmortem tissues from any disease state must be collected under the same circumstances as tissues from control subjects. The selection of control subjects can be difficult, as both premortem and postmortem factors need to be matched. Control tissues should be obtained from patients free of any central nervous system disease. Patients who die in metabolic imbalance, such as diabetic coma or uremic or hepatic failure, should also be excluded. Accidental deaths or immediate deaths from myocardial infarct may provide control brain tissue with optimum conditions for maintaining neurochemical markers. However, many of the patients suffering from the neurological disorders we are interested in often die after prolonged terminal illness such as bronchopneumonia with, no doubt, a degree of cerebral anoxia prior to death. It is therefore also necessary to collect brain tissue from control patients who die from bronchopneumonia without prior neurological disease to serve as another control group. This, therefore, requires a fairly large supply of control tissue, which can be obtained from one of the national tissue resource centers that have now been established in various countries.*

In addition to the clinical state of the patient, medications that a patient may have received need to be considered, especially long-term treatment with psychotropic agents. It is sometimes possible to compare two different neurological disease states treated with similar drugs; i.e., Huntington's disease patients treated with neuroleptics for some years can be compared with schiz-

---

* Information on availability of postmortem specimens for research can be obtained by contacting MRC Brain Bank, MRC Neurochemical Pharmacology Unit, Medical Research Council Centre, Medical School, Hills Road, Cambridge, England, or Dr. E. Bird, Brain Tissue Resource Center, McLean Hospital, Belmont, Massachusetts 02178 U.S.A.

ophrenic patients who have received the same agents for a similar length of time.

It has been shown that many neurochemicals are stable in the postmortem brain for several hours provided the corpse has been stored at 4°C in a mortuary refrigerator and the brain not removed from the skull until just prior to sectioning.[11] Specific studies such as electron microscopy for immunocytochemistry may require prompt removal of the brain from the skull after death for immediate fixation. For most of these studies, plans need to be made prior to death either by having the afflicted person sign a donor card similar to those used in the donations of eyes and kidneys or by having the attending physician discuss the autopsy procedure with the next-of-kin.

In the United States, an organ donation signed by the donor is legal after death, according to the Anatomical Gift Act; however, this witnessed document must have been signed when the patient was mentally competent. If no donor card is in existence, the autopsy should not be carried out until authorization has been signed by the next-of-kin. This can be received by telephone with two recording witnesses at the hospital. When this procedure is planned ahead, the autopsy can be carried out promptly.

In order to shorten the delay between death and autopsy, it is useful to have available, in advance, two protocols:

1. Nurses' Protocol. When an oral agreement regarding a future autopsy has been expressed by the next-of-kin, a protocol for handling the body after death should be placed inside the front cover of the patient's chart so that all staff will be aware that an autopsy is planned. Once death has been certified by a physician, the corpse should be taken promptly to the Pathology Department and placed in a 4°C refrigerator in preparation for autopsy. The attending physician and the pathologist need to be alerted in order that the autopsy can proceed without delay.
2. Pathologists' Protocol. A complete autopsy is always to be carried out, since the information derived from a complete examination may be vital to the interpretation of the neurochemical and histopathological findings in the brain.

The brain is removed promptly along with as much of the pituitary stalk as possible in order to preserve the median eminence. The brain should be sliced sagittally in the midline through the brainstem, cerebellum, and cerebral hemisphere. One-half of the brain is placed in a suitable fixative, usually buffered 10% formalin. The other half can be placed in a plastic container lined with plastic wrap with the flat, medial surface of the brain down in order to maintain the anatomic configuration of the brain when it freezes. The box is sealed and placed in a deep freeze at -20°C immediately and then transferred for more permanent storage at -70°C.

It is important to be aware that frequently the brain is the last organ of the body to be examined, in which case, the brain is removed and placed in a bowl at room temperature until the complete autopsy is done. For the purposes of neurochemistry, the brain should not be removed from the skull until it is to be examined, since certain neurochemical markers deteriorate rapidly after such removal.

Prior to dissection, the brain can be transferred from the $-70°$ freezer to $-15°$ for 24 hr. The frozen brain is sliced into 3-mm coronal sections, and these are placed on a frozen surface. This frozen surface can be made of stainless steel with refrigerated coils beneath or a sheet of thick plate glass on top of a box of dry ice. The surface should be maintained at $-5°$ to $-10°$ in order to keep the whole tissue slice frozen. At this temperature, it is possible to differentiate white from gray matter, making it suitable for dissecting neuroanatomically defined areas.

3. DISSECTION

The dissection of the brain can proceed using standard atlases.[12,13] Starting at the frontal pole of the cerebrum, Brodmann[14] regions of the cortex can be dissected, chopped with a fine blade to produce a homogeneous mix, divided into representative aliquots, and placed into suitable tubes kept frozen in dry ice, and a similar procedure is used for all other dissected regions. The anatomic dissection of the brain proceeds posteriorly until the anterior pole of the striatum appears. As one proceeds more caudally, the internal capsule makes its appearance in the striatum, separating the caudate nucleus from the putamen.

In man, the nucleus accumbens has become enveloped by the anterior striatum, and in many human brain atlases, it is not defined. On gross appearance, it is difficult to differentiate this nucleus from the caudate and putamen. In order to standardize the dissection of this nucleus, it is recommended that Brockhaus's[15] description be used. Brockhaus described histopathological findings that divide the accumbens area into four nuclei.[15] For those wishing to dissect each of these discrete nuclei, we would recommend his detailed description.[15] For others, the entire nucleus accumbens, which includes the four subnuclei, can be removed from respective slices and called the "accumbens area (fundus striati) of Brockhaus" (see Fig. 1a and b).

The accumbens area begins on the first coronal section in which the internal capsule completely divides the putamen from the caudate and continues caudally to the level of the anterior commissure. When the anterior commissure appears on the caudal surface of a slice, the accumbens area can be dissected from the portion of the slice that lies rostral to the anterior commissure. The bed nucleus of the striae terminalis can be found on the dorsal and medial ventricular surface of the caudate where it forms a firm area that can be differentiated from the caudate. At this same level on the coronal slice, the anterior tip of the temporal lobe can usually be seen where cortical Brodmann Area 38 can be taken. Depending on how well the midline sagittal cut was made, the ventral and dorsal parts of the septal nuclei can be dissected.[35]

---------→

Fig. 1. Coronal sections of human brain. a: Section approximately 35 mm posterior of the anteriormost tip of the frontal lobe. b: Section approximately 45 mm posterior of the anteriormost tip of the frontal lobe showing the beginning of the anterior commissure. C, caudate; P, putamen; I.C., internal capsule; Acc, nucleus accumbens; A.C., anterior commissure.

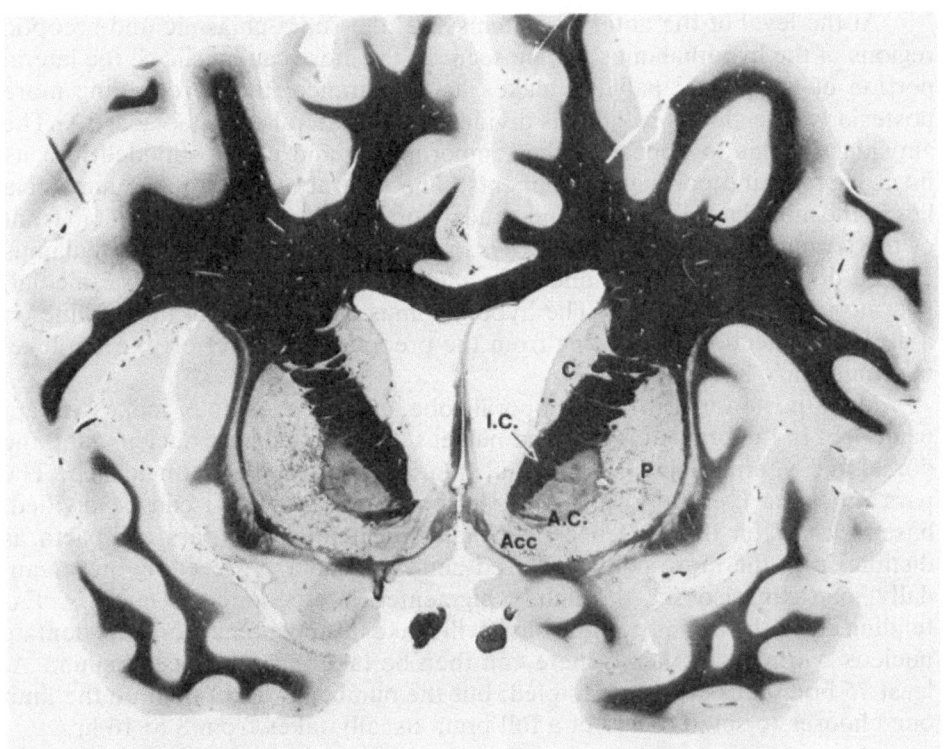

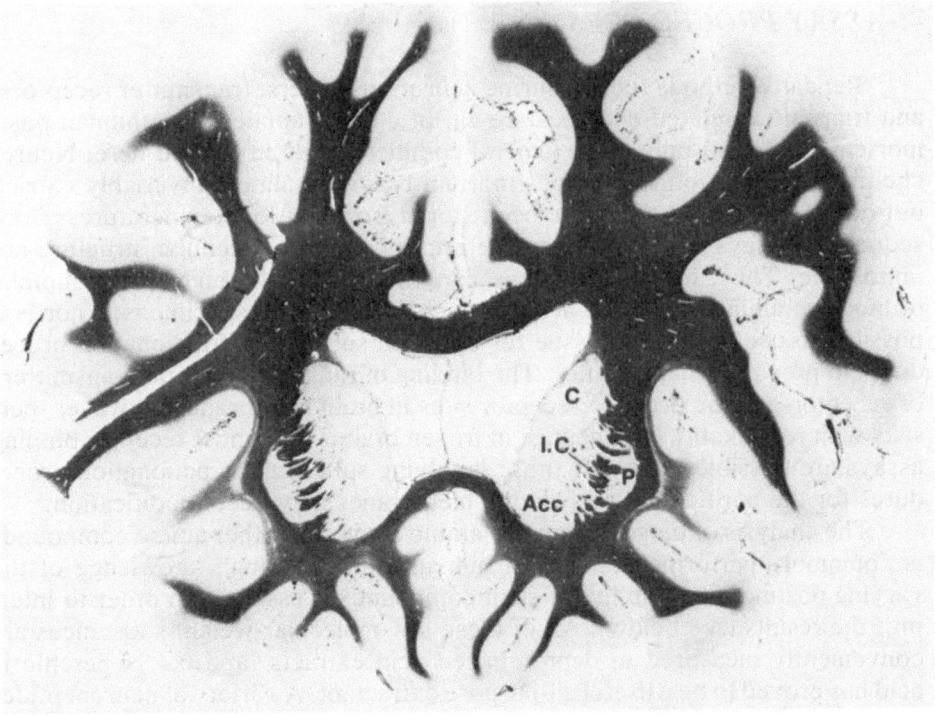

At the level of the anterior commissure, the supra-chiasmic and preoptic regions of the hypothalamus will be seen. In the more caudal slices, the lateral portion of the globus pallidus makes its appearance; then proceeding more posteriorly, the globus pallidus is divided into its lateral and medial areas. The amygdala begins to appear in the temporal lobe and can be divided into its nuclear areas and cortical components. The central nucleus of the amygdala lies in the more lateral, dorsal, and caudal portion of the amygdala, just in front of the termination of the caudate tail. In the midline, regions of the thalamus can be divided into ventral and anteromedial, ventral lateral, dorsal medial, and more posterior regions. The hypothalamus can be seen in the midline on a number of sections extending from the preoptic area anteriorly to the level of mammillary body.

More posteriorly in the temporal lobe, the hippocampus makes its appearance. In the brainstem, the red nucleus, the subthalamic nucleus, and the dorsal and ventral aspects of the substantia nigra can be distinguished. The pars reticulata (ventrally) and the pars compacta (dorsally) can be divided, based largely on the presence of melanin pigment in the pars compacta as distinct from the more lightly colored zona reticulata. Proceeding more caudally, one can also see the lightly pigmented interpeduncular nucleus. Extending down the brainstem, the olive will make its appearance, and the dentate nucleus, vermis, and hemisphere can then be taken from the cerebellum. At least 75 brain areas can be sampled, but the number is dependent on the limit one chooses to set. To dissect a full brain usually takes from 8 to 16 hr.

4. ASSAY PROCEDURES

Standard methods for measuring neurotransmitters, transmitter receptors, and transmitter-related enzymes are all, of course, applicable to human postmortem brain, and only some general comments will be offered here. Neurochemical analyses on human postmortem brain are almost invariably carried out on tissue samples that have been stored frozen at low temperatures. Consequently, assays that depend on the presence of intact cellular structure are impossible. Thus, for example, one cannot measure the high-affinity uptake of biogenic amines or amino acids in such frozen tissue specimens,[16] nor is it possible to use the frozen tissue for standard subcellular fractionation procedures to prepare synaptosomes. The binding of radiolabeled neurotransmitters or receptor-specific drugs to receptor sites in brain membranes, however, persists with remarkably little change in frozen brain,[17] and most receptor binding assays are possible, although those involving subcellular fractionation procedures for the purification of synaptic membranes may need modification.

The analysis of biogenic amines, amino acids, and other amino compounds is commonly performed on frozen human brain, although knowledge of the varying postmortem stability of such compounds is essential in order to interpret the results (see below). All of these low-molecular-weight substances are conveniently measured in deproteinized acid extracts, and 0.4 N perchloric acid has proved to be a useful all-purpose extractant. A variety of neuropeptides can also be measured by radioimmunoassay in human brain extracts. Here, the use of an appropriate extraction procedure is important. To avoid any possible degradation of brain peptides by peptidases during the extraction procedure, boiling is commonly used. We have found that extraction in boiling 1 M acetic acid followed by homogenization, centrifugation, and freeze-drying of supernatant samples is applicable to measurements of most brain peptides, although some such as cholecystokinin may not be adequately soluble in acetic acid and require an additional extraction step, in this case with boiling water.

Many neurotransmitter-related enzymes can be assayed, usually by radiochemical methods, in frozen human brain samples. For enzyme assays, the tissue is usually homogenized in distilled water or in dilute assay buffer, and the assay may also include a detergent such as Triton X-100 to insure release of any occluded enzyme.[18] An important but often overlooked point is that enzymes should be assayed wherever possible under optimal conditions, so that the measured activity approximates as closely as possible V_{max}. The use of subsaturating concentrations of substrate and/or cofactors may often be convenient, particularly in conferring greater sensitivity to a radiochemical enzyme assay, but the results will then not be easy to compare with other published data.[18] This is unfortunately the case for many of the studies carried out so far on neurotransmitter-related enzymes in human brain.

Samples of human brain tissue, especially from the smaller areas of brain, are often precious, and assay procedures can be devised in such a way as to permit the maximum number of different measurements to be made on a single tissue specimen. This may involve, for example, homogenization in water and rapid removal of aliquots for enzyme assays before addition of acid extractants to the remainder of the homogenate to permit amine or peptide assays. Aliquots of the tissue homogenates are also routinely used for protein assay, and neurochemical data should be expressed wherever possible per unit protein rather than per wet weight. The water content of frozen brain samples may vary during storage and because of the possible presence of frozen cerebrospinal fluid.

A problem encountered in long-term studies during which neurochemical data may be accumulated over a period of several years is the necessity of standardizing and checking the various biochemical assay procedures used. Neurochemical results from normal human postmortem brain are inherently variable and cannot be used as a reliable means of standardizing assays. We have found that the routine inclusion of animal brain samples in our assays provides a useful check on the biochemical procedures. Mouse brains from an inbred laboratory strain are used in this way.

5. POSTMORTEM STABILITY

Biochemists are trained to remove animal tissues and to stop ongoing metabolism as rapidly as possible by rapid heating, freezing, or other fixation procedures. The idea that meaningful biochemical data can be obtained from the analysis of human brain after death, when the interval between death and removal of the organ from the body may be several days, seems at first sight impossible. However, the experience in several laboratories that have had the temerity to undertake such studies has shown that many of the chemical markers associated with neurotransmitter mechanisms in brain are remarkably stable after death, so that their measurement is practicable even in tissue samples subjected to normal delays in coming to autopsy. Generalizations are impossible, however, and a high degree of postmortem stability is not found for all of the neurochemical markers that one might wish to employ. Each has to be carefully investigated to determine what postmortem changes are likely to occur.

5.1. Methods for Assessing Postmortem Stability

A number of methods have been used to assess the postmortem stability of neurochemical markers, and a combination of several different approaches is recommended when dealing with each new problem. Nearly all of the com-

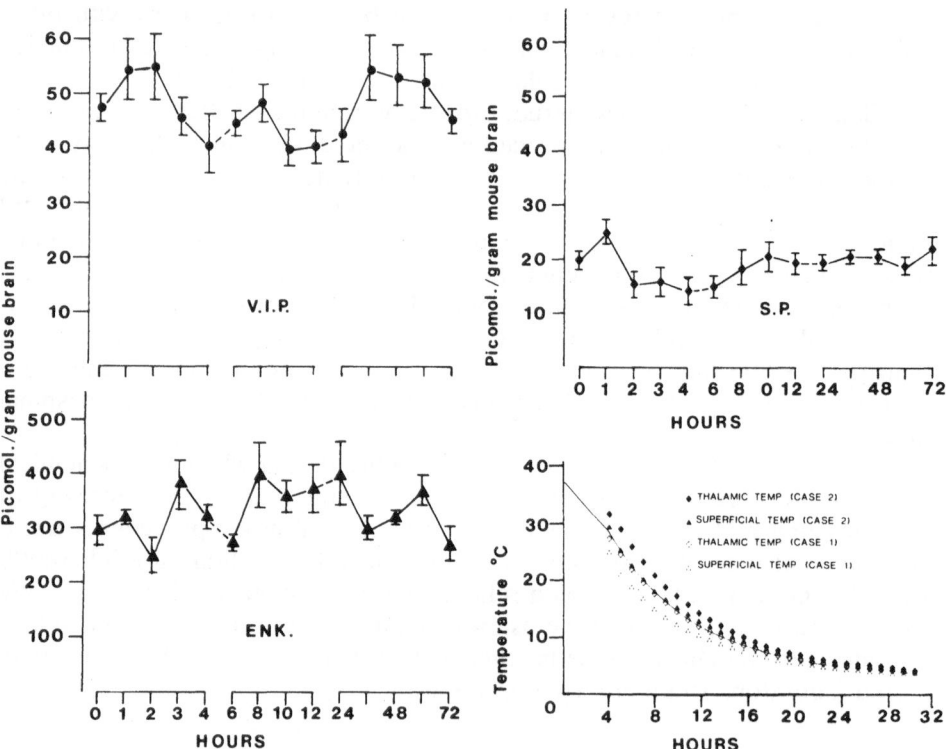

Fig. 2.   Postmortem stability of neuropeptides in mouse brain and cooling curve for normal human brain. Cooling curves of normal human brains (lower right) were obtained by inserting thermoprobes into the brain in two cadavers and monitoring superficial and thalamic temperature after transfer to mortuary refrigerator within 3 hr after death.[19] Neuropeptides were measured in mouse brain at various times after death, using a model in which the animal brain is slowly cooled to simulate the human postmortem conditions. Peptides, measured by radioimmunoassay, were vasoactive intestinal polypeptide (VIP), enkephalin (ENK), and substance P (SP)[65,66]

pounds or enzymes concerned are very stable in deep-frozen tissue samples, and most are relatively stable in brain tissue maintained at or near 4°C. The major postmortem changes, thus, occur during the interval between death and cooling of the brain to mortuary refrigerator temperature. Measurements of the rate of cooling of human brain have been made for both superficial and deep structures by implanting thermoprobes into cadavers.[19] The results (Fig. 2) show a predictably slow fall in brain temperature, with a delay of about 24 hr to reach 4°C.

Postmortem stability studies in either animal or human brain aim to simulate the normal postmortem conditions. With human brain samples, for example, it is useful whenever possible to compare biochemical results obtained in assays of tissue removed as biopsy samples from the living brain with those obtained in tissues subjected to normal postmortem handling and storage. During the surgical excisions of brain tumors, substantial amounts of normal brain tissue have perforce to be removed, and substantial amounts of histologically normal brain may be available, although this is almost always from superficial regions, notably cerebral cortex. Biopsy samples can be rapidly frozen or may be deliberately subjected to conditions designed to simulate the normal postmortem state. The following procedure, described recently by Perry and colleagues,[20] although not suitable for the faint-hearted, will serve to illustrate this approach.

> A portion of each biopsied specimen was immersed in liquid nitrogen within 30 sec of surgical removal from the patient's brain, while two to six additional portions of each biopsy were incubated at approximately 35°C for periods ranging from 10 min to 4 hr before being frozen in liquid nitrogen. This was accomplished simply by placing vials containing the biopsy specimens under the clothing and next to the skin of the person visiting the operating room to collect the specimens.

Most experiments, however, have made use of animal brain as a model for studying the likely postmortem stability of various neurochemicals in human brain. Many studies have involved simply storing animal brain at various temperatures prior to freezing. In our own laboratory, we use a procedure devised by Spokes and Koch[19] in which mice are killed, and the carcasses are slowly cooled in a programmed incubator over a period of up to 48 hr or longer, so that the rate of cooling of the animal brain simulates that observed in normal human brain after death. Brain samples are then removed from the incubator at various intervals and stored frozen until analysis.

5.2. Results Obtained

Some examples of the varying patterns of postmortem stability observed are shown in Figs. 2 and 3, and a summary of the results obtained for several important neurotransmitter-related markers is given in Table I. Results reported recently on the stability of various amino acids and other amino compounds (Table II)[20] illustrate the marked and unpredictable differences that are seen even among chemically similar compounds. Whereas the putative neurotransmitter amino acids (glutamic acid, aspartic acid, glycine, taurine, and GABA) show either no change or an increased content post-mortem, the cat-

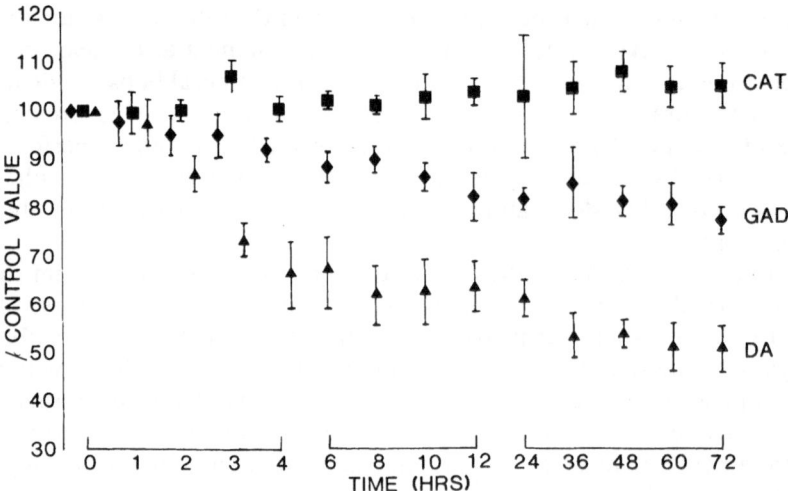

Fig. 3. Postmortem stability of dopamine (DA), choline acetyltransferase (CAT), and glutamate decarboxylase (GAD) in mouse brain, using the model described in Fig. 2. Data from Spokes and Koch.[19]

Table I
Summary of Published Findings on Postmortem Stability[a]

Variable	Increase (+) or decrease (−) post-mortem	Postmortem stability
Monoamines		
Norepinephrine	−	Moderate
Dopamine	−	Poor
5-Hydroxytryptamine	−	Good
Amino acids		
GABA	+	Poor
Glutamate/aspartate	+	Good
Transmitter enzymes		
Choline acetyltransferase	−	Good
Glutamate decarboxylase	−	Moderate
Acetylcholinesterase	−	Good
DOPA decarboxylase	−	Poor
Tyrosine hydroxylase	−	Poor
Dopamine-β-hydroxylase	−	Moderate
Neuropeptides		
Substance P	±	Good
Enkephalins	±	Good
Cholecystokinin	±	Good
Somatostatin	±	Good
Vasopressin	±	Good
Vasoactive intestinal polypeptide	±	Good

[a] Postmortem stability approximately defined: good, no detectable change likely during first 24 hr post-mortem; moderate, less than 50% change during first 24 hr post-mortem; poor, more than 50% change during first 24 hr. Table adapted from Mackay *et al.*[23,27]

Table II

Stability of Some Amino Compounds in Biopsied Human Brain[a]

Compound	Interval between biopsy and freezing	
	0 min (μmol/g)	3 hr (%)
Glycerophosphoethanolamine	0.60 ± 0.04	106 ± 9
Taurine	1.15 ± 0.16	121 ± 6
Phosphoethanolamine	1.55 ± 0.10	117 ± 7
Aspartic acid	1.18 ± 0.08	112 ± 8
Glutathione	2.70 ± 0.13	55 ± 3**
Threonine	0.26 ± 0.04	166 ± 10**
Serine	0.49 ± 0.03	158 ± 7**
Asparagine	0.04 ± 0.01	215 ± 50*
Glutamic acid	10.16 ± 0.51	103 ± 2
Glutamine	5.89 ± 0.47	95 ± 5
Proline	0.05 ± 0.01	551 ± 125**
Glycine	0.85 ± 0.13	213 ± 18**
Alanine	0.49 ± 0.05	382 ± 25**
Valine	0.15 ± 0.01	205 ± 11**
Cystathionine	0.78 ± 0.25	135 ± 17
Isoleucine	0.03 ± 0	438 ± 56**
Leucine	0.09 ± 0.01	282 ± 15**
Tyrosine	0.05 ± 0.01	241 ± 32**
Phenylalanine	0.05 ± 0.01	257 ± 43**
GABA	0.93 ± 0.08	279 ± 15**
Ethanolamine	0.10 ± 0.01	694 ± 109**
Lysine	0.11 ± 0.01	268 ± 22**
Histidine	0.11 ± 0.11	150 ± 10**
Homocarnosine	0.26 ± 0.03	132 ± 15
Arginine	0.07 ± 0.01	230 ± 17**

[a] Values (mean ± S.E.M.) are expressed in μmol/g wet weight for instantly frozen biopsies ($n = 12$) and as percentages of these values for specimens kept at body temperature for 3 hr ($n = 7$).
* $P < 0.05$; ** $P < 0.001$ (as compared with instantly frozen biopsies). Data from Perry et al.[20]

echolamines norepinephrine and dopamine decline markedly during postmortem storage[19-23] (Fig. 3). Studies are currently under way to investigate the postmortem stability of various small peptides present in brain. Many of these display a remarkable degree of postmortem stability in animal and human brain (Fig. 2), although the same peptides are rapidly metabolized by peptidases if added to brain homogenates. The unexpected stability of such substances and other neurotransmitter-related ones in postmortem brain may reflect the manner in which they are normally packaged and protected from metabolism within specialized storage organelles in nerve terminals.

Neurotransmitter-related enzymes differ widely in their postmortem stability. In some cases, such as the cholinergic marker enzyme choline acetyltransferase (CAT),[19] there is virtually no postmortem loss of activity under normal postmortem storage conditions (Fig. 3). With other enzymes, such as DOPA decarboxylase and tyrosine hydroxylase (T-OH), the postmortem instability may be so great as to make reliable measurements virtually impossible.[24-27]

A surprising and useful finding has been that a majority of neurotransmitter and drug receptor binding sites are very stable in postmortem brain and can thus be assayed by radioligand-binding procedures.[17]

In retrospect, the instability of neurochemical markers in postmortem brain has proved to be a much less serious technical limitation than originally feared. For most chemicals, the most important postmortem changes take place within the first 12 hr after death, during the period in which the brain cools to about 10°C. After this, further changes tend to be much slower and approach stable plateau levels. Unless particularly labile substances are being assayed, it may thus be desirable to avoid collecting brain samples with very short postmortem delays and preferable instead to work within the normal autopsy delay range of 24–72 hr, by which time brain chemistry has reached a more or less stable postmortem state.

6. COMPILATION OF CONTROL DATA

Control data for catecholamines, 5-HT, GABA, certain neuropeptides, neurotransmitter-related enzymes, and receptors are summarized in Tables III to XII. This is by no means a comprehensive compilation, and most of the data shown derive from our own series of control subjects, so that the results are

Table III
Dopamine and Norepinephrine Concentrations in Normal Human Brain[a]

Brain region	Dopamine (μg/g protein)	Norepinephrine (μg/g protein)
Putamen	23.2 ± 1.1 (77)	0.9 ± 0.1 (76)
Caudate nucleus	18.5 ± 0.9 (72)	0.8 ± 0.1 (72)
Lateral pallidum	10.6 ± 0.9 (37)	0.8 ± 0.1 (36)
Medial pallidum	3.8 ± 0.4 (36)	1.2 ± 0.1 (35)
Nucleus accumbens	12.2 ± 1.0 (50)	1.3 ± 0.1 (40)
Substantia nigra pars compacta	5.0 ± 0.4 (29)	1.3 ± 0.1 (29)
Substantia nigra pars reticulata	2.6 ± 0.3 (29)	1.0 ± 0.1 (29)
Red nucleus	1.4 ± 0.2 (39)	1.7 ± 0.1 (39)
Anterior perforated substance	1.9 ± 0.3 (33)	0.8 ± 0.1 (33)
Septal nuclei	1.4 ± 0.1 (35)	4.2 ± 0.7 (35)

[a] Values expressed as mean ± S.E.M. Number of patients in parentheses. Data from Spokes.[35,39] See also refs. 4,23,27,42,56,77–79.

Table IV

Distribution of 5-Hydroxytryptamine (5-HT) and 5-Hydroxyindolylacetic Acid (5-HIAA) in Human Brain[a]

Area	5-HT [mean ± S.E. (*n*)]	5-HIAA [mean ± S.E. (*n*)]
Caudate (by definition)	100	100
Substantia nigra	308 ± 72 (6)	568 ± 142 (6)
Pons	322 ± 108 (4)	660 ± 298 (4)
Midbrain	371 ± 86 (7)	735 ± 242 (7)
Thalamus	134 ± 24 (5)	211 ± 47 (5)
Hypothalamus	175 ± 61 (3)	273 ± 102 (3)
Amygdala	94 ± 18 (4)	85 ± 14 (4)
Orbital frontal cortex	40 ± 28 (8)	49 ± 23 (8)
Convexity frontal cortex	14 ± 3 (8)	8 ± 6 (8)
Cingulate gyrus	38 ± 7 (7)	37 ± 7 (7)
Sensory ortex	27 ± 8 (6)	40 ± 18 (6)
Motor cortex	13 ± 4 (8)	31 ± 12 (8)
Parietal cortex	16 ± 4 (8)	14 ± 4 (8)
Calcarine cortex	29 ± 7 (8)	43 ± 8 (8)
Hippocampal cortex	30 ± 5 (8)	79 ± 22 (8)
Temporal cortex	15 ± 6 (8)	14 ± 5 (8)
Cerebellar cortex	15 ± 3 (7)	14 ± 6 (7)
Nucleus accumbens	12 (1)	73 (1)
Olfactory area	93 (1)	97 (1)

[a] Data from Mackay *et al.*[23] Mean absolute concentration of 5-HT in the caudate was 649 ± 176 ng/g wet weight and for 5-HIAA 797 ± 190 ng/g. See also refs. 56,77,79.

to some extent internally consistent in referring to a single control series with a common dissection protocol. The tables are accompanied by lists of references to other published reports which can be referred to for comparison. These results are intended to serve as a convenient summary and source of reference, and it is not possible to comment on each of the neurochemical markers in detail here.

7. FACTORS INFLUENCING POSTMORTEM NEUROCHEMICAL DATA

Apart from the obvious problems posed by the postmortem instability of certain neurochemicals in brain, many other factors can influence the results obtained in such studies. When attempts are made to compare control and pathological cases, it is important that as many of these factors as possible be recognized and allowed for, in order to avoid the danger that observed differences between control and pathological groups are results of extraneous factors other than the illness itself. Over the past several years we have become aware of many of the pitfalls that await the unwary investigator in this field and have certainly fallen into some of them ourselves. Some of the recognized factors are the following.

Table V
Distribution of Choline Acetyltransferase (CAT) in Normal
Human Brain[a]

Brain region	CAT (μmol/hr per g protein)
Putamen	346.5 ± 12.0 (76)
Caudate nucleus	270.9 ± 11.7 (72)
Nucleus accumbens	226.3 ± 10.6 (50)
Anterior perforated substance	54.6 ± 8.5 (33)
Lateral pallidum	49.1 ± 4.8 (35)
Septal nuclei	17.8 ± 1.7 (28)
Medial pallidum	16.2 ± 3.1 (35)
Hippocampus	13.6 ± 0.8 (47)
Substantia nigra pars compacta	10.3 ± 0.9 (29)
Red nucleus	9.3 ± 0.8 (39)
Motor cortex	5.8 ± 0.4 (26)
Substantia nigra pars reticulata	5.2 ± 0.5 (29)
Cerebellar cortex	4.6 ± 0.4 (32)
Olive	4.6 ± 0.4 (29)
Dentate nucleus	4.0 ± 0.2 (35)

[a] Values expressed as mean ± S.E.M. Number of samples in parentheses. Data from Spokes[39]; see also refs. 11,27,35,42,45,80–83.

7.1. Age-Related Trends

A number of studies have reported a progressive loss of monoamine and amino acid neurotransmitter systems from various regions of the brain with aging (for review, see Pradhan[28]). The findings in human brain, however, have not always been consistent. Thus, although several groups have reported a significant decline in dopamine and norepinephrine and the associated marker enzyme tyrosine hydroxylase from basal ganglia with age,[29-34] we have failed to find such correlations (Fig. 4) except for T-OH activity in nucleus accumbens (Table XIII).[35]

Similarly, a decline in the cholinergic marker enzyme CAT with age has been reported,[29,36,38] but we have failed to observe any significant reduction in CAT activity with age in any of the 15 or more brain regions assayed.[35] We

Table VI
Distribution of GABA in Control Human Brain

Brain region	GABA (μmol/g protein)[a]
Lateral pallidum	76.5 ± 3.3 (23)
Nucleus accumbens	59.7 ± 2.4 (37)
Dentate nucleus	44.8 ± 3.5 (19)
Putamen	43.4 ± 2.8 (52)
Subthalamic nucleus	33.6 ± 1.8 (22)
Caudate nucleus	26.2 ± 3.7 (17)
Ventrolateral thalamus	20.4 ± 1.8 (18)
Premotor cortex	18.1 ± 1.1 (21)
Hippocampus	15.3 ± 1.4 (25)
Cerebellar cortex	12.2 ± 0.8 (21)
Amygdala	19.7 ± 1.3 (28)

[a] Values expressed as mean ± S.E.M. Number of samples in parentheses. Data from Spokes *et al.*[40]; see also refs. 11,27,35,42,45.

do, however, find quite marked and widespread reductions in glutamate decarboxylase (GAD) activity and GABA content in several brain regions with age (Table XIII).[35,39,40] The absence of any obvious age-related trends in the control series does not mean, however, that age can be disregarded as an important variable. Paradoxical age-related trends that are not seen in controls may occur in a pathological group. Thus, in our studies on patients dying with

Table VII
Glutamic Acid Decarboxylase (GAD) Activity in Control Human Brain[a]

Brain region	GAD (μmol/hr per g protein)	Brain region	GAD (μmol/hr per g protein)
Substantia nigra, pars compacta	109.1 ± 10.0 (18)	Dentate nucleus	48.6 ± 3.0 (24)
Substantia nigra, pars reticulata	84.2 ± 9.2 (18)	Cerebellar cortex	32.7 ± 3.2 (20)
Medial pallidum	97.2 ± 8.1 (22)	Anterior perforated	32.5 ± 3.9 (23)
Lateral pallidum	92.9 ± 8.5 (22)	Hippocampus	26.2 ± 2.3 (32)
Nucleus accumbens	84.4 ± 6.3 (36)	Red nucleus	21.4 ± 2.0 (26)
Putamen	57.9 ± 3.9 (51)	Olivary nucleus	18.6 ± 2.1 (20)
Motor cortex	53.4 ± 3.9 (15)	Septal nuclei	16.4 ± 2.2 (19)
Caudate nucleus	48.0 ± 3.1 (49)		

[a] All values expressed as mean ± S.E.M. Number of samples in parentheses. Because GAD is reduced in patients dying after prolonged terminal illness, all values refer to patients who suffered a sudden death. Data from Spokes.[35,39] See also refs. 27,29,30,45,81,84,85.

Table VIII
Substance P and Met-Enkephalin Immunoreactivities in Different Regions of Human Brain[a]

Region	Substance P (pmol/g tissue)	Met-enkephalin (pmol/g tissue)	Met:Leu ratio
Frontal cortex (Brodmann Area 10)	14 ± 10 (6)	42 ± 17 (16)	191
Corpus callosum		N.D.	—
Caudate nucleus	138 ± 14 (13)	116 ± 40 (20)	4.48
Putamen	112 ± 29 (10)	200 ± 71 (22)	4.48
Globus pallidus (lateral)	197 ± 99 (10)	1163 ± 216 (40)	6.27
Globus pallidus (medial)	877 ± 253 (10)	675 ± 168 (40)	—
Anterior perforated substance	77 ± 27 (5)	43 ± 18 (3)	—
Septum	89 ± 23 (5)	64 ± 17 (4)	—
Nucleus accumbens	159 ± 21 (6)	1086 ± 240 (5)	—
Amygdala	25 ± 12 (3)	26 ± 10 (10)	4.28
Hippocampus	104 ± 29 (6)	56 ± 16 (3)	—
Dentate nucleus	N.D.	N.D.	—
Subiculum	—	26 ± 14 (3)	—
Claustrum	—	8 ± 2 (3)	—
Thalamic nuclei			
Anterior lateral nucleus	N.D.	24 ± 13 (3)	—
Anterior medial nucleus	N.D.	34 ± 6 (3)	—
Ventral anterior nucleus	N.D.	10 ± 9 (3)	—
Ventral lateral nucleus	N.D.	6 ± 2 (3)	—
Hypothalamus	112 ± 19 (6)	141 ± 22 (18)	3.48
Median eminence	205 ± 97 (3)	—	—
Pineal gland	N.D.	—	—
Substantia nigra			
Pars compacta	1264 ± 239 (27)	557 ± 103 (15)	—
Pars reticulata	1535 ± 177 (30)	661 ± 145 (15)	—
Interpeduncular nucleus	83 ± 25 (3)	—	—
Superior colliculus	—	55 ± 4 (3)	—
Central gray	378 ± 130 (5)	143 ± 46 (3)	—
Locus coeruleus	310 ± 72 (5)	79 ± 18 (3)	—
Raphe-pons	234 ± 23 (3)	70 ± 25 (3)	—
Basal-pons	5 ± 1 (3)	10 ± 2 (3)	—

[a] Values are mean ± S.E.M.; number of determinations in parenthesis. Met:Leu-enkephalin ratios were determined after separation by TLC. N.D., not detected. Data from Emson et al.[64,65,86]

a hospital diagnosis of schizophrenia, we have observed significant age-related trends for dopamine, GABA, and angiotensin-converting enzyme which either do not exist in controls or are of opposite direction.[40–43] It is clear that all comparisons of control and pathological series should utilize carefully age-matched groups.

7.2. Circadian Changes

The possibility that circadian fluctuations may exist in the amount of neurotransmitters, enzyme activity, or receptors adds a further complicating factor to human postmortem studies. The importance of such phenomena, however, remains unclear. Circadian changes in GAD and CAT have been reported,[9] but we were unable to confirm these findings in our own control series.[35] Where marked circadian fluctuations exist normally, as in the rate of synthesis of melatonin in the pineal gland, these may still be apparent from postmortem measurements. Thus, circadian changes have been observed in the activity of the melatonin-synthesizing enzymes in human pineal.[44]

7.3 Agonal State

An important but poorly understood premortem factor is the clinical state of the patient in the period immediately before death. The McGeers[45] and Bowen and colleagues[46] were among the first to report that prolonged terminal illness, in which death is often associated with coma, can adversely affect subsequent postmortem biochemical measurements. In particular, reduced activities of the enzymes T-OH, DOPA decarboxylase, and GAD were reported in all brain regions of patients dying after prolonged terminal illness.[9,45,46] The importance of the agonal state, particularly in determining postmortem GAD activity, has since been amply confirmed. In our own series, a consistent difference was observed between GAD activities in different brain regions from patients dying after prolonged illness and those suffering sudden deaths, the GAD activities in the sudden death group being about twice as high in most brain regions.[35] The nature of this premortem influence, however, remains obscure. Despite the markedly different GAD activities measured in sudden versus prolonged death cases, there is, surprisingly, no significant differences in GABA concentrations in the same brain samples.[47] Furthermore, no sig-

Table IX
Regional Distribution of Somatostatin in Human Brain[a]

Region	Somatostatin (pmol/g tissue)	Region	Somatostatin (pmol/g tissue)
Cerebral cortex		Lateral pallidum	60.0 ± 14.7 (8)
Brodmann area 10 (prefrontal)	65.0 ± 4.8 (23)	Medial pallidum	16.0 ± 3.8 (8)
Brodmann area 4 (motor)	86.4 ± 5.8 (8)	Hypothalamus	308.9 ± 59.8 (10)
		Substantia nigra (pars compacta)	58.2 ± 4.9 (8)
Brodmann area 38 (anterior temporal)	103.2 ± 9.5 (8)	Substantia nigra (pars reticulata)	62.0 ± 15.2 (8)
Brodmann area 32 (medial frontal)	124.5 ± 13.0 (8)	Thalamus (anteromedial nucleus)	112.0 ± 21.0 (8)
Brodmann area 17 (visual)	29.9 ± 3.3 (8)	Thalamus (ventral anterior nucleus)	36.7 ± 11.9 (7)
Amygdala	304 ± 31.6 (23)	Cerebellar cortex	1.4 ± 0.3 (8)
Hippocampus	82.4 ± 7.3 (8)	Dentate nucleus	1.0 ± 0.1 (8)
Nucleus accumbens	223.5 ± 34.0 (8)	Periaqueductal gray	211.7 ± 30.0 (8)
Caudate nucleus	107.5 ± 14.8 (8)	Spinal cord (dorsal horn)	15.2 ± 9.6 (3)
Putamen	113.4 ± 9.7 (23)	Spinal cord (ventral horn)	10.7 ± 6.6 (3)

[a] Values are means $\pm$ S.E. for number of patients indicated in brackets. Data from Emson *et al.*[87]

Table X

The Regional Distribution of Cholecystokinin (CCK-8 Equivalents) and Vasoactive Intestinal Polypeptide in Normal Human Brain[a]

Region	CCK (pmol/g)	VIP (pmol/g)	Region	CCK (pmol/g)	VIP (pmol/g)
Cerebral cortex			Caudate nucleus	79 ± 23	4.6 ± 1.3
Brodmann area 4 (motor)	159 ± 12	7.9 ± 1.4	Putamen	67 ± 6	2.3 ± 0.5
Brodmann area 6 (premotor)	93.4 ± 6	15.7 ± 2.0	Globus pallidus (medial)	21 ± 4	2.4 ± 1.4
Brodmann area 8	60.1	10.2 ± 1.4	Globus pallidus (lateral)	N.D.	1.9 ± 0.2
Brodmann area 10 (prefrontal)	137 ± 45	17.3 ± 2.3	Hypothalamus	N.D.	23.1 ± 6.0
Brodmann area 11 (orbital)	108	22.8 ± 2.6	Substantia nigra (pars compacta/ pars reticulata)	65 ± 10	1.9 ± 1.3
Brodmann area 17 (visual)	N.D.	8.9 ± 1.2	Anteromedial thalamus	3 ± 1	0.3 ± 0.08
Brodmann area 32 (medial frontal)	197	21.4 ± 3.1	Central gray	11.7	2.8 ± 0.5
Brodmann area 38 (ant. temporal)	76.1 ± 5.3	10.9 ± 0.4	Cerebellum	1.0	0.4 ± 0.1
Brodmann area 21 (middle temporal)	N.D.	14.4 ± 1.7	Olive	2.53	—
Hippocampus	109.6	10.7 ± 1.0			
Amygdala	96.5	20.8 ± 8.9			

[a] Values are means ± S.E.M. for at least six separate determinations; where fewer than six determinations have been carried out, a mean value is given. N.D., not determined. Data from refs. 64–66,88.

Table XI
Regional Distribution of Angiotensin-Converting Enzyme Activity in Human Brain[a]

Region	Mean activity ± S.E.M. (pmol His-Leu/min per mg tissue)	n
Caudate nucleus	258.8 ± 17.9	24
Putamen	203.1 ± 14.3	24
Nucleus accumbens	147.5 ± 12.5	24
Lateral pallidus	149.1 ± 5.5	36
Medial pallidus	137.0 ± 6.1	33
S. nigra reticulata	103.4 ± 5.4	40
S. nigra compacta	56.2 ± 5.8	32
Subthalamic nucleus	19.3 ± 2.7	22
Red nucleus	18.8 ± 2.7	16
Raphe nuclei (pons)	41.2 ± 3.7	6
Locus coeruleus	36.3 ± 5.2	6
Basal pons	11.1 ± 1.0	13
Olivary nucleus	18.3 ± 3.1	6
Cortex (area 38)	18.0 ± 3.2	6
Cerebellar cortex	5.5 ± 0.4	21
Dentate nucleus	28.3 ± 5.7	21
Amygdala	24.1 ± 2.0	7
Hippocampus	22.5 ± 4.8	6
Septal nuclei	54.3 ± 12.0	6
Central gray (thal.)	31.2 ± 4.2	6
Ventral anterior thal.	24.2 ± 3.1	6
Ventral lateral thal.	18.4 ± 2.7	6
Anteromedial thal.	22.5 ± 1.9	6
Anterolateral thal.	19.7 ± 4.2	6
Choroid plexus	56.2 ± 6.1	5
Pineal gland	47.1 ± 4.2	5

[a] Data from Arregui *et al.*[43,62]

nificant differences are seen in catecholamine concentrations or CAT activities between sudden death and prolonged death cases.[35,42]

7.4. Postmortem and Storage Interval

The importance of postmortem delays will depend, of course, on the postmortem stability of the neurochemical markers being measured. Careful records have to be kept of the interval between death and transfer to a mortuary refrigerator and of the subsequent interval before autopsy. The data can then be analyzed retrospectively to determine the influence, if any, of these variables. Our experience has been that there is remarkably little difference in neurochemical data obtained from patients in which the postmortem delays were less than 48 hr and those in which autopsy was delayed for more than 48 hr after death.[35,42] Long-term storage of frozen tissue also seems possible. There were no detectable deteriorations in catecholamines, CAT, or GAD in human or animal brain samples stored for up to 8 months at $-70°C$.[35] Human brain samples cannot, nevertheless, be stored safely for indefinite periods

Table XII

Regional Distribution of Neurotransmitter and Drug Receptor Binding in Human Brain[a]

Region	Flunitrazepam (benzodiazepine receptor) (n = 1–4)	Binding (fmol/mg protein)					
		GABA	5-HT (serotonin receptor)	Dihydroalprenolol (β-adrenoreceptor)	Haloperidol (dopamine receptor)	Naloxone (opiate receptor)	QNB (quinuclidinyl benzilate; muscarinic ACh receptor)
Cerebral cortex							
Pre- and postcentral gyrus	81–83	78	98	57	—	50	—
Frontal cortex and pole	97–116	63	152	77	—	93	294
White matter corpus callosum	8	<9	<15	<40	<10	12	—
Limbic forebrain							
Hippocampus	63	76	107	90	—	16	196
Amygdala	80	30	103	140	—	36	298
Midbrain tegmentum	43	16	48	115	—	19	—
Thalamus	19–32	34	31	84	—	38	182
Basal ganglia							
Caudate	24–36	36	66	105	24	35	480
Putamen	34	16	71	89	35	34	472
Globus pallidus	13–19	14	87	104	<10	19	60
Substantia nigra	33	10	70	115	<10	8	18
Cerebellum–lower brainstem							
Cerebellar hemisphere	60	311	<15	70	—	23	0
Anterior cerebellar vermis	58	133	<15	79	—	10	<5
Posterior cerebellar vermis	58	130	<15	84	—	16	<5
Dentate	8	11	<15	174	—	5	0
Inferior olive	14	23	33	194	—	9	<5
Pons							
Base	6	25	<15	40	—	5	33
Tegmentum	9	12	38	193	—	15	16
Medulla							
Tegmentum	6	<9	17	261	—	7	—

[a] Neurotransmitter and drug receptor binding was assayed with various radioligands. Each value is the mean of 3–5 experiments, each performed in triplicate. A dash indicates that assays were not performed. Data from Enna et al.[17]; the flunitrazepam data are from Speth et al.[89]

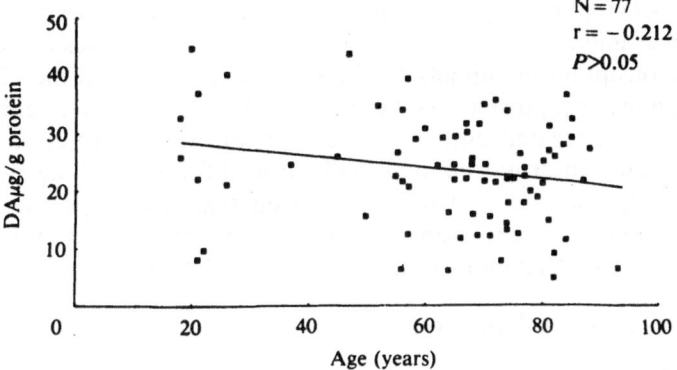

Fig. 4. Absence of decrease in human caudate nucleus dopamine content with age in 77 control cases. Data from Spokes.[35]

before biochemical analysis, and pathological specimens should always be matched by comparison with control material having a similar storage history.

7.5. Drug Treatment

Most patients will have been receiving medication before death, but we have been unable to identify any obvious effects of treatment with commonly prescribed antihypertensive agents, cardiac glycosides, thiazide diuretics, or antibiotics on postmortem measurements of catecholamines, CAT, or GAD in brain.[35,42] Patients dying from drug overdose or those receiving opiates or psychotropics have been deliberately excluded from our control series. When dealing with a pathological group, however, it may not be possible to exclude patients who have been treated for many years prior to death with psychoactive drugs. This problem has been a particularly difficult feature of neurochemical

Table XIII
Statistically Significant Correlations with Age[a]

Brain region	Substance	N	Correlation coefficient	Percent reduction 25–85 yrs
Central nucleus of amygdala	DA	17	−0.608**	82
	GAD	11	−0.92***	69
Septal nuclei	NE	36	−0.374*	64
Substantia nigra pars reticulata	NE	29	−0.380*	50
	GAD	18	−0.466*	54
Red nucleus	GAD	26	−0.562**	57
Cerebellar cortex	GAD	20	−0.745***	58
Hippocampus	GAD	32	−0.538***	60
Nucleus accumbens	B_{max}	16	−0.533*	67
	GAD	50	−0.307*	69
	T-OH	24	−0.449*	55

[a] N is the number of samples. *$P < 0.05$; **$P < 0.01$; ***$P < 0.001$. DA, dopamine; GAD, glutamate decarboxylase; NE, norepinephrine; B_{max}, DA receptors measured by [^{3}H]spiperone binding; T-OH, tyrosine hydroxylase. Data from refs. 35,41 and unpublished.

research on schizophrenic patients in whom the long-term use of neuroleptic drugs is almost universal. Such treatment, for example, causes a proliferation of dopamine receptors in animals,[48-50] and we have observed increased numbers of dopamine receptors in postmortem brain samples from schizophrenic patients who were treated until the time of death with neuroleptic drugs but not in those in whom drug treatment had stopped for a month or more before death.[41] This suggested to us that the increased density of dopamine receptors observed in the postmortem brains from most schizophrenic patients was probably a result of drug treatment rather than of the illness.

7.6. Tissue Volume Changes

One of the most difficult problems in interpreting neurochemical data from postmortem human brain relates to the fact that marked changes in the normal tissue volume of various brain regions may occur in various degenerative brain diseases. There may be in addition changes in the proportion of neuronal and glial cells in the tissue. In comparing neurochemical markers, when results are expressed per unit wet weight of tissue or per unit tissue protein, this clearly poses some difficult problems. If tissue atrophy occurs equally in all cellular components, there may be no change in various neurochemical markers, even in a severely shrunken tissue. Conversely, shrinkage of some but not other components can lead to an apparent increase in the unaffected components. A possible example of the latter phenomenon was observed by Spokes[39] who found significant increases in the dopamine concentrations in basal ganglia in patients dying with Huntington's disease. This change probably reflects the shrinkage of striatal tissue, leaving a more densely packed afferent dopaminergic innervation. Many similar situations exist, often making interpretation of the neurochemical data difficult.

7.7. Other Factors

No differences have ever been reported between male and female brains, and so far there have been no reports of ethnic differences. Despite the clear lateralization of various functions in the human brain, there have been surprisingly few studies of possible neurotransmitter differences between right and left hemispheres. One report suggested that norepinephrine was asymmetrically distributed between right and left thalamic nuclei, but we were unable to identify any significant left–right differences in other brain regions for this or other markers.[52]

8. DISEASE STATES

8.1. Degenerative Disorders

8.1.1. Parkinson's Disease

In the 19th Century the term "paralysis agitans" was used for a condition characterized by rigidity of limbs, trunk, and face with a rapid tremor of the hands. Parkinson (1817) wrote an *Essay on the Shaking Palsy*[53] as it was often called, and since then, this progressive disorder, with its onset in the sixth or seventh decade, has been called Parkinson's disease. No known cause for the typical "paralysis agitans" has been found, although symptoms and signs of Parkinson's disease can be seen secondary to an antecedent viral encephalitis, vascular disease, or manganese intoxication.

It is interesting that Lewy[54] described concentric hyaline cytoplasmic cell inclusions in the substantia innominata well before attention was directed to the substantia nigra (SN). Tretiakoff[55] found the "Lewy" bodies in the pigmented cells of the substantia nigra but also noted a marked reduction of these cells, with decreased SN pigmentation, which is the most significant finding in this disorder. Similar changes are found in other melanin-pigmented cells, the locus coeruleus, the reticular formation, and the dorsal nucleus of the vagus nerve. Lesions can also be seen in the globus pallidus, which is not surprising in view of the large nigro–pallidal projections that are known to exist. In addition to the presence of the Lewy-type inclusions in the globus pallidus, there is frequently rarefaction of nerve cells and a paucity of fine myelinated fibers.

Neurochemical measurements in the various types of Parkinson's disease reveal that in the basal ganglia, the DA concentrations are most markedly reduced in postencephalitic Parkinsonism, less marked in the idiopathic (paralysis agitans) form, and least but still significantly reduced in the arterosclerotic (senile) type.[56] Tyrosine hydroxylase was significantly reduced in the caudate, putamen, SN, and locus coeruleus, and dopamine β-hydroxylase (DBH) was significantly reduced in the locus coeruleus.[57]

8.1.2. Huntington's Disease

Alzheimer[58] was the first to associate the loss of neurons seen in the caudate nucleus and putamen with the familial chorea that had been described so well by Huntington (1872).[59] Others since then have described the enlarged ventricles and the cell loss that occurs in the deeper layers of the cortex. However, it is interesting that little attention has been directed to the substantia nigra. Corsellis[60] has had the opportunity to examine over 300 H.D. cases that we collected in the United Kingdom, and he found that the substantia nigra (SN) is more densely pigmented than normal. The pars reticulata of the SN, on the other hand, is markedly atrophic, with abundant glial cells.

The SN receives a number of projections from other regions of the brain, in particular, a GABAergic pathway from the striatum. This area is also rich in the neuropeptides substance P, cholecystokinin, enkephalin, somatostatin, and neurotensin. It is not surprising, therefore, to find that the marked atrophy seen grossly and microscopically in the pars reticulata is associated with marked reductions in these neuropeptides. The most significant neurochemical changes in the H.D. brain have been found in the most atrophic areas, caudate nucleus, putamen, globus pallidus, and the pars reticulata region of the SN. The small cells in the striatum are thought to be GABAergic interneurons lost early in the disease. Slightly larger striatal cells are also reduced in number, and these may represent in part GABAergic neurons that project to the SN. The marked reduction in the GABA biosynthetic enzyme glutamic acid decarboxylase (GAD)[11] and in GABA itself[5,11] in both the striatum and SN would suggest that the small and medium-sized cells are GABAergic. Large striatal cells tend to be retained in the early course of the disease but are reduced in number when there is severe atrophy; these cells are probably cholinergic. The biosynthetic enzyme for acetylcholine, choline acetyltransferase (CAT), is significantly reduced in the H.D. striatum; however, a small subgroup of cases had normal CAT activity.[11]

The cell bodies in the pars compacta of the SN remain intact in H.D. Dopamine and its biosynthetic enzyme, T-OH, are concentrated in this region of the SN and are increased in the H.D. brain based on wet weight or weight of protein. However, it is likely that these increases are only relative to the loss of other neurons in the region. In addition to decreased GABAergic neurons, there are a number of neuropeptides that are altered in the SN: substance P,[61] enkephalin,[62] somatostatin,[62,63] and cholecystokinin[64,65] (no abnormalities were found in vasoactive intestinal polypeptide levels[66]).

In spite of the atrophy seen in the deeper layers of the cerebral cortex, there have not been any particular neurochemical abnormalities found in this region other than an increase in γ-hydroxybutyrate.[67] Glutamate decarboxylase and CAT activities are always normal in H.D. cerebral cortex. We have found in brain tissues sent to us with a clinical diagnosis of H.D. that where the cortical CAT activity is decreased, histological examination usually reveals the presence of neuropathological features (plaques and neurofibrillary tangles) characteristic of Alzheimer's disease.[68] It is not uncommon for these two disorders to be clinically misdiagnosed, especially where there is a positive family history for a dementing illness with abnormal movements.

8.1.3. Alzheimer's Disease

Alzheimer[69] described the presence of numerous amyloid plaques similar to those originally described by Blocq and Marinesco[70] and neurofibrillary changes in a 51-year-old woman who had a dementia. Kraepelin[71] suggested that this type of presenile dementia be named Alzheimer's disease. It is not recognized that these changes cannot be differentiated from those found in older patients who have a dementia; so Alzheimer's disease would appear to have no age limit and, indeed, is now recognized as a common neuropsychiatric disorder. Alzheimer's disease is characterized by slow progressive mental deterioration, disorientation, confusion, and a failure of memory. Other symptoms include aphasia, agnosia, and apraxia. Myoclonic movements may be present, usually in the autosomal dominant form of the disorder.

There is widespread atrophy, with the brain weight at death often under 1000 g. On coronal sectioning, the ventricles are usually dilated as in Huntington's disease.

Histologically, the two most obvious features of Alzheimer's disease are the argyrophilic plaques found in the cerebral cortex and sometimes in the basal ganglia and neurofibrillary degeneration of nerve cells, particularly in cerebral cortex and hippocampus. Very little attention has been directed in the past to the substantia innominata area in this disorder. However, the probable existence of a cholinergic projection from the substantia innominata to the cerebral cortex[72] and the cholinergic deficit described in postmortem cerebral cortex[7-9] suggest that a closer examination of this area of the Alzheimer's brain is needed.

8.2. Nondegenerative Disorders: Schizophrenia

No consistent histopathological changes have been found in the schizophrenic brain. Although no specific areas of degeneration have been noted, this may need to be reexamined in view of recent CAT scan changes reported in schizophrenia.[73,74] As discussed with regard to the degenerative disorders, neurochemical findings have directed attention to areas of brain not previously examined in detail. Preliminary data indicate that there are significant increases in both dopamine[10,41,75] and norepinephrine[76] in the postmortem schizophrenic brain in anterior striatal and limbic regions, the accumbens area, and the olfactory tubercle. It is conceivable that with newer electron microscopic and immunocytochemical techniques directed to these limbic areas of the schizophrenic brain, very specific abnormalities may be detected.

REFERENCES

1. Blaschko, H., 1939, *J. Physiol. (Lond.)* **96**:13.
2. Carlsson, A., and Waldeck, B., 1958, *Acta Physiol. Scand.* **44**:293–298.
3. Bertler, A., and Rosengren, A., 1959, *Acta Physiol. Scand.* **47**:350–361.
4. Ehringer, H., and Hornykiewicz, O., 1960, *Klin. Wochenschr.* **38**:1236–1239.
5. Perry, T. L., Hansen, S., and Kloster, M., 1973, *N. Engl. J. Med.* **288**:337–342.
6. Bird, E. D., Mackay, A. V. P., Rayner, C. N., and Iversen, L. L., 1973, *Lancet* i:1090–1092.
7. Bowen, D. M., Smith, C. B., White, P., and Davison, A. N., 1976, *Brain* **99**:459–496.
8. Davies, P., and Maloney, A. J. R., 1976, *Lancet* ii:1403.
9. Perry, E. K., Gibson, P. N., Blessed, G., Perry, R. H., and Tomlinson, B. E., 1977, *J. Neurol. Sci.* **34**:247–265.
10. Bird, E. D., Spokes, E. G. S., and Iversen, L. L., 1979, *Brain* **102**:347–360.
11. Bird, E. D., and Iversen, L. L., 1974, *Brain* **97**:457–472.
12. Schaltenbrand, G., and Wahren, W., 1977, *Atlas for Stereotaxy of the Human Brain*, 2nd ed., George Thieme, Stuttgart.
13. Riley, H. A., 1943, *An Atlas of the Basal Ganglia, Brain Stem and Spinal Cord*, Williams & Wilkins, Baltimore.
14. Brodmann, K., 1909, *Vergleichende Lokalisation-lehre der Grosshirnirinde in ihren Prinzipien dargestellt auf Grund des Zellenbaues*, J. A. Barth, Leipzig.
15. Brockhaus, H., 1942, *J. Psychol. Neurol.* **51**:1–55.
16. Weiner, N., Martin, C., Wesemann, W., and Riederer, P., 1979, *J. Neural Transm.* **46**:253–262.
17. Enna, S. J., Bennett, J. P., Jr., Bylund, D. B., Creese, I., Burt, D. R., Charness, M. E., Yamamura, H. I., Simantov, R., and Synder, S. H., 1977, *J. Neurochem.* **28**:233–236.
18. Coyle, J., 1975, *Handbook of Psychopharmacology*, Volume 1 (L. L. Iversen, S. D. Iversen, and S. H. Snyder, eds.), Plenum Press, New York, pp. 71–79.
19. Spokes, E. G. S., and Koch, D. J., 1978, *J. Neurochem.* **31**:381–383.
20. Perry, T. L., Hansen, S., and Gandham, S. S., 1981, *J. Neurochem.* **36**:406–412.
21. Moses, S. G., and Robins, E., 1975, *Psychopharmacol. Commun.* **1**:327–337.
22. Vogel, W. H., Orfie, V., and Century, B., 1969, *J. Pharmacol. Exp. Ther.* **165**:196–203.
23. Mackay, A. V. P., Yates, C. M., Wright, A., Hamilton, P., and Davies, P., 1978, *J. Neurochem.* **30**:841–848.

24. Black, I. B., and Geen, S. C., 1975, *Arch. Neurol. Psychiatry* **32**:47–49.
25. Fahn, S., and Cote, L. J., 1976, *J. Neurochem.* **26**:1039–1042.
26. Grote, S. S., Moses, S. G., Robins, E., Hudgens, R. W., and Croninger, A. B., 1974, *J. Neurochem.* **23**:791–802.
27. Mackay, A. V. P., Davies, P., Dewar, A. J., and Yates, C. M., 1978, *J. Neurochem.* **30**:827–839.
28. Pradhan, S. N., 1980, *Life Sci.* **26**:1643–1656.
29. McGeer, P. L., and McGeer, E. G., 1976, *J. Neurochem.* **26**:65–76.
30. McGeer, E. G., Fibiger, H. C., McGeer, P. L., and Wickson, V., 1971, *Exp. Gerontol.* **6**:391–396.
31. Adolfsson, R., Gottfries, C. G., Roos, B. E., and Winblad, B. 1979, *J. Neurol. Transm.* **45**:81–105.
32. Carlsson, A., and Winblad, B., 1976, *J. Neural Transm.* **38**:271–276.
33. Reis, D. J., Ross, R. A., and Joh, T. H., 1977, *Brain Res.* **136**:465–474.
34. Robinson, D. S., 1975, *Fed. Proc.* **34**:103–107.
35. Spokes, E. G. S., 1979, *Brain* **102**:333–346.
36. Frolkis, V. V., Bezrukov, V. V., and Duplenki, Y. K., 1973, *Gerontologia* **19**:45–47.
37. Meek, J. L., Bertilsson, L., Cheney, D. L., Zsilla, G., and Costa, E., 1977, *J. Gerontol.* **32**:129–131.
38. Davies, P., 1979, *Brain Res.* **171**:319–327.
39. Spokes, E. G. S., 1980, *Brain* **103**:178–210.
40. Spokes, E. G. S., Garrett, N. J., Rossor, M. N., and Iversen, L. L. 1980, *J. Neurol. Sci.* **48**:303–313.
41. Iversen, L. L., and Mackay, A. V. P., 1981, *Lancet* **ii**:149.
42. Bird, E. D., Spokes, E. G. S., and Iversen, L. L., 1979, *Brain* **102**:347–360.
43. Arregui, A., Mackay, A. V. P., Spokes, E. G., and Iversen, L. L., 1980, *Psychol. Med.* **10**:307–313.
44. Smith, J. A., Padwick, D., Mee, T. J. X., Minneman, K. P., and Bird, E. D., 1977, *Clin. Endocrinol.* **6**:219–225.
45. McGeer, P. L., McGeer, E. G., and Wade, J. A., 1971, *Neurology (Minneap.)* **21**:1000–1007.
46. Bowen, D. M., White, P., Flack, R. H. A., Smith, C. B., and Davison, A. N., 1974, *Lancet* **i**:1247–1249.
47. Spokes, E. G., Garrett, N. J., and Iversen, L. L., 1979, *J. Neurochem.* **33**:773–778.
48. Burt, D. R., Creese, I., and Snyder, S. H., 1977, *Science* **196**:326–328.
49. Clow, A., Theodorou, A., Jenner, P., and Marsden, C. D. 1980, *Eur. J. Pharmacol.* **63**:135–144.
50. Owen, F., Cross, A. J., Waddington, J. L., Poulter, M., Gamble, S. J., and Crow, T. J., 1980, *Life Sci.* **26**:55–59.
51. Oke, A., Keller, R., Mefford, I., and Adams, R. N., 1978, *Science* **200**:1411–1413.
52. Rossor, M. N., Garrett, N., and Iversen, L. L., 1980, *J. Neurochem.* **35**:743–745.
53. Parkinson, J., 1817, An essay on the shaking palsy. Reprinted in: *Critchley's James Parkinson* (1955), Macmillan, London.
54. Lewy, F. W., 1913, *Dtsch. Z. Nervenheilkd.* **50**:50.
55. Tretiakoff, C., 1921, *Rev. Neurol.* **37**:592–618.
56. Bernheimer, H., Birkmayer, W., Hornykiewicz, O., Jellinger, K., and Seitelberger, F., 1973, *J. Neurol. Sci.* **20**:415–455.
57. Nagatsu, T., Kato, T., Nagatsu, I., Yukari, K., Inagaki, S., Iizuka, R., and Narabayashi, H., 1979, *Adv. Neurol.* **24**:283–292.
58. Alzheimer, A., 1911, *Z. Ges. Neurol. Psychiatrie* **3**:891–892.
59. Huntington, G. S., 1872, *Med. Surg. Reporter (Philadelphia)* **26**:317–321.
60. Corsellis, J. A. N., 1976, Greenfields Neuropathology, 3rd ed. (W. Blackwood and J. A. N. Corsellis, eds.), Arnold Publishers, London, pp. 822–827.
61. Gale, J. S., Bird, E. D., Spokes, E. G. S., Iversen, L. L., and Jessell, T., 1978, *J. Neurochem.* **30**:633–634.
62. Arregui, A., Emson, P., Iversen, L. L., and Spokes, E. G. S., 1979, *Adv. Neurol.* **23**:517–525.
63. Cooper, P. E., Fernstrom, M. H., Leeman, S. E., and Martin, J. B., 1980, *Neurology (Minneap.)* **31**:64.

64. Emson, P. C., Rossor, M., Hunt, S. P., Clement-Jones, V., Fahrenkrug, J., and Rehfeld, J., 1981, *Transmitter Biochemistry of Human Brain Tissue* (P. Riederer and E. Usdin, eds.), Macmillan, London, pp. 221–234.
65. Emson, P. C., Rossor, M. N., Hunt, S. P., Marley, P. D., Clement-Jones, V., Rehfeld, J. F., and Fahrenkrug, J., 1980, *The Mansell Bequest Symposium.*
66. Emson, P. C., Fahrenkrug, J., and Spokes, E. G. S., 1979, *Brain Res.* **173:**174–178.
67. Ando, N., Gold, B. I., Bird, E. D., and Roth, R. H., 1979, *J. Neurochem.* **32:**617–622.
68. Bird, E. D., 1977, *Neurotransmission and Disturbed Behavior* (H. M. van Praac and J. Bruinvels, eds.), Bohn, Scheltema and Holkemn, Utrecht, pp. 140–149.
69. Alzheimer, A., 1907, *Allg. Z. Psychiatr.* **64:**146–148.
70. Blocq, P., and Marinesco, G., 1892, *Semin Med. (Paris)* **12:**445.
71. Kraepelin, E., 1910, *Klin. Psychiatrie* **2:**624.
72. Emson, P. C., and Lindvall, O., 1979, *Neuroscience* **4:**1–30.
73. Johnstone, E. C., Crow, T. J., Frith, C. D., Hustand, J., and Kreel, L., 1976, *Lancet* **ii:**924–926.
74. Weinberger, D. R., Kleinman, J. E., Luchins, D. J., Bigelow, L. B., and Wyatt, R. J., 1980, *Am. J. Psychiatry* **137:**359–361.
75. Crow, T. J., Baker, H. F., Cross, A. J., Foseph, M. H., Lofthouse, R., Longden, A., Owen, F., Riley, G. J., Glover, V., and Killpack, W. S., 1979, *Br. J. Psychiatry* **134:**249–256.
76. Farley, I. J., Price, K. S. McCullough, E., Deck, J. H. N., Hordyinski, W., and Hornykiewicz, O., 1978, *Science* **200:**456–458.
77. Bertler, A., 1961, *Acta Physiol. Scand.* **51:**97–107.
78. Rinne, U. K., Sonninen, V., and Hyyppä, M., 1971, *Life Sci.* **10:**549–557.
79. Birkmayer, W., Danielczyk, W., Neumayer, E., and Riederer, P., 1974, *J. Neural Transm.* **35:**93–116.
80. Lloyd, K. G., Möhler, H., Heitz, P. H., and Bartholini, G., 1975, *J. Neurochem.* **25:**789–795.
81. Stahl, W. L., and Swanson, P. D., 1974, *Neurology (Minneap.)* **24:**813–819.
82. Bull, G., Hebb, C., and Ratkovic, D., 1970, *J. Neurochem.* **17:**1505–1516.
83. McGeer, P. L., McGeer, E. G., and Fibiger, H. C., 1973, *Neurology (Minneap.)* **23:**912–917.
84. Vogel, W. H., Orfei, V., and Century, B., 1969, *J. Pharmacol. Exp. Ther.* **165:**196–203.
85. Utena, H., Kanamura, H., Suda, S., Nakamura, R., Machiyama, Y., and Takahashi, R., 1968, *Proc. Jpn. Acad. Sci.* **44:**1078–1083.
86. Emson, P. C., Arregui, A., Clement-Jones, V., Sandberg, B. E. B., and Rossor, M., 1980, *Brain Res.* **199:**147–160.
87. Emson, P. C., Rossor, M., and Lee, C. M., 1981, *Neurosci. Lett.* **22:**319–324.
88. Rossor, M. N., Emson, P. C., Iversen, L. L., Mountjoy, C. W., Roth, M., Hawthorn, J., and Fahrenkrug, J., 1980, *The Mansell Bequest Symposium.*
89. Speth, R. C., Wastek, G. J., Johnson, P. C., and Yamamura, H. I., 1978, *Life Sci* **22:**859–866.

Neurological Mutants

Nicole Baumann and François Lachapelle

1. INTRODUCTION

Anyone who has studied the anatomy, embryology, physiology, or biochemistry of laboratory mammals must have been struck by the close similarity of their nervous system to man's in most essentials. The developmental processes of the central nervous system during fetal, neonatal, and early postnatal life do not differ essentially in these species.[1] Laboratory rodents can easily be bred under strictly defined genetic conditions and in any desired number. Furthermore, environmental conditions can be controlled strictly and indeed varied if this is likely to help investigations.

Domestic species, and especially laboratory mice, have produced an array of inherited neurological abnormalities[2,3] similar to or distinct from the ones observed in man, which constitute a wonderful tool for neurochemists for many reasons. On the one hand, these defects, being genetically determined, are stable and stereotyped, which means that they are easily reproducible in large amounts. On the other hand, in a mutant strain, the fate of the animals is predictable, and investigations can thus start before the onset of clinical features. This is obviously advantageous in developmental studies which can be carried out even during early embryonic life. At last, several mutations affecting the same developmental process at various stages allow a dissection of the different component parts of this event. This is especially true for myelination, cerebellar development, and cortical maturation. Actually, in most of these mutants, no causal relationship has been established between the gene defect and the molecular events affecting development. Nevertheless, they have been found to provide a useful experimental approach for investigating the sequential stages of nervous system maturation and the influence of a specific alteration on other developmental processes.

Although these mutations are not identical to those affecting man, they are in many respect comparable and have been used very fruitfully in com-

Nicole Baumann and François Lachapelle • INSERM and CNRS Laboratory of Neurochemistry, Salpetriere Hospital, Paris 13, France.

parative pathology: as in human diseases, mutant's nervous systems have adjusted to their handicap, and functional compensation may have occurred; this is in contrast to experimental lesions which may induce uncontrolled secondary injuries in unrelated systems.

However that may be, mutants remain underutilized in neurochemistry. Three main factors are responsible for this state of affairs. In the first instance, there is a problem of communication; new mutations generally appear in genetics laboratories, and their neurological characteristics are often succinctly described by geneticists in their usual publications. So, neurochemists are simply not yet aware of the material for which they are looking. Second, some people do not know how to obtain the strains that would interest them. At the moment, some periodical publications provide up-to-date information about available mutations and allow continuous communication between "customers" and "users."[4-7] Third, the maintenance of a mutant stock is a complex laboratory technique, often expensive, which requires suitable buildings and material and a minimum of know-how. In the hands of untrained people, many attractive mutations have been lost, sometimes irrevocably, and investigators have been discouraged from using this type of experimental tool.

2. NEUROLOGICAL MUTANTS IN NONRODENT SPECIES

Two main sources, veterinary medicine and laboratory animal colonies, provide hereditary neurological diseases. Veterinary medicine has uncovered several hereditary neurological diseases in large domestic species: for instance, bovine mannosidosis,[8] dysraphic disorders in sheep,[9] bovine or porcine gangliosidoses.[10,11] Unfortunately, these animals remain poorly used for laboratory investigations, perhaps because of the difficulty of developing such models from the spontaneous cases that have occurred. In contrast, in medium-sized species, among various spontaneous models of inherited neurological diseases, some are of high interest for investigators. Four of these, light-sensitive epilepsy in baboon (*Papio papio*), the two feline gangliosidoses, and canine globoid leukodystrophy, have been widely studied. And a caprine β-mannosidosis recently discovered is being maintained and seems especially promising.

2.1. Epilepsy in Baboons

The light-sensitive epilepsy in baboons was first described in 1966 by Killam *et al.*[12] Preliminary statistical studies of the variation in the degree of photosensitivity according to geographical origin, age, and sex of the captured animals indicated that, as in man, there is a lack of penetrance of the gene and a relative predisposition to photosensitive epilepsy in females as opposed to males and in young subjects as opposed to adults.[13] Pharmacological studies suggest a cortical hyperexcitability as the origin of the epilepsy. Photosensitive epilepsy in the baboon, although clinically very similar to the one observed in man, presents certain differences that are probably explainable by phylogenesis and which make schematic extrapolations dangerous. Nevertheless, the animals afford a very useful model for analyses of synaptic interactions, cortical inhibition, neurotransmitter regulation, pharmacological manipulations,[14] and correlations between cerebral and extracerebral blood volume and generalized seizures.[15]

2.2. Inborn Errors of Metabolism

2.2.1. Feline G_{M1} Gangliosidosis

The feline G_{M1} gangliosidosis (for review see ref. 16) first observed in 1971 by Baker *et al.*[17] in siamese cats is an autosomal recessive trait. Clinical features appear at about 4 months with head and hindlimb tremors. Then kittens develop a clumsy wide-based gait which progresses to generalized locomotor ataxia. Morphological alterations are characterized by neuronal degeneration (swelling vacuolization, loss of Nissl bodies, cytoplasmic lamellar inclusions) throughout the nervous system, terminal gliosis, and demyelination. The most peculiar trait is the formation of meganeurites[18] similar to those observed in human gangliosidoses. Vacuolation of cells outside of the nervous system has been limited to liver, acinar cells of the pancreas, and occasionally histiocytes in various organs. Total ganglioside NeuNAc content of cortical brain tissues in diseased cats is approximately 2.5 times that found in normal cat brains. The major ganglioside responsible for this increase is the G_{M1} which accumulates to levels approximately eight times normal. The relative concentrations of other gangliosides are less than or only slightly more than normal. The content of ceramide tetrahexoside is also elevated in gray matter of mutant cats. The G_{M1} ganglioside content of liver is above normal levels. A drastic deficiency of acid β-galactosidase (pH 3.8 to 4.3) is found in brain, liver, kidney, testes, spleen, skin, leukocytes, cultured fibroblasts, and placenta of homozygous cats when tissues are assayed with synthetic substrate. Liver homogenates of diseased cats are completely devoid of β-galactosidase activity. Neutral β-galactosidase and other lysosomal hydrolase activities are normal or slightly elevated. β-Galactosidase activity in tissue of heterozygote animals ranges from 35 to 51% of normal.[19] Comparison of features in feline and human G_{M1} gangliosidoses shows that the feline disease appears to be a virtually exact replica of juvenile human G_{M1} gangliosidosis (type 2).

2.2.2. Feline G_{M2} Gangliosidosis

The feline G_{M2} gangliosidosis first described in 1977 by Cork *et al.*[20] in short-haired domestic cats is also a recessive autosomal trait. Onset of clinical features occurs between 4 and 10 weeks, including tremor, hypermetria, ataxia, and paresis; failure to gain weight, proportional dwarfism, occasional dysphagia, and corneal opacity are associated with neurological symptoms. Neuronal cell body hypertrophies throughout the nervous system, including autonomic ganglia and retina, with absence of Nissl substance and multilamellar round cytoplasmic inclusions are the main morphological features. Finally, a marked terminal gliosis is associated with a mild demyelination. Vacuolated hepatocytes and distended Kupffer cells are seen in the liver. Membrane-bound inclusions are also found in endothelium, smooth muscle cells of vessels, bone marrow cells, splenic macrophages, and renal interstitial cells. In mutants, the total ganglioside content of the brain is two to three times that found in the normal cat brain. The ganglioside responsible for this increase is G_{M2}. The concentration of other gangliosides is reduced. In addition, abnormally high concentrations of the asialo derivative of the G_{M2} ganglioside in brain and liver and a 30-fold increase of globoside in liver are observed. Determinations of β-D-acetylhexosaminidase (pH 4.0 and 4.5) demonstrate that total β-hexosaminidase activity in brain, liver, and fibroblasts from diseased kittens is 0.5 to 2% of normal. The β-galactosidase activity in diseased brain is equal to or slightly higher than the activity of this hydrolase in normal cat brains. Electrophoretic studies on liver extracts of mutant cats show that the two major forms, A and B, of β-hexosaminidases are present but inactive. Cultured skin fibroblasts from parents of affected animals have hexosaminidase activities 30 to 50% that of normals. Patterns of neuronal and visceral glycosphingolipid storage and the total hexosaminidase deficiency suggest that the feline G_{M2} gangliosidosis is analogous to the human G_{M2} gangliosidosis type II (Sandhoff disease).

2.2.3. Canine Globoid Leukodystrophy

Globoid leukodystrophy (GLD) detected in several dog breeds has been most thoroughly studied in the Cairn terrier because a colony of these dogs is maintained for experimental studies (for review see ref. 16). Globoid leukodystrophy is inherited as a single-locus two-allele recessive autosomal character in Cairn terriers.[21] The earliest clinical symptom, pelvic limb incoordination, occurs from 11 to 30 weeks. Pelvic limb paralysis with muscle atrophy generally follows the appearance of thoracic limb hypermetria and head tremor associated with deficits of sensory recognition and visual and mental alterations. Terminal prostration is associated with anorexia. Animals die about 10 weeks after the first symptoms appear. Macrophage hyperproliferation, presence of multinucleated globoid cells, and terminal astrocytosis in place of normal white matter are the main histopathological features. Ultrastructurally disrupted myelin sheaths and axons, excessive extracellular spaces, and abundant astrocytic processes can be observed. Macrophages contain tubular inclusions; dense bodies and lamellar inclusions are present in astrocytes. In the CNS, cerebral and cerebellar white matter are most severely involved. In the PNS, autophagic Schwann cells containing vacuoles, myelin figures, and tubular inclusions are common. Lesion distribution is quite similar in canine and human GLD. Galactosylceramidase activity in GLD dogs is 12% for brain and 18% for leukocytes. Heterozygotes have averaged 51% of normal activity for brain and leukocytes.[22] Less convenient than the Twitcher mouse for *in vivo* radioisotopic or population studies, the canine model is superior for experiments that require large amounts of tissue or dissections of different CNS regions.

2.2.4. Caprine β-Mannosidosis

A rapidly fatal neurovisceral disease, β-mannosidosis has been described in 1979 by Jones and Laine[23] in Nubian goats. This apparently autosomal recessive mutation is expressed at birth by a marked neurological deficit. Microscopic examination revealed demyelination associated with distended axons in the CNS. Membrane-bound cytoplasmic vacuoles are found in neurons. Storage vacuoles are present in oligodendrocytes, renal tubular epithelium, reticuloendothelial cells in liver, spleen, bone marrow, and lymph nodes. Peripheral myelin appears normal despite storage vacuoles in Schwann cells. Occasionally, PNS axons contain dense bodies. Brain contains 2.2 μmol/g of the trisaccharide Man-GlcNAc-GlcNAc. Urine contains elevated levels of both mannose and N-acetylglucosamine. β-Mannosidase activity is reduced (less than 10% of control values) in tissues of affected animals, whereas α-mannosidase and α-fucosidase are 10- to 20-fold elevated over control values.

3. MURINE MODELS

Several neurological mutations have been discovered in various rodent species. They have only been widely used in the mouse.

The mouse is by far the most commonly used laboratory mammal. It owes this unique position to its small size, its high fecundity, and because it resembles man in being largely unspecialized. Well adapted to domestication, its short life-span allows geriatric investigations. At the moment, almost any investigation, even neurophysiological ones, is possible on its nervous system, thus allowing a pluridisciplinary approach. Inbreeding and maintenance of defective mutations are easier than in any other laboratory species. This great genetic flexibility accounts for the overwhelming number of inbred strains which provide varied well-defined genetic backgrounds and some structural or functional peculiarities of the nervous system. At the moment, among almost 700 mutations indexed in the mouse, at least 150 affect the nervous system (for review, see refs. 24, 25) including more than 80 for CNS, PNS, or the neuromuscular system, the others affecting more specifically vision or the inner ear. Some of these mutations exhibit clear-cut defects more or less restricted to a structural or functional abnormality of the nervous system, whereas others have a more complex phenotype with associated alterations in this system as a result of the pleiotropic nature of most of the mutations in vertebrates. This is especially true for some inborn errors of metabolism (similar to or distinct from the ones observed in man) which secondarily affect the nervous system. Here, we try to present a review of some of these genetic alterations. Mutations have been classified relative to their pathological effect. Only mutations widely studied and still existing at the moment are indexed. Mutations affecting only sensory organs are not presented.

Table I

Dysraphic Disorders in Mouse

Symbol	Name	Genetics				Neurological disorders	Other organs	Model for genetic human disease
		Chr	D/R	F/S	V/L/Sl			
my^{26}	Blebs	3	R (VP)		L	Pseudencephaly	Acrania, malformations (eye, skull, skin, kidney)	
Cd^{27}	Crooked	6	$\frac{1}{2}$D (VP)	F	L	Pseudencephaly, exencephaly, or anencephaly	Crooked tail, defects in other organs	
Ds^{28}	Disorganization	14	$\frac{1}{2}$D (VP)		L	Exencephaly, pseudencephaly, anencephaly	Cranioschisis	
ct^{29}	Curly-tail	?	R (VP)		Sl	Lumbosacral spina bifida, sometimes anencephaly	Curly or kinky tail, skeletal abnormalities	Spina bifida
Fu^{30}	Fused	17	$\frac{1}{2}$D		L	Cord malformations (folding, duplication, overgrowth)	Malformations in tail, vertebrae, urogenital tracts	
Fu^{Ki31}	Kinky	17	$\frac{1}{2}$D		L	Same	Same	
Lp^{32}	Loop-tail	1	$\frac{1}{2}$D		L	Open neural plate	Abnormalities of skin, axial skeleton	Spina bifida aperta
Sp^{33}	Splotch	1	$\frac{1}{2}$D		L	Spina bifida, cranial hernia, myeloschisis (overgrowth neural tissue)	Rachischisis, abnormal pigmentation	

[a] Abbbreviations in this and succeding tables: Chr, chromosome; D, dominant; R, recessive; VP, variable penetrance; F, fertile; S, sterile; V, viable; L, lethal; Sl, sublethal; $\frac{1}{2}$D, semidominant.

Table II
Hydrocephalus in the Mouse

Symbol	Name	Genetics			Neurological disorders	Other organs	Model for genetic human disease
		Chr	D/R	F/S	V/L/SL			
cb^{37}	Cerebral degeneration	?	R	S	Sl	Hydrocephalus at birth; white matter degeneration in the cerebral hemispheres		
hpy^{38}	Hydrocephalic polydactyl	6	R	S	Sl	Hydrocephalus	Polydactylia	
ch^{39}	Congenital hydrocephalus	13	R		L	Hydrocephalus with retarded development of the subarachnoid space	Urogenital system	
hy^{340}	Hydrocephalus 3	8	R		L	Hydrocephalus with atrophy of meninges		

3.1. Dysraphic Disorders

This group of mutations (for review see ref. 3) include the main morphological encephalic abnormalities (anencephalies, exencephalies, and pseudencephalies) and several forms of spina bifida[26-33] (Table I). In contrast to man, there are more exencephalies than anencephalies in mouse. Most of these mutations in mice have partial penetrance. Recently, the concept of genetic and environmental factors affecting the teratogenic threshold was highlighted in studies on the curly tail (ct) mouse.[34] The expression of the recessive gene is affected by the maternal genetic background. In this mutant as in the human, the expression of anencephaly, but not of spina bifida, is strongly sex linked although the mutation is autosomal; also, excess of vitamin A, a well-known teratogen in the human, can affect neural tube closure when administered before or during the period of neurulation. These facts give more validity to the model.[35] DNA content as a cell number index measured in the Loop tail mutation reveals no difference between the mutants and their normal littermates.[36] Nevertheless, at the moment, no biochemical approach to the pathological mechanisms involved in these defects has been reported.

3.2. Hydrocephalus

Some hereditary forms of hydrocephalus (for review see ref. 3) have been described in the mouse (Table II). Contrary to most of the genetic human conditions, mouse hydrocephaluses are of the communicating type in which there is no stenosis of aqueducts or foramina. Some of these mutations also involve polycystic kidneys (ch) or polydactylia (hpy). They provide information about the development of the cerebrospinal fluid pathways and possibly about the hydrocephaluses that occur in humans in utero.

3.3. Cerebellar Malformations

In mice, about ten mutations (for review see refs. 2,41–43) have been recognized as leading to morphological changes in cerebellum (Table III). One of them, Swaying (sw), characterized by an absence of most of the vermis, appears to be a true model of the human cerebellar congenital malformation agenesis of vermis. Three cause massive degeneration of granule cells: Reeler (rl), Staggerer (sg), and Weaver (wv). Others such as Nervous (nr) and Purkinje cell degeneration (pcd) lead selectively to Purkinje cell degeneration. Leaner (tg^la) Lurcher (lc), and Ducky (du) lead to more complex disorganization. This group of mutations affecting an integrated developmental process at various steps proved to be very fruitful for a dissectional approach to the mechanisms of cerebellar cortical development.[41]

In the autosomal recessive mutation Reeler (rl), animals manifest reeling ataxia of gait, distonic postures, and tremor after 18 days of age. Animals die at about weaning age, but survival can be prolonged when they are outcrossed. Morphological studies reveal a gross size reduction in cerebellum. The granular layer of the cerebellar cortex contains a reduced number of granule cells. Some

Table III
Cerebellar Malformations in the Mouse

Symbol	Name	Genetics				Neurological disorders	Other organs	Model for genetic human disease
		Chr	D/R	F/S	V/L/SL			
sg^{44}	Staggerer	9	R		SI	Granule cell degeneration by defect in Purkinje cell dendrites		
wv^{45}	Weaver	16	R	S	SI	Absence of granule cells (defect in migration from the external layer)		Congenital cerebellar atrophy of granular layer
pcd^{46}	Purkinje cell degeneration	13	R	S	V	Degeneration of Purkinje cells, mitral cells of the olfactory bulb, and photoreceptors	Structural abnormalities in spermatozoa	
nr^{47}	Nervous	8	R		L	Purkinje cell and photoreceptor degenerations		Congenital cerebral atrophy with loss of Purkinje cell
sw^{48}	Swaying	15	R			Absence of the most of the anterior part of the vermis		Agenesis of vermis
rl^{49}	Reeler	5	R	S	Reduced viability	Abnormal migration of neurons in cerebral and cerebellar cortex		
Lr^{50}	Lurcher	6	½D		L	Absence of Purkinje cells, reduced width of molecular and granular layers		
du^{51}	Ducky	9	R	F	V	Spinocerebellar degeneration		
tg^{la52}	Leaner	8	R	F	V	Degeneration of granule cells, migration abnormalities		

Purkinje cells are in the normal position, but most of them have cell bodies within the granule layer or deep in the underlying white matter and show an abnormal twisted morphology.[53] Cerebral cortex is also affected, with failure of neuronal migration.[54] Chimeric studies of normal–*rl* cerebella show that Purkinje cells of both genotypes are sometimes normal looking and sometimes abnormal, indicating that in this mutation the morphology of cells is specified from outside by the neural environment.[55]

Staggerer (*sg*) is an autosomal recessive mutation. Animals are recognized at 8–12 days of age by mild tremor, impairment of gait, and rare tonic–clonic generalized seizures. Homozygotes weigh less than their littermates at weaning age and remain lighter thereafter. The cerebellum is less than one-third the normal size and shows few fissures. The molecular layer is thin, and granular cells are reduced in number and sometimes degenerated. The number of Purkinje cells is normal, but there are no cell–parallel fiber synapses. There is some indication that in this mutant the granule cell degeneration is caused by a defect in the Purkinje cells which fail to form proper dendritic spines. Analysis and comparison of carbohydrate composition of the surface of cerebellar cells from normal and *sg* mice during early postnatal life show altered cell surface components in this mutant.[56]

Weaver (*wv*) is an autosomal semidominant mutation. Homozygotes appear very small at about 8–10 days of age and nurse poorly. Clinical features are instability of gait, ataxia, hypotonia, tremor, and rare tonic–clonic seizures. Most of the homozygotes die at weaning age. Cerebellar cortex is reduced in volume. Morphological studies show an almost complete absence of granule cells. The Purkinje cells appear relatively normal in number. The defect in the granule cells results from a massive degeneration of external granular layer prior to migration. Bergman glial fibers are hypoplasic.[57] Interestingly, although they do not show behavioral abnormalities, heterozygotes for this mutation have a slightly reduced size of cerebellum with incomplete migration of some granule cells.

The Nervous (*nr*) autosomal recessive mutation is clinically identifiable at about weaning age by hyperactivity and ataxia in mutant mice. Morphological studies show enlargement of mitochondria in Purkinje cells; 90% of Purkinje cells die between 3 and 6 weeks of age. Surviving ones recover normal cytology. Most of the retinal photoreceptors degenerate by 7 weeks.[58,59]

The Purkinje cell degeneration (*pcd*) mutation is an autosomal recessively inherited trait. Animals are identified by a mild ataxia of gait. Morphological studies reveal a total degeneration of Purkinje cells between the 15th day and the third month of age. In addition, the *pcd* mutation produces slow degeneration of mitral cells in the olfactory bulb and photoreceptors and induces structural abnormalities in spermatozoa, leading to sterility.[60,61] Studies of chimeric mice with cerebella containing mixtures of normal and *pcd* cells revealed that all *pcd* cells had atrophied, and all normal cells survived, indicating that, in contrast to reeler, in the *pcd* mutation, the primary defect is cell autonomous.[62]

Correlations between decrease in cyclic nucleotides, putative neurotransmitters, or receptors and the lack of a given cell type give information on cellular localization of these components. For instance, comparison of the cGMP concentration in normal and Nervous cerebellum, and study *in vitro* of the effect of kainic acid on these mutants established the localization of cGMP-dependent protein kinase in Purkinje cells[63] and the role of kainic acid in stimulating Purkinje cells to synthesize cGMP.[64] Regarding the role of glutamate as a neurotransmitter of the granule cells, the 70% decrease in this molecule in cerebellum of Weaver, Staggerer, and Reeler mutants with respect to control animals reflects the high concentration of this amino acid in granule cells.[65–67] Nevertheless, as the same decrease is observed in cerebellar deep nuclei of the Weaver and Staggerer mice[67] whose deep nuclei are morphologically unaffected, the conclusion is not definite.[43] Regarding receptor localization, the reduction to 17% of normal values of the binding of [^{3}H]diazepam in Nervous mice[68] without apparent alteration in receptor activities suggests that the Purkinje cells possess a large number and a high density of diazepam receptors. In contrast, no reduction has been found in Weaver mice, suggesting that the granule cells do not carry benzodiazepine receptors.[69] Measurements of P400 protein content of Nervous and Staggerer cerebellum[70] indicate that this protein is specific for Purkinje cells, but at the moment, although it is clear that this protein is associated with membranes, immunologic techniques do not allow its localization at the ultrastructural level.[70]

3.4. Epilepsies

About 20 neurological mutations (for review see refs. 71,72) are known to display spontaneous or evoked seizures in mice. Many neurological mutants convulse, such as Quaking, Jimpy, and Weaver. Only five appear to be putative models of primary epilepsies[73,77] because of the absence in these mutants of gross anatomic defects (see Table IV).

Two of these, Epilepsy-like (*El*) and Tottering (*tg*), have been electrophysiologically well defined. Biochemical studies on *El* revealed a reduction of cholinesterase activity after seizures, a higher level of choline acetyltransferase activity, and a higher level of acetylcholine,[78] a high level of γ-aminobutyric acid in the cortex,[79] and some distinctive findings in RNA metabolism in the brain.[80]

In the *tg* mouse, a survey of regional cerebral metabolism during focal motor seizures revealed a discrete activation of motor and reticular nuclei. The correlation of the focal nature of the motor seizures with a restricted anatomic substrate for the discharging neuronal circuitry is clearly demonstrated by the lack of increased metabolic activity and hence the absence of seizure spread to surrounding brain regions.[81]

3.5. Myelin Defects

The discovery of murine mutations affecting myelination (for review see refs. 82–84) has provided an effective tool for studying both myelination and pathological disturbances linked with abnormal myelination or secondary demyelinations.

Table IV
Epilepsies

Symbol	Name	Genetics				Neurological disorders	Other organs	Model for genetic human disease
		Chr	D/R	F/S	V/L			
asp[73]	Audiogenic seizure prone	4	R	F	V	Audiogenic seizures, no morphological lesions		
epf[74]	Epileptiform	?	R	F	V	Spontaneous and evoked tonic–clonic seizures no morphological lesions		
El[75]	Epilepsy	?	R	F	V	Evoked focal tonic and/or tonic–clonic seizures, no morphological lesions		Focal epilepsy
lh[76]	Lethargic	2	R	F	V	Focal clonic and/or tonic–clonic seizures sometimes showing typical "jacksonian" march, no morphological examination.		Centralopathic epilepsy
tg[77]	Tottering	8	R	F	V	Spike waves and focal motor seizures, no morphological lesions		

Like cerebellar mutations, those essentially restricted to myelin ontogenesis affect different stages of maturation at various degrees and have contributed to a dissectional approach to this process. Actually, in most of these mutants, no causal relationship has been established between the gene defect and the molecular events affecting myelin. Nevertheless, they have permitted an experimental approach to the various steps underlying myelin formation and to the influence of such events on other developmental processes.

It is clear that several genes are implicated in myelination (Table V), some of which involve, with differing degrees of selectivity, either the central or the peripheral nervous system or both. Some of these mutations exhibit clear-cut defects more or less restricted to the myelination process (Quaking, Jimpy and its allele Jimpy^{msd}, Shiverer and its allele Shiverer^{mld} affecting mainly CNS, and Trembler affecting mainly PNS). Most of the mutations involving secondary demyelinations are more or less generalized metabolic errors. The major ones are related in Section 3.7.

3.5.1. Mutations Affecting Mainly CNS

To a first approximation, mutations affecting CNS appear to block specific steps in the myelination program. For Jimpy, this is the beginning of myelination and results in the persistance of small axons and immature oligodendrocytes, whereas later stages of oligodendrocyte differentiation are affected in Quaking and Shiverer. These three mutations involve to different degrees and at different steps myelin assembly, as can be shown by the dynamics of myelin-specific components, myelin basic protein, and galactosylceramide (galactocerebroside) from their appearance in the oligodendrocyte until their incorporation into the myelin sheath.

The Jimpy (*jp*) mutation is sex-linked recessive. Animals manifest an axial body tremor at 12 days. They develop tonic seizures after 20 days and die at about 25–35 days. The major morphological trait is an almost total lack of myelin in the whole CNS. Privat *et al.*[90] attributed this absence of myelin to a defect of multiplication and maturation of the oligodendrocytes. Recently, it was claimed by Skoff[91] that the earliest defect in the Jimpy mutant is the hyperplasia and hypertrophy of astrocytes which prevent the investment of axons by oligodendrocyte processes. However, recently, by freeze-fracture techniques, Privat *et al.*[92] showed that the investment of the axons by oligodendrocyte processes occurs normally, thus confirming that the defect is primarily in oligodendrocyte maturation. In Jimpy mutants, no major changes in neuronal sizes, number, or arrangement have been described, and axons seem normal, although their growth has been shown to be retarded. Peripheral nerves appear normal.[86]

Since the Jimpy mutant is devoid of mature oligodendrocytes and myelin, it may be assumed that if a component is reduced in this mutant to less than 10% of the level encountered in normal animals, that component is normally synthesized in the oligodendrocytes. This has been shown for the synthesis of myelin sphingoglycolipids, in particular galactocerebrosides, sulfatides (for review see refs. 82,93) and their very-long-chain fatty acids.[94] Myelin enzymes,

Table V
Myelin Defects

Symbol	Name	Genetics				Neurological disorders	Other organs	Model for genetic human disease
		Chr	D/R	F/S	V/Sl/L			
jp^{86}	Jimpy	X	R		Sl	Hypertonic seizures, severe myelin deficiency		
jp^{msd85}	Myelin synthesis deficiency	X	R		Sl	Same		
qk^{86}	Quaking	17	R	S	V	Tonic and tonic–clonic seizures, myelin deficiency in CNS	Sterility in males	
shi^{87}	Shiverer	?	R	F	Sl	Severe hypertonic seizures, myelin deficiency, lack of immunodetectable CNS basic protein		
shi^{mld88}	Myelin deficient	?	R	F	Sl	Same		
Tr^{49}	Trembler	11	D	F	V	Tremor, uncoordination, defect of myelin in PNS		Hereditary hypertrophic neuropathies
dy^{89}	Dystrophia muscularis	10	R	S	Reduced viability	Progressive weakness and paralysis, naked axons in spinal roots and portions of cranial and peripheral nerve		

cholesterol ester hydrolase, and 2',3'-nucleotide phosphodiesterase are also severely reduced in Jimpy (for a review see ref. 95). Myelin proteins, proteolipid protein (PLP), myelin basic protein (MBP), and myelin-associated glycoprotein (MAG) are severely depleted in Jimpy (for review see ref. 96). All subcellular fractions have very low MBP content except the cytosol in which MBP level is similar to that of controls.[97] In Jimpy, the incorporation of MBP into the myelin sheath is drastically reduced, although its synthesis is normal. This fact, correlated with the presence of vesicular and membraneous inclusions in oligodendroglia perikaryon, indicates that the mutation blocks the conversion of precursor membranes into myelin.[98] Astrocytes appear to be reactive (gliosis) in Jimpy; glial Fibrillary acidic protein (GFAP) is considerably increased.[99] The myelin synthesis-deficient (jp^{msd}) mutant allelic with jp has not yet been so extensively studied. Most biochemical findings obtained are similar to those reported in jp.[100]

In the recessive autosomal mutation Quaking (qk), animals show a generalized tremor during locomotor activity. After 35 days, they develop spontaneous and evoked tonic–clonic seizures. The reduction of myelin is appreciably greater in rostral areas and varies in different tracts.[101] The sheaths appear poorly organized, with numerous pockets of oligodendroglial cytoplasm and frequent failure of compaction.[102,103] The oligodendrocyte perikaryon often contains lamellar inclusions, dense bodies, or vacuoles. As in some human genetic and degenerative disorders, morphological alterations have been found in the Purkinje cell axons of Quaking and Shiverer mutations.[104] There are peculiar clusters of densely packed small vesicles in several terminals associated with axonal focal swellings.[105] Although hypomyelination and dysmyelination of the central nervous system are predominant, hypomyelination has been found in the peripheral nervous system: opening of major dense lines, long nodal gaps, variation of Schmidt–Lanterman incisures, naked axolemma segments surrounded only by basal lamina.[106] As in Jimpy, but to a lesser extent, analysis of the brain lipids reveals a drastic deficit of cerebrosides and sulfatides and their very-long-chain fatty acids.[108,109] There are no major changes in total ganglioside level, although G_{M1} appears slightly reduced after 30 days.[110] Activities of enzymes involved in the synthesis of myelin sphingoglycolipids[82,93,94] are reduced about 30–50%, as are the myelin enzymes[95] cited preceedingly (see Jimpy). α-Mannosidase activity is also reduced in the mutant.[111]

The fact that, as in jp, in qk mice so many enzyme activities mainly related to biosynthetic pathways of myelin lipids have been found reduced although these two mutations are located on different chromosomes suggests a feedback control mechanism resulting from intracellular accumulation of end products[82] in the nervous system. This is in agreement with the fact that in organs such as kidney, very-long-chain fatty acids[112] and sulfatides[113] are present and normally synthesized at all times in both mutants. Except for polyphosphoinositide biosynthesis,[114] there appear to be no major disturbances of phospholipid metabolism. On a continuous sucrose gradient, qk myelin distributed in a bell-shaped mode shows an abnormally high density when compared with control myelin.[115,116] Acrylamide gel electrophoresis demonstrates a deficiency of PLP[117] and small basic protein.[118] The synthesis of PLP and MBP occurs at normal rates, but their incorporation into myelin membrane seems blocked.[119] The impairment of myelination could be related to the persistence of large-molecular-weight MAG in myelin of these mutants.[118]

In peripheral nerve, the major specific myelin protein, P0, is normal,[120] whereas a drastic reduction of the two basic protein components, P1 and P2, is observed. In experiments that employ the nerve grafting technique, it has been shown that transplanted Quaking Schwann cells fails to produce sufficient myelin when they ensheath normal axons. Conversely, the grafted normal Schwann cells myelinate Quaking mouse axons normally.[121] These observations suggest that in Quaking, as in Trembler, a primary Schwann cell abnormality is responsible for the peripheral neuropathy. Indirect evidence for a similar alteration in oligodendroglial cells has been provided by the demonstration of normal myelination of CNS axons by *in vitro* cultured Schwann cells from rats transplanted into the spinal cord of Quaking mice.[122] The formation of normal myelin by these exogenous Schwann cells suggests that CNS axons and more general conditions in this mutant are not responsible for the myelin deficit in the Quaking central nervous system.

The Shiverer (*shi*) mutation is an autosomal recessive trait clinically characterized by the appearance at 12–14 days of a generalized fine tremor. Then animals develop strong spontaneous tonic seizures after 35–40 days and die at 3–4 months. Conventional and freeze-fracture electron microscopy studies reveal the absence of the major dense line in myelin sheaths, whereas the intraperiodic line appears normal.[123] The defect is correlated with the almost total absence of myelin basic protein as determined by immunologic techniques.[124] Although the peripheral myelin sheaths appear normal at the ultrastructural level, there is a lack of myelin basic protein.[125] The same neuronal abnormalities as in Quaking have been found in cerebellum.[104] Biochemical studies show drastic reduction of myelin lipids, especially glycolipids.[124,126] Radioimmunoassay of basic protein shows a 100-fold decrease in *shi* brain as compared to controls. Indirect immunofluorescence is negative when done with specific anti-MBP serum. Myelin-deficient (*shi^{mld}*) is an allelic form of the *shi* mutation. The main phenotypic features are similar except with regard to the level of MBP which, although drastically decreased when compared to controls, is nevertheless tenfold higher in *mld* brain than in *shi*.[127,128] Oligodendroglial accumulation of galactocerebroside and sulfatide is observed in *jp* and *shi* mice.[129] In the *jp* mice, MBP is synthesized and deposited at a very low level; this protein is virtually absent in *shi* mice, with a correlated absence of the major dense line; in both mutants galactolipids are found to accumulate in the oligodendroglia. This suggests that there is no direct interference by the MBP on the biosynthesis of galactolipids and that the key regulatory steps must be elsewhere. Brains of Quaking, Jimpy, and Shiverer[131] have already been grown in culture.

3.5.2. Mutations Affecting Mainly PNS

The dystrophic mouse (*dy*) is an autosomal recessive mutation. The first clinical symptoms appear at about 3–4 weeks with progressive dragging of hindlimbs, kyphosis, nodding of the head, gasping, and progressive atrophy of axial limb muscles. The CNS is normal. In PNS, lumbosacral and cervical spinal roots exhibit naked axonal segments and hypomyelination. The striking morphological abnormalities in the spinal roots and portions of cranial and peripheral nerves of this mutant represent the most extreme known abnormality of axonal ensheathment. The naked axonal segments tend to occur in groups so that the membranes of adjacent axons are closely apposed. In addition to the focal absence of Schwann cells, other portions of these fibers are surrounded by inappropriately thin myelin with long nodal gaps or show discontinuities in the basal lamina.[132,133] The abnormality in the spinal roots of dystrophic mice is present at birth and associated with a neonatal deficit of Schwann cell multiplication.[134] Transplantation studies of dystrophic spinal roots have shown that Schwann cells are not implicated in the myelin defect. The lesion could best be explained by an abnormality of extracellular matrix (ECM) within the nerve. This contention derives from evidence that Schwann cell development is retarded if only axonal contact is available without simultaneous contact with appropriate ECM material.[135] The protein composition of the sciatic nerve is normal in this mutant.[136]

The Trembler mouse (*Tr*) is characterized by a dominantly inherited chronic hypertrophic neuropathy. The main clinical features are onset of generalized tremor at 11–13 days, abnormal limb posture, ataxia of gait, incoordination, and stimulus-induced convulsions about weaning age. Tremor alone subsists when animals grow old. They have a normal life-span. There are severe degrees of myelinated fiber loss, hypomyelination, and Schwann cell hyperproliferation associated with "onion bulb" in the PNS. The peripheral nerves of *Tr* mice have abnormally thin, poorly compacted, or totally absent myelin sheaths. In the spinal roots, axonal caliber is nearly half that of the normal. There is evidence for segmental demyelination in *Tr* nerves. The Schwann cell population increases rapidly during the first month after birth to nearly ten times that of control littermates and remains stable in older animals in spite of the persistence of these cell multiplications. The abnormalities are confined to peripheral nerves normally myelinated in nonmutant animals; the morphology and number of Schwann cells in *Tr* nerves composed of unmyelinated fibers are normal.[139]

Experiments using transplantation techniques prove that the defect of myelination resides in a primary Schwann cell disorder rather than in an axonal defect.[140] The lipid level of *Tr* sciatic nerves is diminished, each class of lipid being decreased nearly the same except for the cholesterol esters, the quantity of which is fivefold higher than normal.[141] The specific activity of ceramide galactosyltransferase is highly decreased.[142] In contrast, cerebroside sulfotransferase activities are increased in sciatic nerves of *Tr* mice.[142] Low levels of substrate and high arylsulfatase A activity (218% of normal) could explain the lack of sulfatide accumulation. Nevertheless, this abnormal sulfate metabolism does not seem to affect the incorporation of sulfate in P0. In *Tr* peripheral myelin, P0, LBP, and SBP are diminished.[96] In many respects, the protein composition of the mutant's myelin is similar to that isolated from sciatic nerves undergoing Wallerian degeneration.[96]

3.6. Neuromuscular Disorders

Among several hereditary neuromuscular diseases (for review see refs. 143,144), only four involve a primary motor or sensory neuronal disorder (Table VI). Two of them, Dystonia musculorum (*dt*) and Sprawling (*Swl*), affect primarily sensory neurons with secondary mild effects on skeletal muscles. The two others, Motor end plate disease (*med*) and Wobbler (*wr*), affect motor neurons with strong secondary effects on skeletal muscles.

In *dt* mice, the first clinical features appear about 8–10 days after birth. Then a rapidly progressive incoordination of limb movements follows with absence of voluntary movements. Three allelic forms of this mutation were observed independently showing different life-spans (from 3 weeks up to 6 months). Morphological examinations show that principal sensory nerve fibers develop a typical "neuroaxonal dystrophy": sensory axons appear argyrophilic and filled with masses of neurofilaments, vesicles, or dense bodies. Both peripheral and central branches of sensory neurons are affected. If the animals survive long enough, the sensory roots become severely depleted of axons, and peripheral sensory organs become denervated. Sensory ganglion cells may demonstrate chromatolysis. Myelin degeneration and Wallerian-type degeneration follow. Skeletal muscles are not paralyzed. In young animals, no abnormality of muscle or its innervation is observed except for muscle spindles. Partial motor denervation occurs with longer survival.

The first clinical manifestation of the autosomal dominant mutation Sprawling (*Swl*) appears at 8–10 days of age with a flection of the hind limb to the trunk when animals are lifted by the tail. The animals develop a sensory ataxia and incoordination particularly affecting the hind limb. Microscopic examination shows a severe deficiency in large-diameter sensory fibers primarily in caudal roots and hind limb nerves. Sensory ganglion L4 and L6 cell number is greatly reduced; these cells appear abnormal. They contain excentric nuclei, abundant dense bodies, lipofuscin granules, Golgi complexes, and neurofilaments aggregated near the nucleus. Nissl bodies are absent except at the periphery of the cells. These changes are characteristic of nerve cell reactivity to axonal injury. No sign of degeneration is found. Muscle spindles are deficient in number, especially in distal hind limb muscles. Ventral roots of the *Swl* mice exhibit a striking absence of small-diameter myelinated axons which includes an absence of γ efferents.

The autosomal recessive Motor end plate disease (*med*) mutation causes weakness which begins at about 10–12 days. Animals die at weaning age. Morphological examination shows the appearance at 15 days of sprouting of the motor nerve terminals. The sprouts grow in all directions from end plates, correlated with thin unmyelinated axons lying in the superficial contact with the sarcolemmal membrane which is deficient in postsynaptic folds. Muscles are not affected uniformly: forelimb muscles are worst affected, as are proximal thigh muscles. Muscle fibers become atrophied while their nerve fibers are intact and in contact with them. Electrophysiological studies show a normal axonal conduction, but with increasing age, an increasing number of muscle

Table VI
Neuromuscular Disorders

Symbol	Name	Genetics				Neurological disorder	Other organs	Model for genetic human disease
		Chr	D/R	F/S	V/L/Sl			
mdg[147]	Muscular dysgenesis	?	R		L	Myogenesis retardation	Arthrogryposis	Multiple congenital arthrogryposis or dystrophic neonatal myotonia (Steinert)
med[146]	Motor end plate disease	15	R		Sl	Sprouting, partial functional denervation, myelin and axon abnormalities slowered conduction		
med^J[143]	Motor end plate disease, Jackson	15	R		Sl			
med^Jo[143]	Motor end plate disease, Jolting	15	R		Reduced viability			
wr[149]	Wobbler	?	R		Reduced viability	Motor neuron degeneration in the spinal cord		
dt[145]	Dystonia musculorum	1	R		Reduced viability	Lack of Schwann cells in some fiber tracts		Werdnig–Hoffmann
Swl[148]	Sprawling	?	D	F	V	Selective loss of sensory axons in peripheral nerves and in dorsal roots, sensory ganglion neuron degeneration		

fibers fail to respond to nerve stimulation with an action potential. Muscle fibers display extrajunctional sensitivity to acetylcholine, and action potentials generated by intracellular stimulation are resistant to tetrodotoxin. Muscles of *med* mice appear functionally as opposed to structurally denervated.

In the autosomal recessive mutation Wobbler (*wr*), young mice develop a slight degree of tremulousness, and by the third week, they tend to be smaller than normal and always remain so. Forelimbs become increasingly weak, and animals develop a characteristic "side-to-side" wobbling gait. Many animals die during the second and third month, but some survive up to 1 year or longer. The first morphological abnormalities occur in the perikarya of motor neurons in the brainstem and the spinal cord with enlargement and rounding of the motor neurons. Nissl bodies are not identifiable, but the nucleolus normally stains.

Later, the cytoplasm becomes vacuolated and then disintegrates. No inflammatory reaction is seen. Synapses are identifiable in the neuronal plasma membrane. Abnormalities extend out into neuronal processes, and then myelinated axons in ventral roots of peripheral nerves display Wallerian degeneration. Motor nerve cells in ventral horns of the cord become reduced in number; preterminal axons are observed to branch excessively and to innervate more fibers than normal. Areas of muscles become atrophied and fibrillate. Biochemical studies suggest an abnormally low level of aspartic acid[150] in the spinal cord, a decrease of 80% of cGMP in the cervical spinal cord.[151] and reduced protein synthesis in the spinal anterior horn of the same organ.[152] Electrophysiological investigations have demonstrated that the changes in skeletal muscles are characteristic of partial motor denervation. In fibrillating areas, single muscle fibers exhibit supersensitivity to acetylcholine. Action potentials are resistant to tetrodotoxin. In contrast with *med*, *wr* muscles are unfunctional secondarily to the loss of motor nerve fibers. Leestma[153] suggests that the Wobbler mouse appears to bear a close similarity with the Werdnig–Hoffmann human disease, although neuronal vacuolation is rarely observed in the human syndrome.

3.7. Inherited Metabolic Disorders

Several inborn errors of metabolism in mice involve neurological disorders. Among the best characterized ones, the Crinkled mutation (*cr*) and four allelic forms at the Mottled locus are known to be alterations of copper metabolism. Another, the Twitcher (*twi*) mutation, is a true model of globoid-cell leukodystrophy (Table VII).

The Crinkled (*cr*) mutation is a recessive autosomal trait. Homozygotes are characterized by absence of guard hairs, zigzag hairs in the coat related to an absence of follicle, reduced aperture of the eyelids, respiratory disorders, and a modification of the agouti patterns. Viability and breeding performances are reduced. Morphological examination of the CNS reveals swollen myelin sheaths in whole brain and especially in cerebellum; myelin stains appear much

Table VII

Inherited Metabolic Diseases Involving Nervous System Abnormalities

Symbol	Name	Genetics			Neurological disorder	Other organs	Model for genetic human disease
		Chr	D/R	F/S	V/SI/L			
cr[154]	Crinkled	13	R	F	Reduced viability	Light demyelination in CNS, especially in cerebellum	Abnormal coat color and structure, reduced aperture of eyelid, respiratory disorder, decrease of copper concentration in liver	
Mo[br155]	Brindled	X	½D		SI	Neuronal degeneration in cerebral cortex and thalamic nuclei	Devoid of pigments except in eyes and ears, strongly curled vibrissae and weavy coat, abnormally high copper concentration in kidney	Menkes disease
twi[156]	Twitcher	?	R		V	Generalized tremor and progressive paralysis, demyelination in CNS and PNS	Absence of galactosylceramidase and lactosylceramidase I activity in liver	Globoid cell leukodystrophy

paler in spongy myelin areas. Lipid analysis shows an increase of sulfatides in 21-day-old *cr* brain when compared with controls of the same age. In 1-year-old mouse brain, cholesterol esters are detected, although they are absent in control mouse brains.[157] Developmental patterns of copper in liver and brain show a decrease of liver copper concentration in *cr* mice when compared to controls between 18 days of gestation and 20 days after birth. In *cr* mice older than 20 days of age, liver copper concentration is similar to that of the control mice. Brain copper concentration is similar in both at all times tested.[158] The effect of the mutant allele *cr* has been shown to be partially prevented by dietary copper supplementation of the mother during gestation and lactation.[159]

The Mottled locus located on the X chromosome in the mouse provides several alleles: Tortoise (*To*), Dappled (*Mo^{dp}*), Brindled (*Mo^{br}*), Viable brindled (*Mo^{vbr}*), and Blotchy (*Mo^{blo}*), producing coat color and structure modifications associated with copper metabolism alterations. The most widely studied is the Brindled allele because of its strong similarity with the human Menkes disease.*Mo^{br}* Males are characterized by curly whiskers and absence of fur pigments. They become inactive, losing weight at around 10–12 postnatal days. They usually die in an emaciated state around 15–16 postnatal days. Brain weight is about three-fourths of that of normal littermates. Increasing widespread neuronal degeneration is noted in cerebral cortex and thalamic nuclei after the 12th postnatal day. Cortical neurons contain enlarged mitochondria with tubular or vesicular cristae. Intracytoplasmic vesicular vacuoles and electron-dense matrix containing myelin figures are observed.[160] Abnormally high copper concentration is observed in the kidney.[161] Other studies revealed the presence of an abnormal renal copper thionein.[162] Copper therapy is effective in these mutants.[163]

The Twitcher (*twi*) mutation recently described by Duchen *et al.*[156] appears to be a true model of human globoid-cell leukodystrophy. This apparently autosomal recessive syndrome is clinically identified by a marked generalized tremor which appears after 30 days of age. Other signs include wasting and weakness progressing to almost complete paralysis, especially of the hindlimbs. Degenerating myelin sheaths and degenerating axons are formed in all areas of cerebrum, cerebellum, brainstem, and spinal cord. Evidence for remyelination is also present. Rounded often multinucleated macrophagelike cells are found in demyelinated areas. Various inclusions are present in macrophages and glial cells as well as neurons; the most interesting are multiangular bodies and twisted tubules lying within membrane-bound spaces. Abnormal glial cells are also observed. In peripheral nerves, axons are widely separated from each other by abnormal cells similar to those observed in the CNS. Little or no evidence of axonal degeneration is seen in the PNS. As in the CNS, there are many demyelinated axons, indicating a continuous processing of myelin breakdown and reformation.[164] Studies on galactosylceramidase and lactosylceramidase I activities in brain and liver of *twi* homozygotes revealed an almost total absence of activity for these two enzymes in comparison with control littermates. No significant differences in the activities of other lysosomal hydrolases were detected.[165]

3.8. Problems Related to the Production and Use of Neurological Mutants

During the last 10 years, our attention has been drawn to the difficulties of obtaining reproducible and comparative results when biochemical or immunochemical studies are performed on neurological mutants. This is in part correlated with an absence of genetic uniformity among the different mutant strains or sometimes within a given strain. Because of the various origins of these mutations and the zootechnical difficulties linked to their transfer and inbreeding, very few are maintained on a genetically homogeneous line. The use of mutants maintained on similar genetically defined backgrounds appears obviously recommendable. In order to fulfill zootechnical and scientific requisites, maintenance of these mutations on two different inbred strains bred in parallel and production of mutants by intercrossing parents of these strains seems the best breeding system. Indeed, the F_1 offspring offer the advantages of heterosis and small variability. Regarding the choice of strains, one must consider the possibility of an eventual interference with the experiment. Indeed, it is known that variable penetrance in some mutations is in part dependent on the genetic environment (i.e., dysraphic mice). On the other hand, phenotypic expression of a given mutation may be influenced by the carrying line: some strains are known to carry specific mono- or polygenic characters affecting specifically the structure, development, or physiology of the nervous system. For instance, the DBA2 strain carries together the *asp* mutation and a polygenic determinant of sound-induced seizure susceptibility.[73] Likewise, differences in catecholaminergic tracts have been described in the BALB/C and C57BL/6 strains.[166] Finally, the ICR strain has been shown to present a delay in the timing of myelination.

Another problem is the use of genetic markers. These markers allow an early recognition of mutants before the onset of clinical feature. The use of repulsion intercrosses is highly recommended in order to avoid any interference with the mutation. For instance, in the case of the *jp* mutation and the Tabby *(Ta)* marker, one should prefer the following type of cross: $Ta + / + jp \times + + /Y$ producing $+jp/Y$ hemizygotes rather than the $+ + /Tajp \times + + /Y$ type producing *Tajp*/Y hemizygotes, as the latter animals cannot be used for comparative studies with other non-*Ta* dysmyelinating mutants.

Most of these mutations are concerned with key mechanisms of development of the nervous system. Genetic markers, tissue culture, and new development in molecular biology have only quite recently been used. They have now advanced to a point where rapid progress can reasonably be expected with neurological mutations as biological tools.

REFERENCES

1. Davison, A. N., and Dobbing, J. (eds.), 1968, *Applied Neurochemistry*, F. A. Davies, Philadelphia.
2. Benirschke, K., Garner, S. N., and Jones, T. C., 1978, *Pathology of Laboratory Animals*, Volume 1, Springer-Verlag, New York, pp. 1983–1993.

3. Cummings, J. F., 1978, *Spontaneous Models of Human Diseases*, Volume 2 (E. J. Andrews, B. C. Ward, and N. H. Altman eds.), Academic Press, New York, pp. 108–178.
4. CNRS, 1968–1982, *Animaux de Laboratoire (Bibliographic Review)*, CNRS Paris.
5. Institute of Laboratory Animal Resources, 1957–1982, *ILAR News*, National Research Council, Washington.
6. Festing, M. F. W. (ed.), 1975, *International Index of Laboratory Animals*. Medical Research Council, Laboratory Animals Center, Carshalton.
7. Gilford, L. P. (ed.), 1945–1982, *Mouse News Letters*. MRC Laboratories, Carshalton.
8. Hocking, J. D., Jolly, R. D., and Batt, R. D., 1972, *Biochem. J.* **126**:69–78.
9. Dennis, S. M., 1975, *Aust. Vet. J.* **51**:385–388.
10. Read, D. H., Harrington, D. D., Keenan, T. W., and Kinsman, E., 1976, *Science* **194**:442–445.
11. Pierce, K. R., Kosanke, S. O., Bay, W. W., and Briges, C. H., 1976, *Am. J. Pathol.* **83**:419–421.
12. Killam, K. F., Killam, E. K., and Naquet, R., 1966, *C.R. Acad. Sci. [D] (Paris)* **262**:1010–1012.
13. Naquet, R., 1979, *Wissensch. Z. Ernst Moritz Ardnt Univ. Greifswald.* **28**(3):239–244.
14. Naquet, R., and Meldrum, B. S., 1972, *Experimental Models of Epilepsy* (D. P. Purpura, ed.), Raven Press, New York, pp. 376–406.
15. Ancri, D., Naquet, R., Menini, C., Meldrum, B., Stutzmann, J. M., and Basset, J. Y. 1981, *Electroencephalogy Clin. Neurophysiol.* **51**:91–103.
16. Baker, H. J., Mole, J. A., Russel, J., and Creel, R. M., 1976, *Fed. Proc.* **35**:1194–1201.
17. Baker, H. J., Lindsey, J. R., McKhann, G. M., and Farrel, D. F., 1971, *Science* **174**:828–839.
18. Purpura, D. P., and Baker, H. J. 1977, *Nature* **266**:553–554.
19. Blakemore, W. F., 1972, *J. Comp. Pathol.* **82**:179–185.
20. Cork, L. C., Munnel, J. F., Lorenz, M. D., Murphy, J. V., Baker, H. J., and Rattazi, M. C., 1977, *Science* **196**:1014–1017.
21. Suzuki, Y., Austin, J., Armstrong, D., Suzuki, K., Schlenker, J., and Flechter, T., 1970, *Exp. Neurol.* **29**:65–75.
22. Suzuki, Y., Miyataki, T., Flechter, T. F., and Suzuki, K., 1974, *J. Biol. Chem.* **249**:2109–2112.
23. Jones, M. Z., and Laine, P. W., 1980, *Fed. Proc.* **39**:2521.
24. Sidman, R. L., Green, M. C., and Appel, S. H., 1965, *Catalog of the Neurological Mutants of the Mouse*, Harvard University Press, Cambridge.
25. Green, M. C., 1966, *Biology of the Laboratory Mouse*, 2nd ed. (E. L. Green, ed.), McGraw-Hill, New York, pp. 87–150.
26. Little, C. C., and Bagg, H. J., 1923, *Am. J. Roentgenol.* **10**:975–989.
27. Morgan, W. C., 1954, *J. Genet.* **52**:354–373.
28. Hummel, K. P., 1958, *J. Exp. Zool.* **137**:389–423.
29. Grüneberg, H., 1954, *J. Genet.* **52**:52–67.
30. Rheiler, K., and Glueksohn-Waelsch, S., 1956, *Anat. Rec.* **125**:83–104.
31. Caspari, E., and David, P. R., 1940, *J. Hered.* **31**:427–431.
32. Lyon, M. F., and Meredith, R., 1965, *Mouse News Lett.* **32**:38.
33. Russell, W. L., 1947, *Genetics* **32**:102.
34. Embury, S., Seller, M. J., Adinolf, M., and Polani, P. E., 1979, *Proc. R. Soc. Lond. [Biol.]* **206**:85.
35. Morris, G., 1980, *Nature* **284**:121–122.
36. Wilson, D. B., and Gonzales, L. W., 1978, *Brain Res.* **140**:354–359.
37. Deol, M. S., and Truslove, G. M., 1963, *Proceedings XI International Congress of Genetics*, Volume 1, Pergamon Press, New York, pp. 183–184.
38. Hollander, W. F., 1966, *Am. Zool.* **6**:588–589.
39. Grüneberg, H., 1943, *J. Genet.* **45**:1–21.
40. Grüneberg, H., 1943, *J. Genet.* **45**:22–28.
41. Caviness, V. S., and Rakic, P., 1978, *Annu. Rev. Neurosci.* **1**:297–326.
42. Quinn, W. G., and Gould, J. L., 1979, *Nature* **278**:19–23.
43. Mallet, J., 1980, *Curr. Top. Dev. Biol.* **15**:41–65.
44. Sidman, R. L., Lane, P. W., and Dickie, M. M., 1962, *Science* **137**:610–612.
45. Lane, P. W., 1964, *Mouse News Lett.* **30**:32.

46. Mullen, R. J., Eicher, E. M., and Sidman, R. L., 1976, *Proc. Natl. Acad. Sci. U.S.A.* **73**:208–212.
47. Sidman, R. L., and Green, M. C., 1970, *Les Mutants Pathologiques chez l'Animal, leur Intérêt dans la Recherche Biomédicale* (M. Sabourdy, ed.), CNRS, Paris, pp. 69–79.
48. Lane, P. W., 1967, *Mouse News Lett.* **36**:40.
49. Falconer, D. S., 1951, *J. Genet.* **50**:192–201.
50. Phillips, R. J. S., 1960, *J. Genet.* **57**:35–42.
51. Snell, G. D., 1955, *J. Hered.* **46**:27–29.
52. Dickie, M. M., 1962, *Mouse News Lett.* **27**:37.
53. Mariani, J., Crepel, F., Mikoshiba, K., Changeux, J. P., and Sotelo, C., 1977, *Philos. Trans. R. Soc. Lond. [Biol.]* **28**:1–28.
54. Caviness, V. S., 1977, *Soc. Neurosci. Symp.* **2**:27–46.
55. Mullen, R. J., 1976, *Approaches to Cell Biology of Neurons* (M. J., Cowan, and J. A. Ferendelli, eds.), Society for Neuroscience, Bethesda, pp. 47–66.
56. Hatten, M. E., and Messer, A., 1978, *Nature* **276**:504–506.
57. Sotelo, C., 1975, *Brain Res.* **94**:19–44.
58. Lanois, S., 1973, *J. Cell Biol.* **57**:582–597.
59. Mullen, R. J., and Lavail, M. M., 1975, *Nature* **258**:528–530.
60. Landis, S. C., and Mullen, R. J., 1978, *J. Comp. Neurol.* **177**:125–143.
61. Mullen, R. J., Eicher, E. M., and Sidman, R. L., 1976, *Proc. Natl. Acad. Sci. U.S.A.* **73**:208–212.
62. Mullen, R. J., 1977, *Nature* **270**:245–247.
63. Mai, C. C., Guidotti, A., and Costa, E., 1975, *Brain Res.* **83**:516–519.
64. Schmidt, M. J., and Nadi, N. S., 1977, *J. Neurochem.* **29**:87–90.
65. Hudson, D. B., Valcana, T., Bean, G., and Timiras, P. S., 1976, *Neurochem. Res.* **1**:73–81.
66. Mc Bride, W. J., Aprison, M. H., and Kusano, K., 1976, *J. Neurochem.* **14**:465–472.
67. Roffler-Tarlou, S., and Sidman, R. L., 1978, *Brain Res.* **142**:269–283.
68. Lippa, A. S., Sano, M. C., Coupet, J., Klepner, C. A., and Beer, B., 1978, *Life Sci.* **23**:2213–2218.
69. Skolnic, P., Syapin, P., Pouger, B., and Paul, S., 1979, *Nature* **277**:397–398.
70. Mallet, J., Huchet, M., Pougeois, R., and Changeux, J. P., 1975, *FEBS Lett.* **52**:216–220.
71. Purpura, D. P., Penry, J. K., Tower, D. B., Woodbury, D. M., and Walter, R. D. (eds.), 1972, *Experimental Models of Epilepsy*, Raven Press, New York.
72. Noebels, J. L., 1979, *Fed. Proc.* **32**:2405–2410.
73. Collins, R. L., and Fuller, J. L., 1968, *Science* **162**:1137–1139.
74. Hare, J. E., and Shirashare, A., 1979, *J. Hered.* **70**:417–420.
75. Imazumi, K., Ito, S., Kutsukaki, G., Takizawa, T., Fujiwara, F., and Tsuchikawa, K., 1959, *Exp. Anim. (Tokyo)* **8**:6–10.
76. Dickie, M. M., 1964, *Mouse News Lett.* **30**:31.
77. Green, M. C., and Sidman, R. L., 1962, *J. Hered.* **53**:233–237.
78. Kurchawa, M., Naruse, H., and Kato, M., 1966, *Prog. Brain Res.* **12A**:112–130.
79. Naruse, H., Kato, M., Kurochawa, M., Haba, R., and Yabe, T., 1960, *J. Neurochem.* **5**:359–369.
80. Muri, K., Kirike, N., Yamagami, S., Ohnishi, H., and Kawakita, Y., 1974, *Bull. Jpn. Neurochem. Soc.* **13**:100–103.
81. Noebels, J. L., and Sidman, R. L., 1979, *Science* **204**:1134–1136.
82. Hogan, E. L., 1977, *Myelin* (P. Morell, ed.), Plenum Press, New York, pp. 489–520.
83. Baumann, N. A. (ed.), 1980, *Neurological Mutations Affecting Myelination (INSERM Symposium 14)*, Elsevier/North-Holland Biomedical Press, Amsterdam.
84. Baumann, N. A., 1980, *Trends Neurosci.* **22**:82–85.
85. Meier, H. and Mc Pike, A. D., 1970, *Exp. Brain Res.* **10**:512–525.
86. Sidman, R. L., Dickie, M. M., and Appel, S. H., 1964, *Science* **144**:309–311.
87. March, E., 1973, *Mouse News Lett.* **48**:24.
88. Doolittle, D. P., and Schweikart, M., 1977, *J. Hered.* **68**:331–332.
89. Michelson, A. M., Russel, E. S., and Harman, P. J., 1955, *Proc. Natl. Acad. Sci. U.S.A.* **41**:1079–1084.

90. Privat, A., Robain, O., and Mandel, P., 1972, *Acta Neuropathol.* **21**:282–295.
91. Skoff, R. P., 1976, *Nature* **264**:560–562.
92. Privat, A., Drian, M. J., and Escaig, J., 1979, *Acta Neuropathol.* **45**:129–131.
93. Baumann, N., 1978, *Proceedings of the European Society for Neurochemistry* (V. Neuhof, ed.), Verlag Chemie, Weinheim, pp. 48–63.
94. Bourre, J. M., Paturneau-Jouas, M. Y., Daudu, O. L., and Baumann, N. A., 1977, *Eur. J. Biochem.* **72**:41–47.
95. Suzuki, K., 1980, *Neurological Mutations Affecting Myelination* (N. Baumann, ed.), Elsevier/North-Holland, Amsterdam, pp. 333–347.
96. Matthieu, J. M., 1980, *Neurological Mutations Affecting Myelination* (N. Baumann, ed.), Elsevier/North-Holland, Amsterdam, pp. 275–298.
97. Zimmerman, T. R., Jr., and Cohen, S. R., 1979, *J. Neurochem.* **32**:1437–1446.
98. Barbarese, E., Carson, J. H., and Braun, P. E., 1979, *J. Neurochem.* **32**:1437–1446.
99. Jacque, C., Lachapelle, F., Collier, P., Raoul, M., and Baumann, N., 1980, *J. Neurosci. Res.* **5**:379–385.
100. Billings-Gagliardi, S., Adcock, L., and Wolf, M., 1980, *Brain Res.* **194**:325–338.
101. Friedrich, V. L., 1974, *Brain Res.* **82**:168–172.
102. Berger, B., 1971, *Brain Res.* **25**:35–53.
103. Wisniewski, H., and Morell, P., 1971, *Brain Res.* **29**:63–71.
104. Friedrich, V. L., Iconiecki, D. L., and Massa, P. T., 1980, *Neurological Mutations Affecting Myelination* (N. Baumann, ed.), Elsevier/North-Holland, Amsterdam, pp. 141–146.
105. Suzuki, K., and Zagoren, J. C., 1975, *Brain Res.* **85**:38–43.
106. Suzuki, K., and Zagoren, J. C., 1977, *Brain Res.* **106**:146–151.
107. Suzuki, K., and Zagoren, J. C., 1977, *J. Neurocytol.* **6**:71–84.
108. Baumann, N. A., Jacque, C. M., Pollet, S. A., and Harpin, M. L., 1968, *Eur. J. Biochem.* **4**:340–344.
109. Jacque, C. M., Harpin, M. L., and Baumann, N. A., 1969, *Eur. J. Biochem.* **11**:218–224.
110. Baumann, N., Harpin, M. L., and Jacque, C., 1980, *Neurological Mutations Affecting Myelination*, (N. Baumann, ed.), Elsevier/North-Holland, Amsterdam, pp. 257–268.
111. Kurtz, D. J., and Kanfer, J. N., 1970, *Science* **168**:259–260.
112. Bourre, J. M., Daudu, O., and Baumann, N., 1975, *J. Neurochem.* **24**:1095–1097.
113. Sarlieve, L. L., Neskovic, N. M., and Mandel, P., 1971, *FEBS Lett.* **19**:91–95.
114. Hauser, G., Heichberg, J., and Jacobs, S., 1971, *Biochem. Biophys. Res. Commun.* **43**:1072–1080.
115. Bourre, J. M., Jacque, C., Delassalle, A., Nguyen-Legros, J., Dumont, O., Lachapelle, F., Raoul, M., Alvarez, C., and Baumann, N., 1980, *J. Neurochem.* **35**:458–464.
116. Waehneldt, T. V., Fagg, G. E., Matthieu, J. M., Baumann, N. A., and Neuhoff, V., 1979, *J. Neurochem.* **32**:1679–1688.
117. Greenfield, S., Norton, W. T., and Morell, P., 1971, *J. Neurochem.* **18**:2119–2128.
118. Matthieu, J. M., Koelreutter, B., and Joyet, M. L., 1978, *J. Neurochem.* **30**:788–790.
119. Greenfield, S., Williams, N. I., White, M., Brostoff, S. W., and Hogan, E. L., 1979, *J. Neurochem.* **32**:1647–1651.
120. Matthieu, J. M., 1978, *Biochem. J.* **173**:989–991.
121. Aguayo, A. J., Bray, G. M., and Perkins, C. S., 1979, *Ann. N.Y. Acad. Sci.* **317**:512–531.
122. Duncan, I. D., Aguayo, A. J., Bunge, R. P., and Wood, P. M., 1979, *Soc. Neurosci. Abstr.* **5**:510.
123. Privat, A., Jacque, C., Bourre, J. M., Dupouey, P., and Baumann, N., 1979, *Neurosci. Lett.* **12**:107–112.
124. Dupouey, P., Jacque, C., Bourre, J. M., Cesselin, F., Privat, A., and Baumann, N., 1979, *Neurosci. Lett.* **12**:113–118.
125. Kirschner, D. A., and Ganser, A. L., 1980, *Nature* **283**:207–210.
126. Bird, T. D., Farell, D. F., and Sumi, S. M., 1978, *J. Neurochem.* **31**:387–391.
127. Lachapelle, F., Debaecque, C., Jacque, C., Bourre, J. M., Delassalle, A., Doolittle, D. P., Hauw, J. J., and Baumann, N., 1980, *Neurological Mutations Affecting Myelination* (N. Baumann, ed.), Elsevier/North-Holland, Amsterdam, pp. 147–152.

128. Ginalski, H., Friede, R. L., Cohen, S. R., and Matthieu, J. M., 1980, *Neurological Mutations Affecting Myelination* (N. Baumann, ed.), Elsevier/North-Holland, Amsterdam, pp. 147–152.

129. Zalc, B., Monge, M., and Baumann, N., 1980, *Neurological Mutations Affecting Myelination* (N. Baumann, ed.), Elsevier/North-Holland, Amsterdam, pp. 263–268.

130. Billings-Gagliardi, S., Adcock, L. H., Schwing, G. B., and Wolf, M. K., 1980, *Brain Res.* **200:**135–140.

131. Hauw, J. J., Boutry, J. M., and Jacque, C., 1980, *Neurological Mutations Affecting Myelination*, Elsevier/North-Holland, Amsterdam, pp. 475–480.

132. Madrid, R. E., Jaros, E., Cullen, M., and Bradley, W. G., 1975, *Nature* **257:**319–321.

133. Jaros, E., and Bradley, W. G., 1978, *Neuropathol. Appl. Neurobiol.* **5:**133–147.

134. Bray, G. M., Perkins, C. S., Peterson, A. C., and Aguayo, A. J., 1977, *J. Neurol. Sci.* **32:**203–212.

135. Bunge, R. P., and Bunge, M. B., 1978, *J. Cell Biol.* **72:**943–950.

136. Wiggins, R. C., and Morell, P., 1978, *J. Neurochem.* **31:**1101–1105.

137. Ayers, M. M., and Anderson, R., McD., 1973, *Acta Neuropathol.* **25:**54–70.

138. Low, P. A., 1977, *Neuropathol. Appl. Neurobiol.* **3:**81–82.

139. Aguayo, A. J., Bray, G. M., and Perkins, C. S., 1979, *Ann. N.Y. Acad. Sci.* **317:**512–531.

140. Aguayo, A. J., Attiwell, M., Trecarten, J., Perkins, S., and Bray, G. M., 1977, *Nature* **265:**73–75.

141. Larrouquère-Regnier, S., Boiron, F., Darriet, O., Cassagne, C., and Bourre, J. M., 1979, *Neurosci. Lett.* **15:**135–139.

142. Matthieu, J. M., Reigner, J., Costantino-Ceccarini, E., Bourre, J. M., and Rutti, M., 1980, *Brain Res.* **200:**457–465.

143. Sidman, R. L., Cowen, J. S., and Eicher, E. M., 1979, *Ann. N.Y. Acad. Sci.* **317:**497–505.

144. Duchen, L. W., 1979, *Ann. N.Y. Acad. Sci.* **3:**506–511.

145. Duchen, L. W., and Strich, S. J., 1964, *Brain Res.* **83:**367–378.

146. Duchen, L. W., 1970, *J. Neurol. Neurosurg. Psychiatry* **33:**238–250.

147. Gluecksohn-Waelsch, S., 1963, *Science* **142:**1269–1276.

148. Duchen, L. W., 1975, *Neuropathol. Appl. Neurobiol.* **1:**89–101.

149. Duchen, L. W., Falconer, D. S., and Strich, S. J., 1968, *J. Neurol. Neurosurg. Psychiatry* **31:**535–542.

150. Lewkowicz, S. J., 1974, *Excerpta Medica Int. Cong. Ser.* **334:**129.

151. Brooks, B. B., Cust, D. W., Andrews, J. M., and Engel, W. K., 1978, *Arch. Neurol.* **35:**590–591.

152. Murakami, T., Mastaglia, F. L., and Bradley, W. G., 1980, *Exp. Neurol.* **67:**423–432.

153. Leestma, J. E., 1980, *Am. J. Pathol.* **100:**821–824.

154. Falconer, D. S., Fraser, A. S., and King, J. W. B., 1951, *J. Genet.* **50:**324–344.

155. Fraser, A. S., Sobey, S., and Spicer, C. C., 1953, *J. Genet.* **51:**217–221.

156. Duchen, L. W., 1979, *Mouse News Lett.* **61:**47.

157. Theriault, L. L., Dungan, D. D., Simons, S., Keen, C. L., and Hurley, L. S., 1977, *Proc. Soc. Exp. Biol. Med.* **155:**549–553.

158. Keen, C. L., and Hurley, L. S., 1979, *J. Inorg. Biochem.* **11:**269–277.

159. Hurley, L. S., and Bell, L. T., 1975, *Proc. Soc. Exp. Biol. Med.* **149:**830–834.

160. Yajima, K., and Suzuki, K., 1979, *Acta Neuropathol.* **45:**17–25.

161. Evans, G. W., and Reis, B. L., 1978, *J. Nutr.* **108:**554–560.

162. Prins, H. W., and Van der Hamer, C. J. A., 1980, *J. Nutr.* **110:**151–157.

163. Mann, J. R., Camakaris, J., Danks, D. M., and Walliczek, E. G., 1979, *Biochem. J.* **180:**605–612.

164. Duchen, L. W., Eicher, E. M., Jacobs, J. M., Scaravilli, F., and Texeira, F., 1980, *Neurological Mutations Affecting Myelination* (N. A. Baumann, ed.), Elsevier/North-Holland, Amsterdam, pp. 107–113.

165. Kobayashi, T., Scaravilli, F., and Suzuki, K., 1980, *Neurological Mutations Affecting Myelination* (N. A. Baumann, ed.), Elsevier/North-Holland, Amsterdam, pp. 253–262.

166. Berger, B., Hervé, D., Dolphin, C., Barthelemy, C., Gay, M., and Tassin, J. P., 1979, *Neuroscience* **4:**877–888.

Analytical Aspects of the Pharmacokinetics of Psychotropic Drugs

Thomas B. Cooper

1. INTRODUCTION

The present interest in the pharmacokinetics of psychotropic drugs was initially stimulated by the work of Brodie and his various collaborators.[1] By definition, the science of pharmacokinetics is concerned with the transfer of drug molecules within the body, but in medicine, it is often regarded as the study of drug and drug metabolites in plasma.

In psychiatry, considerable effort has been expended in the examination of blood plasma drug concentrations and clinical efficacy and side effect relationships. It has been clearly shown that patients treated with identical doses of tricyclic antidepressant or antipsychotic medication show great interindividual differences in their steady-state plasma concentrations of the parent compound and metabolites (Figs. 1,2). The correlation, therefore, between dose ingested and drug plasma level is very weak, and it is probable that plasma drug levels will be more strongly correlated with drug concentration at the receptor site. There is a considerable body of evidence indicating that much of this variation is genetically determined.[2,3] It has been suggested that such constitutional variations in the rate of drug metabolism may help explain therapeutic failure and side effects.[4,5] In other branches of medicine, this variation in dosage requirement is often controlled by dosage titration to physiologically quantifiable endpoints such as bleeding time, clotting time, etc. In 1964, Brodie[6] suggested that in psychiatry the administration of antipsychotic medication in a dose just sufficient to elicit extrapyramidal system side effects followed by a slight reduction in dosage to just below this threshold was an attempt by the

Thomas B. Cooper • Analytical Psychopharmacology Laboratory, Rockland Research Institute, Orangeburg, New York 10962.

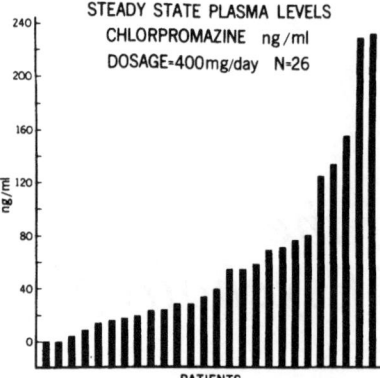

Fig. 1.   Variation in steady-state chlorpromazine levels in patients treated with identical doses of medication. (Reprinted with permission from the Editors, *Clinical Pharmacokinetics*).[48]

skilled physician to correct for the known five- to tenfold variation in the interindividual metabolism of such drugs. Such procedures have not gained wide acceptance. May and Van Putten[7] have stated that ". . . At best, antipsychotic drugs are prescribed in a pragmatic trial and error basis, at worst they are prescribed by rote." It is possible, indeed probable, that many patients deemed "nonresponders" are receiving too little or too much medication.

In other branches of medicine, the value of plasma level measurement of drugs in monitoring therapy has been well documented (e.g., digitalis, anti-

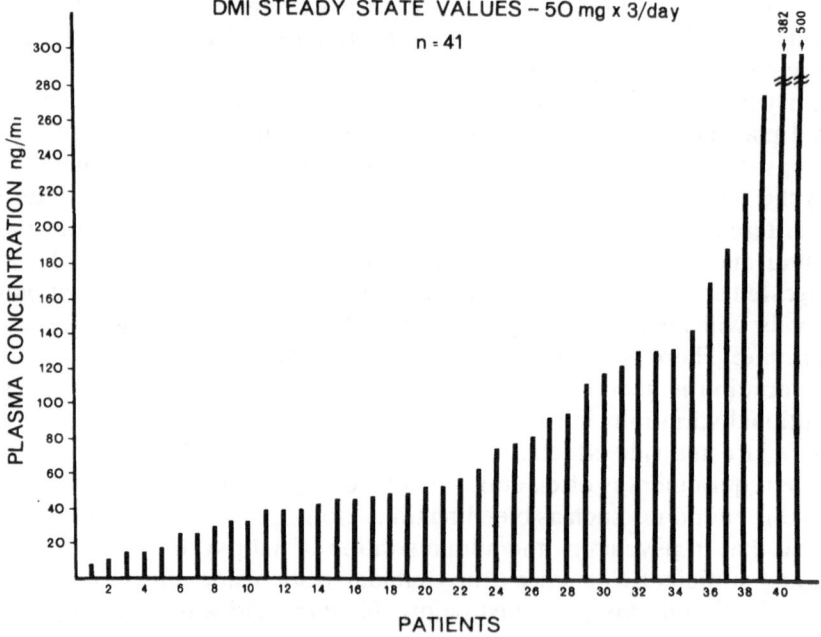

Fig. 2.   Variation in steady-state desmethylimipramine levels in patients treated with identical doses of medication.

convulsants). In psychiatry, with the exception of lithium, the techniques available for quantitation of plasma drug levels have been, until quite recently, totally inadequate in that they lacked sensitivity and/or specificity for accurate measurement of the very low levels of drug found in the plasma of patients receiving antipsychotic drugs (*ca.* 10^{-7}–10^{-12} g/ml). The major emphasis of this chapter is to detail the analytical techniques that have been developed to overcome these problems.

First, however, the rationale for assuming that the blood drug concentration is correlated to the drug concentration in the biophase surrounding the receptor site must be examined. Animal studies have shown a strong correlation between plasma level and brain tissue concentrations of these drugs, with the brain concentration being 5 to 20 times that of plasma.[8-10] In man, two studies have examined the relationship between plasma and CSF levels of drug and major metabolites using GC/MS with deuterated internal standards.[11,12]

The inhibition of NE uptake in isolated adrenergic neurons incubated in the plasma of patients treated with various antidepressant drugs has been shown to be correlated with the plasma concentration of the drug.[13-17]

Tyramine, an indirect-acting sympathomimetic amine, when injected in an intravenous bolus will cause a transient increase in blood pressure. This action is blocked by the administration of TCAs, presumably by blockade of the reuptake mechanism of the noradrenergic neurons.[14-17]

Blockade of this tyramine pressor response in patients chronically treated with antidepressants has been shown to be strongly correlated to the plasma concentration of the drug.[18-21] In a novel approach, Ghose *et al.*[22] has used the tyramine pressor blockade to measure the "pharmacodynamic" half-life of amitriptyline and metabolites, i.e., measuring the decrease in pharmacological blockade rather than the decrease in blood concentration.

Current interest has turned to the role of the hydroxylated metabolites of tricyclic antidepressant drugs because they are now known to be psychoactive, to cross the blood–brain barrier, and to be present in plasma in quantities the

Table I
Inhibition of Norepinephrine Uptake[a] in Adrenergic Neurons Incubated in Plasma Drawn from Patients Treated with Various Antidepressants

Drug treatment (number of patients)	Drug (metabolite) measured in plasma	Correlation between drug plasma level and inhibition of uptake		References
		r	p	
Nortriptyline (14)	Parent drug	0.83^b	$p < 0.001$	Borga *et al.*[14]
Nortriptyline (12)	Parent drug	0.82^b	$p < 0.01$	Hamberger and Tuck[15]
Amitriptyline (6)	Metabolite nortriptyline	0.71–0.88	$p < 0.025$	Tuck *et al.*[16]
Imipramine (14)	Metabolite desmethylimipramine	0.92	$p < 0.0005$	Tuck *et al.*[17]

[a] In percent of control.
[b] No significant correlation between uptake inhibition and the dose of nortriptyline.

Table II
Effect of Metabolites of Tricyclic Antidepressants on the Neuronal
Uptake of Norepinephrine[a]

Drug or metabolite	EC_{50} (nmol/liter)	Relative to DMI	Relative to NT
Imipramine (IMI)	250	0.22	—
2-OH-IMI	180	0.31	—
Desipramine (DMI)	55	1.00	—
2-OH-DMI	61	0.90	—
Amitriptyline (AT)	380	—	0.20
Nortriptyline (NT)	75	—	1.00
E-10-OH-AT	670	—	0.11
Z-10-OH-AT	720	—	0.10
E-10-OH-NT	160	—	0.47
Z-10-OH-NT	160	—	0.47
Desmethyl-NT	1000	—	0.08
AT-N-oxide	7300	—	0.01

[a] Imipramine series data from Potter *et al.*[24]; amitriptyline series data from Bertilsson *et al.*[25]

same order of magnitude as the parent compound.[23–25] The relative potency of NE blockade of these metabolites is shown in Table II.

Similar observations have been made with the antipsychotic drugs. Alfredsson *et al.*[12] have shown that CPZ levels in plasma and CSF are highly correlated ($r = 0.91$). These investigators have also unequivocally shown that 7-OH chlorpromazine is present in the plasma of patients treated with CPZ using the highly specific GC/MS with deuterated internal standards. This metabolite is known to have dopamine-blocking activity approximately equal to that of the parent compound[26,27] and to release prolactin.[28]

The preceding data appear to give a reasonable justification for the assumption that the blood or blood plasma concentration of a drug in steady state is strongly correlated with the drug concentration in the biophase surrounding the receptor. Furthermore, based on this assumption, studies investigating clinical efficacy and side effect relationships have scientific validity.

2. PHARMACOKINETIC BACKGROUND

The individual characteristics of the various drugs in these classes are presented separately under their respective headings. The main pharmacokinetic characteristics are the following:

1. A large volume of distribution of about 5–40 liter/kg which results in blood levels of 10^{-7}–10^{-12} g/ml.
2. A high degree of tissue binding.
3. Efficient elimination via hepatic metabolism.

4. For some antipsychotic drugs, an experimentally demonstrated metabolism in the gastrointestinal lumen or gut wall prior to entering the hepatic portal system.
5. A pharmacokinetic profile from oral and intravenous studies showing a multiphasic elimination which has in some cases resulted in estimation of a plasma elimination (B-phase) half-life utilizing a biexponential model (early and late half-life of elimination) and in others a half-life of elimination calculated for data during a single dosing interval which gives intermediate values, clearly a hybrid of the various phases.
6. Autoinduction of enzymes by antipsychotic drugs (hepatic and/or intestinal wall) has been demonstrated, further complicating pharmacokinetic analysis (i.e., a nonlinear or nonstationary system).
7. Patients given a fixed dosage of an antipsychotic or antidepressant drug will reach pseudo-steady-state plasma levels in 5–10 days, but this steady-state level will vary considerably (cf. Figs. 1,2).
8. Many of these drugs have complex metabolic pathways which can result in a multitude of metabolites, several of which may be psychoactive or have other pharmacological activity. The complexities range from haloperidol, with a relatively uncomplicated pathway in which only the parent compound and one metabolite (reduced haloperidol) are known to be active, to the prototypic phenothiazine chlorpromazine with more than 150 theoretical metabolites, several with demonstrated central activity, the majority never tested.

3. ANALYTICAL PROCEDURES

3.1. Glassware for Collection of Samples and Extraction Procedures

All glassware must be maintained scrupulously clean. This is self-evident but cannot be overemphasized. In kinetic studies in which single pulse doses are given, drug concentration levels can be as low as 10^{-12} g/ml fluid or g/g tissue. Contamination of a sample from a previous assay is therefore an ever-present danger. Similarly, extreme care must be taken to insure that tissue homogenizers do not cross contaminate samples (i.e., multiple rinses are mandatory). In the latter case, where it is often known which tissues have a higher concentration, they should be homogenized in ascending order of concentration. In our laboratory, all glassware used for GLC or HPLC procedures is washed in an area reserved exclusively for this purpose. These items of glassware are never used for any other laboratory procedures, nor are the washing facilities used for any other laboratory glassware. The use of dichromate cleaning fluid followed by neutralizing in dilute ammonia solution and extended heat (160°C overnight) drying is a routine procedure from which we never deviate. Similar procedures have been recommended by Spirtes[29] when he found a contaminating peak in his CPZ assay to be CPZ from previous assays and by Flint et al.[30] as good laboratory practice.

3.2. Sample Collection

It is common practice to use a vacuum collection system for blood samples. The use of Vacutainers® (Becton Dickinson & Co.) causes a spurious lowering of plasma drug concentration. The tris(butoxyethyl) phosphate used as a plasticizer in the container stoppers displaces the drug from the α_1 acid glycoprotein in plasma, thus increasing the "free" fraction of drug in plasma. This increased "free" fraction then redistributes between red cells and plasma with a resultant spurious lowering of the plasma levels of the drug.[31] This artifact has been observed with several basic lipophilic drugs including the TCAs and CPZ.[32–34] We have confirmed the interaction of Vacutainer® stopper and plasma AMI and NT level in a blood sample from an amitriptyline-treated patient (Table III). Other investigators have reported similar findings but with considerably less inhibition of protein binding.[36]

In animal and human experimentation, it is common practice to heparinize the animal or to use a heparin lock for blood collection. Recent studies in man have demonstrated that as little as 50 units of heparin can cause a 30% increase in "free" (i.e., pharmacologically active) drug concentration. The mechanism of action is thought to be heparin-induced activation of lipoprotein lipase activity with consequent degradation of lipoprotein.[37] These basic drugs bind to lipoproteins, and the resultant increased catabolism of these proteins increases levels of unbound drug. This increased level of unbound drug will cause changes in a number of the parameters derived from pharmacokinetic analysis.[38,39] In contrast, Naranjo *et al.*[40] suggest that the above finding resulted, at least partially, from the fact that subjects were in a nonfasting state, which may be the major determinant of activation of lipoprotein lipase.

Clearly, these pitfalls, which have been confirmed in several centers, must be avoided in future studies, and published studies must be subject to some question unless it is clearly stated that none of the above pertains.

3.3. Storage of Samples

Tissues and biofluids containing tricyclic antidepressant drugs can be stored frozen for at least 6 months at $-20°C$.[41] We routinely use 0.1 N HCl or H_2SO_4 for tissue homogenate preparation and then store aliquots of this homogenate at $-20°C$. Plasma samples do not need to be acidified, but urine samples should be made acidic before freezing.

The antipsychotic drugs may present more difficult problems in that some are photosensitive and may require precautions such as collection in light-shielded containers[42] and protection from fluorescent lighting during extraction and preparation for GLC and HPLC assay.[43] Studies on the stability of these drugs have usually been done in aqueous solution and only rarely in the presence of blood plasma.

Thus, Glassman and Perel suggest that an antioxidant be used in sample collection and storage of antipsychotic drug samples because of their experience with trifluoroperazine.[44] Curry[45] has suggested that storage of CPZ samples may result in increased levels of CPZ brought about by reduction of CPZ-

Table III

Effect of Vacuum Tube Containers on AMI and NT Levels in Whole Blood, Plasma, and Red Blood Cells[a]

Stopper	Whole blood (ng/ml)				Plasma (ng/ml)				RBC/plasma ratio	
	AMI	% Change	NT	% Change	AMI	% Change	NT	% Change	AMI	NT
Glass control	121		112		151		84		0.57	1.72
Blue Vacutainer®	121	0	110	−1.8	98	−35.1	65	−22.6	1.51	2.51
Red Vacutainer®	119	−1.7	105	−6.3	107	−29.1	69	−17.9	1.24	2.13
Green Vacutainer®	125	+3.3	115	+2.7	99	−34.4	63	−25.0	1.57	2.79
Violet Vacutainer®	124	+2.5	111	−0.9	100	−33.8	62	−26.2	1.52	2.70
Orange Vacutainer®	122	+0.8	114	+1.8	96	−36.4	62	−26.2	1.59	2.80

[a] AMI and NT data obtained when 60 ml whole blood was collected via all glass syringe from AMI-treated patient. Blood was mixed with oxalate anticoagulant in a glass stoppered flask and divided into 10-ml aliquots in acid-washed tubes. The various colored stoppers were removed from fresh Vacutainer® tubes and inserted into the tubes containing the 10-ml whole blood aliquots and mixed (10 inversions). Duplicate 1-ml samples of whole blood and plasma were then analyzed by GLC for AMI and NT content. (Reprinted with permission from the Editor, *Psychopharmacology*, 1979.[35])

n-oxide, and Friedel[46] suggests that storage of CPZ samples may result in increased levels of CPZ because of reduction of the sulfoxide to CPZ. Traficante *et al.*[47] have demonstrated that rapid *in vitro* sulfoxidation of CPZ by a heme catalyst in hemoglobin is completely inhibited by the presence of a proteinlike factor in human plasma. Other investigators have mentioned the protection afforded by plasma both in preventing oxidation and absorption to glassware during extraction procedures.[48,49] Until carefully controlled studies are performed, extreme care in the handling of these samples must be practiced. There is a clear need for a detailed examination of the claim cited above and for recommendations to be made from a large data base.

3.4. Extraction Procedures

These tricyclic drugs are highly lipophilic bases (with pK_a 8.5–10 or higher), and therefore, at high alkaline pH, the free base can be extracted into a relatively nonpolar organic phase and, at pH <3.0, can be reextracted from this organic phase into an aqueous phase. Plasma and tissue extraction procedures generally involve either one or three steps depending on the method of analysis. Thin-layer chromatography and some HPLC techniques use a single alkaline extraction step into an organic solvent and application of this extract (usually after concentration) to the TLC plate or HPLC injector port. Other methods require a three-step procedure (organic–acid–organic) for further sample clean up and possible derivatization of the compound and/or one of its metabolites before introduction into the analytical instrument.

3.5. Internal Standards

The ideal internal standard is a compound with essentially identical structure, partition coefficient, vapor pressure, adsorption to glass, and chromatographic performance. The nearest to this ideal is the use of a site-specific stable-isotope-labeled drug. The use of a [^{2}H]- or [^{13}C]-labeled compound (the isotope dilution method) meets all of the above requirements but requires a mass spectrometer to determine the ratio of the labeled to unlabeled compound. Because of the expense and requirement for highly trained staff, such sophisticated instrumentation is only available to specialized centers but is gradually becoming used as a reference method in cross validation studies.

For TLC, HPLC, and GLC procedures, it is common practice to use an analogue or closely related compound as an internal standard. When mixtures of secondary and tertiary compound are to be analyzed, then, ideally, two internal standards reflecting each structure should be used. Thus, if a derivative of the secondary amine is needed, then the secondary amine internal standard will control the derivative step. When a derivative step is not included, and the method has demonstrated linearity using a tertiary amine as internal standard, then this additional refinement, although desirable, is not absolutely necessary. When two internal standards are used, it is strongly recommended that the ratio between these internal standards be routinely calculated and recorded. This simple step guards against unusual matrix effects in which the tertiary and secondary amine partition coefficients may change.

4. THIN-LAYER CHROMATOGRAPHY

This technique is widely used in toxicology for rapid screening and semiquantitative analyses. This procedure has been used in the past to quantitate the tricyclic drugs, but the introduction of the nitrogen-specific detector in GLC and HPLC with a variety of detectors has slowed development of this technique considerably. Nevertheless, in skilled hands, this procedure can be used with reasonable precision and accuracy. In recent years, the use of single- and double-beam scanning spectrofluorodensitometers or scanning UV detectors has been advocated. Thus, Kaul *et al.*[50] have described the quantitation of 11 metabolites of CPZ using a dansylation procedure with subsequent fluorometric detection. Lehr and Kaul[51] have also described a novel technique with picomole sensitivity which utilizes quaternization of the tertiary amine of chlorpromazine with 9-bromomethylacridine followed by separation via thin-layer chromatography and photolysis to generate measurable fluorescence. A similar procedure has been developed to quantitate amitriptyline.[52]

Chan and Gershon[53] have developed a TLC procedure using a double-beam scanning spectrofluorodensitometer which has been utilized in a number of clinical studies of chlorpromazine and its metabolites.[54,55] Nagy and Treiber[57] developed a TLC assay for imipramine and metabolites involving oxidation after TLC and scanning densitometry at 405 nm. This procedure has a lower limit of sensitivity of 5 ng/ml (5-ml sample). Using this technique, these investigators have reported on plasma levels of imipramine and metabolites in man and plasma and brain levels of imipramine and chlorimipramine in the rat after different routes of administration.[8,58]

Gram[59] and Faber *et al.*[60] have used a modification of this technique to study OH metabolites of imipramine and amitriptyline and nortriptyline, respectively. Breyer and Villumsen[61] used TLC with UV reflectance densitometry without chemical manipulation of the compounds; i.e., the compounds were extracted, chromatographed, and scanned without derivative formation, oxidation, or any other chemical reaction. These workers quantitated perazine, perazine sulfoxide, perazine-*n*-oxide, clozapine, clozapine-*n*-oxide, amitriptyline, nortriptyline, imipramine, and desmethylimipramine. Fenmore *et al.*[62] and Haefelfinger[63] claim even higher sensitivity and precision using high-performance thin-layer chromatography.

The TLC procedures described have sufficient sensitivity to quantitate patients' drug plasma concentrations when they are receiving therapeutic dose levels. Although the coefficient of variation is somewhat higher than one would desire, these procedures have generated data that are comparable to data generated by GLC, HPLC, and GC/MS techniques; e.g., Gram[59] found OH metabolite levels of approximately 15–20% IMI level and 50% DMI level, figures very much in agreement with the more recent HPLC procedures,[64,65] and in addition was able to quantitate IMI-*n*-oxide (5–10% IMI), a metabolite that has been used in clinical practice but which has received little attention in blood level studies to date.

5. GAS CHROMATOGRAPHY OF TRICYCLIC ANTIDEPRESSANTS AND ANTIPSYCHOTICS

Work prior to 1975 was limited to the use of GC with a flame ionization detector or electron-capture detector. Although the GC/MS was used to some extent, its high cost and limited availability in these years meant that it was utilized but rarely.

5.1. Flame Ionization Detection

The flame ionization detector is severely limited because of lack of sensitivity in terms of the routine monitoring of therapeutic drug levels. Several investigators have published methods using this detector,[66-73] and when these procedures are used carefully, they are adequate for the measurement of steady-state plasma levels of patients treated with normal therapeutic doses of tricyclic antidepressants. They have the disadvantage of requiring relatively large volumes of plasma (3–10 ml/assay), and none are suitable for the analysis of single-dose pharmacokinetic studies.

5.2. Electron-Capture Detection

The electron capture detector was used extensively in these early years for quantitation of halogenated tertiary amines and secondary amines because they could be derivatized to form highly electrophilic compounds which were then detectable at a picomole level.[74-77] The conversion of a tertiary amine compound to one with electrophilic properties has more recently been developed (conversion to anthroquinone or trichlorethylcarbamate by Hartvig *et al*.[78,79]). With these procedures, acceptable plasma assays have been developed for a variety of antidepressant and antipsychotic drugs.[74-84] Single-dose pharmacokinetic studies are easily performed using this detector, and relatively low plasma volumes are required for these assays (1–3 ml plasma).

5.3. Gas Chromatography with Nitrogen–Phosphorous Detection (NPD)

The introduction of the nitrogen-phosphorous detector (alkali flame ionization) operated in the nitrogen mode proved to be a major breakthrough in that it has made possible monitoring of plasma levels of tricyclic drugs in the routine as opposed to specialist laboratory. In addition, the high sensitivity coupled with selectivity made feasible the use of 1-ml plasma samples for steady-state level monitoring. A further major advantage of this detector was that many methods developed for the FID mode could be transferred to the NPD with little or no modification. One of the disadvantages of the NPD is that it is particularly sensitive to halogenated solvents which rapidly cause deterioration in performance. Therefore, many solvents normally used in gas chromatography cannot be used in these procedures.

Two attempts have been made to overcome this: one is to turn off the bead when the sample is injected, to wait until the solvent front has passed through the detection chamber, and then to reheat the bead[85]; the second procedure is to place a bypass valve between the column and detector chamber in which an identical flow of carrier gas maintains stable detector conditions while the column is vented so that the solvent front does not pass through the chamber. After a set time, the bypass valve is actuated, and the column effluent then enters the detector chamber.[86] In these ways, reasonable chromatography can be expected with extended bead life even when using these halogenated solvents.

Since 1975, the number of publications describing analytical procedures for the analysis of these compounds has been increasing at an exponential rate. The basic technique breaks down into extraction followed by derivatization of the secondary amines and/or other metabolites and chromatography[87–89] or extraction without derivatization of the secondary amines followed by chromatography.[90–92] No attempt is made here to cite all references in the literature. For a detailed review, the work of Scoggins *et al.*[93] is recommended. The widespread use of the NPD has enabled the routine measurement of all of the tricyclic antidepressant drugs and, where appropriate, their demethylated metabolites.

In this author's laboratory, the 10-OH metabolites of AMI and NT are routinely measured by the use of the NPD after reduction of the E and 2 forms of these 10-OH metabolites to 10–11 dienes followed by conversion of the secondary amine to an amide using heptafluorobutyric anhydride.[94] A similar approach for reduction of the enantiomeric hydroxy group of NT was first described by Borga and Garle.[95] The same procedure has been used by Garland *et al.*[96] for the reduction of the 10-OH AMI and 10-OH NT enantiomers before subjecting these to GC/MS analysis. The rationale for this reduction is that these isomers have essentially identical pharmacological activity.[25,97] Rather than spending considerable time and effort in achieving separation of these isomers, the conversion to the diene was selected to give a single peak on the chromatogram. Dysken *et al.*[98] have recently demonstrated that subnanogram sensitivity for a fluphenazine assay can be achieved by the use of trifluoroacetic anhydride which not only improves the symmetry of the chromatographic peak but which appears to enhance the response of the detector. The latter observation was reported by Dekirmenjian *et al.*[99] for a variety of psychotrophic drugs. Nitrogen–phosphorus detection assays have been described for haloperidol,[100] fluphenazine,[101] chlorpromazine,[102,103] and other antipsychotic agents.

5.4. Gas Chromatography/Mass Spectrometry

There is little disagreement that isotope dilution procedures using stable-isotope-labeled standards and GC/MS separation with simultaneous ion monitoring comprise one of the most powerful analytical tools available. The high cost and technical sophistication required to operate such equipment make it an analytical tool more suited for research purposes, as a reference procedure against which routine procedures such as GC, HPLC, and RIA are cross validated, and as a quality control check, e.g., 1% of samples confirmed by GC/MS assay. Care must be taken to guard against possible isotope effects when selecting a labeled compound and in the derivative used to enable selection of a suitable fragmentation when operating in the El mode. All of these details are discussed elsewhere in this volume (D. A. Durden and A. A. Boulton, this volume).

Drug assays have been developed that utilize the isotope dilution method and others that do not; yet another subgroup uses a deuterated internal standard for the tertiary amines but an unlabeled compound for quantitation of the secondary amine metabolites.[42,106-123]

Examples of the value of such instrumentation in drug kinetics are as follows. Heck *et al.*[122] used site-specific labeled imipramine to compare a genetic formulation of this drug to the standard (reference) compound Tofranil.® Usually single-dose studies involving 20–30 subjects in a crossover design are required to achieve the statistical power required to satisfy the FDA that these two drugs do or do not differ in their bioavailability by more than 20%. Heck and his associates were able to demonstrate that less than 10 subjects were required to achieve the same statistical power because of the increased precision of the GC/MS assay. Alfredsson *et al.*[12] essentially resolved the controversy that had arisen when investigators using similar GC/ECD techniques were unable to agree whether 7-OH chlorpromazine was present in the plasma of patients treated with CPZ. Analysis of the plasma using deuterated 7-OH chlorpromazine gave unequivocal proof of the existence of this metabolite in all samples analyzed. Finally, of great importance, the GC/MS used as a mass fragmentograph can confirm the identity of a peak thought to be a specific drug or metabolite because of retention time characteristics on a GC or HPLC column. The latter is particularly important in psychiatric patients on whom polypharmacy is widely practiced.

A whole area in pharmacokinetic analysis that is only now becoming fully recognized is the use of site-specific labeled compounds *in vivo* for the determination of drug metabolism profiles and for the examination of single-dose pharmacokinetic profiles when subjects are in steady state. This allows one to determine if changes in the pharmacokinetics of a drug take place after chronic administration or with the coadministration of another drug, etc. Briefly, this is achieved by giving a patient a single dose of medication and collecting blood samples over an appropriate time interval to achieve accurate pharmacokinetic parameters; the subject is then given medication each day until he has achieved steady state, and then one dose of the drug is changed to an identical quantity of a mixture of labeled and unlabeled drug. Blood samples collected over the same time period as in the original single-dose study will give data on the steady-state kinetics (all drug quantified) and single-dose kinetics (labeled drug quantified). Such studies have been recommended for a number of years[125] but only now are becoming feasible as resources for the synthesis of stable-isotope-labeled compounds become generally available.

6. HIGH-PERFORMANCE LIQUID CHROMATOGRAPHY

The introduction of high-performance liquid chromatography (HPLC) has been a major breakthrough in technology that has direct and immediate application to psychotropic drug quantitation. The extraordinary efficiency of commercially available 5-μm columns (greater than 50,000 theoretical plates per meter), the fact that the separation procedure and analysis are a nondestructive process, and the introduction of reverse-phase columns and a variety of highly sensitive and selective detector systems have staggering implications for future research in this and other areas.

Primary use of this technique to date has focused on drug monitoring of the tertiary and secondary amine antidepressant and antipsychotic drugs.[126-136] In these procedures, the hydroxylated metabolites are either excluded via the extraction procedure or rapidly eluted by judicious choice of the mobile phase using reversed-phase columns. It must also be remembered that rapid gains in column efficiency and chemical procedures such as new ion-pair reagents will mean that separations will be more readily achieved in less time; e.g., fast LC columns are already available reducing chromatography time to one-third of that previously attainable yet still retaining base line-to-base line resolution. It is also interesting to speculate on just how long current data reduction systems will have a high enough sampling rate for the "ultrafast" LC of the near future.

The HPLC has not yet been extensively used for the quantitation of hydroxy metabolites of these compounds. This is surprising in that the technique is most suitable for such analyses. The hydroxylated metabolites of IMI and DMI (2-hydroxylated) have been quantitated using normal-phase (silica) columns and spectrofluorometric detection[64] and by reverse-phase ion-paired chromatography with amperometric detection.[65] Both of these methods have adequate sensitivity for single-dose pharmacokinetic studies using 1 ml of plasma. Amitriptyline, unfortunately, is not quite so straightforward in that it does not fluoresce, nor is it readily oxidized; thus, neither of the above methods is applicable. An additional problem is that amitriptyline is hydroxylated in the 10 position (a chiral center) and therefore exists in the E and Z enantiomeric forms. Two procedures for the separation of these enantiomers are in press at this time. Both use reversed-phase paired-ion chromatography with UV detection.[137,138]

Major efforts have been made to develop a general procedure for the separation of all currently marketed antidepressant drugs and their desmethyl metabolites. This is a major importance to routine clinical laboratories in that the system will not require mobile phase changes or detector changes throughout the working day.

A fully automated (i.e., including extraction) "fast LC" method has recently been published which can routinely assay seven separate antidepressant drugs and their desmethyl metabolites.[139]

7. RADIOIMMUNOASSAY

Many of the second-generation antipsychotic drugs are high-potency, low-dose compounds which result in extremely low levels of circulating drug and metabolites (10^{-9}–10^{-12} g/ml). In many cases, this has promoted the development of radioimmunoassay (RIA) procedures simply because other analytical procedures lack the sensitivity to accurately quantitate these subnanogram levels in reasonable sample volumes.

The RIA procedures are technically simple, require minimal sample manipulation, and enable high sample turnover rates. These advantages and the fact that most routine clinical laboratories have an RIA capability in terms of basic instrumentation make this approach to therapeutic drug level monitoring most attractive. However, it is only recently that such procedures have been developed for psychotropic drugs. Initial investigators experienced difficulty in obtaining high-specific-activity tracers, a problem that has now been resolved.[140] A number of techniques for tricyclic antidepressant drugs with adequate sensitivity for therapeutic drug level monitoring have been described, but many lack the necessary specificity to enable use without prior sample extraction and workup.[141-148] Several of these studies have included reports of cross validation against specific procedures such as GC, HPLC, or GC/MS.[142,143,146,147]

Considerable effort and success have been involved in development of specific antisera for the antipsychotic drugs. As previously stated, the high-potency drugs have proven most difficult to measure, but it now appears that we have adequate sensitivity and specificity for measurement of fluphenazine,[149,150] perphenazine,[151] haloperidol,[152,153] trifluoroperazine,[154] and flupenthixol.[155] The sensitivity of many of these antisera is such that unextracted plasma volumes of as little as 50–200 μl can be assayed. An example of the exquisite sensitivity of these procedures is to be found in the paper of Greenberg and Weiss[156] in which the desmethylimipramine content of a single pineal gland was determined using the method of Brunswick *et al.*[143]

The disadvantages of RIA procedures are as follows: antisera must be carefully checked for cross reaction against a large number of interfering compounds. Validation and quality control checks on these antisera should be repeated at frequent intervals using a more selective procedure. Radioactive isotope usage is necessary with the precautions and associated bookkeeping problems. If one wishes to look at metabolite patterns, then additional specific antisera need to be developed. However, these disadvantages appear to be outweighed by the advantages: large volume throughput, technical simplicity, low sample volumes, and little sample manipulation are compelling factors to the clinical chemist or toxicologist.

8. RADIORECEPTOR ASSAYS

Creese and Snyder[157] have described a simple sensitive assay for neuroleptic drugs. This procedure utilizes the fact that all currently marketed antipsychotic (neuroleptic) drugs block the stereospecific binding of tritiated haloperidol to dopamine receptors at concentrations that correlate directly with clinical potency.[158] This neuroleptic radioreceptor assay (NRRA) is based on the principle that the neuroleptics in the patient's serum compete for binding of tritiated haloperidol or spiroperidol to dopamine receptors in membranes from rat or calf caudate. An even more preliminary report has recently investigated a similar procedure for antidepressant drugs.[157]

This assay may overcome two major problems in neuroleptic plasma level monitoring: (1) the technique is applicable to all neuroleptics, thus avoiding the multiplicity of procedures necessary at this time, and (2) it overcomes the necessity of determining the identification and pharmacological activity of each metabolite and the development of analytical procedures for each found to be "active" (blocks dopamine receptors); i.e., NRRA assays any metabolite that competes for binding without the necessity of identification of same.[158]

It is too early to determine whether this procedure will hold to its early promise and intuitive appeal. Preliminary studies are encouraging,[159-171] but studies with large numbers of patients are required, and it is also necessary to demonstrate that the technical procedure can be readily incorporated into a busy clinical reference laboratory as a routine. It is surprising to this author to see that a commercial "kit" for this assay has been marketed when so little hard data are available to facilitate interpretation of the results generated by such a kit.

9. CONCLUSION

In psychiatry, lack of analytical methodology has been the oldest and most vexing problem in the study of plasma drug levels and clinical efficacy/side effect relationships. These data indicate that for many drugs the technology is now more than adequate for investigation of these relationships and in some cases is ready to be used in routine as opposed to specialist laboratories. It should be emphasized that most of these procedures can be used for the analysis of tissues other than blood.

The clinical and experimental investigations that have used these relatively new tools have made a major contribution to our understanding of the pharmacokinetics of psychotropic drugs and, although to a lesser extent, the pharmacokinetic and pharmacodynamic interrelationships.

Clinical studies are most difficult to perform because of the complexities involved in experimental design which must account for factors related to heterogeneity of diagnostic category, age, sex, previous drug exposure, chronic versus acute patient population, compliance, suitable numbers of patients, adequate rating scales, etc. The pitfalls of such trials have been well documented by May and Van Putten[7] in a critical review of studies of the antipsychotic drug chlorpromazine. Many of their criticisms apply to other studies of both the antipsychotic and antidepressant drugs.

Recent reviews of this field include papers by Gram,[172] Cooper,[48] Risch et al.,[173,174] Sjoqvist,[13] Scoggins et al.,[93,175] and Dahl.[176]

Detection of noncompliance, undermedicated and overmedicated patients, and drug interactions are now feasible. Drug level monitoring is generally regarded as of clinical value in the overall evaluation of a treatment-resistant patient. However, despite a considerable number of controlled clinical studies, there are insufficient data to warrant the routine monitoring of all patients receiving these medications.

REFERENCES

1. Brodie, B. B., 1967, *J.A.M.A.* **202**:600–609.
2. Green, D. E., Forrest, I. S., Forrest, F. M., and Serra, M. T., 1965, *Exp. Med. Surg.* **23**:278–287.
3. Alexanderson, B., Evans, D. A., and Sjoqvist, F., 1969, *Br. Med. J.* **4**:746–768.
4. McQueen, E. D., 1976, *Drug Treatment*, 2nd ed. (G. S. Avery, ed.), Adis Press, Sydney, pp. 161–192.
5. Price-Evans, D. A., 1969, *The Present Status of Psychotropic Drugs* (A. Cerletti and F. J. Bove, eds.), Excerpta Medica, Amsterdam, pp. 111–117.
6. Brodie, B. B., 1964, *Absorption and Distribution of Drugs* (T. Binns, ed.), Churchill Livingstone, Edinburgh, pp. 199–251.
7. May, P. R. A., and Van Putten, T., 1978, *Arch. Gen. Psychiatry* **35**:1081–1087.
8. Nagy, A., 1977, *J. Pharm. Pharmacol.* **29**:104–107.
9. Dingell, J. V., Sulser, F., and Gillette, J. R., 1964, *J. Pharmacol. Exp. Ther.* **143**:14–22.
10. Jori, A., Bernardi, D., Muscettola, G., and Garratini, S., 1971, *Eur. J. Pharmacol.* **15**:85–90.
11. Muscettola, G., Goodwin, F. K., Potter, W. Z., Claeys, M., and Markey, S. P., 1978, *Arch. Gen. Psychiatry* **5**:621–625.
12. Alfredsson, G., Wode Helgodt, B., and Sedvall, G., 1976, *Psychopharmacology* **48**:123–131.
13. Sjoqvist, F., 1979, *Prog. Neuropsychopharmacol.* **3**:201–210.
14. Borga, O., Azarnoff, D. L., Plym Forshell, G., and Sjoqvist, F., 1969, *Biochem. Pharmacol.* **18**:2135–2143.
15. Hamberger, B., and Tuck, J. R., 1973, *Eur. J. Clin. Pharmacol.* **5**:229–235.
16. Tuck, J. R., Hamberger, B., and Sjoqvist, F., 1972, *Eur. J. Clin. Pharmacol.* **4**:212–216.
17. Tuck, J. R., Kahan, E., and Siwers, B., 1973, *Acta Pharmacol. Toxicol. (Kbh.)* **32**:304–313.
18. Freyschuss, U., Sjoqvist, F., and Tuck, D., 1970, *Pharmacol. Clin.* **2**:72–78.
19. Ghose, K., Gifford, L., Turner, P., and Leighton, M., 1976, *Br. J. Clin. Pharmacol.* **3**:334–337.
20. Ghose, K., 1980, *Neuropharmacology* **19**:1251–1254.
21. Mulgirigama, L. D., Pare, C. M. B., Turner, P., Wadsworth, J., and Witts, D. J., 1977, *Postgrad. Med. J.* **53**(Suppl. 4):30–34.
22. Ghose, K., 1980, *Eur. J. Clin. Pharmacol.* **18**:151–157.
23. Alvan, G., Borga, O., Lind, M., Palmer, L., and Siwers, B., 1977, *Eur. J. Clin. Pharmacol.* **11**:219–224.
24. Potter, W. Z., Calil, H. M., Manian, A. A., Zavadil, A. P., and Goodwin, F. K., 1979, *Biol. Psychiatry* **14**:601–613.
25. Bertilsson, L., Mellstrom, B., and Sjoqvist, F., 1979, *Life Sci.* **25**:1285–1292.
26. Nyback, H., and Sedvall, G., 1972, *Psychopharmacologia* **26**:155–160.
27. Miller, R. J., and Iverson, L. L., 1974, *J. Pharm. Pharmacol.* **26**:142–144.
28. Meltzer, H. Y., Fang, V. S., Simonovich, M., and Paul, S. M., 1977, *Eur. J. Pharmacol.* **41**:431–436.
29. Spirtes, M. A., 1972, *Clin. Chem.* **18**:317–318.
30. Flint, D. R., Ferullo, C. R., Levandoski, P., and Hwang, B., 1971, *Clin. Chem.* **17**:830.
31. Borga, O., Piafsky, K. M., and Nilsen, O. G., 1977, *Clin. Pharmacol. Ther.* **22**:539–544.
32. Fremstad, D., and Bergerud, K., 1976, *Clin. Pharmacol. Ther.* **20**:120.
33. Brunswick, D. J., and Mendels, J., 1977, *Commun. Psychopharmacol.* **1**:131–134.
34. Veith, R. C., Raisys, V. A., and Perera, C., 1978, *Commun. Psychopharmacol.* **2**:491–494.
35. Robinson, D. S., Cooper, T. B., Ravaris, C. L., Ives, J. O., Nies, A., Bartlett, D., and Lamborn, K. R., 1979, *Psychopharmacology* **63**:223–231.
36. Cochran, E., Carl, J., Hanin, I., Koslow, S., and Robins, E., 1978, *Commun. Psychopharmacol.* **2**:495–503.
37. Krauss, R. M., Levy, R. J., and Fredrickson, D. S., 1974, *J. Clin. Invest.* **54**:1107–1124.
38. Wood, M., Shand, D., and Wood, A. J. J., 1979, *Clin. Pharmacol. Ther.* **25**:103–107.
39. Desmond, P. V., Roberts, R. K., Wood, A. J. J., Dunn, G. D., Wilkinson, G. R., and Schenker, S., 1980, *J. Clin. Pharmacol.* **9**:171–175.
40. Naranjo, C. A., Abel, J. G., Sellers, E. M., and Giles, H. G., 1980, *Br. J. Clin. Pharmacol.* **9**:103–105.

41. Orsulak, P. J., and Gerson, B., 1980, *Ther. Drug Monit.* 2:233–242.
42. May, P. R. A., Van Putten, T., Jenden, D. J., and Cho, A. K., 1978, *Arch. Gen. Psychiatry* 35:1091–1097.
43. Rivera-Calimlin, L., Castaneda, L., and Lasagna, L., 1973, *Clin. Pharmacol. Ther.* 14:978–986.
44. Glassman, A., and Perel, J., 1980, *Clin. Pharmacol. Psychiatry Newslett.* 2:2.
45. Curry, S. H., 1978, *Psychol. Med.* 8:177–180.
46. Friedel, R. O., 1976, *Psychopharmacol. Bull.* 12:63–64.
47. Traficante, L. J., Sakalis, G., Siekierski, J., Rotrosen, J., and Gershon, S., 1979, *Life Sci.* 24:337–346.
48. Cooper, T. B., 1978, *Clin. Pharmacokinet.* 3:14–38.
49. Midha, K. K., Loo, J. C. K., Hubbard, J. W., Rowe, M. L., and McGilveray, I. J., 1979, *Clin. Chem.* 25(1):166–168.
50. Kaul, P. N., Conway, M. W., Clark, M. L., and Huffine, J., 1970, *J. Pharm. Sci.* 59:1745–1749.
51. Lehr, R., and Kaul, P. N., 1975, *J. Pharm. Sci.* 64:950–956.
52. Kaul, P. N., Whitfield, L. R., and Clark, M. L., 1978, *J. Pharm. Sci.* 67:60–62.
53. Chan, T. L., and Gershon, S., 1973, *Mikrochim. Acta* 3:435–452.
54. Sakalis, G., Chan, T. L., Gershon, S., and Park, S., 1973, *Psychopharmacologia* 32:277–284.
55. Schooler, N. B., Sakalis, G., Chan, T. L., Gershon, S., Goldberg, S. C., and Collins, P., 1976, *Pharmacokinetics of Psychoactive Drugs: Blood Levels and Clinical Response* (L. A. Gottschalk and S. Merlis, eds.), Spectrum, New York, pp. 199–219.
56. Sakalis, G., Chan, T. L., Sathananthan, G., Schooler, N., Goldberg, S., and Gershon, S., 1977, *Commun. Psychopharmacol.* 1:157–166.
57. Nagy, A., and Treiber, L., 1973, *J. Pharm. Pharmacol.* 25:599–603.
58. Nagy, A., and Johansson, R., 1975, *Naunyn Schmiedebergs Arch. Pharmacol.* 290(2–3):140–160.
59. Gram, L. F., 1978, *Commun. Psychopharmacol.* 2:373–380.
60. Faber, D. B., Mulder, C., and Man in't Veld, W. A., 1974, *J. Chromatogr.* 100:55–61.
61. Breyer, U., and Villumsen, K., 1976, *Eur. J. Clin. Pharmacol.* 9:457–465.
62. Fenmore, D. C., Meyer, C. J., Davis, C. M., Hsu, F., and Zlatkis, A., 1977, *J. Chromatogr.* 142:399–409.
63. Haefelfinger, P., 1978, *J. Chromatogr.* 145:445–451.
64. Sutfin, T. A., and Jusko, W. J., 1979, *J. Pharm. Sci.* 68:703–705.
65. Suckow, R., and Cooper, T. B., 1981, *J. Pharm. Sci.* 70:257–261.
66. Braithwaite, R. A., and Whatley, J. A., 1970, *J. Chromatogr.* 49:303–307.
67. Hucker, H. B., and Stauffer, S., 1974, *J. Pharm. Sci.* 63:296–297.
68. Corona, G. L., and Bonferoni, B., 1976, *J. Chromatogr.* 124:401–404.
69. Norman, T. R., McGuire, K. P., and Burroughs, G. D., 1977, *J. Chromatogr.* 134:524–528.
70. O'Brien, J. E., and Hinsvark, O. N., 1976, *J. Pharm. Sci.* 65:1068–1069.
71. Bruderlien, H., Kraml, M., and Dvornik, D., 1977, *Clin. Biochem.* 10:3–7.
72. Nyberg, G., and Martensson, E., 1977, *J. Chromatogr.* 143:491–497.
73. Burch, J. E., Raddats, M. A., and Thompson, S. G., 1979, *J. Chromatogr.* 162:351–366.
74. Sisenwine, S. F., Knowles, J. A., and Ruelins, H. W., 1969, *Anal. Lett.* 2:315.
75. Borga, O., and Garle, M., 1972, *J. Chromatogr.* 68:77–88.
76. Ervik, M., Walle, T., and Ehrsson, H., 1970, *Acta Pharm. Suic.* 7:625–634.
77. Walle, T., and Ehrsson, H., 1971, *Acta Pharm. Suic.* 8:27–38.
78. Hartvig, P., Strandberg, S., and Naslund, B., 1976, *J. Chromatogr.* 118:65–74.
79. Hartvig, P., and Naslund, B., 1977, *J. Chromatogr.* 133:367–371.
80. Geiger, U. P., and Rajagopalant, G., and Riess, W., 1975, *J. Chromatogr.* 114:167–173.
81. Vereczkey, L., Bianchetti, G., Rovei, V., and Frigerio, A., 1976, *J. Chromatogr.* 116:451–456.
82. Larsen, N. E., and Marinelli, K., 1978, *J. Chromatogr.* 156:335–339.
83. Cooper, S. F., Albert, J. M., and Dugal, R., 1975, *Int. Pharmacopsychiatry* 10:78–88.
84. Curry, S. H., 1968, *Anal. Chem.* 40:1251–1255.
85. Tracor Instruments, Houston, Texas.
86. Winnett, G., and Silver, B., 1981, *J. Chromatogr. Sci.* 19:52–53.
87. Jorgensen, A., 1975, *Acta Pharmacol. Toxicol. (Kbh.)* 36:79–90.

88. Bailey, D. N., and Jatlow, P. I., 1976, *Clin. Chem.* **22:**1697–1701.
89. Rosseel, M. T., Bogaert, M. G., and Claeys, M., 1978, *J. Pharm. Sci.* **67:**802–805.
90. Cooper, T. B., Allen, D., and Simpson, G. M., 1975, *Psychopharmacol. Commun.* **1:**445–454.
91. Vasiliades, J., and Bush, K. C., 1976, *Anal. Chem.* **48:**1708–1711.
92. Hucker, H. B., and Stauffer, S. C., 1977, *J. Chromatogr.* **138:**437–442.
93. Scoggins, B. A., Maguire, K. P., Norman, T. R., and Burrows, G. D., 1980, *Clin. Chem.* **26:**5–17.
94. Cooper, T. B., 1981, *Clinical Pharmacology in Psychiatry* (E. Usdin, ed.), Elsevier/North Holland, Amsterdam, pp. 35–42.
95. Borga, O., and Garle, M., 1972, *J. Chromatogr.* **68:**77–88.
96. Garland, W. A., Muccino, R. R., Min, B. H., Cupano, J., and Fann, W. E., 1979, *Clin. Pharmacol. Ther.* **25:**844–856.
97. Hyttel, J., Christensen, V., and Fjalland, B., 1980, *Acta Pharmacol. Toxicol. (Kbh.)* **47:**53–57.
98. Dysken, M. N., Javaid, J. I., Chang, S. S., Schaffer, C., Shahid, A., and Davis, J. M., 1981, *Psychopharmacology* **73:**205–210.
99. Dekirmenjian, H., Javaid, J. I., Duslak, B., and Davis, J. M., 1978, *J. Chromatogr.* **160:**291–296.
100. Bianchetti, G., and Morselli, P. L., 1978, *J. Chromatogr.* **153:**203–209.
101. Franklin, M., Wiles, D. H., and Harvey, D. J., 1978, *Clin. Chem.* **24**(1):41–44.
102. Linnoila, M., and Dorrity, F., 1978, *Acta Pharmacol. Toxicol. (Kbh.)* **42:**264–270.
103. Bailey, D. N., and Guba, J. J., 1979, *Clin. Chem.* **25**(7):1211–1215.
104. Alfredsson, G., Wiesel, F. A., Fyro, B., and Sedvall, G., 1977, *Psychopharmacologia* **52:**25–30.
105. DeRidder, J. J., Koppens, P. C., and van Hal, H. J. M., 1977, *J. Chromatogr.* **143:**289–297.
106. Claeys, M., Muscatolla, G., and Marckay, S. P., 1976, *Biomed. Mass Spectrom.* **3:**110–116.
107. Jindal, S. P., Lutz, T., and Vestergaard, P., 1980, *J. Pharm. Sci.* **69:**684–687.
108. Crampton, E. L., Glass, R. C., Marchant, B., and Reese, J. A., 1980, *J. Chromatogr.* **183:**141–148.
109. Gaskell, S. J., 1980, *Postgrad. Med. J.* **56**(Supp. 1):90–93.
110. Vasiliades, J., Asahawneh, T. M., and Owens, C., 1979, *J. Chromatogr.* **164:**457–470.
111. Alkalay, D., Volk, J., and Carlsen, S., 1979, *Biomed. Mass Spectrom.* **6:**200–204.
112. Hornbeck, C. L., Griffiths, J. C., Neborsky, R. J., and Faulkner, M. A., 1979, *Biomed. Mass Spectrom.* **6**(10):427–430.
113. Jenkins, R. G., and Friedel, R. O., 1978, *J. Pharm. Sci.* **67:**17–23.
114. Chinn, D. M., Jennison, T. A., Crouch, D. J., Peat, M. A., and Thatcher, G. W., 1980, *Clin. Chem.* **26:**1201–1204.
115. Biggs, J. T., Holland, W. H., Chang, S., Hipps, P. P., and Sherman, W. R., 1976, *J. Pharm. Sci.* **65:**261–268.
116. Midha, K. K., Charette, C., Cooper, J. K., and McGilveray, I. J., 1980, *J. Anal. Toxicol.* **4:**237–243.
117. Garland, W. A., 1977, *J. Pharm. Sci.* **66:**77–81.
118. Borga, O., Palmer, L., Linnarsson, A., and Holmstedt, B., 1971, *Anal. Lett.* **4:**837–849.
119. Frigerio, A., Pantarotto, C., Franco, R., Gomeni, R., and Morselli, P. L., 1977, *J. Chromatogr.* **130:**354–360.
120. Belvedere, G., Burti, L., Frigerio, A., and Pantarotto, C., 1975, *J. Chromatogr.* **111:**313–321.
121. Dubois, J. P., Kung, W., Theobald, W., and Wirz, B., 1976, *Clin. Chem.* **22:**892–897.
122. Heck, H. d'A., Flynn, N. W., Buttrill, S. E., Dyer, R. L., and Anbar, M., 1978, *Biomed. Mass Spectrom.* **5:**250–257.
123. Johansson, R., Borg, K. O., and Gabrielsson, M., 1976, *Acta Pharm. Suec.* **3:**193–200.
124. Heck, H. d'A., Buttrill, S. E., Flynn, N. W., Dyer, R. L., Anbar, M., Cairns, T., Dighe, S., and Cabana, B., 1979, *J. Pharmacokinet. Biopharm.* **7**(3):233–248.
125. Jenden, D. J., 1978, *Psychopharmacology: A Generation of Progress* (M. A. Lipton, A. DiMascio, and K. Killam, eds.), Raven Press, New York, pp. 879–886.
126. Tjaden, C. R., Lankelma, J., and Poppe, H., 1976, *J. Chromatogr.* **125:**275–286.
127. Watson, I. D., and Stewart, M. J., 1975, *J. Chromatogr.* **110:**389–392.
128. Brodie, R. R., Chasseaud, L. F., and Hawkins, D. R., 1977, *J. Chromatogr.* **143:**535–539.

129. Emanuelsson, B., and Moore, R. G., 1977, *J. Chromatogr.* **146**:113–119.
130. Vandermark, F. L., Adams, R. F., and Schmidt, G. J., 1978, *Clin. Chem.* **24**:87–91.
131. Thoma, J. J., Bondo, P. B., and Kozak, C. M., 1979, *Ther. Drug Monit.* **1**:335–358.
132. Fakete, J., del Castilho, P., and Kraak, J. C., 1981, *J. Chromatogr.* **16**:319–327.
133. Goldstein, S. A., and Van Vunakis, H., 1981, *J. Pharmacol. Exp. Ther.* **217**:36–43.
134. Kabra, P. M., Mar, N. A., and Marton, L. J., 1981, *Clin. Chim. Acta* **111**:123–132.
135. Midha, K. K., Cooper, J. K., McGilveray, I. J., Butterfield, A. G., and Hubbard, J. W., 1981, *J. Pharm. Sci.* **70**:1043–1046.
136. Wallace, J. E., Shimek, E., Stavchansky, S., and Harris, S. C., 1981, *Anal. Chem.* **53**:960–962.
137. Bock, J., Gray, S., Geller, E., and Jatlow, P., 1981, *Clin. Chem.* **27**:1101.
138. Suckow, R. F., and Cooper, T. B., 1982, *J. Chromatogr.* (in press).
139. Bannister, J., van der Wal, S., Dolan, J. W., and Snyder, L. R., 1981, *Clin. Chem.* **27**:849–855.
140. Buchman, O., and Pri-Bar, I., 1978, *J. Label. Comp. Radiopharmaceut.* **14**:263–269.
141. Spector, S., Spector, N. L., and Almeida, M. P., 1975, *Psychopharmacol. Commun.* **1**:421–429.
142. Brunswick, D. J., Needelman, B., and Mendels, J., 1978, *Life Sci.* **22**:137–146.
143. Brunswick, D. J., Needelman, B., and Mendels, J., 1979, *Br. J. Clin. Pharmacol.* **7**:343–348.
144. Heptner, W., Badian, M. J., Baudner, S., Christ, O. E., Fraser, H. M., Rupp, W., Weimer, K. E., and Wissmann, H., 1977, *Br. J. Clin. Pharmacol.* **4**:123S–127S.
145. Read, G. F., and Riad-Fahmy, D., 1978, *Clin. Chem.* **24**:36–40.
146. Mould, G. P., Stout, G., Anerne, G. W., and Marks, V., 1978, *Ann. Clin. Biochem.* **15**:221–232.
147. Midha, K. K., Loo, J. C. K., Charette, C., Rowe, M. L., Hubbard, J. W., and McGilveray, I. J., 1978, *J. Anal. Toxicol.* **2**:185–192.
148. Maguire, K. P., Burrows, G. D., Norman, T. R., and Scoggins, B. A., 1978, *Clin. Chem.* **24**(4):549–554.
149. Wiles, D. H., and Franklin, M., 1978, *Br. J. Clin. Pharmacol.* **5**:265–268.
150. Midha, K. K., Cooper, J. K., and Hubbard, J. W., 1980, *Commun. Psychopharmacol.* **4**:107–114.
151. Midha, K. K., Mackonka, C., Cooper, J. K., Hubbard, J. W., and Yeung, P. K. F., 1981, *Br. J. Clin. Pharmacol.* **1**:85–88.
152. Suzuki, H., Minaki, Y., Iwaisaki, M., Sekine, Y., Kagemoto, A., Utsui, Y., Hashimoto, M., Yagi, G., and Itoh, H., 1980, *J. Pharmacol. Dynam.* **3**:250–257.
153. Rubin, R. T., Forsman, A., Heykants, J., Ohman, R., Tower, B., and Michiels, M., 1980, *Arch. Gen. Psychiatry* **37**:1069–1074.
154. Midha, K. K., Hubbard, J. W., Cooper, J. K., Hawes, E. M., Fournier, S., and Yeung, P., 1981, *Br. J. Clin. Pharmacol.* **12**:189–193.
155. Jorgensen, A., 1978, *Life Sci.* **23**:1533–1542.
156. Greenberg, L. H., and Weiss, B., 1979, *J. Pharmacol. Exp. Ther.* **211**:309–316.
157. Creese, I., and Snyder, S., 1977, *Nature* **270**:180–182.
158. Seeman, P., Lee, T., Chau-Wong, M., and Wong, K., 1976, *Nature* **261**:717–719.
159. Innis, R. B., and Snyder, S. H., 1980, *Psychopharmacol. Bull.* **16**(3):80–82.
160. Cohen, B. M., Herschel, M., and Aoba, A., 1979, *Psychiatry Res.* **1**:199–208.
161. Rosenblatt, J. E., Pert, C. B., Colison, J., van Kammen, D. P., Scott, R., and Bunney, W., 1979, *Commun. Psychopharmacol.* **3**:153–158.
162. Rosenblatt, J. E., Pary, R. J., and Bigelow, L. B., 1980, *Psychopharmacol. Bull.* **16**(3):78–80.
163. Calil, H. M., Avery, D. H., Hollister, L. H., Creese, I., and Snyder, S. H., 1979, *Psychol. Res.* **1**:39–44.
164. Meyers, B., Tune, L. E., and Coyle, J. T., 1980, *Am. J. Psychiatry* **137**(4):483–484.
165. Cohen, B. M., Lipinski, J. F., Pope, H. G., Jr., Harris, P. Q., and Altesman, R. I., 1980, *Psychopharmacology* **70**:191–193.
166. Tune, L. E., Creese, I., Depaulo, R., Slavney, P. R., Coyle, J. T., and Snyder, S. H., 1980, *Am. J. Psychiatry* **137**(2):187–190.
167. Campbell, A., Herschel, M., Cohen, B. M., and Baldessarini, R. J., 1980, *Life Sci.* **27**:633–640.
168. Rosenblatt, J. E., Pary, R. J., and Bigelow, L. B., 1980, *Psychopharmacol. Bull.* **16**(3):78–80.

169. Tune, L. E., Creese, I., Coyle, J. T., Pearlson, G., and Snyder, S., 1980, *Am. J. Psychiatry* **137**(1):80–82.
170. Kurland, A. A., Nagaraju, A., Hanlon, T. E., Wilkinson, E. H., and Ng, K. T., 1981, *J. Clin. Pharmacol.* **21**:42–47.
171. Kurland, A. A., Nagaraju, A., and Hanlon, T. E., 1980, *J. Clin. Pharmacol.* **20**:553–559.
172. Gram, L. F., 1977, *Clin. Pharmacokinet.* **2**:237–251.
173. Risch, S. C., Leighton, Y. H., and Janowsky, D. S., 1979, *J. Clin. Psychiatry* **40**:4–16.
174. Risch, S. C., Leighton, Y. H., and Janowsky, D. S., 1979, *J. Clin. Psychiatry* **40**:58–69.
175. Scoggins, B. A., Maguire, K. P., Norman, T. R., and Burrows, G. D., 1980, *Clin. Chem.* **26**(7):805–815.
176. Dahl, S. G., 1979, *Neuropsychopharmacology* (B. Saletu, P. Berner, and L. Hollister, eds.), Pergamon Press, New York, pp. 567–575.

Perfusion of the Isolated Brain

David D. Gilboe

1. RATIONALE FOR USE OF ISOLATED BRAINS

Customarily, studies of brain metabolism and physiology have employed both *in vitro* methods utilizing simplified systems containing cerebral tissue, cells, or organelles and *in vivo* techniques employing the intact animal. Although a considerable amount of useful information has been obtained from studies with these *in vivo* and *in vitro* systems, a number of difficulties are inherent in such preparations. For example, the usual membrane barriers are disrupted during preparation of most *in vitro* systems; consequently, some enzymes are placed in an atypical milieu which may contain abnormal concentrations of activators, inhibitors, or substrates. Furthermore, the anoxia encountered during preparation of brain slices, homogenates, and the various cells and organelles introduces another serious disadvantage, since similar periods of anoxia are known to produce adverse effects and irreversible changes in the higher centers of the brain. Thus, one is confronted with the overwhelming task of correcting the data obtained from the *in vitro* system for artifacts induced by the preparation and relating the results to metabolism in the intact organ. These problems could be avoided entirely by the use of *in vivo* preparations were it not for physiological and metabolic interference from other tissues and the difficulty in controlling blood flow and composition.

When care is exercised to maintain a physiological environment during and following the surgical preparation of the isolated brain, it is possible to isolate a tissue with an undamaged blood–brain barrier which has the histological, metabolic, and physiological characteristics of the intact brain. Because arterial blood flows exclusively to the brain, any change in venous blood composition may be attributed directly to the organ itself. Although isolated brain preparations simplify arteriovenous difference studies, their principal advantage is that they allow control of both perfusate composition and flow rate. Moreover, it is possible to effect frequent and abrupt changes in blood flow

---

David D. Gilboe • Departments of Neurosurgery and Physiology, University of Wisconsin Center for Health Sciences, Madison, Wisconsin 53706.

and/or composition during each study. With respect to blood composition, constant blood drug levels can be achieved when drugs are infused at a steady rate without recirculation of the perfusate.

Therefore, most of the disadvantages of *in vitro* and *in vivo* systems can be circumvented by use of an isolated brain preparation. Consequently, the isolated brain preparation is an excellent vehicle for the study of those aspects of cerebral metabolism, transport, and vascular response to agonists for which control of perfusate composition and flow is important.

2. REVIEW OF PROCEDURES EMPLOYED TO ISOLATE THE BRAIN

2.1. Isolated Heads

The history of investigations utilizing isolated heads goes back to the corneal and cephalic muscular reflex survival studies of Brown-Sequard[1] who perfused rabbit heads with blood from a syringe. deSomer and Heymans[2] anastomosed the carotid arteries and external jugular veins of the isolated head to the corresponding vessels of a donor dog to achieve the first successful extracorporeal perfusion of a head. Slightly more than 100 years after the report of Brown-Sequard, Gilboe, Cotanch, and Glover[3] described the use of a mechanical support system to maintain viability of dog brains with blood supplied to the carotid arteries of an isolated head from either the femoral artery of a donor dog or a pump oxygenator system. More recently, Thompson *et al.*[4] developed a simple procedure for extracorporeal brain perfusion in which perfusate was supplied to the head and neck of the rat via the aortic arch. Although flow of perfusate to the head and neck was measured during a variety of acid–base disorders, studies of arteriovenous differences could not be satisfactorily made because of retrograde flow through collateral arteries and loss of perfusate into the animal. In general, preparations of this type are difficult to utilize for either metabolic or physiological studies because of the inability to control blood flow to the brain.

The objective of more recent studies has been to effect vascular isolation of the brain so as to eliminate interference from extracerebral tissue. An early attempt at extracorporeal perfusion of the cerebral arteries of the dog was made by Schmidt[5] who reduced blood flow to extracerebral tissues by clamping the carotid arteries, constricting small arteries in adjacent tissues with epinephrine, and pumping blood through the vertebral arteries at higher than aortic pressure. This preparation was used to study the respiratory response to cerebral blood flow. The system devised by Moss[6] for selective perfusion of bovine brain was similar in concept to that of Schmidt in that carotid artery pressure was maintained 15–20 mm Hg higher than systemic pressure to exclude arterial blood from sources other than the carotid arteries. The system was designed primarily for studies of brain tumor and chemotherapy; little metabolic or physiological data have come from this preparation. It is inviting to simplify or eliminate surgery; however, such procedures frequently have the potential for extracerebral tissue perfusion. Consequently, they are seldom useful for quantitative studies.

2.2. Partially Isolated Brains

Chute and Smyth[7] adopted a more radical approach to the problem of vascular isolation when they surgically isolated the cat brain from the trunk of the animal and from part of the extracerebral tissue. Geiger and Magnes[8] devised a similar preparation, but by failing to sever the spinal cord and decapitate the cat, they left open the possibility that systemic blood could mix with perfusion blood via the spinal artery. Both of these groups believed it important to retain the eyes and adjacent structures for the purpose of assessing brain viability during extracorporeal perfusion. Although Geiger and Magnes took measures to limit the venous blood contribution from extracerebral tissue, neither the Geiger and Magnes nor the Chute and Smyth arteriovenous difference studies unequivocally represent changes caused only by brain metabolism. Andjus et al.[9] studied some of the metabolic characteristics of a relatively simple isolated rat brain preparation in which both orbital contents and maxilla remain intact. This preparation is relatively free of metabolically active extracerebral tissue, and, because of the relative simplicity of the preparation, it has been frequently used by other investigators. Michenfelder et al.[10] described a method in which the brain was supplied with arterial blood from the systemic circulation but venous samples were restricted to a circumscribed portion of the brain drained by the sagittal sinus. Studies with this preparation suggest that a greater than normal proportion of the sampled blood is derived from cerebral cortex; consequently, the data are not identical to those for whole brain. Although the latter preparation does not allow for flexibility in manipulating perfusate composition, the changes that occur in venous blood are clearly the result of brain metabolism.

2.3. Completely Isolated Brains

Suda et al.[11] reported the complete isolation of feline brain, and shortly thereafter, White et al.[12] described a comparable procedure for the isolation of the monkey brain. Although primarily interested in the study of cerebrovascular physiology, Bouckaert and Jourdan[13] were the first to surgically isolate canine brain from metabolically active extracerebral tissue. Gilboe et al.[14] using a modification of the dissection described by Bouckaert and Jourdan, developed an isolated dog brain preparation for use in studies of cerebral metabolism. With these preparations, some[11,12] or all[14] of the brain case is retained to protect cerebral tissue from injury resulting from manipulations during the experiment. Because bone metabolism and vascular reactivity are relatively low compared to brain tissue, the presence of the skull does not detract from the usefulness of such preparations for studies of cerebral metabolism and vascular physiology.

3. CRITERIA OF VIABILITY

Whem employing isolated brain preparations for metabolic and physiological research, one must present convincing evidence that the organ is alive
in order to validate the studies. The methods currently used to evaluate brain
viability in the human involve estimation of cerebral blood flow which is usually
normal in the isolated brain or evaluation of reflexes which are totally absent
in most isolated brain preparations because of the extensive surgery.

Some of the techniques used to evaluate cerebral viability can yield mis-
leading information regarding the status of the brain. For example, Blomquist
and Gilboe[15] elicited normal auditory and somatic sensory evoked potentials
in several isolated brain preparations; however, it was subsequently established
that the brains were hypoxic at the time of the study. In their pioneering studies
in the late 1940s and early 1950s, Geiger and Magnes[8] relied heavily on the
presence of reflexes to establish the viability of their isolated cat brain preparation. Evidence suggests that their brain preparations also had normal reflexes when the brains were in fact not metabolically normal. The intrinsic
problem these examples show with most methods of determining cerebral viability is that of quantification, since it is clearly not possible to determine the
degree of functional impairment from the types of information obtained in such
studies.

A prominent characteristic of the brain is the need for a large and continuous supply of oxygen to support oxidative phosphorylation and prevent cellular damage. It is clear that periods of complete ischemia result in damage to
a certain percentage of the brain cells (probably neurons) and a disruption of
their mitochondria with a consequent decrease in oxidative phosphorylation.
This decrease in oxidative phosphorylation is manifested by a change in the
cerebral metabolic rate for glucose ($CMRO_2$). Because $CMRO_2$ probably reflects the overall status of the mitochondria, it shows considerable promise as
an indicator of cerebral viability in isolated brains[16] where the technical difficulties of measuring arterial and venous oxygen content together with cerebral
blood flow are much less than *in vivo*. Fitzpatrick *et al.*[17] reported that elec-
troencephalographic (EEG) frequency and amplitude correlate with $CMRO_2$,
thus providing a means of estimating $CMRO_2$ in real time.

4. FLUIDS USED TO PERFUSE THE BRAIN

An acceptable perfusate for support of an isolated organ should: (1) have
the capacity to transport adequate oxygen to the organ, (2) be capable of
transporting CO_2 from the organ, (3) contain those substrates needed to maintain normal organ function, (4) include sufficient protein or plasma expander
to balance hydrostatic and tissue osmotic pressures, (5) maintain a physiological pH and buffer organic acids that efflux from the tissue, and (6) contain
physiological concentrations of the principal plasma ions. Whole blood would
be the perfusate of choice for all studies with isolated organs were it not for
the increase in blood vasoconstrictor concentration during extracorporeal cir-

culation, the finite nature of the blood supply, the presence of the metabolically active erythrocyte, and the óccasional need for a completely defined perfusion fluid. Although some investigators utilized whole blood in their initial studies,[6,7,14] most seek an alternate perfusate. Only the studies of Michenfelder *et al.*,[10] which involve no extracorporeal circulation of blood, are compatible with the use of whole blood.

Some investigators, principally Gilboe *et al.*[18] and White *et al.*,[12] have chosen to utilize diluted whole blood as the perfusate. Dilution increases the volume of perfusate, and when dextran is used as the plasma expander, the viscosity of the perfusate is markedly reduced. The flow properties of the diluted blood are further improved when platelets that contain serotonin and platelet aggregates which obstruct the microvasculature are removed from the blood by filtration.[19,20] Glucose and oxygen are the only metabolites whose concentration is controlled in diluted blood. Although the concentrations of most other metabolites and ions tend to vary slightly from their usual level in plasma, the data suggest that preparations perfused with diluted blood are metabolically similar to the brain *in situ*.

Having used diluted whole blood in their initial studies, Geiger *et al.*[21] adopted a perfusate containing bovine erythrocytes, bovine serum albumin, and glucose in a modified Ringer's solution (Table I). They reported that "simplified blood" causes lower than normal glucose uptake by the brain cells; consequently, brain glucose levels fall, and lactate accumulation increases. This problem could apparently be avoided by passing the simplified blood through a liver or adding a crude extract of liver to the perfusate.[22] Eventually cytidine and uridine[25] were added more or less empirically to the perfusion fluid and said to improve the utilization of glucose. In a more detailed study of glucose utilization by the isolated cat brain perfused with simplified blood, Allweis and Magnes[26] reported that lactate accumulated in the tissue at the rate of 0.17 μmol/g per min, reaching a concentration of 19.7 μmol/g following 60 min of perfusion. Approximately 70% of the glucose entering the brain was converted to lactate and either remained in the brain or was transported into the perfusion fluid. Only 20% of the glucose was oxidized to CO_2 and water. Added cytidine and uridine did not change this pattern of glucose utilization.

Krieglstein *et al.*[27] used an albumin and erythrocyte mixture in a Krebs–Hensleit buffer (Table I) which differed only slightly in composition from that used by Geiger and Magnes to perfuse an isolated rat brain preparation. Although the brain glucose levels were slightly below normal[24] (1.41 $\pm$ 0.06 S.E. μmol/g) when blood glucose levels were elevated (7.3 mM), the tissue lactate was higher than normal (4.07 $\pm$ 0.17 S.E. μmol/g) and tended to vary directly with blood and tissue glucose levels. Thus, excessive glycolysis continued to be a problem; however, it seemed to be less severe in the Krieglstein preparations.

Andjus *et al.*[9] reported that lactate efflux from isolated rat brains perfused with simplified blood (Table I) was equivalent to 1.4 times the glucose uptake. The metabolic problems associated with use of simplified blood[9] may have been aggravated by ischemia resulting from a low cerebral blood flow and hematocrit, because the proportion of total glucose influx represented by lac-

Table I
Concentration of Various Anions, Cations, and Albumin in Different Perfusates

Source	Values expressed in mEq/liter								
	Na^+	K^+	Ca^{2+}	Mg^{2+}	Cl	HCO_3^-	SO_4^{2-}	PO_4^{3-}	BSA
Geiger et al.[22] (cat brain)	143	5.8	5.0	2.2	133	24.8	2.2	1.1	7%
Gilboe and Betz[23] (dog brain)	147 ± 1	4.3 ± 0.1	4.7 ± 0.0	1.4 ± 0.1	117 ± 1	19.6	—	—	—
Krieglstein et al.[24] (rat brain)	143	8.8	4.9	2.3	135	25.0	2.3	1.2	4%
Andjus et al.[9] (rat brain)	143	5.8	5.0	2.2	133	24.8	2.2	1.1	7%

tate efflux fell to approximately 35% in a subsequent study.[28] Thus, the data from the various laboratories suggest that a factor that is usually present in whole blood and essential for the control of cerebral glucose metabolism is absent from simplified blood.

The data of Gilboe and Betz (Table I) are presented for comparison and indicate that there are unimpressive differences between the composition of the simplified blood and the electrolyte and bicarbonate content of diluted blood, where excessive glycolysis does not appear to be a problem. Although the chloride levels appear to be abnormally high in all simplified blood perfusates, it is doubtful that electrolytes are responsible for the atypical utilization of glucose. Because the factor must cross the blood–brain barrier to influence metabolism, it is unlikely to be a large molecule like insulin. If there is a factor that promotes normal glucose utilization by the brain, it is probably a low-molecular-weight compound.

Sloviter and Kamimoto[28] utilized a fluorocarbon as the oxygen carrier and emulsified it in a perfusate containing albumin in a Krebs–Ringer bicarbonate buffer vehicle. The obvious advantage of the fluorocarbon emulsion is that it eliminates metabolic interference from erythrocytes, a factor that may be important in certain metabolite balance studies. The metabolic data suggest that brains perfused with these mixtures were more hypoxic than the controls perfused with simplified blood containing erythrocytes. It is difficult to achieve an emulsion containing more than 15–20% fluorocarbon; consequently, the fluorocarbon may not be present in a sufficiently high concentration to properly oxygenate the brain at normal rates of blood flow. This seems to be confirmed by recent studies of the Krieglstein group[29] which indicate that cerebral lactate levels remain within normal limits when the brain is perfused with a 20% fluorocarbon emulsion in Krebs–Hensleit buffer at a flow rate of 3.0 ml/min.

Gilboe and collaborators have made extensive use of a diluted blood perfusate.[30] Although the plasma Na^{+}[23] tended to be slightly elevated because of addition of $NaHCO_3$ to control pH, glucose and the remaining electrolytes were usually in the normal range (Table I). Amino acid levels were slightly depressed as the result of plasma dilution[30] (Table X), and it is reasonable to assume that most other plasma components were present in lower than normal concentrations. Unlike studies made with simplified blood, the tissue lactate[31] tended to be in the normal range. This suggests that glucose metabolism is proceeding normally in these preparations, possibly because of the presence of some unmeasured diffusable component of blood.

Michenfelder and Theye[32] utilized whole blood in their studies. Although it must be assumed that the levels of all blood components were normal, Michenfelder did not supply data regarding blood levels of either electrolytes or metabolites other than O_2 and glucose. The brain lactate levels were in the normal range and comparable to those observed in the isolated dog brain perfused with diluted blood.[31]

The conditions under which various isolated brain preparations were perfused are shown in Table II. Although hematocrit was reduced to the same level with most preparations, there was considerable variation in temperature,

Table II
Perfusion Conditions in Various Preparations

Source	Temperature (°C)	Hct (%)	Flow (ml/100 g per min)	pH	P_{CO_2} (mm Hg)	Glucose concn. (mM)
Geiger and Yamasaki[25] (cat brain)	—	30	180	7.3	—	6.67
Kintner et al.[33] (dog brain)	36.5	30	80	7.40	40	5.56
Krieglstein et al.[24] (rat brain)	<30	30	100	7.4	—	7.00
Magnes et al.[34] (cat brain)	37	36–38	118–130	7.40	26–30	5.56
Michenfelder and Theye[35] (dog brain)	37	—	—	7.40	35–45	—
Sloviter and Kamimoto[28] (rat brain)	25	20	—	—	32–40	3.33–11.11
Hostetler et al.[36] (Monkey brain)	31–32.5	27	24–36	—	—	5.56

blood flow rate, glucose concentration, and P_{CO_2}. Only blood pH seems to have been regularly within the normal range.

5. METABOLIC STUDIES

5.1. Metabolite Uptake and Efflux with Various Preparations

It is common practice to quantify the uptake and efflux of cerebral metabolites by arteriovenous difference. In practice, the precision of available analytical methods has limited such studies to those metabolites in which significant arteriovenous differences can be measured during brain perfusion. Fortunately, those metabolites that are of immediate importance, such as glucose, oxygen, lactate, and carbon dioxide, fit this criterion, and representative values for the cerebral metabolic rate for glucose (CMRglu), the cerebral metabolic rate for oxygen (CMRO$_2$), lactate efflux, and carbon dioxide efflux reported by the laboratories of Geiger,[37] Gilboe,[31] Magnes,[26] Michenfelder,[32]

Table III
Metabolite Uptake and Efflux in Various Preparations

	Cat brain[8]	Dog brain[31]	Dog brain[26]	Dog brain[32]	Rat brain[38]	Cat brain[39]
CMRglu (μm/g per min)	0.33	0.30 ± 0.01	0.51	0.40 ± 0.04	0.245	0.41 ± 0.03
Lactate efflux (μm/g per min)	—	0.0	0.54	0.03 ± 0.01	0.170	0.29 ± 0.02
CMRO$_2$ (μm/g per min)	1.58	2.21 ± 0.05	2.19	2.04 ± 0.08	—	2.32 ± 0.37
Carbon dioxide efflux (μm/g per min)	—	—	—	—	—	2.41 ± 0.01

Sloviter,[38] and Watanabe[39] for their particular preparations are seen in Table III.

There is a 100% difference between the highest and lowest CMRglu values; however, this difference is explainable. The low CMRglu reported by Sloviter and Kamimoto[38] was obtained from isolated brains maintained at 25°C, whereas the higher value reported by Michenfelder and Theye[32] was from a preparation that included more than the normal fraction of metabolically active cerebral cortex. Those isolated brains perfused with simplified blood[26,38,39] tend to take up glucose and return about 40% of it to the blood as lactate. For the reasons discussed above, Michenfelder's preparation exhibits a larger CMRglu than that of Gilboe; however, neither preparation releases or takes up significant quantities of lactate.

In contrast to the observations with glucose, the $CMRO_2$ values with the various preparations are in close agreement. This is somewhat unexpected, since those preparations perfused with simplified blood appear to derive a considerable amount of their high-energy phosphate from glycolysis; yet they continue to take up normal amounts of oxygen. One wonders if simplified blood causes uncoupling of oxidative phosphorylation in the isolated brain. The values for CO_2 efflux reported by Watanabe *et al.*[39] suggest that the respiratory quotient of these isolated brains is near 1.0. However, this seems unlikely in view of the conversion of 35% of the glucose to lactate in this preparation. Some of the CO_2 may arise from the buffering of lactate by tissue bicarbonate. Fitzpatrick *et al.*[17] have promoted measurement of $CMRO_2$ as a means of assessing cerebral viability. Such a method would appear to be unreliable when used with preparations perfused with artificial blood.

Michenfelder and Theye[32] reported that $CMRO_2$ and CMRglu dropped approximately 35% when the brain temperature was decreased by an average of 5°C. Although neither hypocapnia ($Pa_{CO_2} < 13$ mm Hg) nor moderate anemia (8 g/100 ml) caused a change in $CMRO_2$, the combination of hypocapnia and extreme anemia (5 g/100 ml) did cause a significant fall in $CMRO_2$.[40] Conversely, hypocapnia, moderate anemia, and hypocapnia with extreme anemia caused CMRglu to increase significantly. Watanabe *et al.*[41] reported that 40 min following a 40–50% reduction in cerebral blood flow, $CMRO_2$ and CO_2 efflux dropped while CMRglu and lactate efflux remained constant. This suggests that oxidative metabolism may slow during slower perfusion; however, the metabolic fate of glucose is not clear, since cerebral lactate could increase under these conditions.

From arteriovenous difference studies with an isolated rat brain preparation, Zivin and Snarr[42] observed that D(−)3-hydroxybutyrate and glucose uptake were linearly related to their respective blood concentrations. Increasing the plasma D(−)3-hydroxybutyrate caused the rate of glucose uptake to decrease and permitted the brain to maintain electrical activity at lower plasma glucose concentrations. Although D(−)3-hydroxybutyrate uptake increased in fasted rats, its contribution to the total energy supply was relatively small.

Arteriovenous difference determinations of blood amino acid flux into and out of the isolated dog brain were reported in detail by Drewes *et al.*[30] These studies are discussed in Section 6.2.

Table IV
Glycolytic Intermediate Levels (nmol/g) in Intact and Isolated Brains

Compound	Intact brains	Isolated brains					
	Dog[45]	Dog[45]	Rabbit[43]	Rat[24]	Rat FC43[29]	Rat[51]	Cat[22]
Glycogen	5160 ± 140	4700 ± 150	3250	1180 ± 180	—	860	—
Glucose	2190 ± 160	1720 ± 140	2340	1440 ± 60	3350 ± 590	1140	2830 ± 352
Glucose-1-PO$_4$	1.4 ± 0.2	4.7 ± 0.4	—	—	—	4.1	—
Glucose-6-PO$_4$	58 ± 3	52 ± 2	76.0	69 ± 4.0	93 ± 21	58	—
Fructose-6-PO$_4$	16.8 ± 1.1	17.2 ± 0.5	—	—	20 ± 5	13	—
Fructose-P$_2$	47.0 ± 3.0	62.6 ± 3.5	93.8	102 ± 3	—	134	—
Glyceraldehyde-PO$_4$	2.3 ± 0.3	3.9 ± 0.3	—	—	—	36	—
Dihydroxyacetone-PO$_4$	10.7 ± 1.2	20.2 ± 1.6	49.4	43 ± 4	—	46	—
3-Phosphoglycerate	55.6 ± 2.4	74.0 ± 4.0	—	—	—	38	—
2-Phosphoglycerate	3.8 ± 0.3	6.2 ± 0.5	—	—	—	15	—
Phosphoenolpyruvate	13.3 ± 1.0	17.6 ± 0.8	—	—	—	12	—
Pyruvate	65 ± 4	102 ± 5	3.6	205 ± 5	57 ± 25	61	—
Lactate	944 ± 58	1950 ± 100	2140	4070 ± 170	1200 ± 740	3100	20,180 ± 1602

5.2. Tissue Metabolism in Various Preparations under Control Conditions

Glucose is the principal substrate for energy metabolism in the brain. Consequently, various phases of glucose metabolism have been studied in isolated brain preparations to provide an index for the efficiency of glucose oxidation and to detect possible metabolic abnormalities in its utilization.

Values for glycolytic intermediate concentrations in isolated brain preparations have been reported by Geiger et al.,[22] Gercken and Roth,[43] Gilboe and co-workers[16,31,44–46] Krieglstein and co-workers,[24,47–50] and Sloviter and co-workers.[51]·Representative values from these authors are presented in Table IV. Brain glucose and glycogen levels in the preparations of Krieglstein and Sloviter were depressed even when the perfusate contained normal levels of gluose. The high brain glucose concentration reported by Geiger was caused by raising the plasma glucose to approximately twice the normal level. Although the levels of glycolytic intermediates varied, they seemed to exhibit some changes that are related to the perfusate used, since the values that were reported by Krieglstein and Sloviter were closely correlated in most instances. Lactate was above normal in all preparations, particularly those that were perfused with simplified blood. Although the preparations and perfusates utilized by Krieglstein and Sloviter were basically the same, there was a significantly lower lactate level in the Sloviter preparation, possibly because of the low-temperature perfusion (25°C) used in those studies.

It should be noted (Table IV) that Krieglstein and associates[29] have recently reported on the use of an artificial blood containing a fluorocarbon (FC43) and high levels of glucose for rat brain perfusion. This perfusate appears to yield more normal tissue metabolite values, particularly of lactate, that other such artificial bloods, possibly as the result of better tissue oxygenation. The high-energy phosphate levels for various preparations are compared in Table V. Although the values reported by Krieglstein and co-workers[27,29] are nearer normal than most comparable data, it is clear that ATP is depressed while ADP and AMP are elevated in the Krieglstein preparations as well as the preparations of Sloviter and co-workers[51] and Gercken and Roth.[43] The phosphocreatine concentration was severely depressed in the preparations of Sloviter and

Table V
High-Energy Phosphate Levels (nmol/g) in Intact and Isolated Brains

	Intact brains	Isolated brains				
	Drewes and Gilboe[44]	Fitzpatrick and Gilboe[16]	Gercken and Roth[43]	Krieglstein et al.[27]	Dirks et al.[29]	Ghosh et al.[51]
PCr	—	4487 ± 126	970	2820 ± 120	3710 ± 530	1750
ATP	2121 ± 30	2522 ± 127	1720	2370 ± 60	2190 ± 190	1590
ADP	318 ± 17	420 ± 6	625	710 ± 20	650 ± 140	1030
AMP	44 ± 5	62 ± 9	246	190 ± 10	220 ± 50	470
Energy charge[a]	0.92	0.91	0.78	0.83	0.82	0.68

[a] Energy charge = ([ATP] + 0.5 [ADP])/([ATP] + [ADP] + [AMP]).

Gercken. The brain phosphocreatine levels reported by Krieglstein were better; however, they were also significantly lower than normal. Despite the more physiological levels of lactate in brains perfused with fluorocarbon-containing artificial blood, the high-energy phosphate compounds remain abnormally low in the isolated rat brain. The high-energy phosphate values reported by Fitzpatrick and Gilboe[16] appear to be more normal.

Several investigators have identified labeled products of [U^{14}C]glucose metabolism in the isolated cat brain. Geiger and co-workers[52-56] and Allweis and Magnes[26] used this system extensively and reported that 20–30% of the labeled glucose was oxidized to $^{14}CO_2$. Most of the remaining CO_2 was derived from metabolism of endogenous nonglucose substrates.[26] However, Otsuki et al.[57] achieved a 55% conversion of [U^{14}C]glucose to $^{14}CO_2$ when the "simplified blood" perfusate was supplemented with an amino acid, cytidine, and vitamin mixture. Oxygen consumption, when measured, was depressed, and all investigators reported that a large percentage of the glucose carbon was incorporated into lactate. Although these findings are of interest, they must be evaluated with caution because of the previously mentioned metabolic irregularities associated with brain perfusions that employ simplified blood.

Barkulis et al.[54] studied ^{14}C incorporation into brain amino acids and reported that total amino acid radioactivity rose for the first 50 min of exposure to [U^{14}C]glucose. Barkulis et al.[54] and later Otsuki et al.[57] observed that some of the ^{14}C label was present in glutamate, aspartate, and glutamine. However, Watanabe et al.[58] found that the accumulation of these amino acids was depressed when cerebral blood flow was reduced. The appearance in brain free amino acid pools of label from [U^{14}C]glucose was more rapid than the incorporation of ^{14}C into cerebral protein.[52] Most of the labeled protein was said to be found in the mitochondrial fraction (probably including polysome)[54]; however, much of the protein radioactivity could be extracted as acid–ethanol-soluble protein, probably the result of incorporation of glutamic and aspartic acids.[56] Margolis et al.[55] reported that ^{14}C from glucose appeared in N-acetyl-L-aspartic acid at only 1–5% of the rate at which it was incorporated into aspartic acid.

Based on a study of ^{14}C incorporation into glycogen utilizing [2-^{14}C]glucose, Hostetler et al.[59] concluded that 5–8% of the labeled glucose was metabolized in an isolated monkey brain via the pentose cycle. This value was significantly lower than that reported by Moss for the calf brain.[60] However, there was no question of extracranial tissue interference when the isolated monkey brain was used.

Otsuki et al.[61] reported that ^{14}C was incorporated into free aspartate, glutamate, and glutamine as well as a small amount of lactate during isolated cat brain perfusion with blood equilibrated with NAH$^{14}CO_3$–$^{14}CO_2$. The specific activities were consistent with the assumption that ^{14}C incorporated by CO_2 fixation was distributed equally between 1-C and 4-C of aspartate or oxaloacetate. Gombos et al.[62] perfused brains with "simplified blood" containing [4-^{14}C]- and [3-^{14}C]-DL-aspartate and [U^{14}C]-L-aspartate. The relative specific activities of free aspartate and lactate suggested that 10% of the pyruvate was derived from oxaloacetate or malate by decarboxylation.

Allweis *et al.*[20] incorporated bovine serum albumin-bound [U¹⁴C]palmitic acid into the simplified blood perfusate and observed that the relative specific activity of the CO_2 rose to about 2.8% of the specific activity of the palmitic acid. When glucose was absent, there was a perceptible increase in the relative specific activity of CO_2 produced by the brain. Although the brain can oxidize palmitic acid, the process is slow compared to glucose oxidation.

Drewes and Gilboe[63] reported that large quantities of ¹⁴CO_2 appeared in the perfusion blood when [1-¹⁴C]leucine, [1-¹⁴C]isoleucine, or [1-¹⁴C]valine was present in the perfusate, thus demonstrating that these branched-chain amino acids are decarboxylated in the isolated dog brain.

Perfusion of brains isolated from immature monkeys with [³H]androstenedione resulted in the formation in hypothalamic and limbic tissue and identification in the blood[64] of free and conjugated [³H]estradiol.

5.3. Metabolite Changes in Blood and Tissue during Hypoxia

The isolated canine brain was utilized in studies of anoxia ($P_{aO_2} \leq 10$ mm Hg), hypoxia (P_{aO_2} 20, 30, or 40 mm Hg), and ischemia (no blood flow for 30 min).[16,31,33,44–46,65] The procedure was to isolate brains in the usual manner, perfuse the organ for 30 min at the experimental P_{aO_2}, and then reoxygenate the tissue until the termination of the experiment. Because the blood flow was completely stopped for 30 min, the ischemic brains were held in a box thermostated at 37°C while the flow was interrupted.

The data presented in Table VI[31,33] indicate that 0 to 67% of the brain's normal oxygen requirements were met as a result of the various experimental treatments. If $CMRO_2$ fell below 20% of normal during the experimental period, it would recover to approximately 60% of normal during reoxygenation. This suggests that a group of vulnerable cells may be irreversibly damaged or destroyed when oxygen consumption is limited to 20% of normal for 30 min. As more oxygen became available during the experimental period (P_{aO_2} 30 and 40 mm Hg),[33] posthypoxic oxygen consumption was also higher, indicating that less cell damage had occurred under such conditions. The CMRglu did not appear to change during anoxia or severe hypoxia; however, it did increase significantly when the brains were perfused with P_{aO_2} 30 or 40 mm Hg blood. Apparently these brains also meet their energy requirement by both aerobic and anaerobic processes.

5.4. Tissue Metabolites and Electrolytes during and following Hypoxia

Glycolytic intermediates were measured and reported for the isolated canine brain at several time intervals not shown in Table VII.[31,33] The values, which are expressed as percent of normal, illustrate some interesting trends. For example, tissue glucose was low in the P_{aO_2} 10 and 20 mm Hg brains at the end of the experimental period but was elevated in the P_{aO_2} 40 mm Hg brains because glycolysis was being controlled at the hexokinase level. The low values for F6P and FDP at 0 min recovery suggest that glycolytic flux was

Table VI
Cerebral Oxygen and Glucose Uptake before, during, and after Hypoxia

P_aO_2	$CMRO_2$ (μmol/g per min)			CMRglu (μmol/g per min)		
	Control	Experimental (28 min)	Recovery (60 min)	Control	Experimental (28 min)	Recovery (60 min)
Ischemia[a]	2.20 ± 0.03	—	1.35 ± 0.03	0.294 ± 0.016	—	0.170 ± 0.022
10[a]	2.08 ± 0.03	0.09 ± 0.03	1.27 ± 0.03	0.276 ± 0.013	0.321 ± 0.039	0.219 ± 0.03
20[b]	2.21 ± 0.05	0.45 ± 0.05	1.19 ± 0.05	0.297 ± 0.013	0.285 ± 0.030	0.207 ± 0.03
30[b]	2.15 ± 0.07	1.05 ± 0.05	1.69 ± 0.16	0.347 ± 0.007	0.468 ± 0.031	0.349 ± 0.008
40[b]	2.14 ± 0.06	1.51 ± 0.03	2.22 ± 0.07	0.377 ± 0.008	0.382 ± 0.010	0.345 ± 0.004

[a] Reference 31.
[b] Reference 33.

Table VII

High-Energy Phosphate and Glycolytic Intermediate Levels before, during, and after Various Degrees of Hypoxia

| | Time after reoxygenation (min) (% of normal)[a] | | | | | | | | |
| | Kintner et al.[31] P_{O_2} 10 mm Hg | | | Kintner et al.[31] P_{O_2} 20 mm Hg | | | Kintner et al.[33] P_{O_2} 40 mm Hg | | |
Substrate	0	15	45	0	15	45	0	15	45
Glucose	0.	53.3	45.2	20.	112.9	136.4	167.4	100.7	101.3
Glucose-6-PO_4	57.3	694.2	442.3	48.5	509.6	199.0	117.8	105.8	107.8
Fructose-6-PO_4	24.9	352.3	276.4	107.0	372.1	166.5	141.2	96.0	113.0
Fructose di-PO_4	31.9	81.5	70.9	13.7	17.9	73.8	82.2	103.2	101.6
Phosphoenol pyruvate	48.3	113.1	103.4	51.7	110.8	106.2	91.3	92.4	96.7
Pyruvate	66.0	492.2	310.7	62.7	43.3	39.9	176.0	155.0	98.0
Lactate	1754.	1697.	1682.0	874.9	938.5	312.8	361.0	101.	103.7
ATP	4.6	49.9	39.7	48.3	63.0	68.0	89.0	93.2	99.1
ADP	81.8	115.4	109.6	89.7	93.5	95.1	111.9	117.	112.4
AMP	441.8	171.6	171.6	88.1	171.6	114.9	353.2	290.	102.1
PCr	—	—	—	—	—	—	4.3	88.7	98.5

[a] Control values were: glucose, 1720 ± 140; G6P, 52 ± 1; F6P, 17.2 ± 0.5; FDP, 62.6 ± 3.5; Pyr, 102 ± 5; Lac, 1950 ± 100; ATP, 2522; ADP, 420; AMP, 62; and PCr 4847. PEP = 17.6 ± 1.2. Control values for P_{O_2} = 40: Glu, 1728 ± 25; G6P, 50.1 ± 2.4; F6P, 17.7 ± 0.5; FDP, 62.0 ± 1.7; PEP, 18.4 ± 0.8; and Pyr, 100.0 ± 2.5.

not impeded at Pa_{CO_2} 10 mm Hg; however, at Pa_{CO_2} 20 mm Hg, F6P was normal, suggesting improved regulation of glycolytic flux. Control at the FDP level was present at Pa_{CO_2} 40 mm Hg as can be seen by the slight rise in F6P that occurred in response to the modest increase in brain lactate. The PEP–pyruvate control point appears to be functioning in the Pa_{CO_2} 20 mm Hg group after 15 min of reoxygenation; however, it was not working in the Pa_{CO_2} 10 mm Hg group at any time. Lactate was elevated in all brains except those reoxygenated for 15 min or more following the Pa_{CO_2} 40 mm Hg perfusion. Neither the Pa_{CO_2} 10 nor 20 mm Hg group appeared to attain normal levels of high-energy phosphates following hypoxia; however, the Pa_{CO_2} 40 mm Hg group returned to normal within 15 to 30 min. The data suggest that 30 min of severe oxygen deprivation may cause irreversible damage to enzymes involved in mitochondrial respiration.

Changes in percent dry weight and Na and K content of brain gray and white matter were studied under experimental and control conditions.[31,65] The changes in gray matter dry weight were most striking in Pa_{CO_2} 10 mm Hg preparations with a decrease from 21.1 ± 0.2% to 18.2 ± 0.3% after 30 min of perfusion. The dry weight in the Pa_{CO_2} 10 mm Hg group continued to decrease for most of the 2-hr period of reperfusion. There were increases in brain sodium content and decreases in potassium content that corresponded with increases in brain water. Following 30 min of perfusion with blood having a Pa_{CO_2} of 20 mm Hg, there was no significant change in gray matter dry weight. Although the Pa_{CO_2} 20 mm Hg brains were not metabolically normal, they did not swell. Edema formation was observed only in the most severely hypoxic brains (ischemia and Pa_{CO_2} 10 mm Hg).

Severe cerebral hypoxia caused disturbances in glycolysis, oxygen consumption, high-energy phosphate levels, and the ability of cells to maintain electrolyte gradients. It is clear that control of glycolysis and the absence of edema following an episode of hypoxia are not directly related to the capacity of a cell to recover normal function, since these variables return to normal when high-energy phosphate levels and oxygen consumption are depressed. It appears that the ability to utilize normal amounts of oxygen and to restore tissue phosphocreatine levels may be the best indicators of metabolic and physiological recovery in the posthypoxic isolated brain.

6. PHARMACOLOGICAL STUDIES

6.1. Influence of Drugs on Cerebral Metabolism

Krieglstein and associates[24,48,66] used an isolated rat brain preparation similar to that described by Andjus et al.[9] for the study of drug-induced changes in cerebral metabolism. These workers utilized either variations in brain tissue metabolite levels or changes in the EEG spectra to evaluate the magnitude of alterations in cerebral metabolism.

Tissue levels of glucose and glycogen plus several high-energy phosphates and glycolytic intermediates were measured in brains from intact rats and isolated rat brains exposed to various anesthetic drugs for 30 min.[24,48,66] The presence (Table VIII) of drugs other than urethane, the anesthetic used during brain isolation, caused significant increases in brain phosphocreatine. Although Michenfelder[67] made a similar observation with thiopental, he did not provide appropriate control data for comparison. Only phenobarbital[24] and chloral hydrate[66] caused significant slowing of lactate accumulation. Unlike the findings of Krieglstein and Stock,[48] Michenfelder's study[67] suggests that lactate does not usually increase above normal levels when thiopental is present. Brain glucose was significantly higher than control with all drugs except ketamine. Among the metabolites measured, only cerebral glucose increased significantly compared to the control when its blood level was elevated in the presence of thiopental.[68] No specific mechanism was suggested to explain these changes; however, the authors theorized that these drugs cause glucose and phosphocreatine accumulation because they depress metabolism.

The isolated rat brain was also used by Krieglstein and colleagues to study the cerebral metabolic effects of other drugs. For example, the antimetabolite 6-aminonicotinamide was administered intraperitoneally (35 mg/kg) to intact rats 7 hr prior to the start of brain perfusion.[69] The significant increase in cerebral glucose coupled with a significant decrease in cerebral lactate suggests that 6-aminonicotinamide caused glycolytic flux to decrease. Toxic amounts (10^{-4} M) of promazine, monodesmethyl promazine, imipramine, and desipramine also caused accumulation of glucose and phosphocreatine as well.[70] The authors believe these changes to be nonspecific and unrelated to the pharmacological action of these neuroleptic and antidepressant drugs.

The isolated rat brain EEG was recorded in the presence of the drugs

Table VIII

High-Energy Phosphate and Metabolite Levels (μmol/g brain ± S.E.) in the Presence of Various Anesthetic Drugs

	Normal[24]	Perfusion control[24]	1.5 mM Phenobarbital[24]	0.2 mM Thiopental[24]	177 mg/kg Thiopental[67]	0.2 mM Hexobarbital[48]	4 mM α-hydroxybutyric acid[48]	0.05 mM Ketamine[48]	3.5 mM Chloral hydrate[66]	3.5 mM Trichloro-ethanol[66]
P-creatine	2.57 ± 0.18	2.82 ± 0.12	3.52 ± 0.12[a]	3.25 ± 0.15[b]	4.96 ± 0.41	3.07 ± 0.11[b]	2.98 ± 0.11[c]	2.91 ± 0.08[c]	3.17 ± 0.13[b]	3.13 ± 0.09[b]
ATP	2.38 ± 0.05	2.37 ± 0.06	2.46 ± 0.06	2.20 ± 0.09	2.33 ± 0.09	2.04 ± 0.08	2.07 ± 0.09	2.24 ± 0.07	2.37 ± 0.06	2.58 ± 0.09
Glycogen	2.02 ± 0.13	1.18 ± 0.18	1.54 ± 0.10	2.32 ± 0.15	—	2.35 ± 0.11	2.17 ± 0.26	1.69 ± 0.09	1.28 ± 0.24	1.61 ± 0.11
Glucose	1.20 ± 0.08	1.44 ± 0.06[a]	2.06 ± 0.10[a]	1.78 ± 0.11[b]		1.80 ± 0.08[b]	1.62 ± 0.08[b]	1.00 ± 0.10	1.89 ± 0.02[b]	2.21 ± 0.09[b]
Lactate	2.39 ± 0.16	4.07 ± 0.17[d]	2.42 ± 0.44[a]	3.28 ± 0.47	2.17 ± 0.23	3.79 ± 0.47	3.22 ± 0.67	5.32 ± 0.3	3.19 ± 0.22[c]	3.55 ± 0.55

[a] P < 0.01 vs. control.
[b] P < 0.01 vs. control (not shown).
[c] P < 0.05 vs. control (not shown).
[d] P < 0.01 vs. normal.

discussed above.[24,48,71–74] Only ketamine[48] caused an increase in low-voltage fast waves. All other drugs brought about a decrease in frequency and in some instances an increase in amplitude of the EEG. Such changes are consistent with a slowing of metabolism.

Changes in EEG frequency and amplitude were used to identify metabolically active forms of several drugs in the isolated rat brain. Chlorpromazine, 7-hydroxychlorpromazine, and chlorpromazine-N-oxide produced an increase in slow-wave activity that correlated with an increase in homovanillic acid levels in the corpus striatum.[73] The authors suggest that striatal dopaminergic neurons were involved in producing the EEG effects that occurred when the phenothiazines were in the perfusate.

Michenfelder and his associates studied anesthetic and drug effects on $CMRO_2$ in the partially isolated dog brain preparation. The $CMRO_2$ fell to 77% of normal when the methoxyflurane level reached 0.44%,[75] to 66% of normal when the enflurane concentration reached 2.2%,[76] to 70% of normal when the isoflurane concentration reached 2.2%,[77] to 58% of normal when the mean dose of thiopental reached 72 mg/kg,[67] and to 75% of normal when cyclopropane exceeded 13% in animals receiving reserpine or spinal anesthesia.[78] Although the $CMRO_2$ did not change in the presence of 0.3 mg/kg droperidol, it fell to 82% of normal with 0.006 mg/kg fentanyl and to 78% of normal when Innovar® (0.3 mg/kg droperidol and 0.006 mg/kg fentanyl) was administered.[79] This preparation, although not completely isolated, permits one to accurately evaluate the effect of drugs on cerebral oxygen utilization because arterial and venous oxygen content and cerebral blood flow can be measured with precision.

Pentylenetetrazol has been used by most investigators during the development of isolated brain preparations because it can usually activate an otherwise flat EEG, thus giving some assurance of brain viability. Geiger and Magnes reported that pentylenetetrazol induced an increase in cerebral O_2 consumption,[37] CO_2 efflux,[80] and glucose carbon incorporation into proteins[52] coupled with a decrease in glucose oxidation to CO_2 and water. Gilboe and Betz[23] observed no increase in O_2 consumption during pentylenetetrazol seizures. This latter observation was attributed to the fact that cerebral blood flow could not increase during a seizure in the isolated dog brain preparation.

Woods *et al.*[81] found 5-hydroxytryptamine levels in the isolated rat brain to be directly proportional to blood tryptophan levels when a monamine oxidase inhibitor (tranylcypromine) was present in blood. Thus, the isolated rat brain behaved exactly as the *in situ* rat brain.

Betz and Gilboe[82] reported that canine brains isolated under pentobarbital anesthesia had an average $CMRO_2$ of 2.54 ± 0.06 S.E. ml/100 g per min, whereas brains from animals anesthetized with halothane had an average $CMRO_2$ of 3.39 ± 0.12 S.E. ml/100 g per min. This was believed to reflect the slowing of cerebral metabolism because of the residual effects of the barbiturate anesthetic. The effect of sodium pentobarbital anesthesia on amino acid and urea flux was also studied in the isolated dog brain.[82] A significant uptake of isoleucine and leucine with a highly significant efflux of nitrogen (asparagine, glutamine, glutamic acid, phosphoserine, taurine, and urea) occurred in the presence of pentobarbital. The nitrogen efflux was attributed to NH_3 detoxification caused by increased amino acid catabolism following barbiturate–induced slowing of oxidative metabolism. This study also demonstrated that urea could be synthesized in the brain under appropriate conditions. In contrast to these findings, pentobarbital was reported by Barkulis *et al.*[54] to have no influence on free amino acid levels in the isolated cat brain.

Studies of the metabolic consequences of drug administration in the isolated brain have not usually been aimed at investigation of problems peculiar to a specific drug. Instead, most studies are less specific, involving a determination of the drug's influence on oxygen utilization, glycolytic flux, and/or energy metabolism. Although such data are of interest, isolated brain preparations could be more profitably exploited for determination of drug-induced changes in cerebral metabolism that might be more specific to the hypothalmic mode of action of the drug under investigation.

6.2. Drug Effects on Cerebrovascular Resistance

Although Geiger and Sigg[83] observed that epinephrine and norepinephrine (NE) induced vasoconstriction in the isolated cat brain, the report was anecdotal. Lowe and Gilboe[84] noted that cerebrovascular resistance (CVR) increased when phenylephrine, norepinephrine, and epinephrine were administered during studies with the isolated dog brain. However, the change in CVR could be reversed or blocked with the α-adrenergic blocking drug phenoxybenzamine HCL. The CVR decreased in the presence of isoproterenol, but this response was reversed by administration of the β-adrenergic blocking agent propranolol. Although these studies clearly demonstrated that both α- and β-adrenergic receptors were present in the cerebral vascular bed, they were flawed in that, for technical reasons, it was not always possible to demonstrate a vascular response that was in direct proportion to the dose of agonist administered.

Zimmer *et al.*[85] studied changes in cerebral blood flow (CBF) during infusion of 0.2 μg per min of NE and epinephrine into an isolated dog brain. The CBF decreased $10.2 \pm 6\%$ and $4.1 \pm 3.3\%$, respectively. Infusion of isoproterenol at a similar rate caused a $9.3 \pm 3.6\%$ increase in CBF. Norepinephrine-induced changes in CVR were studied in the isolated dog brain by Saffitz *et al.*[86] following injection of boluses containing 0.02 to 10,000 μg of NE or infusions of 0.1 to 8850 μg of NE per ml of blood. The resulting data were fitted to the equation for a rectangular hyperbola using an iterative least-square technique. The maximum response (Δ_{max}) was a $96.6 \pm 7.9\%$ S.E. increase over preinfusion CVR, and the dose at half-maximum response (EC_{50}) was 2.60 ± 0.88 S.E. $\times 10^{-5}$ M. In the presence of 0.1 mg/liter of phenoxybenzamine HCl, a noncompetitive inhibitor, the Δ_{max} was $79.0 \pm 0.6\%$ S.E., and the EC_{50} was 85.1 ± 26.8 S.E. $\times 10^{-5}$ M. Although the EC_{50} for the injections was an order of magnitude higher than the corresponding value for infusion, the Δ_{max} values were similar. These studies clearly demonstrate the presence of adrenergic receptors in the cerebral vascular bed; however, the results have been criticized by those who believe that the ability of the cerebral arteries to respond to catecholamines is related to their denervation during the brain isolation procedure.

The effect on CVR of intracarotid injection of various doses of acetylcholine, nitroglycerin, angiotensin, and 5-hydroxytryptamine was examined by Lowe and Gilboe.[87] Acetylcholine (0.6–80 μg/bolus) and nitroglycerin (0.1–20 μg/bolus) caused CVR to decrease, whereas 5-hydroxytryptamine (0.05–6.6 μg/bolus) and angiotensin (1.0–100 μg/bolus) caused CVR to increase. Although cerebral vascular smooth muscle responded to these vasoactive compounds in a manner similar to that for other vascular smooth muscle, the magnitude of the response was less in the cerebral vascular bed.

Ingenito *et al.*[88] studied the effect of various concentrations of nicotine HCl (1–100 μg/ml of perfusate) on CVR in the isolated cat brain. They reported a mild, transient vasoconstriction mediated largely by the superior cervical ganglion. This effect was diminished if the vagi were intact.

Michenfelder and associates studied CBF in the presence of various anesthetic drugs. The CBF increased by 33% and 63% when enflurane reached 1.4% and 2.4%, respectively.[76] Although cerebral blood flow increased abruptly as the cyclopropane concentration was raised, it decreased during spinal anesthesia or reserpine treatment.[78] This suggests that the changes in CBF are secondary to circulating catecholamine levels. Methoxyflurane caused CBF to fall to 74%[75] of normal, whereas fentanyl and Innovar® precipitated a fall in CBF to 50% of normal.[79]

β-Phenyl-γ-aminobutyric acid is a derivative of γ-aminobutryic acid that suppresses psychomotor excitation. This drug caused an increase in CBF in the isolated cat brain which reached a maximum when the dose of the drug was 5 mg per bolus or above.[89]

6.3. Drug Uptake

Isolated dog and monkey brain preparations were utilized to study the cerebral demethylation and acetylation of aminopyrine, the glucurono conjugation of oxazepam,[90] and their subsequent concentration in the cortex. The uptake of promazine, chlorpromazine, and their respective desmethyl derivatives was studied in the isolated rat brain.[91] Promazine and chlorpromazine crossed the blood–brain barrier in larger quantities than their less lipophilic desmethyl metabolites. Addition of reserpine to the perfusate supplying the isolated rat brain caused a decrease in tissue norepinephrine, dopamine, and 5-hydroxytryptamine and an increase in 5-hydroxyindoleacetic acid levels.[92] Physostigmine in the perfusate caused an increase in cerebral acetylcholine levels, whereas oxotremorine did not.[93] The partially isolated cat brain was perfused with blood containing either reserpine which caused a significant decrease in tissue norepinephrine or hydralazine which caused a significant increase in brain norepinephrine.[94] There was no temporal correlation between the peripheral vascular effects of these drugs and the brain level of norepinephrine. Although none of the work described above has been extremely detailed, the studies clearly demonstrate that isolated brains can be used to study the effect of certain drugs and their metabolism in the central nervous system.

7. METABOLITE TRANSPORT

7.1. Glucose Transport

7.1.1. Net Glucose Flux

The net rate of glucose entry (net flux) into the isolated dog brain was determined by arteriovenous difference and reported by Gilboe and Betz[95] to correlate with the total arterial glucose available to a unit weight of brain per unit time. Data obtained over an arterial glucose concentration range of 0.5–15.6 mM described a complex curve which was subsequently resolved into a straight line with a slope of 0.028, representing simple diffusion of glucose into the brain, and a rectangular hyperbola depicting the apparent kinetic constants (K_m of 1.4 mM and V_{max} of 2.66 μmol/g per min) of the glucose carrier in the cerebral capillaries. Although the latter values resemble the kinetic constants for membrane transport, one must bear in mind that the constants do not describe a single well-defined process. Instead, the data are the resultant of cerebral glucose influx and efflux and probably represent a summation of cerebral metabolic reactions involving glucose.

7.1.2. Unidirectional Glucose Flux

7.1.2a. Preliminary Studies. Betz et al.[96] used an indicator dilution technique with ^{22}Na as the intravascular marker to describe the unidirectional movement of glucose through the luminal membrane of the cerebral capillary endothelial cell. The rate of unidirectional glucose transport, $v = (E - 0.036)AF_p/W$, where E is the fractional extraction of glucose from blood, A is the arterial plasma glucose concentration, F_p/W is the plasma flow rate per unit weight of brain, and 0.036 is a correction for glucose diffusion. The diffusion correction for glucose transport was based on the fractional extraction of [6-^{3}H]-D-fructose over a concentration range of 0.2–6.2 mM. An iterative least-squares technique was utilized to fit v versus $\bar{A}$ (average capillary glucose concentration) directly to the nonlinear Michaelis–Menten equation and calculate K_m and V_{max} with their respective standard errors. The K_m of 8.26 ± 1.12 mM and V_{max} of 1.75 ± 0.11 μmol/g of brain per min were not altered significantly in the presence of either 0.03 mM sodium pentobarbital or 40 μg/liter of insulin.

7.1.2b. The Effect of Inhibitors on Glucose Transport: Structural Analogues of Glucose. Betz and Gilboe[97] demonstrated by kinetic analysis that 3-O-methyl-D-glucose is a competitive inhibitor of cerebral glucose transport with a K_i of 1.8 mM. It was assumed that other glucose analogues are also

Table IX
Effectiveness of Various Monosaccharides as Inhibitors
of Glucose Transport in Brain

Inhibitor	Percent normal brain glucose transport in presence of inhibitor
α-D-Glucose	53
α, β-D-Glucose	58
2-Deoxy-D-glucose	61
3-O-Methyl-D-glucose	62
β-D-Glucose	63
D-Galactose	67
D-Mannose	69
1,5-Anhydro-D-glucitol	77
D-Xylose	80
D-Fucose	85
5-Thio-D-glucose	86
D-Ribose	88
L-Arabinose	90
i-Inositol	91
L-Fucose	94
Sorbitol	97
L-Glucose	100
D-Fructose	101
0.9% NaCl (control)	100

competitive inhibitors of glucose transport; thus, the capacity of an analogue to inhibit glucose transport should be directly related to its affinity for the glucose carrier.[98] Eighteen analogues of glucose were tested for their ability to inhibit glucose transport. The inhibitor concentration in the injectate containing ^{22}Na and [^{3}H]glucose was 500 mM, resulting in a peak blood inhibitor concentration of approximately 2.5 mM. The blood glucose concentration was maintained relatively constant at 4.94 ± 0.06 S.E. mM. Based on these studies (Table IX), it was concluded that glucose is normally transported in the pyranose ring conformation, since transport was inhibited by 1,5-anhydro-D-glucitol and *i*-inositol but not by sorbitol. The studies suggest that no single hydroxyl group is absolutely required for binding, since neither elimination nor epimerization at any single asymmetric carbon completely abolished the ability of the analogue to inhibit glucose transport. The hydroxyl on C-1 is probably involved in binding, since neither elimination (1,5-anhydro-D-glucitol) nor anomerization (α or β) of the hydroxyl at this position decreased the inhibitory potency. The data suggest that binding to the glucose carrier occurs through hydrogen bonding to hydroxyls on carbons 1, 3, 4, and 6.

7.1.2c. Competitive and Noncompetitive Inhibitors of Glucose Transport.
Glucose transport was studied in the presence of 0 to 1.5 mM phlorizin.[98] A kinetic analysis indicated that phlorizin, a phenolic glucoside, is a competitive inhibitor of glucose transport with a K_i of a 1.11 ± 0.12 S.E. mM. Glucose transport was also investigated in the presence of 0 to 0.2 mM phloretin, the

structurally related phenol.[98] Phloretin is a partially competitive inhibitor with a K_i of 0.29 ± 0.04 S. E. mM. Inhibition by phlorizin and phloretin was mutually competitive, indicating that these inhibitors both compete for binding to the same site on the cerebral capillary endothelial cell glucose carrier.

Cytochalasin B, a fungal metabolite, is a potent noncompetitive inhibitor of glucose transport at the blood–brain interface.[99] Both glucose transport into (K_i 6.6 ± 1.9 S.E. μM) and out of the capillary endothelial cell was inhibited; however, inhibition was reversed by perfusion with blood containing no cytochalasin B.

Based on the inhibitory patterns of phlorizin and phloretin, a hydrophobic binding site for phenols is believed to exist in close proximity to the glucose-binding site on the cerebral capillary endothelial cells. Cytochalasin B, which appears to bind at a more distant site, reduces the apparent V_{max} but does not alter glucose binding.

7.1.2d. Effect of Hypoxia on Cerebral Glucose Transport: Hypoxia and Anoxia. After 2 min of anoxia[100] or hypoxia,[31] the rate of unidirectional glucose transport began to decline. Within 15 min of the beginning of the low oxygen perfusion, glucose transport leveled off at a rate that was approximately half the control value. The decrease in glucose flux was not caused by a general change in blood–brain barrier permeability, since neither the diffusion of fructose[100] nor the transport of leucine[101] was affected by anoxia. These observations led to the following experiments in an attempt to explain the phenomenon.

7.1.2e. Accelerative Exchange Diffusion. In some systems, metabolite flux into the cell slows markedly when the cellular metabolite concentration falls. This process, termed accelerative exchange diffusion, is believed to occur in carrier-mediated systems because the mobility of the loaded carrier is greater than that of the unloaded carrier. The conditions for accelerative exchange diffusion appear to be similar to those describing glucose concentration in the hypoxic brain cell. Although accelerative exchange diffusion was apparently demonstrated for glucose transport at the blood–brain barrier,[102] the effect was too small to explain the 50% decrease in glucose uptake during hypoxia. Furthermore, the original study was inaccurate, since the intravascular glucose concentration increased when arterial blood containing relatively low levels of glucose was used to perfuse brains equilibrated with high levels of glucose. Thus, the measured arterial glucose concentration was not truly representative of the actual capillary glucose levels. When $\bar{A}$ (mean capillary glucose) was substituted for A (arterial glucose concentration) in the equation for calculation of v (Section 7.1.2), it was no longer possible to demonstrate accelerative exchange diffusion.[103]

7.1.2f. Effect of Blood pH on Glucose Transport. Because intracellular pH and presumably extracellular pH in the brain tend to fall during anoxia, it seemed possible that the pH change might in some way influence glucose transport across the cerebral capillary endothelial cells. Kinetic analyses of

Table X
Whole Blood Amino Acid Concentrations and Amino Acid Flux under Control Conditions[a]

Amino acid	Arterial amino acid concentrations (μmol/100 ml blood)	Net amino acid uptake (+) or efflux (−) (μmol/100 g brain per min)
Alanine	59.60 ± 3.30	−0.41 ± 0.65
α-aminobutyric acid	2.01 ± 0.23	−0.03 ± 0.00
Ammonia	25.00 ± 0.90	−0.26 ± 0.37
Arginine	8.69 ± 0.75	−0.13 ± 0.14
Asparagine	16.63 ± 2.38	−0.08 ± 0.36
Aspartic acid	2.13 ± 0.32	−0.20 ± 0.09*
Carnosine	0.68 ± 0.07	−0.02 ± 0.02
Citrulline	3.97 ± 0.15	−0.06 ± 0.06
Glutamic acid	11.97 ± 0.84	−1.06 ± 0.23*
Glutamine	3.24 ± 0.33	+0.06 ± 0.08
Glycerophosphoethanolamine	2.18 ± 0.40	−0.22 ± 0.09*
Glycine	16.61 ± 0.59	−0.39 ± 0.15*
Histidine	6.28 ± 0.53	−0.27 ± 0.13*
Isoleucine	3.95 ± 0.31	+0.14 ± 0.05*
Leucine	8.32 ± 0.72	+0.16 ± 0.07*
Lysine	20.31 ± 1.14	−0.07 ± 0.33
Methionine	2.72 ± 0.25	−0.49 ± 0.09
3-Methylhistidine	1.27 ± 0.21	−0.10 ± 0.05*
Ornithine	2.37 ± 0.14	−0.01 ± 0.05
Phenylalanine	3.90 ± 0.28	−0.08 ± 0.05
Phosphoserine	4.95 ± 0.46	−0.32 ± 0.14*
Proline	12.89 ± 0.69	−0.22 ± 0.18
Serine	9.78 ± 0.41	−0.16 ± 0.10
Taurine	14.09 ± 1.96	−0.26 ± 0.14*
Threonine	16.63 ± 1.28	−0.37 ± 0.19
Tyrosine	2.67 ± 0.13	−0.08 ± 0.03*
Urea	345 ± 46	−4.90 ± 3.80
Valine	13.25 ± 0.67	−0.03 ± 0.13

[a] Values are means ± S.E. for 13 different brain preparations. A total of 62 A–V samples were collected during 13 different perfusion experiments and analyzed for amino acids. The F ratio of analysis of variance indicated a significant animal-to-animal variation in amino acid concentrations. Therefore, the mean concentration of each animal was used to calculate the average arterial concentrations. No animal-to-animal variation was detected in the value of (A–V) F/W. Therefore, net uptake or efflux is the mean of all observations. *$P <$ 0.05, significantly different than zero.

glucose transport were made at arterial blood pH of 6.8, 7.0, 7.2, 7.4, 7.5, 7.6, and 7.8.[104] There was no significant change in either K_m or V_{max} until pH 7.6 when V_{max} increased from 1.59 ± 0.24 S.E. to 2.1 ± 0.27 S.E. Because of the lack of an effect when the pH was lower than 7.4, this observation eliminated blood and probably interstitial fluid pH as factors influencing the transport of glucose into the brain during hypoxia. The glucose carrier was obviously modified at alkaline pH; consequently, one of the binding sites on the glucose carrier, probably an amino group, must become deprotonated, thereby altering the V_{max}.

7.1.2g. Kinetics of Glucose Transport following Ischemia. It was impossible to study glucose transport kinetics in anoxic preparations because tissue edema interfered with perfusate flow. A reduction in glucose unidirectional flux was also observed in isolated brains after 30 min of complete ischemia. A kinetic analysis of glucose transport was made in postischemic brains, and the apparent K_m was 17.4 ± 5.5 S.E. mM, and the V_{max} was 2.6 ± 0.6 S.E. μmol/g per min. Thus, unidirectional glucose flux into the brain during and following hypoxia was probably depressed, in part because of a decrease in the affinity of the carrier for glucose.[104]

7.2. Amino Acids

7.2.1. Net Amino Acid Flux

In studies conducted with the isolated dog brain, Drewes *et al.*[30] examined net blood-to-brain transport of 28 amino acids and related compounds for up to 2.5 hr following evaporation of the halothane anesthetic. The amino acid concentrations in perfusate plasma averaged 35% less than those in plasma from fasted dogs, probably because of dilution of the plasma with dextran. Perfusion of the brain with this medium caused a significant efflux of glutamic acid, glycerophosphoethanolamine, glycine, methionine, 3-methylhistidine, phosphoserine, taurine, threonine, and tyrosine and significant uptake of leucine and isoleucine (Table X). The concentrations of aspartic acid, glutamic acid, glycerophosphoethanolamine, methionine, phosphoethanolamine, phosphoserine, and taurine in cerebral cortex decreased during perfusion.

In a companion paper, Drewes and associates[105] reported a significant uptake of histidine and lysine by brain during 30 min of anoxic perfusion (Pa_{CO_2} 10 mm Hg) (Table XI); nonetheless, nitrogen balance appeared to be maintained. Following 30 min of anoxic perfusion, there were an increased brain concentration of five essential amino acids (methionine, histidine, leucine, lysine, and valine) and decreased levels of two essential amino acids (threonine and phenylalanine) (Table XII). The total brain tissue pool of essential amino acids continued to increase during aerobic perfusion and was nearly doubled following 120 min of reoxygenation. With respect to nonessential amino acids and related compounds, brain taurine, γ-aminobutyric acid, alanine, glycerophosphoethanolamine, and phosphoethanolamine increased during anoxic perfusion, whereas aspartic acid concentrations declined (Table XIII). The combined concentrations of asparagine and glutamine, serine, α-aminobutyric acid, and cystathionine also increased during postanoxic perfusion. Only taurine and phenylalanine concentrations returned toward normal. These findings suggest that proteolysis continues to occur during reoxygenation of the anoxic brain.

7.2.2. Unidirectional Leucine Flux

The unidirectional uptake of leucine was studied by Betz *et al.*[101] using an indicator dilution method. The arterial leucine concentration was varied in

Table XI
Whole Blood Amino Acid Concentrations and Amino Acid Flux During Anoxia[a]

Amino acid	Arterial amino acid concentrations (μmol/100 ml blood)	Net amino acid uptake (+) or efflux (−) (μmol/100 g brain per min)
Alanine	60.10 ± 6.29	+0.69 ± 0.43
α-aminobutyric acid	2.69 ± 0.37	+0.02 ± 0.05
Ammonia	25.80 ± 3.44	+1.02 ± 0.45*
Arginine	8.65 ± 0.86	+0.10 ± 0.12
Asparagine	21.75 ± 3.05	+0.01 ± 0.45
Aspartic acid	3.29 ± 0.44	−0.19 ± 0.16
Carnosine	0.83 ± 0.17	+0.03 ± 0.02
Citrulline	5.66 ± 0.26	+0.09 ± 0.09
Glutamic acid	16.56 ± 1.27	−0.34 ± 0.27*
Glutamine	3.62 ± 0.37	+0.10 ± 0.08
Glycerophosphoethanolamine	2.65 ± 0.34	+0.06 ± 0.12
Glycine	19.89 ± 1.55	+0.22 ± 0.19
Histidine	6.00 ± 0.83	+0.22 ± 0.10*
Isoleucine	3.82 ± 0.35	+0.04 ± 0.03
Leucine	7.95 ± 0.84	+0.04 ± 0.05
Lysine	18.65 ± 1.48	+0.50 ± 0.22*
Methionine	3.78 ± 0.29	+0.03 ± 0.06
3-Methylhistidine	1.07 ± 0.08	+0.02 ± 0.03*
Ornithine	3.78 ± 1.27	−0.02 ± 0.13
Phenylalanine	3.89 ± 0.50	+0.02 ± 0.03
Phosphoserine	6.37 ± 1.32	+0.14 ± 0.13*
Proline	14.08 ± 1.89	+0.20 ± 0.18
Serine	12.16 ± 0.99	−0.12 ± 0.18
Taurine	13.46 ± 1.09	−0.02 ± 0.26
Threonine	19.65 ± 1.21	−0.11 ± 0.26
Tyrosine	2.64 ± 0.24	+0.02 ± 0.03
Urea	385 ± 67	+0.92 ± 5.06
Valine	13.39 ± 1.13	+0.20 ± 0.12

[a] Values are means ± S.E. for five dog brain preparations. The data of 50 A–V samples from five different perfusion experiments were subjected to analysis of variance, and the F ratio computed. No animal-to-animal variation was detected in the value of $(A–V) F/W$. Therefore, net uptake or efflux is the mean of all observations.
* $P < 0.05$ when compared to zero.

increments by adding unlabeled leucine to the blood. Valine and isoleucine concentrations were varied independently of leucine to permit evaluation of their effect on leucine transport. A preliminary analysis indicated that both valine and isoleucine were competitive inhibitors. Therefore, all data were fitted to a form of the Michaelis–Menten equation that describes transport kinetics in the presence of two competitive inhibitors. The apparent kinetic constants for leucine transport in the absence of isoleucine and valine were $K_m = 1.58$ mM ± 0.28 S.E. and $V_{max} = 0.323 ± 0.035$ S.E. μmol/g per min. The apparent K_i for inhibition of leucine transport was 1.76 mM ± 0.34 S.E. for valine and 0.73 mM ± 0.14 S.E. for isoleucine. While the leucine level was held constant, indicator dilution injections were made at 1, 5, and 10 min after the start of perfusion with anoxic blood. Unlike glucose transport, the rate of leucine transport was unaffected by anoxia.

Table XII

Effect of Anoxia and Reoxygenation on Concentration of Essential Amino Acids in Cerebral Cortex[a]

Amino acid	Time of anoxia (min)		Time after anoxia (min)
	0 (28)	30 (4)	15 (4)
Methionine	0.065 ± 0.008	0.144 ± 0.020*	0.098 ± 0.013
Arginine	0.178 ± 0.014	0.210 ± 0.008	0.173 ± 0.014
Threonine	0.589 ± 0.053	0.362 ± 0.045*	0.443 ± 0.127
Histidine	0.161 ± 0.015	0.211 ± 0.021	0.279 ± 0.014*
Isoleucine	0.094 ± 0.006	0.172 ± 0.027*	0.130 ± 0.016
Leucine	0.219 ± 0.012	0.304 ± 0.024*	0.287 ± 0.044
Lysine	0.249 ± 0.022	0.326 ± 0.007*	0.421 ± 0.047
Valine	0.244 ± 0.019	0.364 ± 0.034*	0.321 ± 0.030
Phenylalanine	0.222 ± 0.009	0.128 ± 0.010*	0.121 ± 0.012

[a] The values indicated are the amino acid concentrations in μmol/ml H_2O ± S.E. Numbers in parentheses are numbers of observations. *$P < 0.05$ compared to preanoxic control levels.

Table XIII

Effect of Anoxia and Reoxygenation on Concentration of Nonessential Amino Acid Related Compounds in Cerebral Cortex[a]

Compound	Time during anoxia (min)		Level 15 min after 30 min anoxia (4)
	0 (28)	30 (4)	
Taurine	1.414 ± 0.094	1.693 ± 0.030*	1.388 ± 0.133*
GABA	0.620 ± 0.061	1.057 ± 0.140*	0.875 ± 0.087
Aspartic acid	1.654 ± 0.136	1.217 ± 0.053*	1.359 ± 0.122
Asparagine and glutamine	6.454 ± 0.280	7.307 ± 0.459	7.228 ± 0.336
Alanine	1.489 ± 0.242	2.077 ± 0.104*	2.471 ± 0.111*
Glutamic acid	10.151 ± 0.314	10.009 ± 0.333	9.165 ± 0.444
Glycine	0.833 ± 0.041	0.836 ± 0.047	0.986 ± 0.026†
Tyrosine	0.061 ± 0.005	0.074 ± 0.008	0.084 ± 0.010
Serine	0.615 ± 0.045	0.541 ± 0.020	0.529 ± 0.036
α-Aminobutyric acid	0.102 ± 0.012	0.100 ± 0.008	0.094 ± 0.022
Ammonia	0.956 ± 0.021	1.043 ± 0.040	1.048 ± 0.089
Glycerophosphoethanolamine	0.215 ± 0.033	0.484 ± 0.040*	0.398 ± 0.050
Ornithine[b]	0.408 ± 0.061		
Phosphoserine	0.120 ± 0.014	0.082 ± 0.015	0.072 ± 0.005
Phosphoethanolamine	0.728 ± 0.086	1.751 ± 0.217*	1.132 ± 0.052†
Urea	3.355 ± 0.322	3.659 ± 0.575	9.945 ± 1.870†
Cystathionine	0.228 ± 0.024	0.273 ± 0.008	0.256 ± 0.022

[a] The values indicated are the amino acid concentrations in μmol/ml H_2O ± S.E. Numbers in parentheses are numbers of observations. *$P < 0.05$ compared to preanoxic control levels. †$P < 0.05$ compared to preoxygenation (30 min anoxic) levels.

[b] Determined only in control samples.

7.3. Unidirectional Flux of Other Compounds

Unidirectional flux of NE was studied by Saffitz *et al.*[86] using the isolated canine brain and was found not to be saturable when blood NE concentration was varied from 0 to 1200 μg/ml. It was concluded that approximately 7.3% of the blood NE normally crosses the blood–brain barrier by simple diffusion.

The transport of 13 different blood-borne substances was investigated by Drewes and Gilboe[106] at normal and elevated solute concentrations using the indicator dilution technique. Saturable processes were observed for the unidirectional transport of tyrosine, tryptophan, and L-DOPA but not for dopamine. Free palmitic acid was transported, whereas free oleic acid was not. The purines adenine, adenosine, and guanosine together with the pyrimidine thymine were transported into the brain, but the purines guanine and hypoxanthine were not. Neither of the two vitamins folic acid and cyanocobalamine exhibited evidence of transport by carrier-mediated processes.

ACKNOWLEDGMENTS. The author gratefully acknowledges the support of Grant NS-05961 during the preparation of this manuscript and the expert assistance of Mrs. Donna Brackett in typing and assembling the manuscript.

REFERENCES

1. Brown-Sequard, C. E., 1858, *J. Physiol. Homme Anim.* **95**:353–367.
2. deSomer, E., and Heymans, J. F., 1912, *J. Physiol. Pathol Gen.* **14**:1138–1142.
3. Gilboe, D. D., Cotanch, W. W., and Glover, M. B., 1964, *Nature* **202**:399–400.
4. Thompson, A. M., Robertson, R. C., and Bauer, T. A., 1968, *J. Appl. Physiol.* **24**:407–411.
5. Schmidt, C. F., 1928, *Am. J. Physiol.* **84**:202–222.
6. Moss, G., 1964, *J. Surg. Res.* **4**:170–177.
7. Chute, A. L., and Smyth, D. H., 1939, *Q. J. Exp. Physiol.* **29**:379–394.
8. Geiger, A., and Magnes, J., 1947, *Am. J. Physiol.* **149**:517–537.
9. Andjus, R. K., Suhara, K., Sloviter, H. A., 1967, *J. Appl. Physiol.* **22**:1033–1039.
10. Michenfelder, J. D., Messick, J. M., Jr., and Theye, R. A., 1968, *J. Surg. Res.* **8**:475–481.
11. Suda, I., Adachi, C., and Kito, K., 1963, *Kobe J. Med. Sci.* **9**:41–67.
12. White, R. J., Albin, M. S., and Verdura, J., 1963, *Science* **141**:1060–1061.
13. Bouckaert, J. J., and Jourdan, F., 1936, *Arch. Int. Pharmacodyn. Ther.* **53**:523–539.
14. Gilboe, D. D., Cotanch, W. W., and Glover, M. B., 1965, *Nature* **206**:94–96.
15. Blomquist, A. J., and Gilboe, D. D., 1970, *Nature* **227**:409.
16. Fitzpatrick, J. H., Jr., and Gilboe, D. D., 1982, *Israel J. Med. Sci.* **18**:67–73.
17. Fitzpatrick, J. H., Jr., Gilboe, D. D., Drewes, L. R., and Betz, A. L., 1976, *Am. J. Physiol.* **231**:1840–1846.
18. Gilboe, D. D., Andrews, R. L., and Dardenne, G., 1970, *Am. J. physiol.* **219**:767–773.
19. Gilboe, D. D., Cotanch, W. W., Glover, M. B., and Levin, V., 1967, *Am. J. Physiol.* **212**:589–594.
20. Allweis, C., Landau, T., Abeles, M., and Magnes, J., 1966, *J. Neurochem.* **13**:795–804.
21. Geiger, A., Magnes, J., and Geiger, R. S., 1952, *Nature* **170**:754–755.
22. Geiger, A., Magnes, J., Taylor, R. M., and Verralli, M., 1954, *Am. J. Physiol.* **177**:138–149.
23. Gilboe, D. D., and Betz, A. L., 1973, *Am. J. Physiol.* **224**:588–595.
24. Krieglstein, G., Krieglstein, J., and Stock, R., 1972, *Naunyn Schmiedebergs Arch. Pharmacol.* **275**:124–134.
25. Geiger, A., and Yamasaki, S., 1956, *J. Neurochem.* **1**:93–100.
26. Allweis, C., and Magnes, J., 1958, *J. Neurochem.* **2**:326–336.

27. Krieglstein, G., Krieglstein, J., and Urban, W., 1972, *J. Neurochem.* **19**:885–886.
28. Sloviter, H. A., and Kamimoto, T., 1967, *Nature* **216**:458–460.
29. Dirks, B., Krieglstein, J., Lind, H. H., Kleger, H., and Schutz, H., 1980, *J. Pharmacol. Methods* **4**:95–108.
30. Drewes, L. R., Conway, W. P., and Gilboe, D. D., 1977, *Am. J. Physiol.* **233**:E320–E325.
31. Kintner, D., Costello, D. J., Levin, A. B., and Gilboe, D. D., 1980, *Am. J. Physiol.* **239**:E501–E509.
32. Michenfelder, J. D., and Theye, R. A., 1968, *Anesthesiology* **29**:1107–1112.
33. Kintner, D., Fitzpatrick, J. H., Jr., Louie, J. A., and Gilboe, D. D., 1982, *Am. J. Physiol.* (in press).
34. Magnes, J., Allweis, C., and Abeles, M., 1967, *J. Neurochem.* **14**:859–871.
35. Michenfelder, J. D., and Theye, R. A., 1970, *Anesthesiology* **33**:430–439.
36. Hostetler, K. Y., Landau, B. R., White, F. J., Albin, M. S., and Yashon, D., 1970, *J. Neurochem.* **17**:33–39.
37. Geiger, A., and Magnes, J., 1947, *Am. J. Physiol.* **149**:517–537.
38. Sloviter, H. A., and Kamimoto, T., 1967, *Nature* **216**:458–460.
39. Watanabe, S., Kono, S., Nakashima, Y., Mitsunobu, K., and Otsuki, S., 1975, *Folia Psychiatr. Neurol. Jpn.* **29**:67–76.
40. Michenfelder, J. D., and Theye, R. A., 1969, *Anesthesiology* **31**:449–457.
41. Watanabe, S., Otsuki, S., Mitsunobu, K., Sannomiya, T., and Okumura, N., 1970, *J. Neurochem.* **17**:1571–1577.
42. Zivin, J. A., and Snarr, J. F., 1972, *J. Appl. Physiol.* **32**:664–668.
43. Gercken, G., and Roth, E., 1961, *Pfluegers Arch.* **273**:589–603.
44. Drewes, L. R., and Gilboe, D. D., 1973, *J. Biol. Chem.* **248**:2489–2496.
45. Drewes, L. R., and Gilboe, D. D., 1973, *Biochim. Biophys. Acta* **320**:701–707.
46. Drewes, L. R., Gilboe, D. D., and Betz, A. L., 1973, *Arch. Neurol.* **29**:385–390.
47. Krieglstein, J., and Stock, R., 1974, *Psychopharmacologia* **35**:169–177.
48. Krieglstein, J., and Stock, R., 1975, *Biochem. Pharmacol.* **24**:1579–1582.
49. Hein, H., Krieglstein, J., and Stock, R., 1975, *Naunyn Schmiedebergs Arch. Pharmacol.* **289**:399–407.
50. Krieglstein, J., and Stock, R., 1975, *Naunyn Schmiedebergs Arch. Pharmacol.* **290**:323–327.
51. Ghosh, A. K., Mukherji, B., and Sloviter, H. A., 1972, *J. Neurochem.* **19**:1279–1285.
52. Geiger, A., Horvath, N., and Kawakita, Y., 1960, *J. Neurochem.* **5**:311–322.
53. Geiger, A., Kawakita, Y., and Barkulis, S. S., 1960, *J. Neurochem.* **5**:323–338.
54. Barkulis, S. S., Geiger, A., Kawakita, Y., and Aguilar, U., 1960, *J. Neurochem.* **5**:339–348.
55. Margolis, R. U., Barkulis, S. S., and Geiger, A., 1960, *J. Neurochem.* **5**:379–382.
56. Otsuki, S., and Geiger, A., 1963, *J. Neurochem.* **10**:415–422.
57. Otsuki, S., and Watanabe, S., Ninomiya, K., Hoaki, T., and Okumura, N., 1968, *J. Neurochem.* **15**:859–865.
58. Watanabe, S., Otsuki, S., Mitsunobu, K., Sannomiya, T., and Okumura, N., 1970, *J. Neurochem.* **17**:1571–1577.
59. Hostetler, K. Y., Landau, B. R., White, R. J., Albin, M. S., and Yashon, D., 1970, *J. Neurochem.* **17**:33–39.
60. Moss, G., 1964, *Diabetes* **13**:585–591.
61. Otsuki, S., Geiger, A., and Gombos, G., 1963, *J. Neurochem.* **10**:397–404.
62. Gombos, G., Geiger, A., and Otsuki, S., 1963, *J. Neurochem.* **10**:405–413.
63. Drewes, L. R., and Gilboe, D. D., 1973, *Fourth International Meeting of the International Society for Neurochemistry: Abstracts* International Society for Neurochemistry, Tokyo, p. 456.
64. Flores, F., Naftolin, F., Ryan, K. J., and White, R. J., 1973, *Science* **180**:1074–1075.
65. Gilboe, D. D., Drewes, L. R., and Kintner, D., 1976, *Dynamics of Brain Edema* (H. M. Pappius and W. Feindel, eds.), pp. 228–235, Springer-Verlag Berlin, Heidelberg.
66. Krieglstein, J., and Stock, R., 1973, *Naunyn Schmiedebergs Arch. Pharmacol.* **277**:323–332.
67. Michenfelder, J. D., 1974, *Anesthesiology* **41**:231–236.
68. Hein, H., Krieglstein, J., and Stock, R., 1975, *Naunyn Schmiedebergs Arch. Pharmacol.* **289**:399–407.

69. Krieglstein, J., and Stock, R., 1975, *Naunyn Schmiedebergs Arch. Pharmacol.* **290**:323–327.
70. Krieglstein, J., Stock, R., and Rieger, H., 1973, *Naunyn Schmiedebergs Arch. Pharmacol.* **279**:243–254.
71. Gruner, J., Krieglstein, J., and Rieger, H., 1973, *Naunyn Schmiedebergs Arch. Pharmacol.* **277**:333–348.
72. Rieger, H., and Krieglstein, J., 1974, *Psychopharmacologia* **39**:163–179.
73. Krieglstein, J., Rieger, H., and Schutz, H., 1979. *Eur. J. Pharmacol.* **56**: 363–370.
74. Krieglstein, J., Rieger, H., and Schütz, H., 1980, *Biochem. Pharmacol.* **29**:63–67.
75. Michenfelder, J. D., and Theye, R. A., 1973, *Anesthesiology* **38**:123–127.
76. Michenfelder, J. D., and Cucchiara, R. F., 1974, *Anesthesiology* **40**:575–580.
77. Cucchiara, R. F., Theye, R. A., and Michenfelder, J. D., 1974, *Anesthesiology* **40**:571–574.
78. Michenfelder, J. D., and Theye, R. A., 1972, *Anesthesiology* **37**:32–39.
79. Michenfelder, J. D., and Theye, R. A., 1971, *Br. J. Anaesthesiol.* **43**:630–636.
80. Allweis, C., and Magnes, J., 1958, *Nature* **181**:626–627.
81. Woods, H. F., Green, H. R., Youdim, M. B. H., and Grahame-Smith, D. G., 1976, *Biochem. Soc. Trans.* **4**:22–26.
82. Betz, A. L., and Gilboe, D. D., 1973, *Am. J. Physiol.* **224**:580–587.
83. Geiger, A., and Sigg, E. B., 1955, *Trans. Am. Neurol. Assoc.* **80**:117–120.
84. Lowe, R. F., and Gilboe, D. D., 1971, *Stroke* **2**:193–200.
85. Zimmer, R., Lang, R., and Oberdorster, G., 1974, *Stroke* **5**:397–405.
86. Saffitz, M. S., Costello, D. J., and Gilboe, D. D., 1979, *J. Pharmacol. Exp. Ther.* **211**:13–19.
87. Lowe, R. F., and Gilboe, D. D., 1973, *Am. J. Physiol.* **225**:1333–1338.
88. Ingenito, A. J., Barrett, J. P., and Procita, L., 1971, *Stroke* **2**:67–75.
89. Otsuki, S., Watanabe, S., Edamatso, K., Nakashima, Y., Hoaki, T., Ninomiya, K., Mitsunobu, K., Sannomiya, T., and Okumura, N., 1968, *Folia Psychiatr. Neurol. Jpn.* **22**:227–231.
90. Benzi, G., Berté, F., Crema, A., and Frigo, G. M., 1967, *J. Pharm. Sci.* **56**:1349–1351.
91. Krieglstein, J., and Jähnchen, E., 1970, *Naunyn Schmiedebergs Arch. Pharmcol.* **266**:380–381.
92. Niemeyer, D. H., and Krieglstein, J., 1975, *J. Neurochem.* **24**:829–831.
93. Kilbinger, H., and Krieglstein, J., 1974, *Naunyn Schmiedebergs Arch. Pharmacol.* **285**:407–411.
94. Ingenito, A. J., Barrett, J. P., and Procita, L., 1969, *J. Pharmacol. Exp. Ther.* **170**:210–220.
95. Gilboe, D. D., and Betz, A. L., 1970, *Am. J. Physiol.* **219**:774–778.
96. Betz, A. L., Gilboe, D. D., Yudilevich, D. L., and Drewes, L. R., 1973, *Am. J. Physiol.* **225**:586–592.
97. Betz, A. L., and Gilboe, D. D., 1974, *Brain Res.* **65**:368–372.
98. Betz, A. L., Drewes, L. R., and Gilboe, D. D., 1975, *Biochim. Biophys. Acta* **406**:505–515.
99. Drewes, L. R., Horton, R. W., Betz, A. L., and Gilboe, D. D., 1977, *Biochim. Biophys. Acta* **471**:477–486.
100. Betz, A. L., Gilboe, D. D., and Drewes, L. R., 1974, *Brain Res.* **67**:307–316.
101. Betz, A. L., Gilboe, D. D., and Drewes, L. R., 1975, *Am. J. Physiol.* **228**:895–899.
102. Betz, A. L., Gilboe, D. D., and Drewes, L. R., 1975, *Biochim. Biophys. Acta* **401**:416–428.
103. Christensen, O., Lund-Andersen, H., Betz, A. L., and Gilboe, D. D., 1982, *Acta Physiol. Scand.* (in press).
104. Yanushka, J. M., Costello, D. J., and Gilboe, D. D., 1982, *Am. J. Physiol.* (in Press).
105. Drewes, L. R., Conway, W. P., and Gilboe, D. D., 1977, *Am. J. Physiol.* **233**(4):E326–E330.
106. Drewes, L. R., and Gilboe, D. D., 1977, *Fed. Proc.* **36**:166–170.

15

Principles and Application of Positron Emission Tomography in Neuroscience

Jonathan D. Brodie, Nora Volkow, and John Rotrosen

1. INTRODUCTION

Positron Emission Tomography (PET) is a new nuclear medicine technique which evolved from the need to establish the relationship among the functional activity of the brain, its metabolic rate, and its anatomic structure. This technique allows the detection of the three-dimensional distribution within the brain of a previously administered positron-emitting radionuclide. Although similar in principle to quantitative autoradiography, it has the advantage of providing functional information of a structure in a living organism. Positron emission tomography incorporates principles from both transmission computed tomography and the uptake and flow techniques initially developed by Kety and Schmidt[1] but has the advantage of being able to give information for localized regions of the brain. This information is based on the detection of the annihilation energy produced when a positron emitted from a radionuclide-labeled compound interacts with an electron in the tissue.

The four basic radionuclides utilized have been ^{11}C, ^{13}N, ^{15}O, and ^{18}F, with half-lives of 20.4, 9.96, 2.07, and 110 min, respectively.[2] These radionuclides are isotopes of elements that are natural constituents of molecules of biological significance (fluorine can be used as a substitute for hydrogen).

2. METHODOLOGY

2.1. Preparation of the Radionuclide

Positron-emitting radionuclides are unstable isotopes (their nuclei are deficient in neutrons) that stabilize by liberating positrons. Their synthesis re-

Jonathan D. Brodie and Nora Volkow • Department of Psychiatry, New York University School of Medicine, New York, New York 10016. *John Rotrosen* • Psychiatry Service, Veterans Administration Medical Center, New York, New York 10010.

quires a cyclotron to accelerate protons or deuterons to bombard a suitable target material, thereby generating a positron-emitting isotope. There are several nuclear reactions for the production of a particular radionuclide. The stable precursor is a different atom than the isotope: ^{11}C is usually prepared by irradiating nitrogen or boron, ^{18}F is produced using $H_2^{18}O$ as a target, ^{15}O by irradiating nitrogen, and ^{13}N by irradiating either oxygen or carbon.[2] These procedures allow the preparation of radioactive isotopes relatively free of contamination from the stable isotope of the same element. This can be critically important in studies requiring a tracer of very high specific activity, where the number of molecules necessary for appropriate imaging approximates the number of receptors available for binding.

The short half-lives of these radionuclides require the existence of a cyclotron in the immediate environment. Although the complexity of the first cyclotrons delayed the development of the PET technique, the increasing availability of dedicated "baby" cyclotrons with simple operating modes has made the technology feasible.

The wider availability of positron-emitting nuclides coupled to the fact that many of the metabolic processes in the brain are relatively fast with respect to the useful life of these radionuclides makes their short half-lives advantageous. It minimizes the dose of radiation exposure to the subject, and it permits repeated studies in the same subject within a short period of time.

2.2. Detection of the Radioactivity

Once in the brain, a labeled compound will differentially concentrate as a function of uptake, flow, and binding. The amount of annihilation energy liberated in a particular region of the brain will be a scalar function of the amount of radionuclide in the region.

The isotope emits a positron which loses its kinetic energy after having traveled for a certain distance (mean free path 1–6 mm) and collides with an electron, yielding two coincident high-energy photons (511 keV) which travel in antiparallel directions.[3] These high-energy particles have great tissue penetration and can be detected by coincident detectors placed outside the region of interest (Fig. 1). To simultaneously cover a complete transverse section of the brain, detectors are arranged in a circumferential array, and each radiation detector is set coincident with multiple opposing detectors. The number of detectors may vary, and these can be rotated around the imaged object to give the maximum coverage of angles. There are now PET devices that can achieve angular and linear sampling without mechanical motion.[4]

Present systems cannot measure the time difference between the inception and the detection of coincident annihilation photons (time of flight) in a volume element as small as a brain. This time difference is a function of the distances separating the annihilation event from the detector, and its measurement would permit the localization of the event along a coincidence line. To establish the coincidence line, a resolving time is set within which the detection of two events is considered to be coincident.[5] This circuit will only register coincident annihilation photons that are produced within the volume of space limited by the straight lines that join them to a detector pair. Photons that are detected

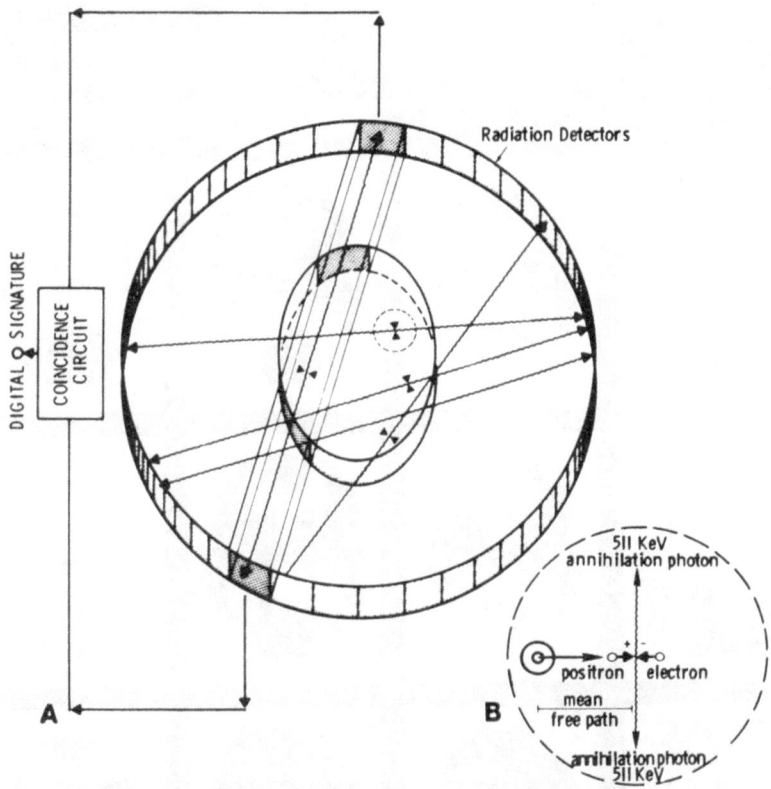

Fig. 1. A. Schematic representation of the arrangement of detectors in PET. The detectors are arranged so that only simultaneous events 180° apart are recorded. To increase the number of coincidence lines, each detector is made coincident with multiple detectors. A coincidence circuit between pairs of detectors records an event only if the events occurred within the resolving time of the instrument. B. The isotope decays by the emission of positrons (+) which travel for a certain distance (mean free path) before interaction with electrons (−). The two particles anihilate each other, and their mass is converted to two anihilation photons traveling at 180° from each other. (Adapted from ref. 11.)

in a noncoincident manner are excluded by the coincidence circuitry and are not registered.

The cross section of this volume element (a function of the aperture of the detectors) and the mean free path of the positron will determine the spatial resolution of the instrument.[5] The spatial resolution of today's PET instruments (8–10 mm) is marginally able to resolve the major structures of the brain.[7] The size of the voxels (volume of tissue discriminated as separate by the instrument) exceeds the volume of most neuroanatomic structures. The current voxel values of tomographic instruments range from values of 2.0 × 2.0 × 2.0 cm to values of 0.5 × 0.5 × 1 cm.[7] However, a PET image can be correlated with the information from a CT scan to help identify the substructures of the brain (Fig. 2).

Different sections of the brain can be obtained sequentially by moving the head with respect to the detectors or simultaneously by arranging the detectors so that they surround the head (Fig. 3). The length of time required to obtain

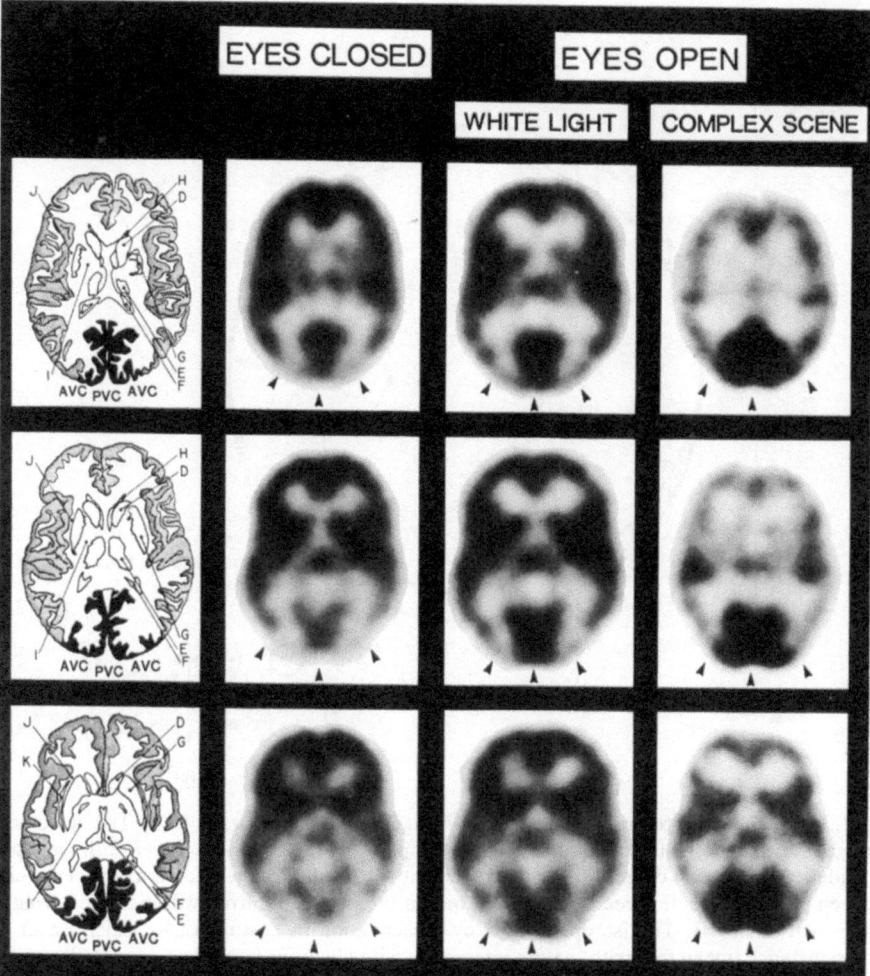

Fig. 2.   Metabolic study with FDG showing response to the primary (PVC) and associative (AVC) visual cortex from eyes-closed control to white light stimulation (570 lux) and to a complex visual scene of a park. The control and the white light stimulation are the same subject, whereas the complex stimulation was a different subject (levels from the third subject did not exactly correspond to the images in columns 2 and 3 and the sketches). Note an increasing metabolic or functional activity of the visual cortex from the quiescent state of eyes closed to increasing complexity of visual stimulation. Also note that in each eyes-closed study, one can clearly see the delineation of posterior parietal cortex from the Brodman area 19 of the associative visual cortex. (Figure and legend courtesy of Phelps, see ref. 21.)

a clear tomographic image of the brain is determined by the time it takes to collect sufficient counts for an adequate reconstruction of the image; PET units currently in operation can collect data for up to seven slices in a few seconds if sufficient radioactivity is administered.

The photon detectors are scintillation crystals; the three most commonly used are cesium fluoride, bismuth germanate, and activated sodium iodide. When these crystals are activated by an annihilation photon, they liberate

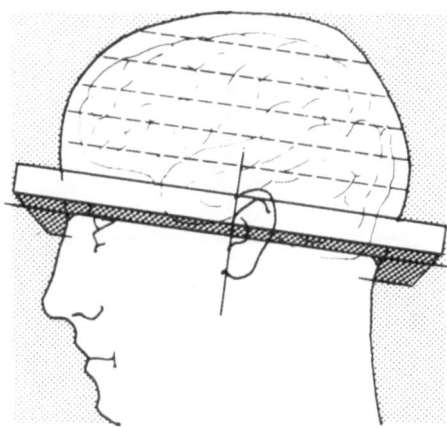

Fig. 3.   Transverse scans parallel to the cantho-meatal line (CM) are taken for the different levels of the brain.

visible photons that decay in a short period of time. Using a scintillation crystal with short decay constant enhances the resolving time of the detector and reduces the time needed for scanning. This diminishes the probability of registering as true events two photons that arrive within the resolving time of the two coincident detectors but from different annihilation sources (random coincidence).[8] Cesium fluoride has a very short decay constant and is therefore used in the new, faster machines[9] which have virtually eliminated random coincidence as a source of noise. The light generated by the scintillator after its interaction with the annihilation photon passes through a photomultiplier tube where it is converted into an electronic signal. This signal is recognized and clocked by a timing coincidence circuit. The ionization event is thus translated into a digital pulse that passes to a fast computer with a large storage capacity.

2.3. Reconstruction Process

The digital pulses are the raw data from which the computer reconstructs the image of the radionuclide distribution in the brain. The acquisition system

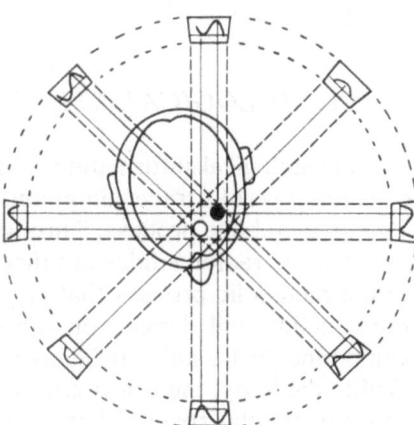

Fig. 4.   Profiles of the radioactivity of 2 points detected by circular array of detectors. (Adapted from ref. 11.)

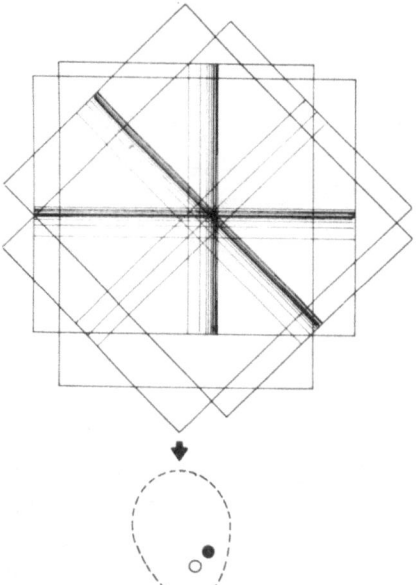

Fig. 5. Backprojection of the profiles and deblurring of the image.

provides the information of a set of projections from different angles around the object (Fig. 4). Each projection represents the sum of radioactivity within the volume of tissue sensed by coincident detectors. The accuracy of the reconstruction will depend on the number of projections detected. If an infinite number of projections could be obtained, it would be possible to obtain an exact reconstruction of the object within the resolution of the instrument.

The circumferential array of detectors provides for each plane a series of straight lines of different angles. These lines are recorded as profiles and are filtered by a mathematical process called convolution to correct for artifactual imagery.[10] The convoluted profiles are then put back through the image space, which gives a superposition of the different projections from the different angles. The array of profiles is added to obtain the image (Fig. 5). The image is then corrected for the attenuation suffered when the photons transversed the skull.[10]

3. PHYSIOLOGICAL STUDIES WITH PET

Physiological information is derived from the tomograph by relating the concentration of a trace compound in a particular site and its significance within a given metabolic process. Through a series of different biological or chemical procedures, radionuclides are incorporated into the trace compound pertinent to the metabolic process that is going to be investigated.[11] The choice of the radionuclide will depend on the characteristics that are needed for a trace compound to be able to behave in a similar way to the natural substance. Within the body, the tracer compound will distribute between the extracellular and intracellular space. Therefore, sequential measures of radioactivity in the peripheral blood are determined throughout the scanning to correct for detected radioactivity that is caused by unmetabolized tracer in the blood compartment of the volume elements. In order to register the activity from the metabolized tracer, it must remain within the area of interest during the measurement period. For this reason, and for limitations of PET technology, studies performed until recently have been limited to physiological phenomena varying slowly as a function of time such as accumulation or equilibrium processes.

Brain glucose metabolism and brain blood volume are well suited for PET measurements and are the most commonly studied. Other parameters including blood flow[12] and oxygen consumption,[12] receptor binding,[13,14] amino acid synthesis,[51] nucleotide synthesis,[16] and lipid turnover[17] are now being investigated in different laboratories.

3.1. Glucose Metabolism

Since, under most circumstances, the brain derives most of its energy from glucose, the measurement of glucose metabolism would reflect cerebral functioning. However, the rate of glucose uptake within one region of the brain is not a measure of any particular function but of the overall activity within that area. Thus, even a large change in any particular function may appear as a subtle variation in the overall glucose metabolic rate. In controlled experiments, PET appears to be sensitive enough to detect the regional differences in glucose metabolism produced by the activation of particular functions.

Two strategies have been utilized to study glucose metabolism: the use of radiolabeled glucose ([^{11}C]glucose) and the use of such radiolabeled glucose analogues as [^{18}F]deoxyglucose (^{18}F-DG) and [^{11}C]deoxyglucose (^{11}C-DG). ^{11}C-Labeled natural glucose is biochemically identical to unlabeled glucose, and corrections need not be made for differences in transport properties and enzyme affinities. As glucose, it crosses the cell membranes through carrier-mediated facilitated diffusion, and within the cell it is phosphorylated to [^{11}C]-glucose-6-phosphate which has a very low membrane permeability. The de-

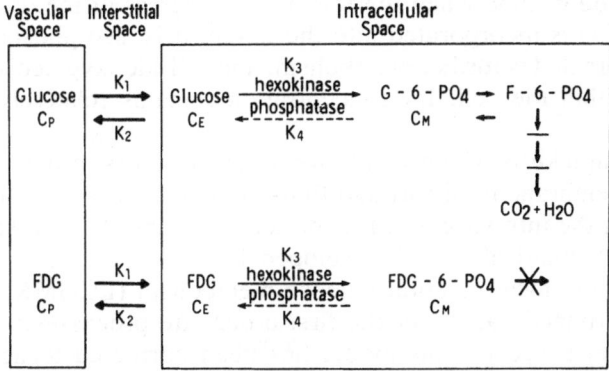

Fig. 6. Simplified schematic representation of cerebral utilization of glucose and 2-fluoro-2-deoxy-D-glucose (FDG). For abbreviations, see Fig. 7.

$$\text{LCMRglu} = \frac{C_p \left(C_i^* (T) - \dfrac{k_1^*}{\alpha_2 - \alpha_1} \cdot \{(k_4^* - \alpha_1)\, e^{-\alpha_1 t} + (\alpha_2 - k_4^*)\, e^{-\alpha_2 t}\} \circledast C_p^* (t)\right)}{LC \left(\dfrac{k_2^* + k_3^*}{\alpha_2 - \alpha_1}\right)(e^{-\alpha_1 t} - e^{-\alpha_2 t}) \circledast C_p^* (t)}$$

$$LC = \frac{\lambda V_m^* K_m}{\phi V_m K_m^*}$$

LCMRglu	Local cerebral metabolic rate of glucose
*	Superscript asterisk denotes symbols that apply to FDG; symbols without asterisk apply to glucose
$\circledast$	Denotes operation of convolution
k_1^* k_2^*	First-order rate constants for FDG forward and reverse capillary membrane transport, respectively (Fig. 6)
k_3^* k_4^*	First-order rate constants for phosphorylation of FDG and dephosphorylation of FDG-6-PO$_4$, respectively (Fig. 6)
LC	Lumped constant: ratio of arteriovenous extraction fraction of FDG to that of glucose under steady-state conditions and when k_4^* is small
λ	Ratio of distribution volume of FDG to that of glucose
K_m	Michaelis Menten constant for phosphorylation of glucose (K_m^* is for FDG)
V_m	Maximum velocity of phosphorylation of glucose (V_m^* is for FDG)
ϕ	Ratio of the difference between the rates of phosphorylation and dephosphorylation of glucose to its rate of phosphorylation
C_i (T)	Cerebral tissue concentration of FDG plus FDG-6-PO$_4$ in region i at a single time (T)
C_p	Capillary plasma glucose concentration (steady state)
C_p^* (t)	Capillary plasma FDG concentration as a function of time
α_1, α_2	Rate constants for the model response to an impulse change in FDG capillary plasma concentration

Fig. 7. Mathematical model for calculation of local cerebral metabolic rate of glucose (see ref. 18).

phosphorylation and oxidative decarboxylation are very low, and for at least the first few seconds the egress of the tracer from the brain is minimal. Therefore, [^{11}C]glucose is a useful tracer for studies of glucose uptake. Glucose-6-phosphate is further metabolized through glycolosis to CO$_2$ (Fig. 6). The measurement of the spatial distribution of [^{11}C]glucose must be completed within the short period of time when there is minimal egress of the tracer.

When FDG is incorporated into the tissue, it is also phosphorylated, but its structure precludes further metabolism, and [^{18}F]deoxyglucose-6-phosphate is trapped within the cell. Its concentration remains relatively constant for about 1 hr.

Cellular uptake of glucose follows a series of first-order rate processes (first-order membrane transport and first-order Michaelis–Menten enzyme kinetics) among the intravascular, the intracellular, and the interstitial compartments. A mathematical model developed by Sokoloff[18] (Fig. 7) allows the estimation of the local cerebral glucose metabolism (LCGMR) using experimentally determined values for the first-order rate processes.[19,20] The operational equation based on this model has been corrected to account for the differences between tracer and tracee (glucose) that occur when radiolabeled analogues are used. It can also be corrected to account for the dephosphorylation reaction.[20]

The use of the average rate constants is based on the assumption of a steady state for glucose consumption. This steady state is achieved about 30 min after the injection of glucose. When FDG is used as a tracer, this problem can be obviated by delaying counting until steady state is attained. With [^{11}C]-glucose, measurement must be made during the presteady state, when the model is sensitive to the exact values of the rate constants. Therefore, sequential measurements of blood must be obtained to calculate the absolute glucose concentration and specific activity rate constants. The method has been replicated both with [^{11}C]glucose and ^{18}F-DG by several authors[19-23] who have obtained consistent results similar to those obtained with the Kety–Schmidt technique.[24] Average rate constants have been obtained from a group of normal subjects, and these values may be affected in pathological states. This must be borne in mind when patients with brain diseases are studied. The measurement of these constants under different abnormal conditions will enable an accurate extrapolation of the model to neuropathology.

3.2. Oxygen Metabolism

Oxygen consumption gives a measure of the rate of aerobic glycolosis and for this reason has also been used to assess metabolism. Brain tissue oxygen consumption cannot be measured utilizing the strategy for glucose metabolism because the positron-emitting radionuclide ^{15}O is promptly converted to [^{15}O]-labeled water which egresses rapidly from the tissue. The technique that has been utilized with PET is the one originally described by Jones and others.[12,25-27] The method involves the continuous inhalation of ^{15}O to label the subject's blood as [^{15}O]oxyhemoglobin. The ^{15}O is liberated by the erythrocyte and, within the tissue, enters the respiratory chain where it is reduced and leaves the cytochrome system as $H_2^{15}O$ of metabolism. The water of metabolism is the endproduct of respiration, and its measurement is used to assess oxygen consumption. The counting of $H_2^{15}O$ radioactivity is delayed until a steady state of radioactivity is reached. This steady state represents a balance between the formation of $H_2^{15}O$ of metabolism through utilization of ^{15}O and the disappearance of the respiration product because of radioactive decay and tissue washout by flow. However, the interpretation of the data is complicated by the recirculation of water of metabolism. To correct for this, a sequential procedure that involves inhalation of ^{15}O-labeled CO_2 to equilibrium is performed. Inhalation of $CO^{15}O$ labels the circulating water pool in blood because of the presence of carbonic anhydrase in the red cell which exchanges O_2 in blood water with $^{15}O_2$ from $CO^{15}O$. The resulting brain image simulates the recirculation of metabolic water to the brain. The regional oxygen fractional extraction can then be computed by dividing the steady-state image of metabolic water by the image of circulating water.

Within the normal range of blood perfusion, the $H_2^{15}O$ production is linearly dependent on the oxygen extraction rate by the tissue. When perfusion rate diminishes, the $H_2^{15}O$ production has a linear dependence on blood flow and oxygen extraction.[28] These relationships must be considered when changes in oxygen extraction are interpreted in physiological states in which perfusion may be deranged.

3.3. Brain Hemodynamics

3.3.1. Cerebral Blood Volume

The quantitative image of the distribution of a vascular tracer was one of the first measurements accomplished with PET. Cerebral blood volume (CBV) can be satisfactorily quantitated by labeling the red blood cells with [^{11}C]carbon monoxide.[29] The regional cerebral blood volume is determined by comparing the equilibrium tomography image of [^{11}C]carboxyhemoglobin activity in the brain with the amount of [^{11}C]Hb in the total blood pool as determined from samples of venous blood taken concurrently with the emission scan. The ability to measure regional cerebral blood volume provides the means to determine differential concentration of a substance in the vascular and in the extravascular compartment.

3.3.2. Cerebral Blood Flow

Accurate quantitative measurement of cerebral blood flow has not yet been accomplished with present emission tomographic systems. The reason has been that the techniques for the measurement of blood flow depend on the analysis of dynamic tracer data.[32] Until recently, PET instruments did not have the capability of collecting sufficient data rapidly enough to make a dynamic tracer measurement. The multislice technique, the use of cesium fluoride as a detector, and the incorporation of the time of flight information into the reconstruction process are different designs being used among others to diminish the scanning time of the instrument. Positron emission tomography has been applied to fast dynamic studies of the brain and of the heart.[30,33] The measurement of CBF has focused on techniques that insure equilibrium imaging. One approach has been the measurement of the equilibrium image of ^{15}O-labeled water obtained by the continuous inhalation of ^{15}O-labeled carbon dioxide.[26] The equilibrium activity of ^{15}O-labeled water in a region of brain is a function of the blood flow of that region. The computation is very sensitive to uncertanties in the brain tracer concentration, and the probability of error is high.

Another approach has been the tomographic imaging of ^{13}N-labeled ammonia (^{13}NH$_3$) at equilibrium.[34] Ammonia has been used for its ability to be efficiently trapped by the tissue as it passes through the microcirculation. The pH of both the blood and the interstitial fluid affects ^{13}NH$_3$ uptake. The effects of cerebral pathology on ammonia uptake are not known, and the technique must be interpreted with these uncertainties in mind.

3.3.3. Blood–Brain Barrier

With an intact blood–brain barrier, gallium-68 remains in the vascular space. If there is a breakdown in the blood–brain barrier, ^{68}Ga will leak into the tissue. The positron-emitting [^{68}Ga]EDTA (ethylene diamine tetraacetic acid) can be used for tomographic identification of cerebral lesions with blood–brain barrier defects.[30,31]

3.4. Brain Chemical Composition

Once CBV is established, PET can provide quantitative measurement of brain-to-blood partition coefficients for various substrates. It can be used to estimate the chemical composition of tissues by selecting appropriately labeled compounds that rapidly distribute between the blood and the tissue of interest. This strategy has been utilized to determine brain carbon dioxide content using ^{11}C-labeled carbon dioxide as tracer. The theoretically expected relationships between arterial carbon dioxide tension and tissue CO_2 content were confirmed.[29] This method can be used in the study of regional acid–base chemistry in the living brain. This principle can be extended to explore tissue response to drugs such as anticonvulsants. It will also allow the anatomic mapping of specific brain receptors and transmitters as a function of pharmacological intervention.

4. FUNCTIONAL STUDIES

Since it is thought that under most conditions glucose metabolism in a given region in the brain reflects its functional activity, measurement of the regional metabolism should make it possible to map the areas of the brain activated by different stimuli.[35,36]

Using the FDG technique to assess glucose metabolic rate, PET has been utilized to study the normal functioning of the brain. Most of the studies have been focused on brain processing of visual and auditory information. The metabolic response of the human visual cortex to stimulation with light and different patterns of images has yielded data that conform well to what is known about this system from clinical anatomic and electrophysiological studies. The tomographic image of glucose metabolism on base-line conditions is bilaterally symmetrical in the striate cortex.[35] Binocular as well as monocular stimulation increases the glucose metabolic rate of the visual cortex symmetrically. The stimulation of the nasal hemifield increases the glucose metabolic rate of the contralateral calcarine cortex.[35,37] These data confirm that each eye provides half of the input to each visual cortex.

Through the use of visual stimuli of increasing complexity, these studies have been able to show differential increases in activity between the primary visual cortex (PVC) and the associative visual cortex (AVC).[38,39] Simple visual stimulation with white light produces a small but measurable metabolic response in the PVC and an even smaller response in the AVC. As the complexity of the visual scene is increased, a concomitant increase in metabolic response of the visual cortex occurs (Fig. 2). This increase is larger for the AVC than for PVC and suggests greater involvement of the AVC for the interpretation of complex visual scenes. The metabolic rate of the visual cortex is also sensitive to visual imagery. Visual hallucination can considerably increase the activity in the visual processing areas of the brain.[40]

The functional activation of the auditory system has been studied using verbal and musical stimuli. Under base-line conditions, there is minimal asymmetry in the temporal lobes, especially in the areas of the auditory cortex. The tomographic image of the brain when subjects are stimulated either with a verbally meaningful story or an unintelligable story spoken in a language they do not know shows in both cases an increase in glucose metabolism in the right temporal cortex independent of the ear to which the stimuli were presented.[35] These results are contradictory to the currently held notion that the left hemisphere is dominant for speech and language functions and the right hemisphere for nonverbal visual spatial processing. This discrepancy may reflect additional levels of performance required by the test other than language perception. A recent publication by the same authors reports that monoaural stimulation with a factual story caused an increase in glucose metabolism in the auditory cortex in the hemisphere contralateral to the stimulated ear.[37]

The interhemispheric differences in information processing are illustrated in a study of a group of subjects who were asked to discriminate whether two groups of musical tones were different or similar. The subgroup of subjects who tried to remember the melody showed an increased activation of the right hemisphere, whereas those who mentally plotted the notes on a music staff showed an increased activation of the left hemisphere.[40] These preliminary data throw light on the enormous potential of PET to help understand the complex interaction of processes within different brain structures.

5. CLINICAL APPLICATIONS

Positron emission tomography has entered the clinical world. Disciplines that have already begun to profit from its advances include neurology, psychiatry, and cardiology. The method has been applied to studies of local changes in glucose metabolism in different brain diseases: epilepsy,[41] stroke,[42] dementia,[43,44] schizophrenia,[45,46] Huntington's chorea,[47] Parkinson's disease (D. E. Kuhl, unpublished data), aphasia,[48] cerebral tumors,[49,50] and subdural hematomas.[31] Although these studies are still preliminary, some have already shown the unique capabilities of the PET technique in providing insight into the pathophysiology of some of these entities.

5.1. Epilepsy

Studies done in patients with partial epilepsy have shown, during the interictal periods, tomographic images with local regions of decreased CMRglu (20–50%) as assessed by the FDG technique. These hypometabolic zones correlated anatomically with the focal EEG abnormalities but appeared normal by CT scan (Fig. 8). Positron emission tomographic images during ictal periods showed foci of increased (80–130%) glucose metabolism which coincided temporally and anatomically with ictal EEG spike foci and corresponded to zones of interictal hypometabolism. These also showed other areas of hypermetabolism which may represent the pathways involved in the propagation of ictal activity and the activation of the zones mediating the behavioral seizures.[34]

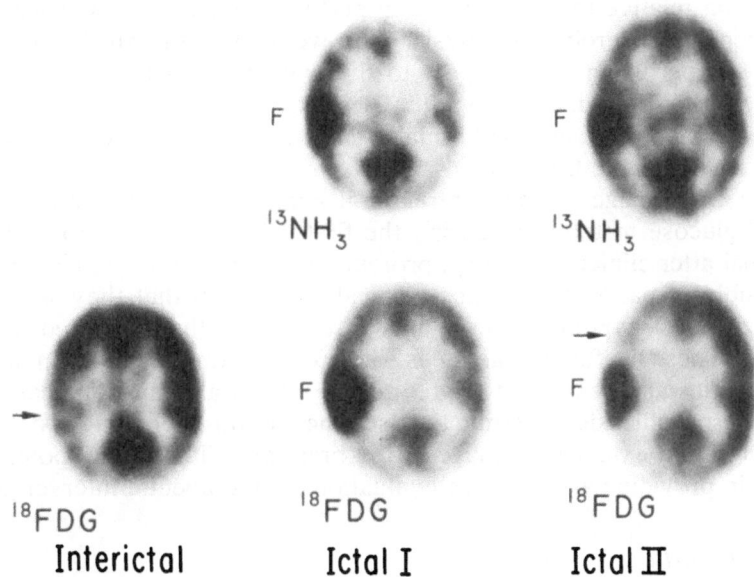

Fig. 8. Interictal and ictal scans of a 5-year-old boy who had a 10-month history of right focal motor seizures. The CT, arteriography, and pneumoencephalography were normal. The interictal ^{18}FDG scan shows hypometabolism in the left temporoparietal cortex (dotted arrow). At this same site, two separate ictal studies show marked focal hypermetabolism and hyperperfusion (F) coincidental with right facial twitching and epileptiform EEG spike activity in the left frontotemporal region. In one study, hypometabolic cortex (solid arrow) is seen adjacent to the hypermetabolic focus, suggesting a "surround inhibition" process. Ictal mean CMR$_{glu}$ of the right hemisphere was normal. The CT showed no abnormalities at the time of these scans. (Figure and legend courtesy of Kuhl *et al.*, see ref. 41.)

Perfusion as assessed by the ^{13}NH$_3$ technique paralleled changes in FDG uptake in abnormal zones during the interictal and ictal periods.

The PET scan can confirm the primary epileptogenic focus by identifying areas of most severe hypometabolism during the interictal period. Since June, 1979, all patients at the UCLA Medical Center who are candidates for surgical resection of the epileptogenic focus receive a PET scan to locate these presumed primary foci. Further studies will undoubtedly help to elucidate the pathophysiology of the different types of epilepsy in man and will provide a powerful tool to assess the effects of anticonvulsant medication.

5.2. Stroke

In patients suffering from stroke, PET has been able to evaluate the derangement of metabolism and perfusion throughout the evolution and recuperation of the episode. The ^{18}FDG and ^{13}NH$_3$ techniques were used to assess, respectively, glucose metabolism and the perfusion rate within and outside the lesion.[35] Soon after cerebrovascular occlusion, there is an increase in metabolic activity relative to the perfusion rate within the infarct. This increase in glucose metabolism relative to the decrease in perfusion appears to be caused by enhancement of anaerobic glycolysis. One week after the stroke, there is an increase in perfusion with respect to the hypometabolism of the zone. This "luxury perfusion" disappears after 2 weeks, and a residual deficit in both metabolism and perfusion will remain. This zone is correlated with an image of low density in the CT scan.

The PET image of the contralateral hemisphere shows moderately depressed glucose metabolism during the first week after the stroke, returning to normal after clinical recovery; profoundly depressed metabolism is seen in irreversible coma. These studies were also useful in that they were able to detect zones of hypometabolism outside the lesion that appeared normal on CT scan. These dysfunctional zones were correlated with clinical neurological abnormalities apparently not related to the structural change indicated by CT scan. These studies demonstrate the usefulness of this technique in determining the degree and extent of damage in the brain and illustrate its potential usefulness in providing an objective evaluation for therapeutic interventions.

5.3. Dementia

Preliminary data with senile dementia indicate an important diffuse diminution in glucose metabolic rate in comparison with elderly normal subjects. Furthermore, the degree of diminution in glucose metabolism was highly correlated with the degree of cognitive impairment. Studies comparing Alzheimer dementias with multiinfarct dementias failed to show any significant difference in patterns of blood flow and oxygen metabolism. In both groups, these parameters were diminished with respect to age-matched controls. The degree of depression in CBF and $CMRO_2$ correlated with the severity of the dementia. Clinical deterioration of the patients after 6 months of reevaluation was also reflected in a further diminution of CBF and $CMRO_2$.[44]

5.4. Huntington's Chorea

Huntington's chorea patients show tomographic images with diminished activity in the caudate and the putamen. These abnormally low levels of activity are also shown by some symptom-free subjects who have a family history of the disease.[47] It is not known whether these are the individuals who will go on to develop Huntington's disease. Until now, means were not available to detect patients who will eventually develop Huntington's chorea. Thus, PET may be useful in the early and accurate detection of this disease and in identifying its pathogenetic defects. As such, it may provide useful data for genetic counseling.

5.5. Cerebral Tumors

The studies done on brain neoplasms have focused mainly on gliomas. Blood flow, oxygen metabolism, and glucose uptake have been determined in tumor patients with different histological grading. Positive correlations have been reported between altered glucose metabolism and malignancy.[49,50] Blood flow within the tumor appears as heterogeneous areas of both increased and decreased perfusion. The tissue surrounding the tumors shows a parallel decrease in blood flow and glucose metabolism. The effects of radiotherapy and chemotherapy on the metabolism of brain neoplasms are now being monitored with PET.

5.6. Schizophrenia

The pattern of glucose metabolism by the brain as assessed by the ^{18}F-DG technique appears to be deranged in a group of schizophrenics studied with PET. This group of patients shows a significant decrease in the relationship between frontal and occipital cortical metabolism. In normal subjects, the frontal cortex shows greater activity than the occipital cortex. The schizophrenic patients show a diminished metabolism in the frontal cortex with respect to the temporal and occipital cortex and relative to the values reported in the frontal cortex of normal subjects.[46]

In this chapter we have reviewed theoretical methodological aspects of PET and very preliminary clinical studies that have been reported to date. Although this field is in its infancy, its potentialities are already starting to appear. The information being obtained with the PET suggests that this will be a powerful tool with far-ranging research and clinical applications.

ACKNOWLEDGMENTS. Partial support was provided by USPHS grants RO NS 15638 from NINCDS and Research Scientist Development Award MH00137 from the NIMH. We thank Dr. A. P. Wolf for helpful discussions and Phyllis Kirshen and Linda Fernhoff for their assistance in preparation of the manuscript.

REFERENCES

1. Kety, S. S., and Schmidt, C. F., 1948, *J. Clin. Invest.* **27**:476–483.
2. Wolf, A. P., 1981, *Semin. Nucl. Med.* **11**:2–12.
3. Brownell, G. L., Correia, J. A., and Zamenhay, R. G., 1978, *Recent Advances in Nuclear Medicine*, Volume 5 (J. N. Lawrence and T. F. Budinger, eds.), Grune & Stratton, New York, pp. 1–47.
4. Derenzo, S. E., Banchero, P. G., and Cahoon, J. L., 1978, *IEEE Trans. Nucl. Sci.* **25**:341–349.
5. Ter-Pogossian, M. M., Mullani, N. A., Ficke, D. C., Markham, J., and Snyder, D. L., 1981, *J. Comput. Assist. Tomogr.* **5**:227–239.
6. Raichle, M. E., 1979, *Brain Res. Rev.* **1**:47–68.
7. Mazziota, J. C., Phelps, M. E., Plummer, D., and Kuhl, D. E., 1981, *J. Comput. Assist. Tomogr.* **5**(5):734–743.
8. Ter-Pogossian, M. M., 1981, *Semin. Nucl. Med.* **11**:13–23.
9. Mullani, N. A., Ficke, C., and Ter-Pogossian, M. M., 1980, *IEEE Trans. Nucl. Sci.* **27**:572–575.
10. Budinger, T. F., 1978, *A Primer on Reconstruction Algorithms*, World Federation of Nuclear Medicine and Biology, pp. 1–38.
11. Ter-Pogossian, M. M., Raichle, M. E., and Sobel, B. E., 1980, *Sci. Am.* **243**:170–181.
12. Lenzi, G. L., Jones, T. M., McKenzie, C. G., Buckingham, P. D., Clark, J. C., and Moss, S., 1978, *J. Neurol. Neurosurg. Psychiatry* **41**:1–10.

13. Comar, D., Maziere, M., Godot, J. M., Berger, G., and Soussaline, F., 1979, *Nature* **280**:329–331.
14. Tewson, T. J., Raichle, M. E., and Welch, M. J., 1980, *Brain Res. Rev.* **193**:291–295.
15. Gelbard, A., 1981, *J. Label. Compd. Radiopharm.* **18**:933–945.
16. Shive, C. Y., Fowler, J. S., and Mac Gregor, R. R., 1979, *Radiopharmaceuticals* **11**:259–264.
17. Knust, E. J., Kupfernagel, C., and Stocklin, G., 1979, *J. Nucl. Med.* **20**:1170–1175.
18. Sokoloff, L., Reivich, M., Kennedy, C., Des Rosiers, M. H., Patlak, C. S., Pettigrew, K. D., Sakurada, O., and Shinohara, N., 1977, *J. Neurochem.* **28**:897–916.
19. Huang, S. C., Phelps, M. E., Hoffman, J. E., Sideris, K., Selin, C. J., and Kuhl, D. E., 1980, *Am. J. Physiol.* **238**:69–82.
20. Phelps, M. E., Huang, S. C., Hoffman, E. J., Selin, C., Sokoloff, L., and Kuhl, D. E., 1979, *Ann. Neurol.* **6**:371–388.
21. Phelps, M. E., 1981, *Semin. Nucl. Med.* **11**:32–49.
22. Raichle, M. E., Welch, M. J., Grubb, R. L., Higgins, C. S., Ter-Pogossian, M. M., and Larson, K. B., 1978, *Science* **199**:986–987.
23. Reivich, M., Kuhl, D., Wolf, A., Greenberg, J., Phelps, M., Ido, T., Casella, V., Fowler, J., Hoffman, E., Alavi, A., Som, P., and Sokoloff, L., 1979, *Circ. Res.* **44**:127–137.
24. Scheinberg, P., and Stead, E. A., 1949, *J. Clin. Invest.* **28**:1163–1171.
25. Jones, T., Chesler, D. A., and Ter-Pogossian, M. M., 1976, *Br. J. Radiol.* **49**:339–343.
26. Subramanyam, R., Alpert, N. M., Hoop, B., Brownell, G. L., and Taveras, J. M., 1978, *J. Nucl. Med.* **19**:48–53.
27. Raichle, M. E., 1981, *Fed. Proc.* **40**:2331–2334.
28. Raichle, M. E., Grubb, R. L., Gado, M. H., Eichling, J. O., and Ter-Pogossian, M. M., 1976, *Arch. Neurol.* **33**:523–526.
29. Raichle, M. E., Grubb, R. L., and Higgins, C. S., 1979, *Brain Res.* **166**:413–417.
30. Yamamoto, Y. L., Thompson, C. J., Meyer, E., Robertson, J. S., and Feidel, W., 1977, *J. Comput. Assist. Tomogr.* **1**:43–56.
31. Ericson, K., Bergstrom, M., and Erikson, L., 1980, *J. Comput. Assist. Tomogr.* **4**:737–745.
32. Frackowiak, R. S., Lenzi, G. L., Jones, T., and Heather, J., 1980, *J. Comput. Assist. Tomogr.* **4**:727–736.
33. Ter-Pogossian, M. M., Klein, M. S., and Markham, J., 1980, *Circulation* **61**:242–255.
34. Phelps, M. E., Hoffman, E. J., and Raybaud, C., 1977, *Stroke* **8**:694–702.
35. Alavi, A., Reivich, M., Greenberg, J., Hand, P., Rosenquist, A., Rintelmann, W., Christman, D., Fowler, J., Goldman, A., Mac Gregor, R., and Wolf, A., 1981, *Semin. Nucl. Med.* **11**:24–31.
36. Sokoloff, L., 1981, *Fed. Proc.* **40**:2311–2316.
37. Greenberg, J. H., Reivich, M., Alavi, A., Hand, P., Rosenquist, A., Rintelmann, W., Stein, A., Tusa, R., Dann, R., Christman, D., Fowler, J., MacGregor, B., and Wolf, A., 1981, *Science* **212**:678–680.
38. Phelps, M. E., and Kuhl, D. E., 1981, *Science* **211**:1445–1448.
39. Phelps, M. E., Mazziotta, J. C., Kuhl, D. E., Nuwer, M., Packwood, J., and Metter, J., 1981, *Neurology (Minneap.)* **31**:517–529.
40. Miller, J. A., 1981, *Sci. News,* **119**:76–78.
41. Kuhl, D. F., Engel, J., Phelps, M. E., and Selin, C., 1980, *Ann. Neurol.* **8**:348–360.
42. Kuhl, D. E., Phelps, M. E., Kowell, A. P., Metter, J. E., Selin, C., and Winter, J., 1980, *Ann. Neurol.* **8**:47–60.
43. Ferris, S. H., De Leon, M. J., Wolf, A., Farkas, T., Christman, D. R., Reisberg, B., Fowler, J. S., MacGregor, R., Goldman, A., George, A. E., and Rampal, S., 1980, *Neurobiol. Aging* **1**:127–131.
44. Frackowiack, R. S., Pozzilli, C., Legg, N. J., Du Boulay, G. H., Marshall, J., Lenzi, G. L., and Jones, T., 1981, *J. Cerebral Blood Flow Metab.* [*Suppl.*] **1**:S453–S454.
45. Farkas, T., Reivich, M., Alavi, A., Greenberg, J. H., Fowler, J. S., MacGregor, R. R., Christman, D. R., and Wolf, A., 1980, *Cerebral Metabolism and Neural Function* (J. V. Passonneau, R. A. Hawkins, W. D. Lust, and F. A. Welsh, eds.), Williams & Wilkins, Baltimore, pp. 403–408.

46. Buchsbaum, M. S., Kessler, R., Bunney, W. E., Cappelletti, J., Coppola, R., van Kammen, D. P., Nigal, F., Waters, R., Sokoloff, L., and Ingvar, D., 1981, *J. Cerebral Blood Flow Metab. [Suppl.]* **1:**S457–S458.

47. Kuhl, D., Phelps, M., Markham, C., Winter, J., Metter, J., and Rieoe, W., 1981, *J. Cerebral Blood Flow Metab. [Suppl.]* **1:**S459–S460.

48. Metter, J. E., Wasterlain, C. G., Kuhl, D. E., Hanson, R. W., and Phelps, M. E., 1981, *Ann. Neurology* **10:**173–183.

49. Di Chiro, G., De La Paz, R., Smith, B., Kornblith, P., Sokoloff, L., Brooks, R., Blasberg, R., Cummins, C., Kessler, R., Wolf, A., Fowler, J., London, W., and Sever, J., 1981, *J. Cerebral Blood Flow Metab. [Suppl.]* **1:**S11–S12.

50. Ackerman, R. H., Davis, S. M., Correia, J. A., Alpert, N. M., Buonanno, F., Finkelstein, S., Brownell, G. L., and Taveras, J. M., 1981, *J. Cerebral Blood Flow Metab. [Suppl.]* **1:**S575–S576.

51. Bustany, P., Sargent, T., Saudubray, J. M., Henry, J. P., and Comar, D., 1981, *J. Cerebral Blood Flow Metab. [Suppl.]* **1:**S17–S18.

Selected Micromethods for Use in Neurochemistry

Volker Neuhoff

1. INTRODUCTION

Micromethods do not seem to be widely applied in neurochemical analysis, perhaps because of a widespread prejudice that micromethods are only useful for "microminded," technically perfect experimenters. In addition, there may be worries about the reproducibility of microtechniques. A microgel, however, with many clearly separated and well-defined bands, each representing some nanograms of protein (see Fig. 10), is rather impressive when seen for the first time, especially when it is realized that the time taken for separation, staining, and destaining is only about one-tenth of the time necessary for the equivalent procedure using a macroscale method. A saving of experimental time is inherent in most of the micromethods described in some detail in this chapter. Such micromethods are also reproducible if they are performed correctly. The use of micromethods is therefore to be recommended even when the amount of material to be analyzed is not limited. Furthermore, now that it is harder to get sufficient funds for relevant research, micromethods are useful because the equipment needed for the analyses is not expensive.

The correct procedure for a micromethod is no more difficult to learn than is any other method. The time needed to learn any method is strongly dependent on the handiness and experience of the experimenter, and this is the same for learning a micro- or a corresponding macrometltod. Obviously, every experimenter will take some time to be satisfied with the application of a micromethod and to clearly understand its critical features and limitations. But these are points that apply to every method. Generally, there are no arguments against micromethods, and it is surprising that in neurochemical research, where so often only small sample volumes are available, the application of micromethods

Volker Neuhoff • Forschungsstelle Neurochemie, Max-Planck-Institut für Experimentelle Medizin, 3400 Göttingen, Federal Republic of Germany.

is not more widespread. On the other hand, it is not surprising that progress in microscale methods is mostly connected with neurochemical applications. For example, the first application of polyacrylamide gel electrophoresis on the microscale was carried out in 1964 when Pun and Lombrozo[1] fractionated brain proteins. Since then, microscale methods for determination of RNA[2,3] and RNA base composition,[4,5] measurement of enzyme activity and substrates of single neurons,[6–12] mass determination,[13,14] micro-flame photometry,[15,16] picoliter volumes,[17] microchromatography on thin-layer plates for dansyl amino acids,[18–21] phospholipids,[22–24] isoenzymes,[25] glycoproteins,[26] and sugars,[27] as well as other auxiliary methods such as microhomogenization[28] have mostly come from neurochemical laboratories.

This chapter, because of the limited space available, will neither review all micromethods possible nor describe techniques for preparation of minute amounts of defined tissue material or single nerve cells (see ref. 29). The methods selected for description are those with which the author has personal experience and therefore the necessary critical competence to describe. Very seldom are micromethods developments of their own. Usually they are deduced from macro procedures adapted to a micro scale, thereby more or less automatically increasing the sensitivity by a factor of between 10 and 1000. Most normal methods can be rather simply converted to micro versions, although one has to accept that every method will have its special merits and disadvantages.

2. PROTEIN DETERMINATION WITH MICROLITER VOLUMES

The quantitative determination of protein is a common prerequisite for many biochemical analyses. The different methods currently available often have a drawback in that agents used for preparation of biological tissues, e.g., sodium dodecyl sulfate (SDS), Triton X-100, NP 40, mercaptoethanol, urea, interfere with the analysis. In such cases, a protein precipitation step is often required, which is not only time consuming but also uncertain with respect to quantification. The volume routinely used for protein determinations is often rather large, and for microdeterminations, special and often cumbersome modifications are necessary. The method described here avoids all of these disadvantages and furthermore allows quantitative determination of small amounts of protein, even if the volume used (in the range between 1 and 5 μl) is unknown.

The method is, in principle, a spot analysis, a type of analytical method first introduced in 1859 by H. Schiff. Such methods had their peak approximately 40–50 years ago but are nowadays almost forgotten. However, spot methods still have very many advantages.[30] One major point is the fact that by spotting sample solutions onto a suitable layer, the contents will be concentrated over a very small area. During fixation, staining, and washing out of the excess stain, all interfering components not firmly bound to the sample being analyzed, will be almost completely washed out. With this protein determination procedure, as described recently,[31] it is possible to determine the protein concentration in a microliter volume when neither the volume nor the protein content in that volume are known.

2.1. *Performance of the Protein Determination*

Cellulose acetate strips commonly used for electrophoresis are used in this procedure (see Appendix, Scheme 1 for details). Prior to use, the acetate strips are stored in a moist chamber to facilitate the application of the sample. If a series of determinations is to be performed, the sample number may be indicated simply by cutting the corners of the acetate layer for the first five samples or by combining cut corners with arrowlike cuts on the sides. The number of strips that are stained and destained together is not critical. Sample application, staining, and destaining are done at room temperature, and commercially available 0.5-, 1-, 2-, or 5-μl capillaries are used. Cleaning of the capillaries[32] prior to use is recommended. The capillaries are filled by capillary attraction simply by being dipped into the solution to be analyzed. Complete filling has to be confirmed, and for routine determinations, 2-μl capillaries have been found to be the best. The standard deviation for a given protein concentration depends on the capillary volume as follows: 0.5 μl, 6.5%; 1 μl, 2.5%; 2 μl, 1.5%; 5 μl, 2.4%; 10 μl, 7.5%. In most instances, immediately after sample application, the acetate strip is transferred to a petri dish containing the stain dissolved in methanol/acetic acid. Prefixation of a protein spot in methanol/acetic acid before staining results in a loss of 70–80% of the dye-binding capacity. In contrast, drying at room temperature or with a hot air stream has no influence on quantitative staining, thereby allowing for repeated sample application in the case of very dilute protein solutions.

2.1.1. *Evaluation with Densitometry*

If protein spots on acetate paper are round, any densitometer can be used for evaluation of Amido black-stained spots after the sample-containing strip is made transparent.[31] Four typical densitograms of spots, each representing 2 μl of bovine serum albumin (2 μl of a solution containing 1 mg protein/ml), are shown in Fig. 1. If a sample is repeatedly measured ten times[31] and is repositioned in the equipment for measurement each time, the coefficient of variation is 0.9%. The integral of the area recorded is equivalent to the amount of protein in the spot (compare Fig. 2a). In cases in which it is difficult to obtain an ideal round spot during sample application, e.g., if the sample contains high concentrations of sucrose, glycerol, urea, etc. use of Hoechst 2495 (a fluorescent benzoxanthene derivative) is recommended for staining followed

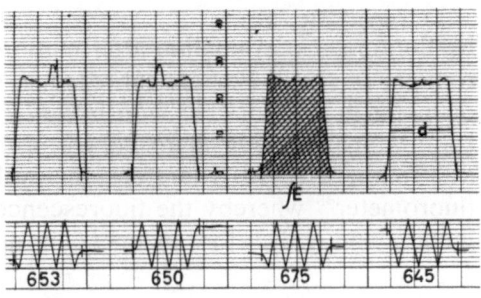

Fig. 1. Densitograms of four spots stained with amido black and representing 2 μg protein each. The densitograms were obtained with a Zeiss Gel Scanner (ZK4) and an adapter for micro-gel evaluation (123). ∫E, integrated absorption; d, spot diameter.

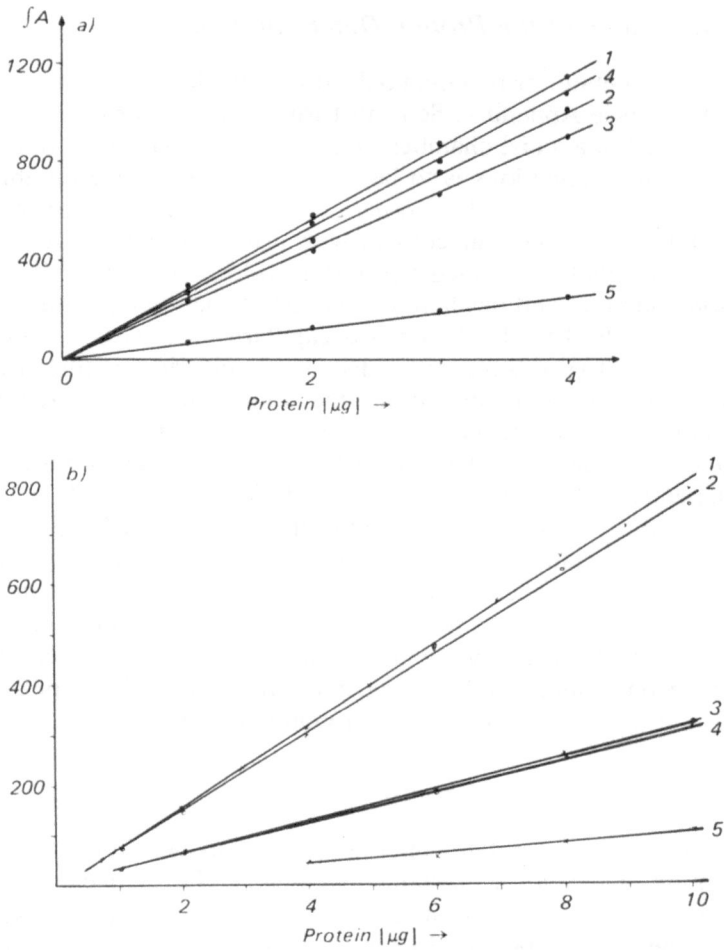

Fig. 2.   Calibration curves for different proteins obtained by densitometric evaluation after amido black staining (a) and staining with Hoechst 2495 and spot fluorometry *in situ* (b). The curves are calculated according to $y = b \cdot x^a$ (log y = log b + $a \cdot$log x) r, correlation coefficient; n, number of spots evaluated per curve; abscissae, µg protein; ordinate in a, integrated absorption; in b, arbitrary units. (1) Bovine serum albumin: (a) r = 0.9989, n = 15; (b) r = 0.9928, n = 41. (2) Human serum albumin: (a) r = 0.9975, n = 15; (b) r = 0.9946, n = 24. (3) Transferrin: (a) r = 0.9989 n = 15; (b) r = 0.9725, n = 23. (4) Globulin: (a) r = 0.9985, n = 15; (b) r = 0.995, n = 24. (5) Bovine myelin: (a) r = 0.9935, n = 15; (b) r = 0.9306, n = 14. Note not only the differences in stainability of the different proteins but also the differences between the two stains.

by fluorometric evaluation after elution of the stain (see below) or spot fluorometry *in situ*.[33] For optical reasons, evaluation by densitometry across the diameter is not possible with irregular spots, and complete scanning of the spots with a scanning microscope is prohibitively expensive and not acceptable as a routine method. Therefore, the evaluation is performed with a simple spot fluorometer[33] whereby the fluorescence of the whole spot is measured as an integral and is therefore independent of the form of the spot.

2.1.2. Photometric and Fluorometric Evaluation

If none of the abovementioned instruments is available for evaluation of stained spots, quantitative determination can easily be performed with a normal spectrophotometer for amido black staining after complete solubilization of the acetate strip containing the stained spot in a suitable volume of dioxane, N,N-dimethylformamide, or dimethylsulfoxide. Instead of a photometer, a fluorometer can be used after staining of the protein spots with Hoechst 2495 and solubilizing with the same three solvents; alternatively, elution of the chromophor Hoechst 2495 from the protein spot with 1% NH_4OH followed by fluorometry is possible.

Fluorescence is measured, in arbitrary units, between the peak maximum

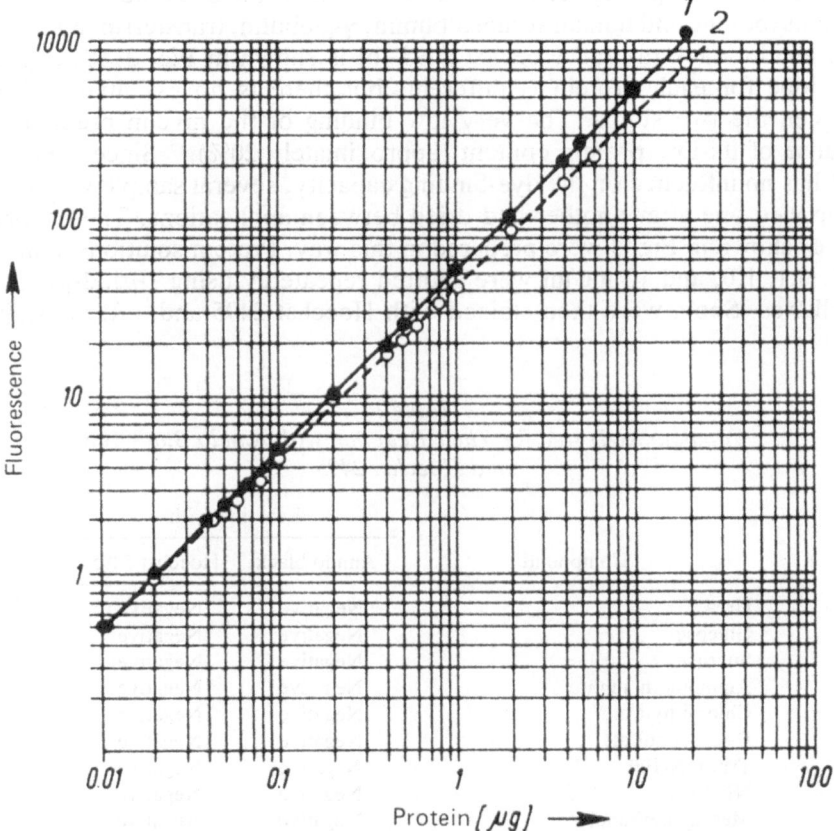

Fig. 3. Calibration curves for bovine serum albumin obtained after staining with Hoechst 2495 and either (1) elution of the stain in 15% NH_4OH or (2) solution of the acetate strip containing the sample in dimethylsulfoxide. Abscissa, μg protein; ordinate, arbitrary units. Fluorescence was measured with a Turner spectrophotometer with an excitation wavelength of 430 nm (for dimethylsulfoxide) or 425 nm (for NH_4OH) and an emission wavelength of 475 nm. The measured volumes were 2 ml for 0.01–2 μg and 5 or 15 ml for the larger amounts. The curves are calculated according to $y = b \cdot x^a$. Curve 1: $\log y = \log 51.8 + 1.01 \log x$; $r = 0.9994$; $n = 36$. Curve 2: $\log y = \log 41.9 + 0.97 \log x$; $r = 0.9995$; $n = 34$.

and the corresponding blank value. If the value is calculated according to the equation

$$Fl = \text{arbitrary units (sample} - \text{blank)} \times \text{volume/amplification}$$

complete independence of volume and amplification of the instrument are reached, and a calibration curve over a wide concentration range can be prepared as shown in Fig. 3. If measurements are to be performed in the lowest concentration range possible (0.01–0.1 mg protein/ml), the precautions for fluorometric measurements described previously[34] should be followed. In this case, the lowest measuring value should be approximately five times that of the blank.

The calibration curves shown in Fig. 2 are obtained with different sample volumes in the range between 0.5 and 5 μl. Further, the dye binding of different proteins (bovine and human serum albumin, γ-globulin, transferrin, and myelin) with amido black (a) and Hoechst 2495 (b) is shown in Fig. 2. As expected, the slope for each protein is different, but there is no essential difference between the two stains. The very low binding of the myelin preparation is because of its low protein content (approximately 20%).[35] Since drying of a spot has no influence on the dye-binding capacity, several sample volumes can be applied one after another and dried between applications. The calibration curves shown in Fig. 3 were prepared in this way. Protein solutions containing 0.01, 0.1, 1.0, and 10 mg/ml were spotted repeatedly using 1-μl, 2-μl, or 5-μl capillaries. Spots were then stained with Hoechst 2495 and, after destaining,

Table I
Compounds Tested for Their Reaction with Amido Black 10B
and Hoechst 2495

Compound	Reaction with	
	Amido black	Hoechst 2495
Urea	Negative	Negative
Glucose	Negative	Negative
Sucrose	Negative	Negative
Ammonium sulfate	Negative	Negative
Chloral hydrate	Negative	Negative
Dodecyl sulfate	Negative	Negative
Triton X-100	Negative	Negative
NP 40	Negative	Negative
Mercaptoethanol	Negative	Negative
Desoxycholate	Negative	Negative
tRNA (2 mg/ml)	Negative	Positive
sRNA	Negative	—
Cholesterol	Negative	Negative
Egg kephalin (saturated solution)	Positive	Negative
Egg lecithin	Negative	Negative
Ampholine	Positive	Positive

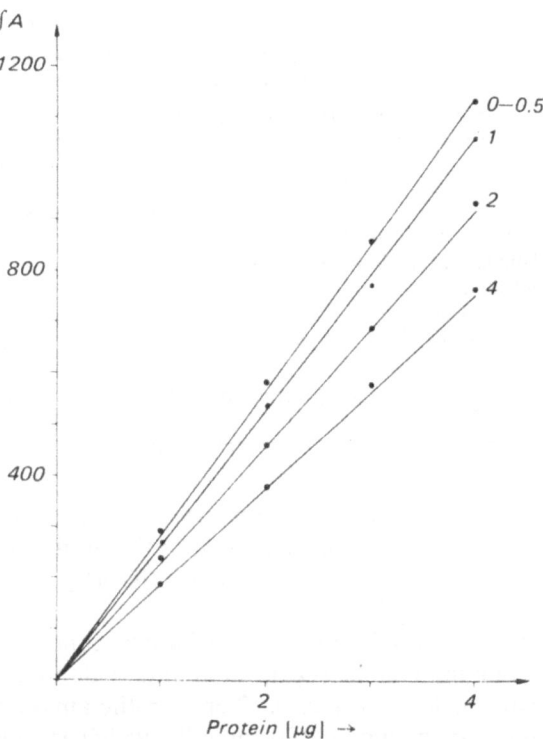

Fig. 4. Relationship between protein concentration and the integrated absorption after amido black staining for different concentrations of dodecyl sulfate. Abscissa, μg protein; ordinate, integrated absorption. Each curve was calculated according to $y = b \cdot x^{a}$ on the basis of 12 evaluated spots. The correlation coefficients are 0.9995, 0.9985, 0.9985, and 0.9979 for 0–0.5%, 1%, 2%, and 4% dodecyl sulfate present in the sample, respectively. Bovine serum albumin was the protein standard. Note that 0.5% dodecyl sulfate has no influence on the stainability with amido black; staining with Hoechst 2495 is also not affected by dodecyl sulfate (results not shown in this Figure).

were solubilized in 2 ml dimethylsulfoxide (Fig. 3, curve 2) or eluted in 2 ml 1% NH_4OH and evaluated fluorometrically (Fig. 3, curve 1).

2.1.3. Interference in the Method

Compounds tested for their influence on the staining procedure are listed in Table I. High concentrations of egg cephalin (saturated solution) can be stained with amido black but not with Hoechst 2495. High concentrations of tRNA (2 mg/ml) are weakly stained with Hoechst 2495 but not with amido black. None of the listed compounds in a protein solution has any influence on the quantitative staining. The influence of dodecyl sulfate on amido black binding to protein is shown in Fig. 4. The presence of 0.5% dodecyl sulfate has no effect, but 1%, 2%, and 4% dodecyl sulfate decreases the staining by 6.2%, 17.4%, and 33.0%, respectively, in comparison with the same protein concentration (bovine serum albumin) without dodecyl sulfate.

A stoichiometric relationship between protein concentration and staining can be observed. This is demonstrated in the calibration curves in Fig. 4. This stoichiometry, on the other hand, may be useful for the determination of an unknown dodecyl sulfate concentration. The presence of mercaptoethanol in a sample solution has no further effect. In contrast to amido black, the binding of Hoechst 2495 to all of the proteins studied so far is not influenced by dodecyl sulfate (0.5–2%) or dodecyl sulfate plus mercaptoethanol (1% each). Thus, turbid protein solutions containing unsolubilized particles or even protein sed-

Table II
Differences for Known Amounts of Different Proteins and Brain Extracts Measured
According to Lowry and with Spot Analysis: Values Are Related to the
Corresponding Standard Calibration Curve Prepared with Bovine Serum Albumin

Protein	Method using amido black[31]	Method of Lowry et al.[36]
Bovine serum albumin	0	0
Human serum albumin	− 14.9%	− 15.6%
γ-Globulin	− 3.3%	+ 15.6%
Transferrin	− 19.6%	− 14.9%
Water-soluble proteins from brain tissue	1.72 mg/ml (1.4% S.D.)	2.12 mg/ml (2.1% S.D.)
Brain total particulate fraction in 1% SDS	2.12 mg/ml (1.72% S.D.)	2.62 mg/ml (2.9% S.D.)

iments can easily be converted to clear solutions suitable for protein determination with Hoechst 2495 by the addition of a suitable volume of 4% SDS/conc. NH_4OH (1:1 v/v), e.g., 100 µl protein solution plus 10 or 20 µl SDS/NH_4OH.

When solutions with a known concentration (1 mg/ml) of bovine serum albumin, human serum albumin, γ-globulin, and transferrin are assayed according to Lowry et al.[36] or with the amido black method described here, and the known amount of protein (by weight) is related to the appropriate calibration curve using bovine serum albumin as a standard, differences are observed for both methods (as shown in Table II). For human serum albumin and transferrin, the negative difference is of the same order, but for γ-globulin, there is a negative difference of 3.3% for the amido black method and a positive difference of 15.6% after determination according to the method of Lowry et al.[36] Similar types of differences are also observed for protein determinations of biological material (Table II).

2.2. Protein Determination with Unknown Volume and Unknown Concentration

For the calculation of protein concentration, the analyzed volume is usually needed, and the amount of protein in that volume is then determined. Both values can be obtained by a single determination with the amido black method. There is a linear relationship between spot size and the applied volume (1–5 µl), as shown in Fig. 5. The spot size is further determined by the protein concentration in that volume, as is shown for protein concentrations of 0.1–2.0 mg/ml. The linear relationship between the amount of protein (0.1–10 µg) and the integrated absorption for different volumes (1–5 µl) is shown in Fig. 6. As shown in Fig. 1, the densitometric evaluation of the stained spot gives both values needed for the determination of a protein concentration in an unknown volume between 1 and 5 µl—the absorption as a measure of the amount of protein present in the spot and the diameter of the spot as a measure of the volume applied. The only prerequisite for the determination of protein concentration in an unknown volume is an absolutely round spot on the acetate strip. This can easily be obtained if the procedures already described[31] are followed. If Triton X-100 and dodecyl sulfate are present in a sample solution, the spot remains round, but the diameter is larger, and this must then be taken into consideration. If a set of calibration curves similar to Figs. 5 and 6 is prepared, a simple iterative procedure may be used[31] to determine the protein concentration, especially if a modest computer program is available. The whole protein determination procedure with unknown volumes has a standard deviation of 4.1%.

Since protein determinations are frequently related to bovine serum albumin, it is useful to calibrate the method described above with this protein. Which stain or which of the evaluation procedures is used may depend on the equipment available in the laboratory. The sensitivity could be improved if a chromophore with better fluorescence intensity is available, e.g., dansyl chloride or fluorescamine, but these compounds are not suitable for this method because they do not permit a staining technique as simple as those with amido black or Hoechst 2495.

The easy handling of this method and especially its independence of or

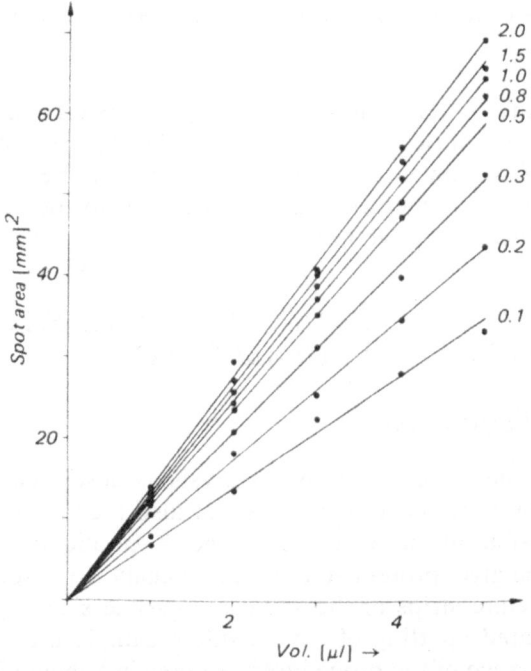

Fig. 5. Spot size on a moistened acetate strip in relation to volume for different protein concentrations evaluated after amido black staining. Bovine serum albumin was the protein standard. Abscissa, μl applied volume; ordinate, spot area (mm^2). The curves are calculated according to $y = b \cdot x^a$ on the basis of 25 evaluated spots per curve. The correlation coefficients for the different protein concentrations are: 2 mg/ml, 0.9985; 1.5 mg/ml, 0.9989; 1 mg/ml, 0.9989; 0.8 mg/ml, 0.9989; 0.5 mg/ml, 0.9989; 0.3 mg/ml, 0.9975; 0.2 mg/ml, 0.9985; 0.1 mg/ml, 0.9955.

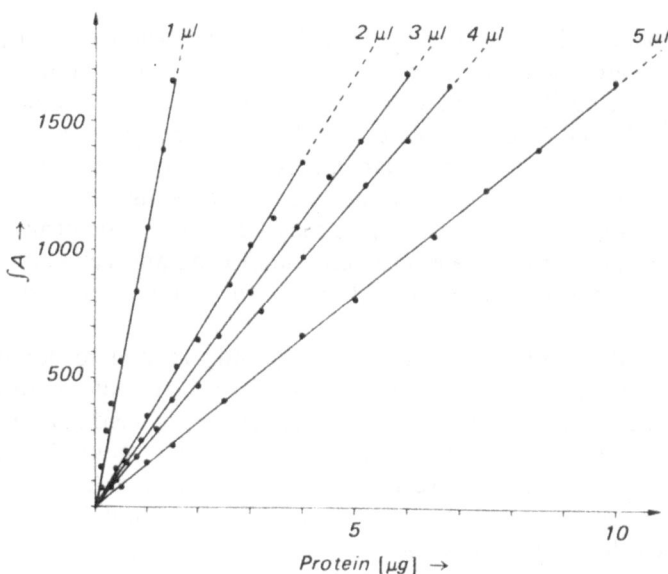

Fig. 6.   Relationship between the amount of protein and the integrated absorption after amido black staining for different volumes. Bovine serum albumin was the standard. Abscissa, μg protein; ordinate, integrated absorption. Each curve is calculated according to $y = b \cdot x^a$ on the basis of 27 evaluated spots. The correlation coefficients are: 5 μl, 0.9995; 4 μl, 0.9989; 3 μl, 0.9915; 2 μl, 0.9989; 1 μl, 9.9969.

correction for other external influences are reasons for using this method in neurochemical laboratories. The same principle of quantitative spot analysis can still be used if volumes smaller than 0.5 μl have to be analyzed. If suitable stains are available, the method may also be used for determination of other biological substance, e.g.,glycoproteins.

3. GLYCOPROTEIN DETERMINATION WITH FITC-LABELED LECTINS AT THE NANOGRAM RANGE

3.1. Practical Performance

Using the same principle of spot analysis as described for protein determination, a calibration and in some respect also a characterization of glycoproteins are possible in the nanogram range.[26] An aliquot of 0.5, 1, or, preferably, 2 μl of the glycoprotein solution to be analyzed is spotted as described on moistened acetate strips of the smallest possible size, e.g., 1 × 1 cm. Air drying and repeated spotting of a very dilute sample are also possible. The practical performance of the procedure[26] is given in Scheme 2 (Appendix). The binding of concanavalin A (Con A) to glycoproteins, with glucose oxidase as an example, is seen in Fig. 7 to be dependent on incubation time and temperature at the pH optimum of 7.4. For practical reasons, a standard incubation time of 80 min at room temperature is recommended, with mild shaking during incubation.

Then, any Con A not firmly bound to the glycoproteins is washed out with PBS buffer or 0.9% NaCl. Longer washing has no effect on elution of Con A from the blank. Some Con-A remains strongly bound to the blank acetate paper. To obtain reproducible results, it is therefore always necessary to use actate strips of exactly the same size for blank and sample. Finally, the Con A–glycoprotein complex is completely dissolved within approximately 45 min in a solution containing 2% SDS and 12% NH_4OH. The fluorescence is measured at 492 nm for excitation and 518 nm for emission. Care has to be taken that the slit on the excitation side is narrow enough to avoid overlapping with the emission side, since the maxima are very close together.

With this rather simple procedure, one can prepare, with FITC-labeled Con A and glucose oxidase as a glycoprotein, an acceptable standard curve ranging from 10 ng to 50 μg. The highest standard deviation is in the most sensitive range (Fig. 8). However, the use of a standard curve for quantitation in this glycoprotein method is not recommended. In addition to the problem of what to use as a reference glycoprotein, there are other reasons against the use of a calibration curve. The problem of the high blank value has already been mentioned. Further, the stability of the Con A solution is diminished by repeated freezing and thawing, and after each incubation of glycoprotein spots in the Con A solution, the Con A concentration will be reduced. This can easily be observed when several spots of high glycoprotein content are incubated successively. To counteract this, it is recommended that a fresh Con A stock

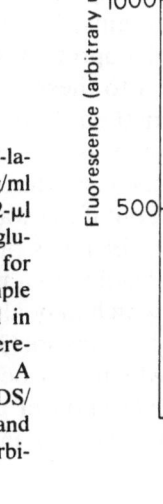

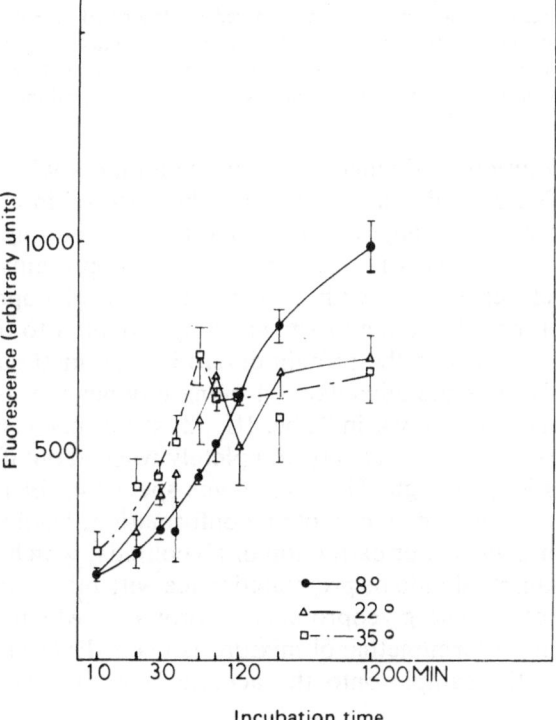

Fig. 7. Incubation in FITC-labeled concanavalin A (0.1 mg/ml in PBS buffer, pH 7.4) of 2-μl spots each representing 10 μg glucose oxidase. After incubation for different time intervals, sample and blank pieces are washed in PBS buffer (or 0.9% NaCl). Thereafter, the glycoprotein–Con A complex is solubilized in SDS/NH_4OH (2%/12% in water), and fluorescence measured in arbitrary units (bars, ±S.D.).

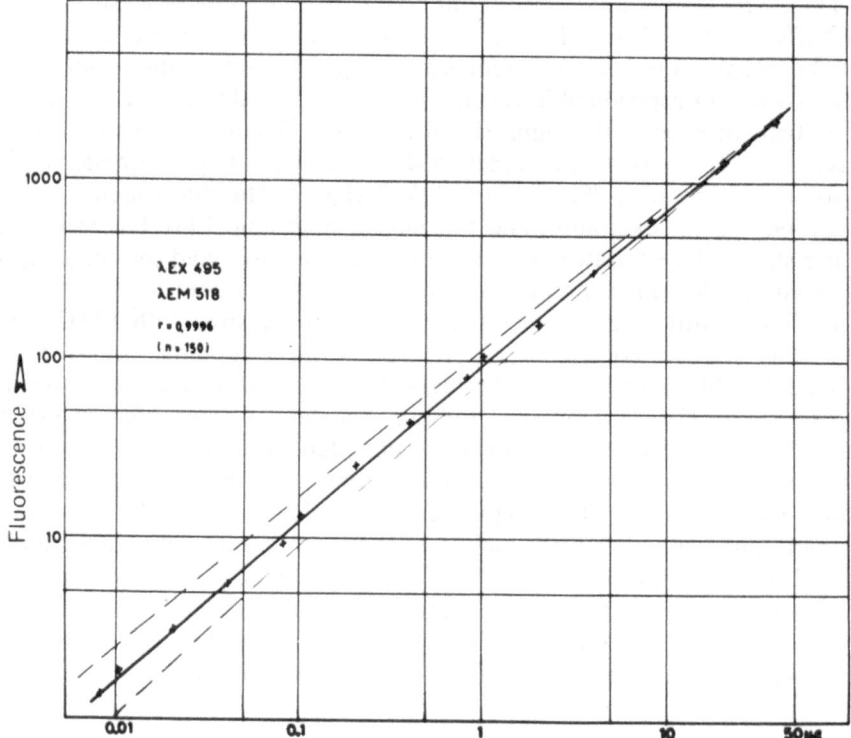

Fig. 8.   Calibration curve prepared as described in Scheme 2 with glucose oxidase and FITC-labeled concanavalin A. The dashed line represents standard deviation. Fluorescence was measured in arbitrary units at 495 nm at the excitation monochromator and 518 nm at the emission monochromator; 150 sample values were evaluated, resulting in a correlation coefficient of $r = 0.996$.

solution be divided into suitable aliquots which are kept frozen until needed. The diluted Con A solution is best stored in a normal freezer at 4°C if further use in a few days is expected; otherwise, storage is at $-20°C$. For determination of an unknown concentration of glycoproteins, it is recommended that some reference spots with concentrations of glycoprotein in the expected range be prepared and the unknown sample related to these.

As with the protein determination method described above, this method also has the advantage of being independent of most common external influences as shown in Table III. Most normally interfering compounds, with the exception of urea, are completely washed out if the washing time in PBS buffer is long enough. That urea gives special problems is presumably because of its influence on glycoprotein conformation, leading to a reduced binding capacity of Con A. For extraction of glycolipids, which may also react with lectins, the sample should be pretreated twice with ether/ethanol (3 : 2). One must be careful not to lose glycoproteins or proteins, which will occur if extraction with a chloroform/methanol mixture is used. Extraction of glycolipids after spotting of the sample onto the acetate layer by washing in ether/ethanol does not

Table III
Effect of Different Detergents on the binding of FITC-Labelled Concanavalin A to Glucoseoxidase: The Detergents Are Added to the Glucose Oxidase Solution Prior to Spotting of 2-μl Samples

Detergent	Washing time	
	1 × 10 min PBS	4 × 10 min PBS
None	100%	100%
2% SDS	78	100
1% SDS	100	100
0.5% SDS	100	100
2% SDS/12% NH₄OH	—	100
1% SDS (5 min, 100°C)	100	100
2% Triton X-100	62	100
1% Triton X-100	68	100
0.5% Triton X-100	80	100
2% NP40	70	100
1% NP40	77	100
0,5% NP40	77	100
2% Mercaptoethanol	100	100
1% Mercaptoethanol	—	100
1% Mercaptoethanol + 1% SDS	—	100
9 M Urea	—	50
6 M Urea	—	44
1% Desoxycholate	—	100

succeed. If a known amount of glucose oxidase is added to a brain tissue homogenate, membrane fraction, or myelin fraction, the 100% recovery demonstrates that there is no interference with compounds in the biological material that may inhibit the binding of Con A to glycoproteins. The binding of three different lectins to three different glycoproteins using spot analysis is shown in Table IV and demonstrates that this is a very simple method for testing the binding capacities of different lectins as well as for determining the influence of different ions, pH, etc.

Table IV
Binding of Different FITC-Labeled Lectins to Different Glycoproteins[a]

FITC-lectin[b] (0.1 mg/ml)	Glucose oxidase (5 mg/ml)	Ovalbumin (5 mg/ml)	Fetuin (5 mg/ml)
Con A	802	140	40
WGA	39	39	15
RCA 120	12	—	—

[a] Experiments are performed as described in Scheme 2 with 2-μl samples. Values in arbitrary units of fluorescence.
[b] Con A, concanavalin A; WGA, wheat germ agglutinin; RCA 120: *Ricinus communis* agglutinin, mol. wt. 120,000.

3.2. Simultaneous Determination of Protein and Glycoprotein

In cases in which the material to be analyzed is very limited or rare, one can perform the protein determination and glycoprotein determination with a single spot. The protein determination is performed with the chromophor Hoechst 2495 as described above. The stain is then eluted from the protein spot with 1% NH_4OH and measured fluorometrically. The normal calibration curve can be used as for the protein determination. Thereafter, the spots are washed first with PBS buffer (twice for 5 min and once for 10 min) and then transferred to the FITC-lectin and further treated as described in Scheme 2 (Appendix). Control experiments have shown that the values obtained by this double determination with a single spot are the same as those found with separate determinations of the protein and glycoprotein concentration using two different spots.

4. DETERMINATION OF SUGARS IN THE PICOMOLE RANGE

For the more detailed analysis of pure glycoproteins, which are often available only in minute amounts from biological material, it may be useful to have a highly sensitive method for the resolution and quantitation of the sugar content. A microprocedure to determine sugars in the picomole range has recently been described.[27] Sugars are reacted with dansyl hydrazine and amino sugars with dansyl chloride, respectively. Two-dimensional microchromatography on 3×4 cm micropolyamide sheets (see Appendix, Scheme 3) is used to resolve the sugar derivatives.

If the reaction is performed with a sugar mixture containing 10 mM each of fucose, ribose, arabinose, xylose, galactose, glucose, mannose, 2-deoxy-D-ribose, N-acetylgalactosamine, N-acetyglucosamine, glucuronic acid, and galacturonic acid, only 0.01–0.02 μl is needed for a microchromatogram. This small volume is applied with a very fine capillary under a stereomicroscope to a corner of the microchromatogram to give a starting point approximately 2 mm in diameter. After drying, several chromatograms are developed together using a special adapter or singly in a closed beaker of suitable size.

A UV photo[37] of such a microchromatogram, representing approximately 5 pmol sugar per single spot, is shown in Fig. 9. Dansyl *l*-proline is the internal standard. The reaction of the sugars and uronic acids is stoichiometric, thereby allowing quantitative determination. After scraping of the spots with a small microknife[38] and complete elution with absolute ethanol, they can be measured in a microcuvette fluorometrically at 340 nm for excitation and 515 nm for emission. Alternatively, the chromatogram can be automatically evaluated by scanning fluorescence microscopy as described below.

The amino sugars glucosamine and galctosamine do not react with dansyl hydrazine under acid reaction conditions, since the amino group is protonated and not the aldehyde group as is necessary for hydrazilation. Both amino sugars easily react with dansyl chloride to form a dansyl compound which can be separated chromatographically in the presence of dansyl amino acids. For the first dimension, acetone/H_2O (25:75) is used; for the second dimension, ethyl

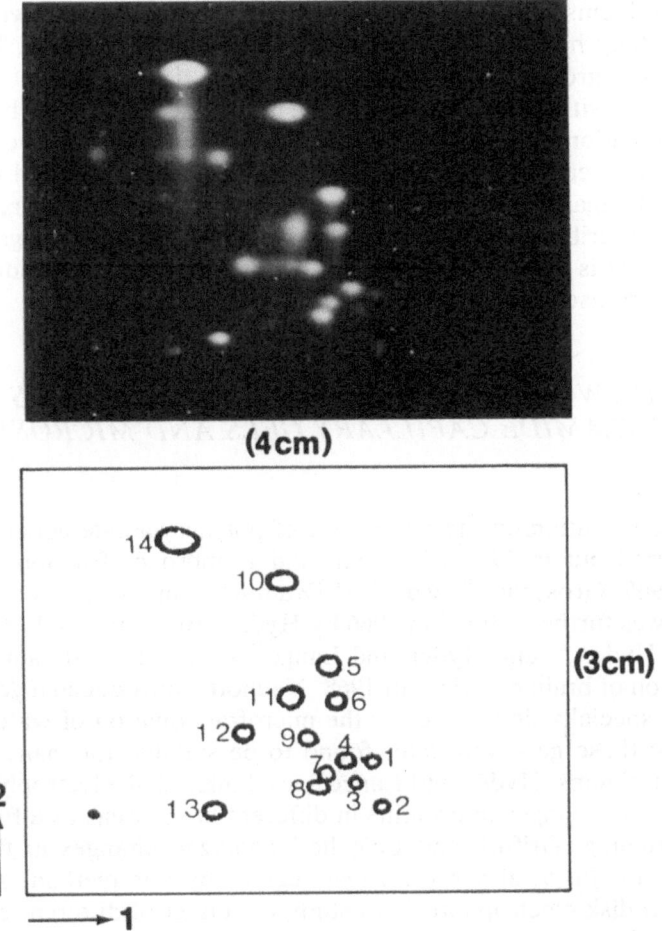

Fig. 9. Two-dimensional microchromatogram on a 3 × 4 cm micropolyamide sheet of dansyl hydrazone sugars and map for identification: 1, N-acetylgalactosamine; 2, galacturonic acid; 3, N-acetylglucosamine; 4, galactose; 5, fucose; 6, arabinose; 7, glucose; 8, mannose; 9, xylose; 10, dansyl-L-proline (internal standard); 11, ribose; 12, 2-deoxy-D-ribose; 13, glucuronic acid; 14, dansylhydrazine. Larger spots represents approximately 5×10^{-12} mol of the respective sugar, and faint spots approximately 5×10^{-13} mol sugar. Photo taken under UV illumination.[37]

acetate/isopropanol (65:35) is used. Under these chromatographic conditions, the dansyl amino acids line up in the first dimension and do not migrate in the second dimension where the two amino sugars are detectable.

N-Acetylamino sugars react readily with dansyl hydrazine but not with dansyl chloride. This is in contrast to amino sugars and occurs because the nitrogen atom in acetylamino sugars is an amide nitrogen which, under acid conditions, is not protonated. Instead, the aldehyde group is protonated and can react with dansyl hydrazine. At the present time, fructose and neuraminic acid present problems, for neither reacts with dansyl hydrazine or with dansyl chloride since they have no amino or aldehyde group. They may, however, through their keto groups be able to react under special conditions which are being investigated with another more reactive hydrazine than dansyl hydrazine. The optimal conditions for quantitative liberation of sugars from glycoproteins or glycolipids in microscale need to be elaborated. Conditions and problems for liberation of sugars from glcyoproteins using different procedures of acid hydrolysis are described elsewhere.[39] Determination of unbound sugars in any biological material is without problems so long as the reaction conditions described above are used.

5. ONE- AND TWO-DIMENSIONAL ELECTROPHORESIS IN POLYACRYLAMIDE CAPILLARY GELS AND MICROSLAB GELS

The first application on the microscale of polyacrylamide gel electrophoresis was carried out in 1964 when Pun and Lombrozo[1] fractionated brain proteins. In 1965, Grossbach[40] used 5-μl Drummond microcaps for this technique, which was further refined in 1966 by Hydén, Bjurstam, and McEwen,[41] McEwen and Hydén,[42] and Hydén and Lange[43] who used 2-μl capillaries for the fractionation of brain proteins. In 1968, Neuhoff[44] introduced a gel mixture that had been specially developed for the microfractionation of water-soluble brain proteins; these gels were later found to be suitable for many different fractionation problems. Hydén and Lange[45] used microdisk electrophoresis for the analysis of the changes in proteins in different brain areas as a function of intermittent training. Griffith and LaVelle[46] analyzed changes in the developmental proteins in facial nerve nuclear regions by this method. Ansorg *et al.*[47] used microdisk electrophoresis to study the effect of different extraction procedures on the pattern of brain proteins and on the heterogeneity of S-100 protein.[48] Nir *et al.*[49] used the method for the analysis of the effect of light on rat pineal proteins.

Microelectrophoresis may already be, or will be in the future, the most widely used micromethod in both biochemistry and neurochemistry, since this microscale method is very flexible and is easily adaptable to almost any problem of separation. Capillary electrophoresis can be used with homogeneous polyacrylamide gels[32,40–51] as well as with gradient gels,[32,52–62] for isoelectric focusing,[25,32,63–69] and in combination with immunoprecipitation.[32,70,71] Furthermore, any system of capillary electrophoresis can be combined for two-dimensional separations with homogeneous or gradient microslab gels.[72,73] All of these methods can be used as such or in suitable combination for enzyme analysis (for review see ref. 74).

5.1. Electrophoresis in Homogeneous Capillary Gels

The performance of these micromethods has been described in full detail previously[38] and in an abridged form by Osborne.[29] Therefore, only the main features of the technique will be described in this chapter. Generally, 5- or 10-μl capillaries (Drummond Microcaps, Brand Intraend) are used, since their inner diameter (0.42 and 0.60 mm, respectively) allows for easy sample application using fine capillary pipettes. Small 2-μl, 1-μl, or even 0.5 μl capillaries can also be used freehand without special equipment, but for sample application into smaller capillaries (0.1 or 0.05 mm inner diameter[40,63]), one needs a micromanipulator as well as a stereomicroscope. Cleaning the capillaries batchwise prior to use[32] is strongly recommended, since there should then be no problems using their capillary attraction for filling, simply by dipping them into a suitable gel mixture. For homogeneous gels, they are filled to approximately two-thirds of their total volume and are then pressed into a plasticine cushion covered with Parafilm.® After careful overlayering of the remaining space in the tube with water, they are kept in a moist chamber for polymerization which normally takes place in 1 hr.

Prior to use, the water layer is aspirated off, and a few millimeters of a rapidly polymerizing 5% collecting gel is introduced and again overlayerd with water. After polymerization of the collecting gel (10–15 min), the water can be exchanged against the sample to be fractionated. The gel concentration and a suitable discontinous or continous buffer system in the gel and at the electrodes can be adapted for any problem of separation, e.g., larger or smaller protein particles, with or without SDS, Triton X-100, urea, etc. Even for the separation of sensitive enzymes, cooling during electrophoresis (routinely performed to 60–120 V at 60–120 μA per capillary gel) is normally not necessary, since the advantageous surface-to-volume ratio in capillary gels creates no heating problems during the short period of 20 to 60 min required for electrophoresis. Bromphenol blue or another suitable dye can be used as front marker to define the endpoint of separation which is normally when the front marker reaches the lower end of the gel.

Any suitable power supply may be adapted for electrophoresis with capillaries, but a power supply that allows independent control of the current for each capillary[32] is recommended (obtainable from E. Schütt, Jr., Göttingen, FRG). For electrophoresis, the single capillaries are fixed in the rubber cap of a funnel and held in position with a suitable stand. When electrophoresis is finished, gels with low acrylamide concentration can be extracted from the capillaries with water pressure.[32] A well-fitting steel wire can be used for homogeneous gels, provided the lower capillary end is first filled with a small piece of plasticine to avoid the mechanical destruction of the gel with its separated protein bands.

Staining in 1% amido black 10 B in 7.5% acetic acid is done for only 10 min, whereas destaining in several baths of fresh acetic acid takes approximately 30 min. With Coomassie brilliant blue, staining and destaining take somewhat longer, even for capillary gels. The question of what dye to use for staining of proteins is often a matter of personal preference. The most commonly used dye, Coomassie brilliant blue, is not necessarily the best or most sensitive stain for proteins, compared with amido black. Coomassie blue may, however, be better for protein molecules of lower molecular weight. The optical evaluation of stained microgels is described in Section 7. Finally, to illustrate

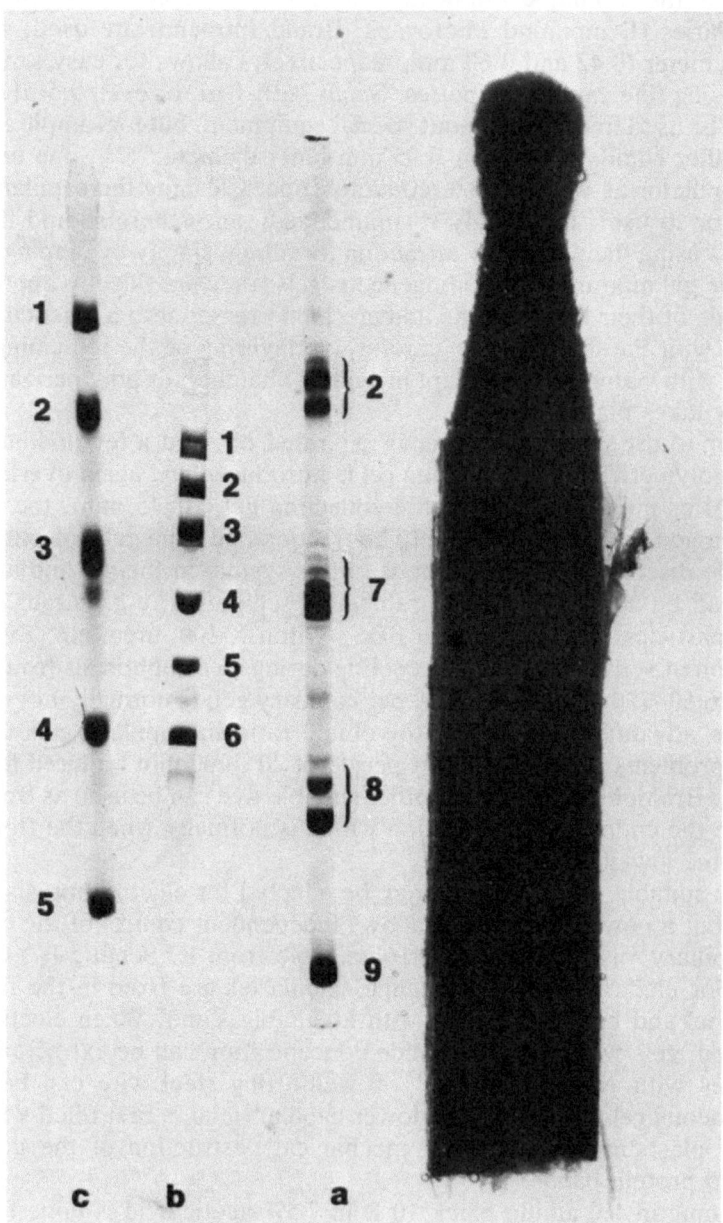

Fig. 10. Ten-microliter gels prepared in capillaries beside a normal matchstick. (a) Isoelectric focusing of a mixture of marker proteins (2 μg total protein) in a 5% polyacrylamide gel containing 9 M urea and 2% carrier ampholytes (pH 2–11); (b) 1–30% acrylamide gradient gel with 0.6 μg of marker proteins; (c) 15% homogeneous polyacrylamide gel with 0.6 μg of marker proteins. The following marker proteins were used: 1, phosphorylase *b;* 2, bovine serum albumin; 3, ovalbumin; 4, carbonic anhydrase; 5, soybeam trypsin inhibitor; 6, α-lactalbumin; 7, β-lactoglobulin; 8, myoglobin (horse); 9, cytochrome *c*.

the dimensions of microgels, a homogeneous gel, a gradient gel, and an IEF-gel, all prepared in 10-µl capillaries, are compared beside a normal matchstick in Fig. 10.

5.2. Microgradient Gels

Protein mixtures are generally more sharply resolved in gradient gels. In addition, gradient capillary gels are more easily prepared than homogeneous gels (with the special 5% collecting gel), since the top of the gradient gel can be made of only 1 or 2% acrylamide which is suitable for stacking of the protein mixture. Microgradient gels can be prepared in capillaries of different diameter or length in batches by using a special apparatus[32] or, according to Rüchel *et al.*,[52] freehand with single capillaries by using capillary attraction. For this purpose, the gel mixture is divided in two parts. One contains only the catalyst ammonium peroxodisulfate in a suitable concentration and buffer. The second part contains the acrylamide, bisacrylamide, and TEMED, again in suitable concentration and buffer. This mixture finally determines the slope of the acrylamide concentration along the linear gradient and can be made up at any suitable concentration according to the problem of separation. The maximal concentration of acrylamide can be as high as 53%, and the slope can further be influenced by adding sucrose to this mixture.[32,52]

To form the acrylamide gradient, the capillary is first half filled by capillary attraction with the ammonium peroxodisulfate solution and finally completely filled up to the rim by dipping in the correct concentrated acrylamide mixture. This mixture is thereby linearly diluted from the top to the bottom of the capillary tube. The catalyst is diluted in the opposite direction, thereby guaranteeing that on top, where the acrylamide concentration has its minimum, the catalyst starting the polymerization of the acrylamide has its maximum. This gradient formation can easily be observed if some stain is dissolved in the concentrated acrylamide mixture. The linearity of the gradient can be densitometrically checked if albumin is incorporated into the acrylamide mixture and is finally stained with amido black 10 B. After the gradient mixture is taken up, the capillary is immediately transferred into a small beaker in the bottom of which is a solution of 50% sucrose and some ammonium peroxodisulfate to insure a complete sealing of the capillary and a complete polymerization of the highly concentrated acrylamide at the lower end of the gradient. Capillary forces are again sufficient to hold the capillary with the gradient in an exactly upright position if the capillary is put onto the wall of a small glass beaker.[32]

After polymerization, the capillaries can be transferred into a closable tube filled with suitable buffer for prolonged storage at 4°C. Prior to use, the unpolymerized upper phase is aspirated away, and sample application and electrophoresis can follow without further manipulations. After electrophoresis, the gels are pulled out with a well-fitting steel wire from the high-concentration end, thereby avoiding damage to the separation area. Since Triton X-100 need not be incorporated into gradient gels, they are excellently suited for all types of SDS electrophoresis and can especially be used to demonstrate the types of artifacts that can easily be produced during SDS electrophoresis[53] in micro-

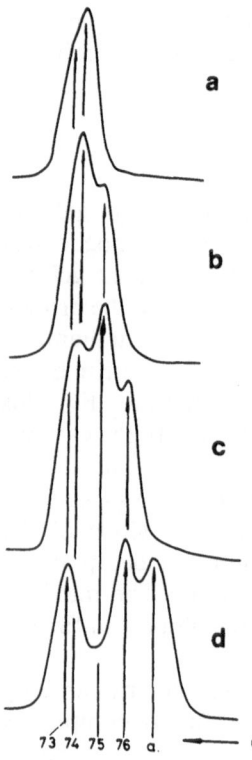

Fig. 11.   Separation of different mixtures of tRNA^{phe}. The numbers of the peaks correspond with the numbers of the 3'- end nucleotide of the tRNA; a indicates the position of the amino acylated tRNA; (a) tRNA^{phe} A$_{73}$ + tRNA^{phe} C$_{74}$; (b) tRNA^{phe} C$_{74}$ + tRNA^{phe}C$_{75}$; (c) tRNA^{phe}C$_{74}$ − tRNA^{phe}C$_{75}$ + tRNA^{phe} A$_{76}$; (d) tRNA^{phe}A$_{73}$ + tRNA^{phe}A$_{76}$ + Phe-tRNA^{phe}C-C-3' NH$_2$A.

as well as in macro gels if the conditions of SDS electrophoresis are not carefully controlled. An example of the very high resolving power of these microgradient gels is shown in Fig. 11. The difference in only one nucleotide is enough for separation of tRNA species in a 1–40% microgradient gel.[75] For further examples with separated proteins from brain tissues, see refs. 2,25,32,41–49, 56,60–62.

5.3. Isoelectric Focusing in Capillary Gels

With the development of ampholytes,[76,77] it became easier to fractionate protein mixtures into single-protein components according to their isoelectric points. Isoelectric focusing (IEF) was initially carried out in saccharose gradients[78] and later in polyacrylamide gels.[79,80] In a comprehensive review, Haglund[81] has described the historical development, the theoretical foundation, and the application on the macro scale of this very important and powerful method for protein fractionation. The micro version of this method, described by Riley and Coleman[82] and Catsimpoolas[83] requires 10 μg of a single protein or 0.2 to 0.4 mg of a protein mixture for a single fractionation. Quentin and Neuhoff,[25] Grossbach,[63] and Gainer,[64] working independently, refined the IEF methods further; today, they are routinely performed in 5- or 10-μl capillaries with the same equipment described for homogeneous or gradient microgels.

Even though one needs only microgram amounts or less of a protein mixture for IEF, it is sometimes advantageous, especially in combination with two-dimensional separations, to separate larger amounts of a protein mixture (up to 20 μg per 10-μl gel), and this can be achieved without danger of overloading with sample. Micro-IEF is, furthermore, excellently suited for fractionation of isoenzymes followed by enzyme activity determinations.[25,32,66–68,74]

Since all technical details of IEF in capillary gels have been extensively described,[32,66–68] only some principles are described here. For IEF, rather soft gels are used, with only 4–7% acrylamide. For cross linking, relativly high concentrations of bisacrylamide (C = 2–4%) or diallyltartrate diamide[84] (DATD, C = 15%) were used to produce so-called nonrestrictive gels. This avoids sieving effects which are one of the main features of homogeneous and especially gradient gels.

Capillaries are filled as usual by capillary attraction with a suitable gel mixture containing 2–4% carrier ampholytes of suitable pH range and are overlayered with water. After polymerization, the water phase is removed, the sample is applied, and this is then carefully overlayered with a cushion of a suitable solution (e.g., 10% glycerine in H_2O) to avoid direct contact of sample with the acidic anolyte. Whether or not carrier ampholines have to be added to the sample has to be determined empirically, since it depends on the protein mixture to be focused. In many instances, this is not required. There is no strict rule for an optimal anolyte or catolyte—most of the published systems can be used. One of the benefits of microelectrophoresis is the very short separation time which, for IEF, is approximately 10 min. Therefore, one can find out easily by trial and error experiments which system is optimal for the separation problem in question.

The current and voltage (normally 200 V) also need to be optimized for the separation problem. Too high a voltage can easily produce artifacts, as has been found, for example, in the fractionation of LDH isoenzymes.[66,67] If well-separated and stained[66] protein separations, as shown in Fig. 10, are not achieved, this is an indication that the optimal conditions still need to be defined.

5.4. One- and Two-Dimensional Microslab Gel Electrophoresis

All three types of microelectrophoresis in capillaries are well suited for reproducible fractionation of complex protein mixtures. All three methods, however, have the same disadvantage in that a direct comparison of different sample separations is not possible in the same way that it is when slab gels are employed. Poehling and Neuhoff[72] have adapted the widely used one-dimensional[85] and two-dimensional[86,87] slab gel techniques to a very flexible and sensitive microsystem.

5.4.1. Preparation of Microslab Gels

The chamber in which microgels are cast is prepared from microscope slides cut to a suitable size (2.5 × 3 cm or 3 × 3.5 cm, Fig. 12). Small strips (approximately 1–2 mm wide) cut from plastic sheets of a suitable thickness are used as spacers. Gel chambers are laterally sealed by special Teflon® clamps (Fig. 12.3), dental wax, or electrical tape. However, chambers sealed with wax are not suitable for long storage and require more careful handling. Sealing of the lower end of the chambers prior to filling with the gel solution is not necessary if the spacer is less than 0.25 mm thick. Capillary forces are sufficient to keep the solution between the plates. Chambers wider than 0.25 mm are sealed by pressing them into a cushion of plasticine coated with Parafilm® (Fig. 12.3). For fixation of the polyacrylamide microslab gels to glass plates (a procedure that is optimal for photometric or autoradiographic evaluation after staining and drying), the glass plates are first silanized according to Radola.[88] The upper buffer reservoir is constructed from Plexiglas,® and the bottom has a suitably sized slit and is covered with a layer of soft rubber (approximately 5 mm thick) in order to hold the gel chamber firmly in a vertical position (Fig. 12.5). The edges between the gel chamber and the soft rubber are then sealed with agarose prior to filling with the electrode buffer. The lower buffer reservoir is a beaker of a suitable size which also supports the upper reservoir (Fig. 12.5). For two-dimensional separations, isoelectric focusing is performed in capillaries as described above.

5.4.2. Homogeneous Microslab Gels

Gel mixtures are often prepared, according to the method of Laemmly,[85] from stock solutions of 30% acrylamide and 0.8% bisacrylamide, but, depending on the separation problems, any other gel system could be used as well. To adapt to the fractionation range required, homogeneous separating gels can be made up with 10, 15, or 20% acrylamide and 375 mM Tris HCl, pH 8.8. Stacking gels are made up from 3% or 6% acrylamide and 125 mM Tris HCl, pH 6.8. Polymerization is achieved with 0.025% TEMED and 0.025% ammonium peroxodisulfate. Chambers with spacers of 0.3 to 1 mm thickness are filled in a vertical position using fine glass pipettes until the separation gel solution fills approximately two-thirds of the chamber. This is then carefully overlayered with water. Chambers with thickness less than 0.3 mm can be filled by capillary attraction and thereafter fixed to a stand. In order to fill several gel chambers simultaneously, a chamber used for preparation of gradient gels (see Fig. 14) may be utilized. After polymerization (approximately 30 min), the water layer is aspirated off, and the stacking gel solution is added up to the rim of the chamber. Into this solution a Teflon® comb is introduced (Fig. 12.1) to form sample wells of 1 mm width 3 mm deep.

After removal of the Teflon® comb, the sample wells are carefully cleaned with a capillary pipette[38] connected to a water vaccum pump and are then completely filled with a suitable buffer (e.g., electrophoresis buffer). The sample, containing 15–20% glycerol and 0.001% bromphenol blue, is carefully underlayered in the sample wells. This step is especially important if wells less than 1 mm in width are used; otherwise, the samples may diffuse out of the wells during application. Unevenly distributed sample solution in the wells reduces the resolution capability of the system. The end sample wells are only

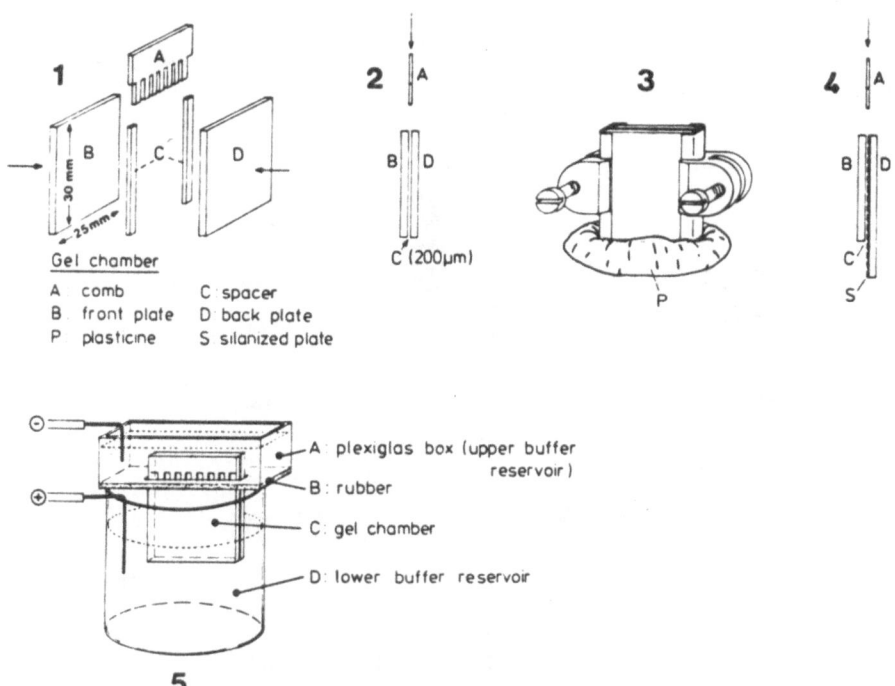

Fig. 12. Schematic drawing of the technical equipment and electrophoresis stand for microslab gels. 1 and 2 show the microgel chamber; 3 the Teflon clamps; 4 the chamber suited for preparation of gradient gels on a silanized glass plate; 5 the electrophoretic stand.

filled with sample buffer or marker proteins, since in these lanes the separation can be impaired if the gel chamber is not completely sealed (compare Fig. 15C). A 50 mM Tris/glycine buffer, pH 8.4, with 0.1% SDS is used as electrode buffer, and electrophoresis is started at low constant voltage. After the sample has stacked, and a sharp bromphenol blue front is formed, electrophoresis is continued at a higher constant voltage. An example of the resolution that can be achieved with a 15% homogeneous SDS microslab gel (3 × 3.3 × 0.03 cm) is shown in Fig. 13. This shows SDS electrophoresis of water-soluble rat brain proteins, purified rat brain myelin (0.3 μg total protein per well), and a mixture of marker proteins (bovine serum albumin, ovalbumin, chymotrypsinogen A, and cytochrome *c*, 0.4 μg total protein per well). Proteins of high molecular weight are very well separated into sharp bands in this system; peaks from proteins of lower molecular weight are more diffuse.

5.4.3. Gradient Microslab Gels

Gradient microslab gels can also be prepared, as are capillary gels, free-hand by using capillary forces to form convex gradients as well as linear gradients.[72] It is, however, easier and more reproducible to use the Plexiglas chamber in Fig. 14 for simultaneous preparation of batches of polyacrylamide gradient gels. For example, several gel chambers (3 × 3.5 cm with variable

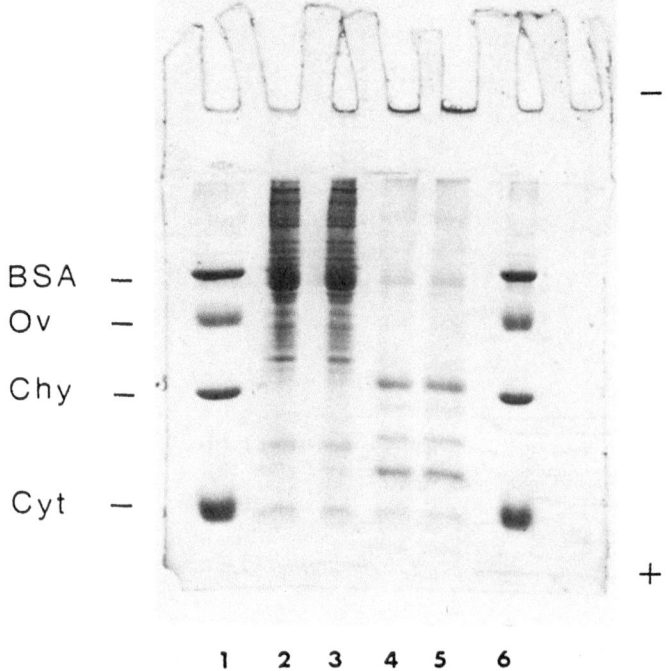

BSA —
Ov —
Chy —
Cyt —

1   2   3   4   5   6

Fig. 13. SDS electrophoresis in a homogeneous polyacrylamide gel prepared in a 3 × 3.5 cm chamber and swollen after staining to 3.5 × 4 cm. Stacking gel: 6% acrylamide, 125 mM Tris HCl, pH 6.8, and 0.1% SDS; thickness of the gel 0.35 mm; electrode buffer 50 mM Tris/glycine, pH 8.4, and 0.1% SDS; electrophoresis: 60 V, 15 min, and 120 V, 50 min. Sample wells 1 and 6 are loaded with a mixture containing 0.1 μg each of bovine serum albumin (BSA), ovalbumin (Ov), chymotrysinogen A (Chy), and cytochrome *c* (Cyt). Wells 2 and 3 are each loaded with 0.6 μg soluble rat brain proteins, and wells 4 and 5 with 0.3 μg rat myelin. All samples contain 1% SDS, 15% glycerol, and 0.001% bromphenol blue.

Fig. 14. Plexiglas® chamber for simultaneous preparation of several polyacrylamide gradient gels. Not shown is the usual gradient mixer connected to a peristaltic pump. The six microgel chambers are sealed with Teflon® clamps (compare Fig. 12). By means of screws connected to Plexiglas® rods, the front plate is held in position and may subsequently be removed to facilitate removal of the polymerized acrylamide block.

spacer width 0.1 to 1 mm) can be sealed with Teflon® clamps which then rest on the Plexiglas® rods in the simple Plexiglas® gradient chamber (Fig. 14). Thereafter, the chamber is filled to the rim with water or 0.1% SDS to completely eliminate capillary forces within the gel chambers when small spacers are used. A suitable preformed gradient mixture is slowly pumped through a central hole in the bottom of the chamber. The ascending gradient mixture displaces the water in the gradient chamber through the overflow pipe (Fig. 14).

After the microgel chambers are filled to a suitable height with the gel mixture, a buffered 50% glycerol solution (375 mM Tris HCl, pH 8.8, containing 0.01% TEMED and 0.1% ammonium peroxodisulfate) is pumped into the bottom of the chamber to a depth of 2 mm (for suitable gel mixtures see ref. 72). Filling through the central hole in the bottom of the gradient chamber leads to a horizontal distribution of the gradient solution if the filling speed is limited so that turbulence in the gradient is minimized. Turbulence can be controlled visually if 0.01% bromphenol blue is incorporated into the acrylamide stock solution. This also provides for densitometric determination of the effective slope and linearity of the final gradient. Preparation of stacking gels, sample wells, sample application, staining, and destaining are as described above. When storage of the slab gels is required, it is recommended that a gradient-forming chamber that can be taken apart easily (Fig. 14) be used. The whole gel block with its enclosed slab gel chambers may be removed and then stored in gel buffer at 4°C. Slabs can be cut out of the block as required.

A fractionation of complex protein mixtures on microslab gels of polyacrylamide gradients is shown in Fig. 15. In the separation in Fig. 15A, a gradient gel with a convex gradient prepared by capillary attraction has been used to fractionate marker proteins, soluble rat brain proteins, and rat brain myelin (prepared according to Norton and Poduslo[89]). The protein peaks are very sharp but narrowly spaced. The same samples were also fractionated in a linear gradient prepared by capillary attraction (Fig. 15B). The reproducibility of fractionation is strongly dependent on the reproducibility of the gradient in the gels. This is difficult to achieve and requires more experience in "handmade" gradient gels.

Separations in gradient gels prepared with a gradient mixer are shown in Fig. 15C and D. In Fig. 15C, the same proteins are separated as in A and B, but to demonstrate the importance of good sealing of the gel chamber, the Teflon® clamps were removed and put on again after polymerization was complete. Contact between gel and spacer was disturbed, with the result that the separations in the outer wells were disturbed. That microgradient slab gels are optimally suited for comparative studies is shown in Fig. 15D. In this gel, with a linear gradient of 6–30% acrylamide, total brain proteins of control rats of different ages (10, 15, and 20 days old) are compared with those of rats with experimental phenylketonuria.[90-92] The appearance of different protein bands according to age and variations caused by the experimentally induced state is clearly visible.

Homogeneous microslab gels can be used for IEF as well (Fig. 16); for technical details consult ref. 72. Recently, ultrathin slab gels for macro-scale

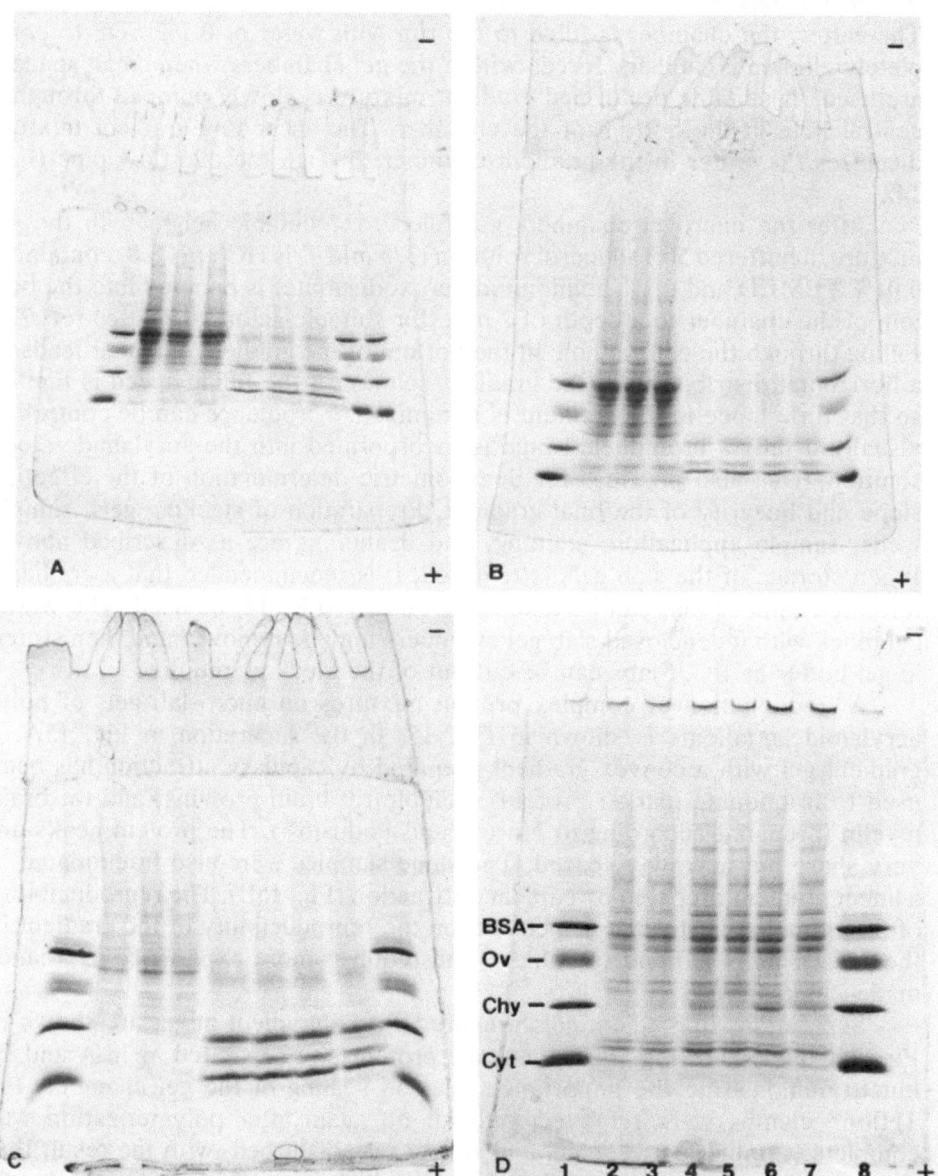

Fig. 15. SDS electrophoresis in different gradient microslab gels. In A and B, the gradients (2.5 × 3.0 × 0.02 cm) were prepared using capillary attraction to form (A) a convex 2–40% and (B) a linear 2–30% acrylamide gradient. The linear 6–25% gradient gels (C and D) (3.0 × 3.5 × 0.5 cm) are prepared with the gradient-forming chamber (Fig. 14). The proteins separated in A, B, and C are the same as in Fig. 13. In D, 0.5 μg/well total brain proteins are separated from control rats of different ages: 10 days old (lane 2), 15 days (lane 4), and 20 days (lane 6); and from rats of the same age having experimentally induced phenylketonuria, lanes 3, 5, and 7. Lanes 1 and 8: marker

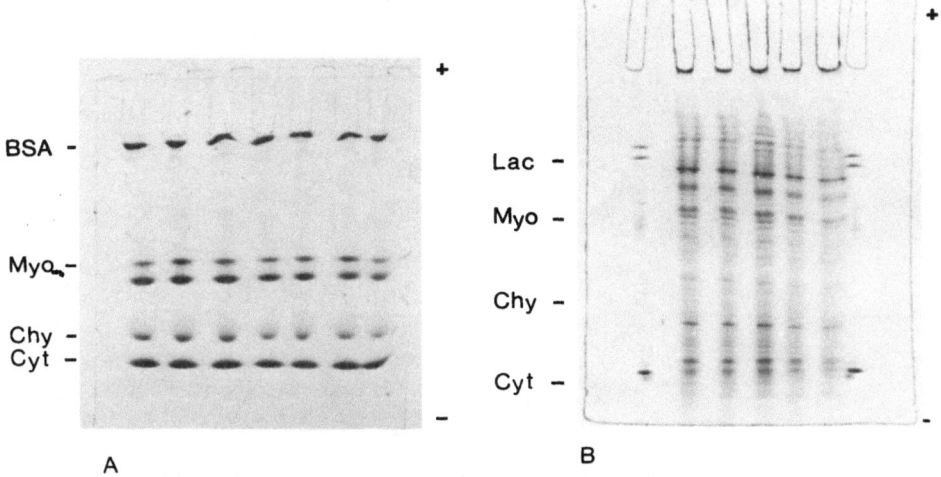

Fig. 16. One-dimensional IEF of (A) marker proteins and (B) total rat brain proteins in 2.5 × 3 × 0.02 cm homogeneous slab gels under different experimental conditions: (A) 7.5% acrylamide, 4% Servalyt®, pH 2–11, prefocusing at 60 V for 15 min, focusing after application of each 0.3 μg of bovine serum albumin (BSA), horse myoglobin (Myo), chymotrypsinogen A (Cyt) at 150 V for 15 min followed by 200 V for 15 min with 0.1% acetic acid as the anolyte and 370 mM Tris as the catholyte; (B) 7.5% acrylamide, 1% Servalyt® (2 parts pH 2–11, 4 parts pH 5–7), 9 M urea, prefocusing at 80 V for 15 min followed by sample application and focusing at 150 V, 15 min, 200 V, 15 min, and 250 V, 15 min. In 1 and 6, 0.1 μg of each marker protein [instead of BSA lactoglobulin (Lac) was used]; 2–5, 0.5 μg of soluble rat brain proteins.

separations by IEF[88,93] have been described. These may easily be converted to the micro scale. For routine use, microslab gels 0.3 to 0.5 mm thick with sample wells of 0.7 to 1 mm width and 3 mm deep are optimal. A suitable sample volume is approximately 1 μl of a solution containing 1 mg/ml protein. An increase in sensitivity is achieved by using both thinner slabs and smaller sample wells to achieve concentration of the proteins into the smallest possible band. However, the absolute detection limit is dependent on the stainability of a protein. Both amido black and Coomassie blue can resolve in the range of 10^{-8}–10^{-9} g per single protein band. Silver staining[73,94–97] is more sensitive than Coomassie blue staining but is not stoichiometric and is very difficult to handle.[97]

5.4.4. Two-Dimensional Microslab Gel Electrophoresis

For the two-dimensional separation of protein mixtures in microslab gels, IEF in microliter capillaries in the first dimension is combined with electrophoresis in microslab gels for the second dimension. Either homogeneous or gradient microslab gels can be used, the most suitable size being 3 × 3.5 cm,

---

proteins, 0.15 μg of each. Note that in C, the uneven separation in the outer lines of the marker proteins is produced artificially in that the Teflon® clamps of the gel chamber were temporarily removed before electrophoresis.

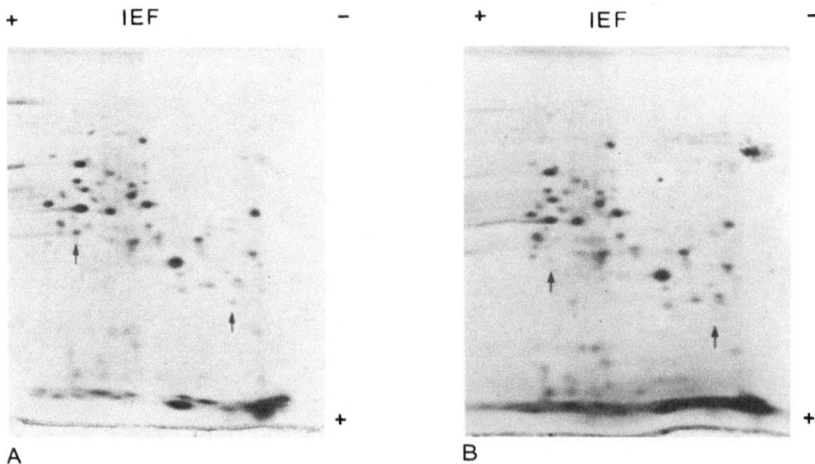

Fig. 17. Two-dimensional electrophoresis of (A) total spinal cord proteins of 30-day-old rats and (B) of rats of the same age but with experimental phenylketonuria. Spinal cords were homogenized, and urea added to a final concentration of 9 M to give a clear solution after centrifugation. Eight micrograms of protein were used for the IEF in the first dimension. The IEF gel was prepared in a 10-μl capillary, and electrophoresis performed with 0.5% phosphoric acid as the anolyte and 0.5% ethylenediamine as the catholyte. Separation in the second dimension was performed in a 3 × 3.5 × 0.075 cm, 15% homogeneous acrylamide gel containing 0.1% SDS. One of the glass plates of the gel chambers was silanized, resulting in a prolonged staining and destaining time of the gel fixed to the glass plate without interfering with the protein separation. The most obvious differences between the tissue extracts from (A) control and (B) rats with phenylketonuria are labeled with arrows. Note that in the homogeneous gel, several proteins migrating with the buffer front are not separated but that the clearly separated protein spots alone amount to approximately 400.

0.25 to 1 mm thick (for suitable gel mixtures see ref. 72). The free space on top of the stacking gel is filled with either an acrylamide gel mixture that polymerizes within 3–5 min or with 0.3% agarose (in 125 mM Tris HCl, pH 6.8, with 0.1–1% SDS). With fine forceps made from elastic spring steel,[38] the IEF gel is transferred into this embedding solution and carefully pushed down onto the stacking gel. If gel chambers with spacers of 0.25 mm are used, the upper inner sides of the two glass plates have to be obliquely ground to form a V-shaped groove. When agarose is used as the embedding gel, which is imperative for the complete electrophoretic elution from IEF gels, [72] the gel chamber, the foreceps, and the agarose have to be warmed to between 40°C and 50°C.

The two-dimensional microgel system has a similar resolving power to that obtained with macrosystems. As an example, the two-dimensional separation of total spinal cord proteins from (1) a 30-day-old control rat and (2) a rat with experimental phenylkotonuria[90–92] is shown in Fig. 17. Independent of the differences caused by the experimental phenylketonuria (two of which are labeled with arrows), the resolving power of the homogeneous 15% acrylamide gels is good enough to separate approximately 400 spots, although, under these conditions, the low-molecular-weight proteins are not separated. It is possible that more spots would be detected if either silver staining[73,94–97] or radioactive labelling[73,98] of proteins were used.

A crucial point in a two-dimensional separation is the embedding of the IEF separation on top of the slab gel. If this is not properly achieved, there may be incomplete electrophoretic elution of the sample out of the first-dimension gel. The most commonly used embedding medium is a fast-polymerizing acrylamide gel in order to avoid diffusion of the separated proteins in the first-dimension gel. The effect of two different embedding media on the final two-dimensional protein pattern is shown in Fig. 18. The two gels illustrated are identical in the two dimensions, but the embedding media were 6% acrylamide with 1% SDS in A and 0.3% agarose with 1% SDS in B. Note that not only is elution incomplete in the former case but also the elution of the proteins is highly variable. Spots are missing on the gel in Fig. 18A (labeled with arrow) that are clearly visible in the other slab gel (Fig. 18B).

The same observation is also made when even higher SDS concentrations are used in the embedding acrylamide gel and/or in the electrode buffer. To gain some insight into the reasons for this observation, control experiments

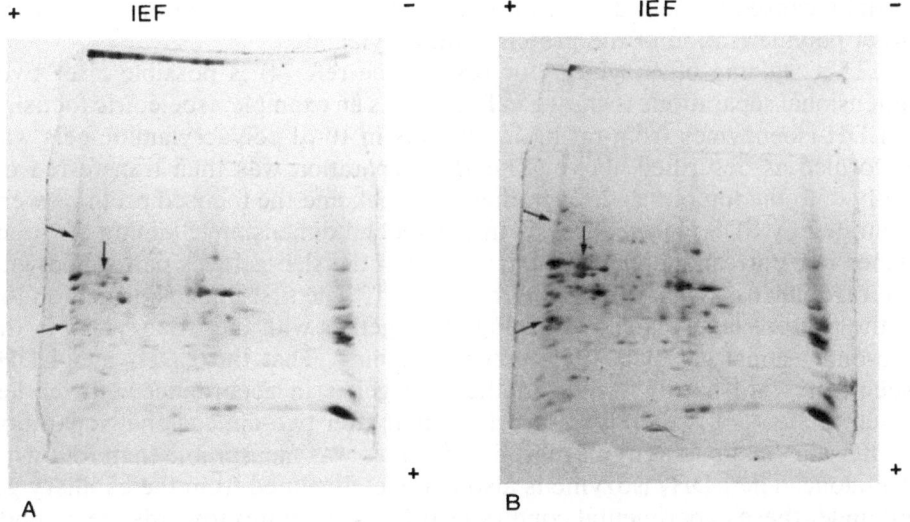

Fig. 18. Two-dimensional separation of 6 μg total proteins from rat brain under identical conditions in the two gels with the exception of embedding the IEF gel (A) in 6% acrylamide with 1% SDS and (B) 0.3% agarose with 1% SDS. For IEF in the first dimension: 10-μl capillaries, 4% acrylamide, 1% Servalyt® (2 parts pH 2–11, 2 parts pH 2–4, 2 parts pH 4–6 and 2 parts pH 5–7), 9 M urea (in gel and sample); anolyte, 0.5% phosphoric acid; catholyte, 0.5% ethylenediamine. Second dimension: 6–25% acrylamide gradient gel (3 × 3.5 × 0.075 cm) electrophoresis for 15 min at 60 V and 40 min at 120 V. Note in A that the IEF gel is not completely eluted because of embedding in fast-polymerizing acrylamide and that, in contrast, in B, the IEF gel is eluted completely because of embedding in agarose. The most obvious differences between the gels are labeled with arrows showing that several proteins are not eluted from the IEF gel embedded in acrylamide.

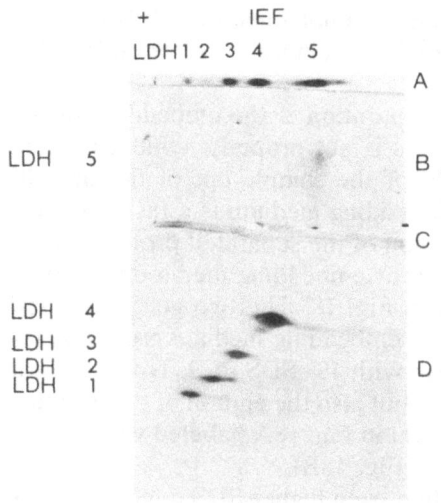

Fig. 19. Two-dimensional separation of soluble rat brain extract stained for LDH with the tetrazolium assay.[32] The IEF for the first dimension was performed in a 10-μl capillary (5% acrylamide, 4% Servalyt®, pH 2–11, 100 V for 45 min, anolyte 0.1% acetic acid, catholyte 370 mM Tris). Second dimension: 2–25% acrylamide gradient gel (3 × 3.5 × 0.075 cm), 40 V for 15 min followed by 80 V for 40 min. The IEF gel C was placed directly on top of the gradient gel D and overlayered with 1% agarose (B) in 125 mM Tris HCl, pH 6.8. The IEF gel A on top of the slab gel is a control gel with the same volume of extract as in gel C, showing the separation of the LDH isoenzymes in the first dimension after staining in the tetrazolium assay for the same time (30 min) as the staining of the slab gel.

were performed with bovine serum albumin.[72] It was found that because of the speed of polymerization in the 6% embedding polyacrylamide gel, up to 33% of the total protein can be firmly bound into the gel. This effect may be partially caused by heat denaturation during rapid polymerization and also by direct polymerization of the protein with acrylamide.

That staining of enzymes (for review see ref. 74) is possible after two-dimensional separations is shown in Fig. 19. As an example, isoelectric focusing of LDH isoenzymes from rat brain extracts in 10-μl polyacrylamide gels was performed as described above. The IEF separation was then transferred directly onto the top of a 2–25% gradient slab gel, and the focused proteins were separated by SDS electrophoresis in the second dimension. Staining of isoenzymes in a corresponding capillary gel and in the slab gel was performed with a tetrazolium assay mixture as described.[25,74,99] The first-dimension gel (on top of the agarose layer) is shown in Fig. 19 together with the isoenzymes in the two-dimensional gel slab after enzyme staining. That the LDH_1 and LDH_2 isoenzymes are hardly visible in the IEF gel is in accordance with earlier observations.[67] It is therefore surprising that after two-dimensional separation greater enzymatic activity of these isoenzymes was measurable than following IEF alone. The LDH_5 isozyme is also completely eluted from the capillary gel but under these experimental conditions (pH 8.4) migrates towards the cathode and is visible as a diffuse spot in the agarose layer. The observation that after two-dimensional separation there are additional spots below LDH_4 and beside LDH_2 requires further investigation.

Automated photometric evaluation of slab gels with the capability of measuring optical densities between 0.5×10^4 and 2.5 O.D. (units of optical density) combined with automated spot detection (see below) is of increasing interest, especially for the two-dimensional separation of proteins in genetic screening. Poehling and Neuhoff have investigated[72] whether the silanizing of one glass plate of the slab gel chamber according to the method of Radola[88] could be adapted to the microslab gel method. The fixation of the gel slab to an optically clear support is of great advantage in the case of photometric evaluation on a high-speed scanning stage. As has been already shown,[88] silanization does not interfere at all with the protein separation. There is, however, a problem with gradient gels in that the gel can dry unevenly because of the high gel concentration at the lower end. This can easily be avoided by having the silanized glass plate approximately 0.5 cm longer than the opposing plate (see Fig. 12.4). Before staining, this extra portion of the gel is scraped off, with the result that after drying one obtains a slab gel well suited for photometric evaluation.

6. DETERMINATION OF AMINO ACIDS AND RELATED COMPOUNDS WITH DANSYL CHLORIDE IN THE PICOMOLE RANGE

Dansyl chloride (dans-Cl, 1-dimethylamino-naphthalene-5-chloride) was first used by Weber in 1952 for the preparation of fluorescent conjugates of albumin. This reagent has subsequently found almost as wide an application as Fischer's naphthalene sulfonyl chloride or Sanger's 2,4-dinitroflurobenzene (for review, see refs. 101–103). Its usefulness stems from the fact that its reaction products with amino acids, amines, peptides, proteins, phenols, imidazoles, and sulfhydryl groups have an intense yellow to yellow–orange fluorescence and can be separated easily with suitable chromatographic systems.[102] Woods and Wang[103] first described the fractionation of dansylated amino acids on polyamide sheets (for review see ref. 104), and Gray and Hartley[105] introduced this method of separation for the determination of end groups and in sequence analysis of proteins and peptides.

It was B. S. Hartley who, in 1969, suggested adapting this technique to the micro-scale. This was immediately successful; the normal 15 × 15 cm polyamide sheets were simply replaced by 3 × 3 cm ones, and the application of the dansylated sample was performed with a very fine capillary under the stereomicroscope. On a 15 × 15 cm polyamide layer, about 10^{-9} mol of each dansylated amino acid is detectable; using 3 × 3 cm polyamide layers, as little as 2×10^{-13} mol can be detected. However, it subsequently became obvious that, in order to obtain consistent data even when only semiquantitative results are required, a number of factors have to be taken into account. This is particularly important when working on the micro scale.

6.1. Practical Procedure

The reaction of the amino acids with dansyl chloride (dans-Cl) depends on a variety of conditions, e.g., concentration of dans-Cl, time of reaction, temperature, and pH. Neadle and Pollit[106] and Seiler[101] have shown that the carboxyl group of the N-dansyl amino acid formed can react further with excess dansyl chloride to give dansylamine with the degradation of the dansyl amino acid. The yield of dansyl amino acid is consistent for individual amino acids, and so is the quantity of dansylamine that is formed simultaneously. If a mixture of seven amino acids is allowed to react with different concentrations of dans-

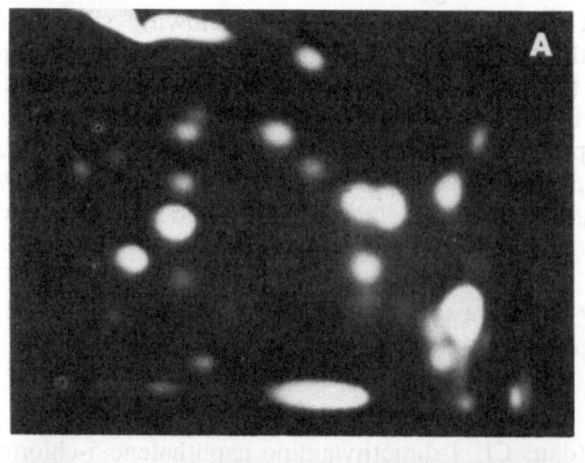

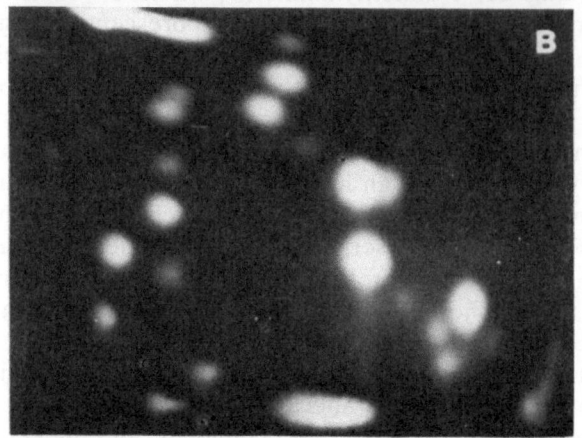

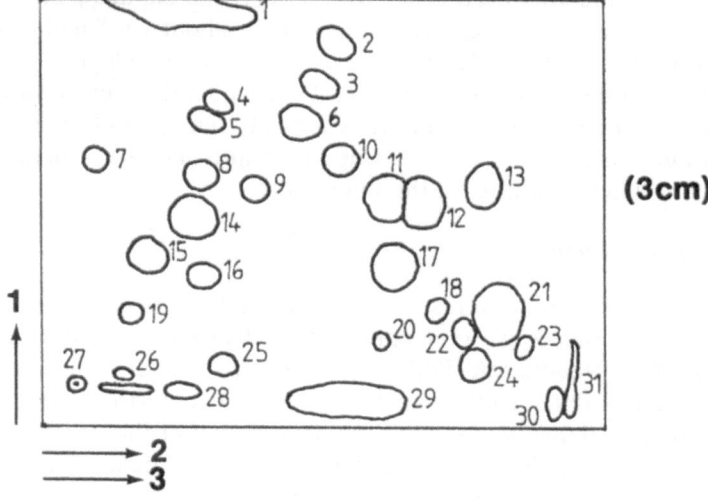

(3cm)

Cl, the best yield of dansyl amino acid is obtained with a 5.9-fold excess per reactive group. Again, differing reactions are observed for the individual amino acids when a mixture of 24 amino acids is treated with dansyl chloride (for details see ref. 107). The dependance of the dansyl reaction on the amino acid pool to be dansylated and the different reaction yield of single amino acids to form the corresponding dansyl compound makes it necessary, prior to an experiment, to find the optimal reaction conditions for the type of tissue to be analyzed. A reaction scheme suitable for practical use with most tissues is given in Scheme 4 (Appendix).

If the dansyl method is to be used for the microdetermination of amino acids from very small biological samples, e.g., isolated nerve cells, very small cores of material from particular regions of the brain, biopsy material, the usual agents for extraction, such as picric acid, ethanol, sulfosalicylic acid, acetone/hydrochloric acid, ethanol/HCl, or perchloric acid, are unsuitable. All of these substances affect the dansyl reaction and interfere with microchromatography, resulting in bad separations. After a single microhomogenization (see Section 9) of weighed pieces of tissue[14] in 0.05 M $NaHCO_3$, 85% of the free amino acids are found in the supernatant. The remaining 15% are extracted almost completely on a second homogenization with $NaHCO_3$; no further dansylamino acids can be detected on microchromatography after a third extraction.[107]

The proteins contained in the supernatant after extraction and centrifugation can interfere with the dansyl reaction, since dansyl chloride also reacts with proteins.[69] In addition, the proteins often remain at the origin and impair the final quality of the microchromatogram. They must therefore be removed if possible before allowing the sample to react with dansyl chloride. Heat denaturation (5 min, 85°C) using a capillary is possible, but in this very drastic method, just as in precipitation with trichloroacetic acid, many of the free amino acids coprecipitate with the denatured material, so that the final yield is reduced by *ca.* 30%.

Precipitation with trichloroacetic acid is also unsuitable because any excess is difficult to remove completely, and the pH for the dansylation must be adjusted accordingly. Also, trichloroacetic acid that has not been removed completely on lyophilization and which is subsequently neutralized causes very bad microchromatograms.

The best method of precipitating the proteins is to add acetone and let the mixture stand at −20°C to complete precipitation. Not all of the proteins are precipitated if a solution contains only 50% acetone (about 90% acetone is necessary to precipitate all the proteins), but enough are precipitated and can

---

Fig. 20. Two-dimensional chromatogram on 3 × 4 cm micropolyamide layer of dansyl/amino acids from human spinal fluid (A) and human serum (B) and map for identification of the spots. 1, Dans-diethylamine; 2, dans-methylamide; 3, dans-proline; 4, dans-isoleucine; 5, dans-leucine; 6, dans-valine; 7, $dans_2$-tyrosine; 8, dans-phenylalanine; 9, dans-methionine; 10, dans-α/γ-aminobutyric acid; 11, dans-alanine; 12, dans-amine, 13, dans-ethanolamine; 14, dans-α-phenylglycine (internal standard); 15, $dans_2$-lysine; 16, $dans_2$-ornithine; 17, dans-glycine; 18, dans-hydroxyproline; 19, dans-tryptophan, 20, dans-glutamic acid; 21, dans-glutamine; 22, dans-threonine; 23, dans-asparagine; 24, dans-serine; 25, dans-tyrosine; 26, $dans_2$-cystine/cysteine; 27, START; 28, dans-taurine; 29, dans-sulfonic acid; 30, dans-amino sugars.

be removed by centrifugation so that good microchromatograms can be obtained after treatment of the remaining solution with dansyl chloride (Fig. 20). The use of acetone as the agent for precipitating the proteins has the additional advantage of providing the right conditions for the reaction with dansyl chloride. If the optimal concentration of dan-Cl is used, the dans-OH formed by reaction with water (easily detectable on its bright blue fluorescence under UV light and its typical position on the chromatogram) is so low that its removal by microcolumn chromatography[107] is unnecessary. If too much dans-OH is present, the chromatography and the quantitative evaluation of the chromatogram can be appreciably impaired.

6.2. Microchromatography of Dansyl Amino Acids

Microchromatography (see Appendix, Scheme 5) is best performed on micro-polyamide sheets,[107,108] since these layers are homogeneous and give sharp separation of the dansyl derivatives (Fig. 20). A mixture of 30–40 dansyl derivatives can be separated on 3 × 3 cm or 3 × 4 cm micro-polyamide layers after two-dimensional chromatography. In microchromatography of dansylated amino acids on polyamide sheets, the quality of the sample applied is of critical importance. Buffers of low molarity or distilled water for extraction of amino acids from biological material are recommended, since those of higher molarity impair the chromatography significantly.[107] Under the stereomicroscope, it is very easy to see if too many foreign substances are present in the sample, since the application spot appears as a microatoll with a plateau in the middle, surrounded by a yellow bank of dansylated products.

The way in which the sample is applied determines the quality of a microchromatogram. Especially fine microcapillaries should be used, and their orifices must be of such a size that the acetone/acetic acid solution can be released by applying slight pressure via the mouthpiece. To apply the sample, the capillary tip is brought carefully into contact with the layer so that the latter is not damaged. If the origin is damaged, the spots of dansyl compounds show some tailing. The point of application should be at one corner of the sheet, 3 mm or, at most, 4 mm from the edges. The diameter of the application point should be no more than 0.5 mm if possible, the maximum permissible being 1 mm.

When this technique is used for the first time, it is a good idea to build up a chromatographic map of the final dansyl amino acids by systematically separating stepwise mixtures of different dansyl amino acids.[107] The typical positions of the dansyl compounds in a microchromatogram are thus observed. Compounds having two reactive groups in the molecule, e.g., serotonin ($-NH_2$ and $-OH$), can easily form doubly dansylated products in addition to the two monodansyl derivatives.[19] This reaction depends mostly on the pH of the mixture, and each bisdansyl compound will show specific migration during two-dimensional chromatography. Identification of compounds with more than two reactive groups (like many of the biologically active amines) as dansyl products[107] is very difficult, since each can form more than three different dansyl products.

Microchromatograms of dansyl compounds are most accurately evaluated by fluorescence scanning microscopy (see Section 7). If the necessary equipment is not available, they can be evaluated fluorometrically after elution of the isolated dansyl spot with ethanol. Complete elution is essential and is not very easy for all dansyl amino acids. If [^{14}C]-labeled dansyl chloride has been used for the reaction, autoradiography can be performed as described.[107] Autoradiography is also well suited for turnover studies if radioactively labeled amino acids are reacted with unlabeled dansyl chloride. Lowest accuracy with variations up to ±10% are obtained after cumbersome scraping off of the spot area with a micro-knife[38] followed by scintillation counting. Many examples of the application of the micro-dansyl procedure to analyze free amino acids from many different biological materials have been described.[29,107,109-129] The micro-dansyl method is also suitable for determining C- and N-terminal amino acids from minute amounts of purified proteins.[107]

7. PHOTOMETRY AND FLUOROMETRY OF MICROGELS AND MICROCHROMATOGRAMS

Photometric evaluation of microgels creates no problem, and any densitometer or photometer fitted with a suitable device for microgels[32,130,131] can be used, even for the evaluation of gels stained for enzyme activity measurements.[32] With all the equipment commonly used for evaluating micro- or macrogels, even under optimal conditions, errors commonly occur, since a single scan is not sufficient to correctly evaluate a whole gel. Special problems occur if the protein bands are not exactly lined up in parallel. A minute change in the relationship between the optical slit and the gel can result in quite a different pherogram. All quantitative figures given on the basis of such an evaluation are therefore doubtful. The only correct procedure is the complete scanning of both one-dimensional (as is described for two-dimensional chromatograms[132-136]) and two-dimensional microgel slabs.[135] The combination of scanning photometry and image analysis[135,137-139] is a sensitive, versatile, and accurate method for the evaluation of slab gels. This method has the high sensitivity of 5×10^{-4} O.D. because it takes advantage of the noise characteristics of photometric signals. Digital image analysis is used to locate and evaluate the individual spot on the two-dimensional electropherogram. A special procedure of segmentation[140] detects the faintest protein spots hardly visible by eye and separates confluent spots reliably and automatically.

Automatic evaluation of two-dimensional thin-layer chromatograms with scanning fluorescence microscopy[132-136,140] demonstrates clearly the applicability of signal processing, image analysis, and pattern recognition to microphotometry by an enlarged dynamic range of the photometer, improved reproducibility through noise reduction, discrimination between objects and background, feature extraction, and, in addition, automatic classification[140] of the results. Using the same principles described for the evaluation of two-dimensional microchromatograms and microgels, a computer-directed scanning device suitable for macrogels as well as for microgels and a suitable program for two-dimensional automatic evaluation of one-dimensional separations are under development. It is likely that this modern type of microphotometry[137] combined with image analysis will be adapted in the near future for the quantitative evaluation of histochemical staining reactions (which are otherwise not quantifiable), and it will permit photometry on the borderline of light microscopy, e.g., a direct scan of chromosome banding.

8. MICROMETHODS FOR ANALYSIS AT THE CELLULAR LEVEL

The methods described so far are not well suited for analysis at the cellular level with the exception of the rather large nerve cells that can be obtained from invertebrates (for review see refs. 29,141,142). A short chronological review of the development of neurochemical methods at the cellular level going back to the pioneer work of Linderstrøm-Lang has been published by Giacobini.[143] Analyses at the cellular level are not as easy to perform as the methods described above, and the equipment required is generally sophisticated and expensive.

Dry mass determination at the cellular level can be performed with the famous quartz-fiber fishpole balance introduced by Lowry,[144–146] using an electron microscope,[147] or by X-ray absorption.[13,148,149] Very many highly sensitive histochemical and radiochemical reactions can also be performed at the cellular level or even on small pieces of single cells.[74,99,142,145,150–155] A very sensitive but rather difficult method can be used for RNA-base analysis of isolated cells.[3–5,17] The determination of phospholipids at the cellular level is relatively easy.[22–24] Quantitative autoradiography at the cellular level is also possible,[156] as are microphotometry[157,158] and cytofluorometry.[159] The cartesian diver technique for measuring enzyme activity in single neurons[6,143] is elegant but difficult to perform. By using microflame photometry,[15,16] determination of K and Na is also possible for single cells.

This list is not a complete review on methods available for microanalytical approaches at the cellular level, but for all of those described, the author has at least some experience. Neurochemists seeking a specific micromethod should look to other disciplines where micromethods have been employed because of the same problem of limited material availability. Many of the above micromethods and some others are used in prenatal diagnosis,[87,160–171] in single-cell analysis of plants,[172] or in connection with immunocytochemistry.[170,174,175]

9. AUXILIARY EQUIPMENT FOR MICROANALYSIS

Ideally, anyone working with microscale methods should have available a small workshop (and handy machinist), since much of the equipment can be easily made or adapted from normal equipment. For example, capillary centrifugation can be performed in a special capillary rotor, but a normal test-tube rotor can also be adapted.[176] Analytical high-speed centrifugation for determination of sedimentation coefficients with volumes less then 1 μl can be achieved with any analytical centrifuge if the usual analytical cell is adapted for capillaries.[176,177] Microdialysis chambers for variable volumes between 15 μl and 500 μl[178] are easily prepared in a workshop or are otherwise commercially available (e.g., from E. Schütt, Jr., Göttingen, FRG).

Microhomogenization is an important step in microprocedures. The use of steel wire for the homogenization of isolated cells in a capillary was first described by Eichner.[179] Easier to handle are nerve-canal drills driven by a dental drilling machine.[28] For volumes larger then 5 or 10 μl, suitable homogenizers can easily be home made.[28] The production of capillary pipettes, repeatedly used in microelectrophoresis, is easy, provided a suitable burner is available.[17,180] Two-dimensional microimmunodiffusion[70,181] is a suitable and easy way for immunoprecipitation with volumes of less then 1 μl of antigen and antibody solutions. Here there is the additional advantage that immunoprecipitates are clearly visible within minutes or hours, in contrast to days with the classical Ouchterlony test. The same procedure, for which one needs only a capillary pipette, a microscope slide, and a stereomicroscope as equipment, can also successfully be used to test complex formation other then immunoprecipitates.[50,182]

10. CONCLUDING REMARKS

The experience of the author over more than 20 years has shown that the use of micromethods requires no more technical expertise and critical understanding than are required by other methods. The same critical appraisal of the results is needed whether they are obtained with microscale or macroscale methods. The present era in neurochemistry is that of the isolated cell. New bulk separation methods for the preparation of neurons, which maintain most of their processes,[182–188] and for oligodendrocytes and astrocytes are available.[187–194] These bulk-prepared cells can be maintained in culture. With such approaches, it should be possible to examine nerve tissue at the single-cell level as well as with homogeneous cell populations. There is a clear need for sensitive micromethods to obtain information from these small cell samples.

Combining currently available experimental and methodological approaches, we can now perhaps begin to tackle all the unanswered questions about the morphological appearance of a nerve cell, its function, and its inherent biochemical makeup. Characterization of these three parameters defining nerve cells in their entirety will undoubtedly reveal them as most fascinating entities.

ACKNOWLEDGMENTS. I am grateful to Dr. Martin Rumsby (York, England) for carefully reading the manuscript and to Dr. G. Huether and Dr. H.-M. Pöhling for valuable discussions.

1. APPENDIX

A.1. Scheme 1: Spot Analysis for Protein Determination

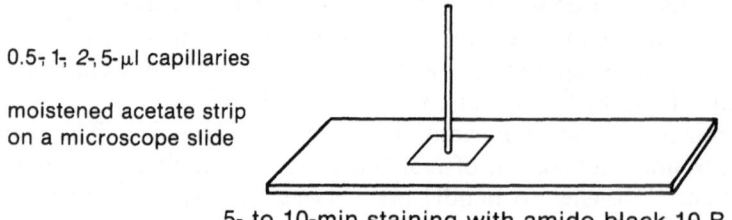

0.5-, 1-, 2-, 5-μl capillaries

moistened acetate strip
on a microscope slide

5- to 10-min staining with amido black 10 B
or Hoechst 2495 in methanol/acetic acid (9:1)

Three times 5- to 10-min destaining in methanol/acetic acid (9:1)

Transparency of the strip
on a microscope slide for
densitometric evaluation
or *in situ* spot fluorometry

Dissolution of the strip
together with the spot in
dioxane, dimethylsulfoxide,
or N',N-dimethylformamide,
for photometric or fluorometric evaluation

Elution of Hoechst 2495
with 1% NH$_4$OH for most
sensitive fluorometric
evaluation.

A.1.1. Reagents and Procedure

Amido black 10 B is 0.5% in methanol/glacial acetic acid (9:1); Hoechst 2495 is a fluorescent benzoxanthene derivative (Serva, Heidelberg, FRG); 500 mg are dissolved in 1 ml H$_2$O, and 20 μl of this solution is dissolved in 25 ml of 9:1 methanol/glacial acetic acid.

Cellulose acetate layers (Membranfilter SM 11200, Sartorius, Göttingen, FRG) are moistened prior to use in a wet chamber to facilitate sample application and to obtain ideal round protein spots in case of densitometric evaluation. Different capillary (Drummond Microcaps or Brand Intraend) volumes can be used, but most reproducible spots are obtained with 2-μl capillaries. They are filled by capillary attraction simply by dipping into the protein solution and are emptied by touching (not pressing) the acetate strip.

A.1.2. Evaluation

For quantitative evaluation of the stained spots, different possibilities exist according to the equipment available.

Densitometric evaluation is performed with any suitable densitometer after making the spots stained with amido black 10 B transparent.[31] For this procedure, the spots have to be perfectly round. For optical reasons, the slit of the measuring instrument must not be larger than one-tenth of the diameter of the spot. A suitable instrument is a photometer together with a gel scanner adapted for evaluation of microgels.[130] This can be used without further technical modification. Densitometric evaluation is also used if the protein concentration in an unknown volume has to be determined, since with a single measurement one obtains the volume applied to the acetate strip from the diameter of the spot and the concentration from the extinction of the stained protein (compare Fig. 1).

In situ spot fluorometry is used if spots stained with Hoechst 2495 are not ideally round. This may be the case, for example, if the sample contains high salt concentrations.

Photometric evaluation can be performed with any photometer if the amido black 10 B-stained spots together with the acetate strips are dissolved in dioxane, dimethylsulfoxide, or N′,N-dimethylformamide. The maximum wavelength for amido black in the three solvents is 600 nm, 630 nm, and 620 nm, respectively. The lowest sensitivity in this case is 1 μg. Hoechst 2495 is not suited for photometric evaluation because its absorption is too low.

Fluorometric evaluation can be performed, as with photometric evaluation, after solubilization of the spots stained with Hoechst 2495 together with the strips. Fluorescence in dioxane (at 427 nm for excitation and 532 for emission) is tenfold less sensitive than for the two other solvents, which have the same fluorescence maxima (430 nm and 475 nm). Highest fluorescence is obtained when the chromophor Hoechst 2495 is eluted from the protein spot with 1% NH_4OH in H_2O, and the fluorescence is measured at 425 nm and 475 nm.

A.2. Scheme 2: Spot Analysis for Glycoprotein Determination

0.5-, 1-, 2-,5-μl capillaries

1 × 1 cm moistened acetate
strip on a microscope slide

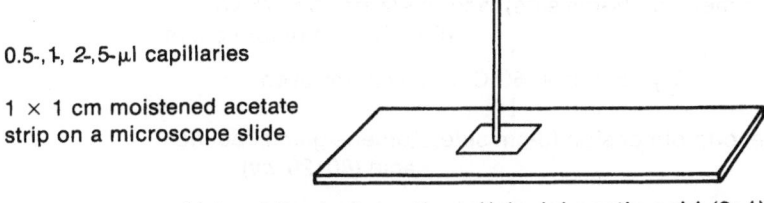

At least 5 min in methanol/glacial acetic acid (9:1)

↓

Two washes in methanol (to remove acetic acid)

↓

60-min preincubation in suitable buffer

↓

80-min (or longer) incubation in FITC-labeled lectin in
buffer together with blank piece of exactly the same size as
the sample piece

↓

One 5-min and two 10-min removals of unbound lectin

↓

Solubilization of the glycoprotein–lectin complex in 2% SDS
in 12% NH_4OH in water

↓

Fluorometric evaluation at 492 nm for excitation and 518 nm at
the emission monochromator of sample and blank piece

A.2.1. Reagents and Procedure

Pretreatment of the acetate strips and sample application is as described for spot analysis of proteins. The strips used for sample application and the blank piece have to be of exactly the same size, since some of the fluorescent-labeled lectin, even after prolonged washing in buffer, remains on the acetate sheet. Highest values of complex formation are obtained with FITC-labeled concanavalin A in PBS buffer (0.1 mg/ml Con A in PBS buffer, pH 7.4; PBS buffer: 6.46 mM Na_2HPO_4, 1.47 mM KH_2PO_4, 136 mM NaCl, 27 mM KCl, 0.68 mM $CaCl_2$, 0.49 mM $MgCl_2$).

A.3. Scheme 3: Determination and Microchromatography of Sugars in Picomole Range

100 µl sample solution
+ 300 µl glacial acetic acid
+ 500 µl dansyl hydrazine (1% in ethanol)
+ 50 µl internal standard solution

(or corresponding aliquots)

↓

10 min at 50° C in a water bath with carefully closed test tubes

↓

Transfer onto ice to stop further reaction

↓

Microchromatography on 3 × 4 cm micropolyamide sheets; first dimension (4-cm side): acetone/water (25:75 v/v)

↓ + 100 mM ammonium acetate

Dry 15 min at 50°C in a vacuum oven

↓

Second dimension (3-cm side): toluene/glacial acetic acid (80:20 v/v)

Dry 15 min at 50°C in a vaccum oven

A.3.1. Reagents and Procedure

Dansyl hydrazine is solubilized in warmed ethanol and stored in the dark at 4°C. Longer incubation at 50°C or higher temperature leads to destruction of the formed dansyl hydrazine compounds. Sample application and chromatography are performed as described for dansyl amino acids (see Scheme 5). Ammonium acetate is added to the first-dimension solvent mixture to sharpen the spots during chromatography. As internal standard, a 0.5% dansyl-L-proline solution in ethanol is used as reference for the evaluation with automated scanning fluorescence microscopy. If the necessary equipment is not available, fluorometric determination can be performed after completely scraping off the spots with a microknife[38] and eluting the dansyl hydrazine sugar with ethanol. Fluorescence is measured at 340 nm for excitation and 515 nm on the emission monochromator. Evaluation using HPLC is possible,[27] but not all dansyl hydrazine sugars separated by microchromatography will be resolved in one run.

A.4. Scheme 4: Determination of Amino Acids as Dansyl Compounds in the Picomole Range

5 mg tissue in 100 μl H$_2$O (or aliquot)

2-min homogenization and centrifugation

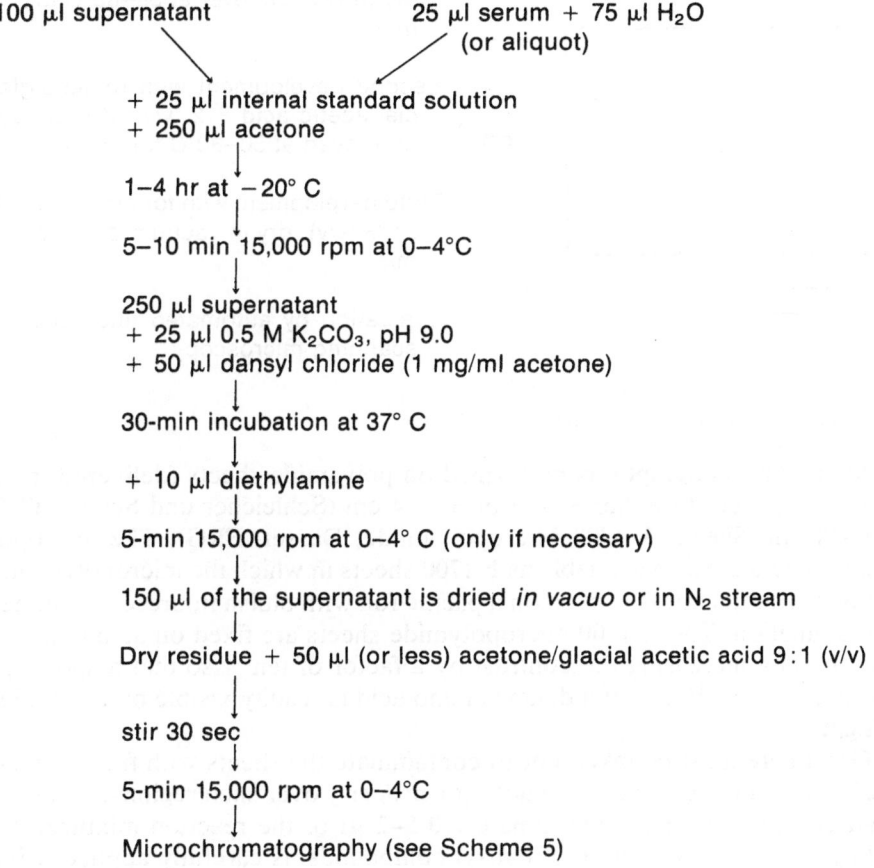

100 μl supernatant   25 μl serum + 75 μl H$_2$O
                                              (or aliquot)

+ 25 μl internal standard solution
+ 250 μl acetone

1–4 hr at −20° C

5–10 min 15,000 rpm at 0–4°C

250 μl supernatant
+ 25 μl 0.5 M K$_2$CO$_3$, pH 9.0
+ 50 μl dansyl chloride (1 mg/ml acetone)

30-min incubation at 37° C

+ 10 μl diethylamine

5-min 15,000 rpm at 0–4° C (only if necessary)

150 μl of the supernatant is dried *in vacuo* or in N$_2$ stream

Dry residue + 50 μl (or less) acetone/glacial acetic acid 9:1 (v/v)

stir 30 sec

5-min 15,000 rpm at 0–4°C

Microchromatography (see Scheme 5)

A.4.1. Reagents and Procedure

Small tissue pieces are homogenized with microhomogenizers of suitable size (see Section 9); centrifugation is performed in a capillary rotor (see Section 9) or a suitable centrifuge at 0–4°C, depending on the volume available. As internal standard, a solution of 4×10^{-4} M α-phenylglycine dissolved in 0.01 N HCl is used. This is stored in suitable portions at 20°C until use. Prior to use 0.06 N KOH in a ratio of 1:1 (v/v) is added for neuralization. The K$_2$CO$_3$ solution is brought to pH 9.0 with gaseous CO$_2$ without increasing salt concentrations. Dansyl chloride of purest quality or after recrystalization is dissolved (1 mg/ml) in acetone. After dansylation, any excess of dansyl chloride is converted immediately to dansyl diethylamine by adding diethylamine to the reaction mixture.

A.5. Scheme 5: Microchromatography of Dansyl Amino Acids

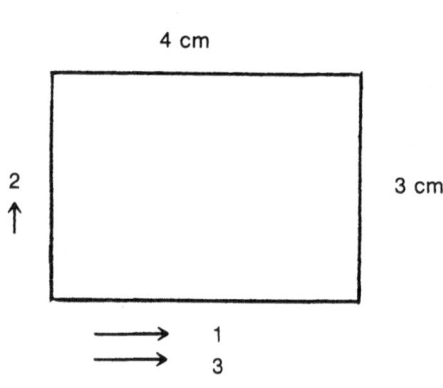

First development with distilled water; dry in vacuum over at 50–80°C for 30 min.

Second development with toluene/glacial acetic acid 8:2 (v/v); dry in vacuum oven at 50–80°C for 15 min.

Third development with formic acid/H_2O, 2:98 (v/v); dry in vacuum oven for 15 min.

Evaluation by automated fluorescence scanning microscopy

A.5.1. Reagents and Procedure

Microchromatography is performed on polyamide sheets (delivered in 15 × 15 cm size) cut to either 3 × 3 or 3 × 4 cm (Schleicher und Schüll, TLC Ready-Plastic Sheets A 1700 Micropolyamide, Dassel, FRG). The micropolyamide sheets are also available as F 1700 sheets in which the micropolyamide layer of 25 μm thickness is fixed on a plastic foil with bluish fluorescence under UV illumination. The A 1700 micropolymide sheets are fixed on an aluminum layer, thereby increasing the contrast by a factor of ten[108] so that a spot representing 5×10^{-13} mol of a dansyl amino acid is readily visible by eye under UV light.

Great care must be taken not to contaminate the sheets with fingerprints, since the developing solvents do not flow properly over such regions, resulting in bad chromatograms. Approximately 0.5–2 μl of the reaction mixture, depending on the concentration of dansyl compounds, is carefully applied using a fine capillary under a stereomicroscope at a point 3–4 mm in from the edges of the chromatogram.

A first development with water up the edge of the layer takes only 4–5 min and is necessary if the sample contains sugars or a high salt content.

Careful drying between developments is essential, since otherwise the separation will be impaired, resulting in uneven spots.

Care must be taken that the solvent mixture never reaches the sample application point when the chromatograms are dipped. Development is performed in a closed beaker. In the first dimension, it takes 4–5 min; in the second dimension, 5 min. Developed chromatograms should be stored in the dark, separately, to avoid contact between individual chromatograms. Contact may result in loss of material through fingerprinting of the spots if two sheets are packed side by side.

REFERENCES

1. Pun, J. Y., and Lombrozo, K., 1964, *Anal Biochem.* **9**:9–20.
2. Hydén, H., 1943, *Acta Physiol. Scand.* [*Suppl.*] **17**(6):1–136.
3. Edstrøm, J.-E., 1953, *Biochem. Biophys. Acta* **12**:361–386.
4. Edstrøm, J.-E., and Hydén, H., 1954, *Nature* **174**:128–129.
5. Edstrøm, J.-E., 1960, *J. Biophys. Biochem. Cytol.* **8**:39–43.
6. Lindstrøm-Lang, K., 1937, *Nature* **140**:108.
7. Lowry, O. H., Roberts, N. R., and Chang, M. W., 1956, *J. Biol. Chem.* **222**:97–107.
8. Giacobini, E., and Zajicek, J., 1956, *Nature* **177**:185–186.
9. Lowry, O. H., Passonneau, J. V., Schultz, D. W., and Rock, M. K., 1961, *J. Biol. Chem.* **236**:2746–2755.
10. Giacobini, E., and Grasso, A., 1966, *Acta Physiol. Scand.* **66**:49–57.
11. Giacobini, E., and Marchisio, P. C., 1966, *Acta Physiol. Scand.* **66**:247–248.
12. Buckley, G., Consolo, S., Giacobini, E., and McCaman, R., 1967, *Acta Physiol. Scand.* **71**:341–347.
13. Brattgard, S.-O., and Hydén, H., 1952, *Acta Radiol.* [*Suppl.*] (*Stockholm*) **94**:1–48.
14. Neuhoff, V., 1971, *Anal. Biochem.* **41**:270–271.
15. Carlsson, B., Giacobini, E., and Hovmark, S., 1967, *Acta Physiol. Scand.* **71**:379–390.
16. Haljamae, H., and Larsson, S., 1968, *Chem. Instrum.* **1**:131–144.
17. Edstrøm, J.-E., and Neuhoff, V., 1973, *Micromethods in Molecular Biology* (V. Neuhoff, ed.), Springer-Verlag, Berlin, Heidelberg, New York, pp. 215–256.
18. Neuhoff, V., von der Haar, F., Schlimme, E., and Weise, M., 1969, *Hoppe Seylers Z. Physiol. Chem.* **350**:121–128.
19. Neuhoff, V., and Weise, M. 1970, *Arzneim. Forsch.* **20**:368–372.
20. Briel, G., Neuhoff, V., and Osborn, N. N., 1971, *Int. J. Neurosci.* **2**:129–136.
21. Briel, G., and Neuhoff, V. 1972, *Hoppe Seylers Z. Physiol. Chem.* **353**:540–553.
22. Schiefer, H.-G., and Neuhoff, V., 1971, *Hoppe Seylers Z. Physiol. Chem.* **352**:913–926.
23. Althaus, H. H., and Neuhoff, V., 1973, *Hoppe Seylers Z. Physiol. Chem.* **354**:1073–1976.
24. Althaus, H. H., Osborne, N. N., and Neuhoff, V., 1973, *Naturwissenschaften* **60**:553–554.
25. Quentin, C.-D., and Neuhoff, V., 1972, *Int. J. Neurosci.* **4**:17–24.
26. Neuhoff, V., Ewers, E., and Huether, G., 1981, *Hoppe Seylers Z. Physiol. Chem.* **362**:1427–1434.
27. Seiler, P., Neuhoff, V., and Thorn, W., 1982, *Hoppe Seylers Z. Physiol. Chem.* (in press).
28. Neuhoff, V., 1973, *Micromethods in Molecular Biology* (V. Neuhoff, ed.), Springer-Verlag, Berlin. Heidelberg, New York, pp. 399–402.
29. Osborn, N. N., 1974, *Microchemical Analysis of Nervous Tissue*. Pergamon Press, New York, Oxford.
30. Feigl, F., 1960, *Tüpfelanalyse, II. Organischer Teil*. Akademische Verlagsgesellschaft MBH, Frankfurt/Main.
31. Neuhoff, V., Philipp, K., Zimmer, H.-G., and Mesecke, S., 1979, *Hoppe Seylers Z. Physiol. Chem.* **369**:1657–1670.
32. Neuhoff, V., 1973, *Micromethods in Molecular Biology*, (V. Neuhoff, ed.), Springer-Verlag, Berlin, Heidelberg, New York, pp. 4–5.
33. Zimmer, H.-G., Kiehl, F., and Neuhoff, V., 1979, *Hoppe Seylers Z. Physiol. Chem.* **360**:1671–1672.
34. Neuhoff, V., 1973, *Micromethods in Molecular Biology*, (V. Neuhoff, ed.), Springer-Verlag, Berlin, Heidelberg, New York, pp. 149–178.
35. Norton, W. T., 1977, *Myelin* (P. Morell, ed.), Plenum Press, New York, pp. 161–199.
36. Lowry, O. H., Rosebrough, N. I., Farr, A. C., and Randall, R. I., 1951, *J. Biol. Chem.* **193**:265–275.

37. Zimmer, H.-G., and Neuhoff, V., 1977, *GIT Fachz. Lab.* **21**:104–105.
38. Neuhoff, V., 1973, *Micromethods in Molecular Biology* (V. Neuhoff, ed.), Springer-Verlag, Berlin, Heidelberg, New York.
39. Marshall, R. D., and Neuberger, A., 1972, *Glycoproteins*, Volume 5, Part A (A. Gottschalk, ed.), Elsevier, Amsterdam, pp. 224–290.
40. Grossbach, U., 1965, *Biochim. Biophys. Acta* **107**:180–182.
41. Hyden, H., Bjurstam, K., and McEwen, B., 1966, *Anal. Biochem.* **17**:1–15.
42. McEwen, b., and Hydén, H., 1966, *J. Neurochem.* **13**:823–833.
43. Hydén, H., and Lange, P. W., 1968, *J. Chromatogr.* **35**:336–351.
44. Neuhoff, V., 1968, *Arzneim. Forsch.* **18**:35–39.
45. Hydén, H., and Lange, P. W., 1972, *Proc. Natl. Acad. Sci. U.S.A.* **69**:1980–1984.
46. Griffith, A., and LaVelle, A., 1971, *Exp. Neurol.* **33**:360–371.
47. Ansorg, R., Dames, W., and Neuhoff, V., 1971, *Arzneim. Forsch.* **21**:699–710.
48. Ansorg, R., and Neuhoff, V., 1971, *Int. J. Neurosci.* **2**:151–160.
49. Nir, I., Dames, W., and Neuhoff, V., 1973, *Arch. Int. Physiol. Biochem.* **81**:607–616.
50. Neuhoff, V., Schill, W.-B., and Sternbach, H., 1970, *Biochem. J.* **117**:623–631.
51. Cupello, A., and Hydén, H., 1975, *Neurobiology* **5**:129–136.
52. Rüchel, R., Mesecke, S., Wolfrum, D. I., and Neuhoff, V., 1973, *Hoppe Seylers Z. Physiol. Chem.* **354**:1351–1368.
53. Rüchel, R., Mesecke, S., Wolfrum, D. I., and Neuhoff, V., 1974, *Hoppe Seylers Z. Physiol. Chem.* **355**:997–1020.
54. Wolfrum, D. I., Rüchel, R., Mesecke, S., and Neuhoff, V., 1974, *Hoppe Seylers Z. Physiol. Chem.* **355**:1415–1435.
55. Rüchel, R., Richter-Landsberg, C., and Neuhoff, V., 1975, *Hoppe Seylers Z. Physiol. Chem.* **356**:1283–1288.
56. Endou, H., and Neuhoff, V., 1975, *Hoppe Seylers Z. Physiol. Chem.* **356**:1381–1396.
57. Reichel, W., Wolfrum, D., Weber, M., Scheler, F., and Neuhoff, V., 1975, *Contrib. Nephrol.* **1**:109–118.
58. Poehling, H. M., Wolfrum, D. I., and Neuhoff, V., 1976, *Entomologia Exp. Appl.* **19**:271–286.
59. Peter, R., Wolfrum, D. I., and Neuhoff, V., 1976, *Comp. Biochem. Physiol.* **55B**:583–589.
60. Fagg, G. E., Waehneldt, T. V., and Neuhoff, V., 1978, *Myelination and Demyelination* (J. Palo, ed.), Plenum Press, New York, pp. 135–145.
61. Fagg, G. E., Schipper, H. I., and Neuhoff, V., 1979, *Brain Res.* **167**:251–258.
62. Tauber, H., Waehneldt, T. V., and Neuhoff, V., 1980, *Neurosci. Lett.* **16**:235–238.
63. Grossbach, U., 1972, *Biochem. Biophys. Res. Commun.* **49**:667–672.
64. Gainer, H., 1973, *Anal. Biochem.* **51**:646–650.
65. Bispink, G., and Neuhoff, V., 1976, *Hoppe Seylers Z. Physiol. Chem.* **357**:991–997.
66. Bispink, G., and Neuhoff, V., 1977, *Electrofocusing and Isotachophoresis* (R. J. Radola, and D. Graesslin, eds.), Walter de Gruyter, Berlin, pp. 135–146.
67. Gustke, H. H., and Neuhoff, V., 1978, *Hoppe Seylers Z. Physiol. Chem.* **359**:1481–1489.
68. Gustke, H. H., and Neuhoff, V., 1979, *Hoppe Seylers Z. Physiol. Chem.* **360**:605–608.
69. Neuhoff, V., and Poehling, H.-M., 1980, *Hoppe Seylers Z. Physiol. Chem.* **361**:77–78.
70. Neuhoff, V., and Schill, W.-B., 1968, *Hoppe Seylers Z. Physiol. Chem.* **349**:795–800.
71. Neuhoff, V., and Mesecke, S., 1977, *Hoppe Seylers Z. Physiol. Chem.* **358**:1623–1637.
72. Poehling, H.-M., and Neuhoff, V., 1980, *Electrophoresis* **1**:90–102.
73. Poehling, H.-M., Wyss, U., and Neuhoff, V., 1980, *Electrophoresis* **1**:198–200.
74. Huether, G., and Neuhoff, V., 1981, *Histochem. J.* **13**:207–225.
75. Sprinzel, M., Wolfrum, D. I., and Neuhoff, V., 1975, *FEBS Lett.* **50**:54–56.
76. Kolin, A., 1954, *J. Chem. Phys.* **22**:1628–1629.
77. Svensson, H., 1961, *Acta Chem. Scand.* **15**:325–341.
78. Vesterberg, O., and Svensson, H., 1966, *Acta Chem. Scand.* **20**:820–834.
79. Dale, G., and Latner, A., 1968, *Lancet* **1**:847–848.
80. Wrigley, C. W., 1968, *J. Chromatogr.* **36**:362–365.
81. Haglund, H., 1971, *Methods in Biochemical Analysis* (D. Glick, ed.), Volume 19, Wiley–Interscience, London, pp. 1–104.
82. Riley, R. F., and Coleman, M. K., 1968, *J. Lab. Clin. Med.* **72**:714–720.

83. Catsimpoolas, N., 1968, *Anal. Biochem.* **26**:480–482.
84. Baumann, J., and Chrambach, A., 1976, *Anal. Biochem.* **77**:216–225.
85. Laemmly, U. K., 1970, *Nature* **227**:680–685.
86. O'Farrel, P. H., 1975, *J. Biol. Chem.* **250**:4007–4021.
87. Klose, J., Blohm, M., and Gerner, L., 1977, *Methods in Prenatal Toxicology* (D. Neubert, H. J. Merker, and T. Kwasigroch, eds.), Georg Thieme, Stuttgart, PP. 303–313.
88. Radola, B. J., 1980, *Electrophoresis* **1**:43–56.
89. Norton, W. T., and Poduslo, S. E., 1973, *J. Neurochem.* **21**:749–757.
90. Lane, J. D., and Neuhoff, V., 1980, *Naturwissenschaften* **67**:227–233.
91. Lane, J. D., Schöne, B., Langenbeck, U., and Neuhoff, V., 1980, *Biochim. Biophys. Acta* **627**:144–156.
92. Huether, G., and Neuhoff, V., 1981, *J. Int. Metab. Dis.* **4**:67–68.
93. Görg, A., Postel, W., and Westermeier, R., 1978, *Anal. Biochem.* **89**:60–70.
94. Switzer, R. C., Merrill, C. R., and Shifrin, S., 1979, *Anal. Biochem.* **98**:231–237.
95. Allen, R. C., 1980, *Electrophoresis* **1**:32–37.
96. Merrill, C. R., Goldman, D., Sedman, S. A., and Ebert, M. H., 1981, *Science* **211**:1437–1438.
97. Poehing, H.-M., and Neuhoff, V., 1981, *Electrophoresis* **2**:141–147.
98. O'Farrell, P. H., 1975, *J. Biol. Chem.* **250**:4007–4021.
99. Cremer, T., Dames, W., and Neuhoff, V., 1972, *Hoppe Seylers Z. Physiol. Chem.* **353**:1317–1329.
100. Gray, W. R., 1967, *Methods Enzymol.* **17**:139–151.
101. Seiler, N., 1970, *Methods Biochem. Anal.* **18**:259–337.
102. Seiler, N., and Wiechmann, M., 1970, *Progress in Thin-Layer Chromatography and Related Methods* (A. Niederwieser and G. Pataki, eds.), Volume III, Ann Arbor Science Publications, Ann Arbor, pp. 95–144.
103. Woods, K. R., and Wang, K. T., 1967, *Biochim. Biophys. Acta* **133**:369–370.
104. Wang, K. T., and Weinstein, B., 1972, *Progress in Thin-Layer Chromatography and Related Methods* (A. Niederwieser and G. Pataki, eds.), Volume III, Ann Arbor Science Publications, Ann Arbor, pp. 177–231.
105. Gray, W. R., and Hartley, B. S., 1963, *Biochem. J.* **89**:59.
106. Neadle, D. J., and Pollit, R. J., 1965, *Biochem. J.* **97**:607–608.
107. Neuhoff, V., 1973, *Micromethods in Molecular Biology* (V. Neuhoff, ed.), Springer Verlag, Berlin, Heidelberg, New York, pp. 85–148.
108. Zimmer, H.-G., Neuhoff, V., and Schulze, E., 1976, *J. Chromatogr.* **124**:120–122.
109. Leonard, B. E., Neuhoff, V., and Tonge, S. R., 1974, *Z. Naturforsch.* **29c**:184–186.
110. Osborne, N. N., Wu, P. H., and Neuhoff, V., 1974, *Brain Res.* **74**:175–181.
111. Osborne, N. N., and Neuhoff, V., 1974, *Brain Res.* **74**:366–369.
112. Quentin, C.-D., Behbehani, A. W., Schulte, F. J., and Neuhoff, V., 1974, *Neuropediatrie* **5**:138–145.
113. Behbehani, A. W., Quentin, C.-D., Schulte, F. J., and Neuhoff, V., 1974, *Neuropediatrie* **5**:258–270.
114. Quentin, C.-D., Behbehani, A. W., Schulte, F. J., and Neuhoff, V., 1974, *Neuropediatrie* **5**:271–278.
115. Neuhoff, V., Behbehani, A. W., Quentin, C.-D., and Prinz, H., 1974, *Hoppe Seylers Z. Physiol. Chem.* **355**:891–894.
116. Leonard, B. E., Neuhoff, V., and Tonge, S. R., 1974, *Z. Naturforsch.* **29c**:767–772.
117. Osborne, N. N., and Neuhoff, V., 1974, *Brain Res.* **80**:251–264.
118. Behbehani, A. W., Quentin, C.-D., and Neuhoff, V., 1975, *Neurobiology* **5**:52–59.
119. Neuhoff, V., Behbehani, A. W., Quentin, C.-D., and Briel, C., 1975, *Neurobiology* **5**:254–261.
120. Richter-Landsberg, C., and Neuhoff, V., 1975, *Naturwissenschaften* **62**:491.
121. Leonard, B. E., Neuhoff, V., and Tonge, S. R., 1975, *J. Neurosci. Res.* **1**:83–92.
122. Schulze, E., and Neuhoff, V., 1976, *Hoppe Seylers Z. Physiol. Chem.* **357**:593–600.
124. Stenzel, K., and Neuhoff, V., 1976, *J. Neurosci. Res.* **2**:1–9.
125. Osborne, N. N., Stahl, W. L., and Neuhoff, V., 1976, *J. Chromatogr.* **123**:212–215.
126. Osborne, N. N., and Neuhoff, V., 1977, *J. Chromatogr.* **134**:489–496.
127. Poehling, H.-M., Wyss, U., and Neuhoff, V., 1980, *Physiol. Plant Pathol.* **16**:59–61.

128. Ulmar, G., and Neuhoff, V., 1980, *Exp. Neurol.* **69**:99–109.
129. Meyer, W., Poehling, H.-M., and Neuhoff, V., 1980, *Comp. Biochem. Physiol.* **67C**:83–86.
130. Zimmer, H.-G., and Neuhoff, V., 1975, *GIT Fachz. Lab.* **19**:481–484.
131. Lane, J. D., Zimmer, H.-G., and Neuhoff, V., 1979, *Hoppe Sylers Z. Physiol. Chem.* **360**:1405–1408.
132. Zimmer, H.-G., and Neuhoff, V., 1977, *Informatik-Fachberichte* (W. Brauer, ed.), Springer-Verlag, Berlin, Heidelberg, New York, pp. 12–20.
133. Kronberg, G., Zimmer, H-G., and Neuhoff, V., 1978, *Fresenius Z. Anal. Chem.* **290**:133–134.
134. Kronberg, H., Zimmer, H.-G., and Neuhoff, V., 1979, *Microsc. Acta* **82**:223–228.
135. Zimmer, H.-G., 1979, *J. Microsc.* **116**:365–372.
136. Kronberg, H., Zimmer, H.-G., and Neuhoff, V., 1980, *Electrophoresis* **1**:27–32.
137. Zimmer, H.-G., and Neuhoff, V., 1981, *Naturwissenschaften* **68**:464–470.
138. Zimmer, H.-G., Kronberg, H., Berstein, R., and Neuhoff, V., 1981, *Pattern Recogn.* **13**:79–82.
139. Zimmer, H.-G., Kronberg, H., and Neuhoff, V., 1980, *Microsc. Acta [Suppl.]* **4**:217–221.
140. Kronberg, H., and Neuhoff, V., 1978, *Informatik-Fachberichte*, Volume 17 (E. Triendle, ed.), Springer-Verlag, Berlin, Heidelberg, New York, pp. 334–337.
141. Giacobini, E., 1975, *J. Neurosci. Res.* **1**:1–18.
142. Giacobini, E., 1968, *Neurosci. Res.* **1**:1–202.
143. Giacobini, E., 1977, *Biochemistry of Characterised Neurons* (N. N. Osborne, ed.), Pergamon Press, Oxford, New York, pp. 3–17.
144. Lowry, O. H., 1953, *J. Histochem. Cytochem.* **1**:420–428.
145. Lowry, O. H., and Passonneau, J. V., 1972, *A Flexible System of Enzymatic Analysis*, Academic Press, New York, pp. 236–249.
146. Lehrer, G. M., 1973, *Micromethods in Molecular Biology* (V. Neuhoff, ed.), Springer-Verlag, Berlin, Heidelberg, New York, pp. 285–296.
147. Bahr, G. F., 1973, *Micromethods in Molecular Biology* (V. Neuhoff, ed.), Springer-Verlag, Berlin, Heidelberg, New York, pp. 257–284.
148. Hydén, H., and Rosengren, B., 1962, *Biochim. Biophys. Acta* **60**:638–640.
149. Hydén, H., and Larsson, S., 1960, *Proceedings Second International Symposium on X-Ray Microscopy and X-Ray Microanalysis*, Elsevier, Amsterdam, pp. 51–55.
150. Giacobini, E., 1970, *Biochem. Psychopharmacol.* **2**:9–64.
151. Giacobini, E., 1962, *J. Neurochem.* **9**:169–177.
152. Larsson, S., 1972, *Anal. Biochem.* **50**:245–254.
153. Hydén, H., 1959, *Nature* **4684**:433–435.
154. Cummins, J., and Hydén, H., 1962, *Biochim. Biophys. Acta* **60**:271–283.
155. Hydén, H., Lange, P. W., and Larsson, S., 1980, *J. Neurol. Sci.* **45**:303–316.
156. Dörmer, P., 1973, *Micromethods in Molecular Biology* (V. Neuhoff, ed.), Springer-Verlag, Berlin, Heidelberg, New York, pp. 347–394.
157. Hydén, H., and Larsson, S., 1956, *J. Neurochem.* **1**:134–144.
158. Zimmer, H. G., 1973, *Micromethods in Molecular Biology* (V. Neuhoff, ed.), Springer-Verlag, Berlin, Heidelberg, New York, pp. 297–328.
159. Ruch, F., and Lehmann, U., 1973, *Micromethods in Molecular Biology* (V. Neuhoff, ed.), Springer-Verlag, Berlin, Heidelberg, New York, pp. 329–346.
160. Galjaard, H., Niermeijer, M. F., Hahnemann, N., Mohr, J., and Sørensen, S. A., *Clin. Genet.* **5**:368–377.
161. Galjaard, H., Hoogeveen, A., Keijzer, W., De Wit-Verbeek, E., and Flek-Noot, C., 1974, *Histochem. J.* **6**:491–509.
162. Jongkind, J. F., Ploem, J. S., Reuser, A. J. J., and Galjaard, H., 1974, *Histochemistry* **40**:221–229.
163. Galjaard, H., Hoogeveen, A., De Wit-Verbeek, Keijzer, W., and Reuser, A. J. J., 1975, *Histochem. J.* **7**:499–501.
164. Galjaard, H., Hoogeveen, A., Van Der Veer, A., and Kleyer, W. J., 1976, *Excerpta Med. Int. Congr. Ser.* **411**:194–206.
165. Van Der Veer, E., Kleijer, W. J., de Josselin de Jong, J. E., and Galjaard, H., 1978, *Hum. Genet.* **40**:285–292.

166. Galjaard, H., 1979, *Ann. Clin. Biochem.* **16**:343–353.
167. Galjaard, H., 1980, *Trends in Enzyme Histochemistry and Cytochemistry*, Excerpta Medica, Amsterdam, pp. 161–180.
168. Aitken, D. A., Kleijer, W. J., Niermeijer, M. F., Herbschleb-Voogt, E., and Galjaard, H., 1980, *Clin. Genet.* **17**:293–298.
169. Galjaard, H., 1980, *Trends Biochem. Sci.* **5**:201–203.
170. De Josselin de Jong, J. E., Jongkind, J. F., and Ywema, H. R., 1980, *Anal. Biochem.* **102**:120–125.
171. Galjaard, H., 1980, *Genetic Metabolic Diseases. Early Diagnosis and Prenatal Analysis*, Elsevier/North-Holland, Amsterdam.
172. Outlaw, W. H., Jr., 1980, *Annu. Rev. Plant Physiol.* **31**:299–311.
173. Hydén, H., and Rönnbäck, L., 1975, *Brain Res.* **100**:615–628.
174. Sternberger, L. A., 1979, *Immunocytochemistry*, John Wiley & Sons, New York.
175. Chang, J. Y., Brauer, D., and Wittmann-Liebold, B., 1978, *FEBS Lett.* **93**:205–214.
176. Neuhoff, V., 1973, *Micromethods in Molecular Biology* (V. Neuhoff, ed.), Springer-Verlag, Berlin, Heidelberg, New York, pp. 205–214.
177. Neuhoff, V., and Rödel, E., 1973, *Hoppe Sylers Z. Physiol. Chem.* **354**:1541–1549.
178. Neuhoff, V., 1973, *Micromethods in Molecular Biology* (V. Neuhoff, ed.), Springer-Verlag, Berlin, Heidelberg, New York, pp. 395–398.
179. Eichner, D., 1966, *Experientia* **22**:620.
180. Neuhoff, V., 1973, *Micromethods in Molecular Biology* (V. Neuhoff, ed.), Springer-Verlag, Berlin, Heidelberg, New York, pp. 407–409.
181. Neuhoff, V., 1973, *Micromethods in Molecular Biology* (V. Neuhoff, ed.), Springer-Verlag, Berlin, Heidelberg, New York, pp. 179–204.
182. Hydén, H., and Rönnbäck, L., 1978, *J. Neurol. Sci.* **39**:241–246.
183. Althaus, H. H., Hutter, W. B., and Neuhoff, V., 1977, *Hoppe Sylers Z. Physiol. Chem.* **358**:773–775.
184. Althaus, H. H., Neuhoff, V., Huttner, W. B., Monzain, B., and Shahar, A., 1978, *Hoppe Sylers Z. Physiol. Chem.* **359**:773–775.
185. Althaus, H. H., Gebicke-Härter, P., and Neuhoff, V., 1979, *Naturwissenschaften* **66**:117.
186. Huttner, W. B., Meyermann, R., Neuhoff, V., and Althaus, H. H., 1979, *Brain Res.* **171**:225–237.
187. Fewster, M. E., and Blackstone, S., 1975, *Neurobiology* **5**:316–328.
188. Chao, S. W., and Rumsby, M. G., 1977, *Brain Res.* **124**:347–351.
189. Poduslo, S. E., 1978, *Adv. Exp. Med. Biol.* **100**:71–94.
190. Szuchet, S., Arnason, B. G. W., and Polak, P. E., 1978, *Biophys. J.* **21**:51a.
191. Kennedy, P. G. E., and Lisak, R. P., 1980, *Neurosci. Lett.* **16**:229–233.
192. McCarthy, K. D., and de Vellis, J., 1980, *J. Cell Biol.* **85**:890–902.
193. Szuchet, S., Stefansson, K., Wollmann, R. L., Dawson, G., and Arnason, B. G. W., 1980, *Brain Res.* **200**:151–164.
194. Gebicke-Härter, P. J., Althaus, H. H., Schwartz, P., and Neuhoff, V., 1981, *Dev. Brain Res.* **1**:497–518.

Mass Spectrometric Analysis of Some Neurotransmitters and Their Precursors and Metabolites

David A. Durden and Alan A. Boulton

1. INTRODUCTION

During the past decade, mass spectrometry (MS) has been applied to the identification and quantitation of several neurotransmitter substances and their precursors and metabolites in a variety of nervous and other tissues and in complex body fluids such as urine, cerebrospinal fluid (CSF), whole blood, and plasma. It is a very sensitive and relatively specific procedure and frequently is used in conjunction with a suitable separative method such as gas (GC), thin-layer (TLC), or liquid chromatography (LC). These chromatographic–mass spectrometric methods vie in sensitivity with other analytical procedures such as fluorimetry, radioimmunologic assays, and radioenzymatic assays; in general, however, they provide a much greater specificity.

Perhaps the greatest impetus in the use of MS for the analysis of neurotransmitters came from the reports of the successful interfacing of GC with MS[1,2] and the exploitation of the GC–MS method for the determination of acetylcholine[3] and other biogenic amines.[4] Others developed procedures using high-resolution mass spectrometry (HRMS) to identify biogenic compounds after separation by paper chromatography (PC) or TLC.[5] Since then, there has been a rapid expansion in the application of MS to neurobiological problems, and many techniques have been developed for handling all classes of neurotransmitter compounds, their acidic and alcoholic metabolites, and their precursor amino acids. In this chapter we discuss, in the main, only those methods that have been developed to identify and quantitate neurotransmitters in brain or other nervous tissue, blood, or csf. The very large field of urinary analyses by GC–MS is not included, although some procedures concerning substances present in low concentration have been mentioned.

David A. Durden and Alan A. Boulton • Psychiatric Research Division, University Hospital, Saskatoon, Saskatchewan S7N OXO, Canada.

As can be seen above and throughout this chapter, in the interests of brevity and in order to minimize the need for continuous repetition, we include a large number of acronyms and abbreviations to describe the various types of mass spectrometers and their linkages as well as the extensive number of substances analyzed and their various derivatives (Appendix). We hope that this has improved the readability of the text and not done the converse.

2. MASS SPECTROMETRY

2.1. The Process of Mass Spectrometry

A MS converts the molecules in the sample into gaseous ions which may then fragment to produce other ions representative of parts of the molecule and then separates the ions according to their mass to charge (m/z) ratio. As a consequence, two items of information are provided, the m/z ratios of the different ions and their abundances, represented by their relative intensities. A recording of the intensity against m/z of all of the ions produced from a compound is called the mass spectrum (see Fig. 1).

Mass spectrometers may be divided into two groups according to their ability to separate ions of different m/z values. Low-resolution devices separate ions that differ by unit mass; possible structures of the ions are determined assuming integer mass numbers for the elements. High-resolution machines

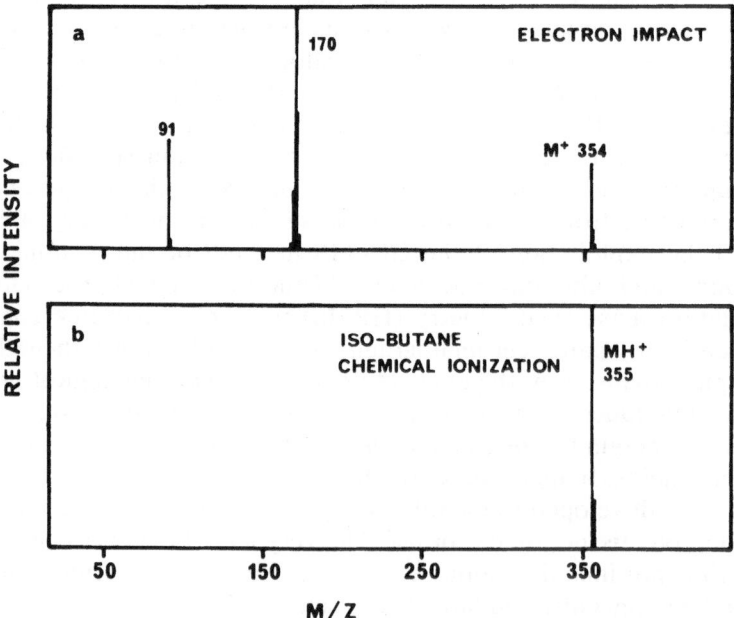

Fig. 1. Low-resolution mass spectra of 1-dimethylaminonaphthalene-5-sulfonamidophenylethylamine (DNS-PE). a: Produced by electron-impact ionization. b: Produced by chemical ionization with isobutane reagent gas.

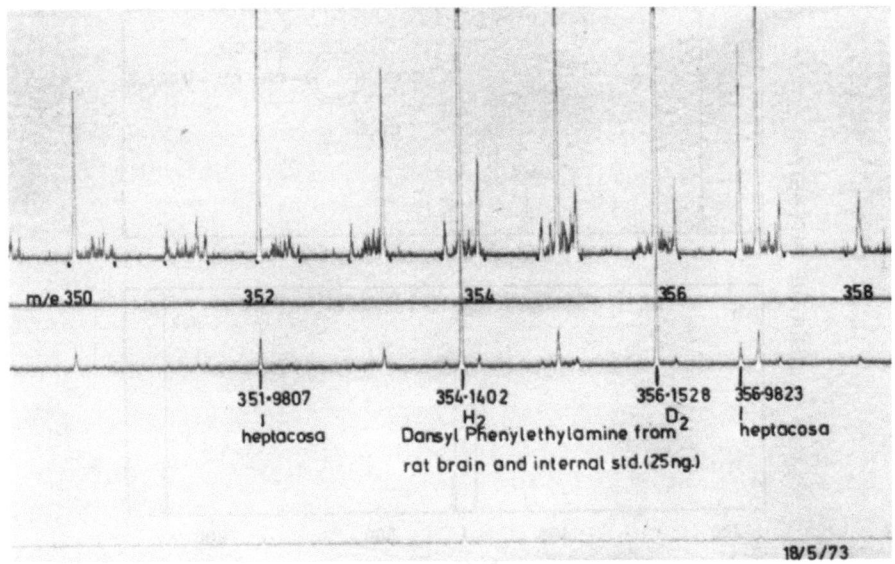

Fig. 2. Molecular ion region of a high-resolution mass spectrum of a derivatized extract from a rat brain (after MAO inhibition). The major peaks at m/z 354.1402 and 356.1528 are caused by DNS-PE and the internal standard, DNS-PE-d_2. Note that ions of several different elemental compositions (and hence exact m/z values) are present at m/z 354 and that reference mass ions derived from heptacosafluorotri-n-butylamine are also present in the spectrum.

are capable of separating ions of different fractional mass values and of measuring the precise ionic masses (to within a few parts per million) from which the elemental compositions can be determined. A portion of a high-resolution spectrum of a derivatized extract of rat brain containing 1-dimethylaminonapthalane-5-sulphonamidophenylethylamine (DNS-PE) is shown in Fig. 2.

In biomedical analyses, essentially three types of mass spectrometers have been used. These are the low-resolution single-focusing magnetic or quadrupole devices and the high-resolution double-focusing magnetic instruments.

2.1.1. Low-Resolution Mass Spectrometers

In the magnetic single-focusing MS, ions accelerated by a high voltage (1 to 8 kV) pass through a defining slit and are deflected by a magnetic field. By adjustments of either the magnetic field or the accelerating voltage, each m/z value can be detected, and although the mass scale under magnetic scanning conditions is nonlinear, the results can be linearized by computer methods. The magnetic MS usually displays a relatively constant sensitivity over a wide mass range (m/z of 1 to perhaps 4000). The present generation of magnetic instruments, capable of resolving powers of perhaps 3000 to 5000 and able to scan very rapidly (down to about 0.2 sec per decade in mass), are suitable for use with GC capillary columns.

Quadrupole mass spectrometers resolve the ions according to their m/z values using radio-frequency and DC electric fields generated on four cylin-

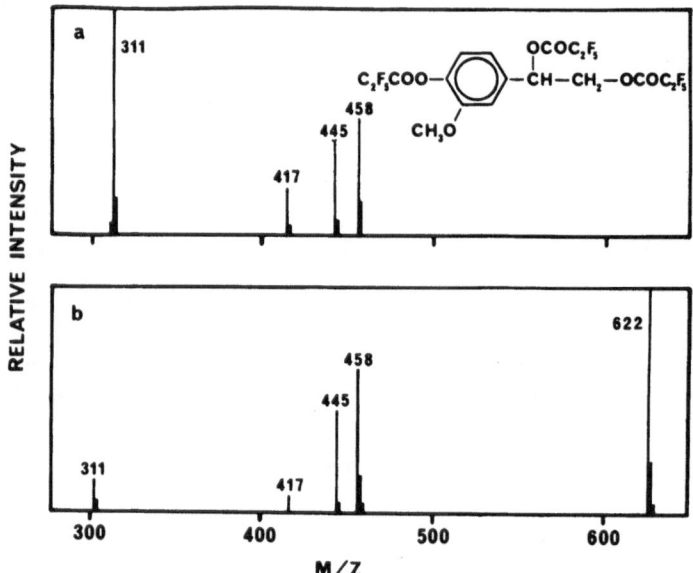

Fig. 3. Low-resolution mass spectra of 3-hydroxy-4-methoxyphenylethylene glycol (HMPG) derivatized with pentafluoropropionic anhydride; as recorded by a quadrupole (a) and a magnetic sector (b) mass spectrometer. Redrawn from data in refs. 96 and 4, respectively.

drical rods. Their major advantages are a linear mass scale and the possibility of very rapid scanning over the mass range. In comparison with the magnetic instrument, however, their mass range is more restricted (m/z of 1 to less than 1000), their sensitivity tends not to be constant over this range, and their resolving power is also usually lower. Figure 3 illustrates a rather extreme example of the differences in spectra produced by magnetic and quadrupole mass spectrometers. Because of the quadrupole instrument's insensitivity to high-mass ions, the molecular ion was not detected, and m/z 311 was recorded as the base peak.

2.1.2. High-Resolution Mass Spectrometers

The conventional high-resolution MS is a double-focusing device in which an electric sector energy analyzer is placed before or after the magnetic analyzer in the ion trajectory. By choosing a narrow band of ion energies, resolutions of up to 300,000 may be attained. Since resolution and intensity are inversely proportional, the double-focusing instrument may be operated either at low resolution and high sensitivity (equal to or better than that of quadrupole or single-focusing magnetic instruments) or at high resolution and lower sensitivity for increased specificity of mass measurement. Thus, the double-focusing mass spectrometer becomes the most versatile of instruments in that it possesses the capabilities of the low-resolution instruments plus the ability to isolate a precise mass value (hence a specific elemental composition).

2.2. Formation of Ions

2.2.1. Electron Impact

The most common way of producing ions, electron impact (EI), is also technically the simplest. In EI, a beam of electrons passes through the ion source in which the sample gas is present at a pressure of less than 10^{-4} mb. Ions are formed primarily by the removal of an electron, thus producing a positively charged excited molecular ion. The molecular ion itself may be detected, or smaller ionic fragments which are formed from it and which provide information on its structure. One important limitation of electron impact spectra is that many compounds, especially biogenic metabolites, produce no molecular ions or very low abundances of them. This can be overcome by using other ionization techniques.

2.2.2. Chemical Ionization

Chemical ionization (CI) is a technique in which the sample ions are formed through ion–molecule reactions from a reagent gas. If a reagent gas is admitted to the ion source at a pressure greater than 10^{-4} mb (preferably about 1 mb), it is preferentially ionized by the electron beam, and the ions lose energy by colliding with the remaining gas molecules. These "thermalized" reagent ions ionize the sample molecules, usually by transfer of a portion of themselves to produce "pseudomolecular" ions of mass greater (e.g., one unit in the case of a proton transfer) than the molecular weight of the sample. The resultant spectrum is frequently very simple, being composed primarily of "pseudomolecular" ions and few fragment ions. Chemical ionization thus complements EI, as is illustrated in the spectra in Fig. 1. Chemical ionization can also be used to produce negatively charged ions and thus become a very specific method for the analysis of electrophilic compounds.

Although most of the ionization of the sample is in the form of "pseudomolecular" ions, and these are increased greatly in intensity relative to the fragment ion intensities, the overall sensitivity of EI and CI are about the same, since in the CI procedure the ion source exit aperture must be reduced considerably in order to maintain the higher pressure.

2.2.3. Field Desorption

Field desorption (FD) is yet another procedure used to produce high molecular ion intensities from biological materials. The sample is deposited on a fine emitter wire which possesses a dense coating of carbon microneedles. When the emitter is placed in a high electric field in the ion source, the sample evaporates from the surface and is ionized by the intense electric field surrounding each microneedle to produce ions, with low internal energies, that do not fragment extensively. Although this technique holds great promise, it is currently technically rather difficult and has not yet been much used in the analysis of neurotransmitter substances.

2.3. Mass Spectrometric Methods

2.3.1. Qualitative

Complete spectra of compounds are obtained by scanning over a wide mass range and recording all m/z values against intensities (as shown in Fig. 1). Identification may then be determined either *a priori* from the intensities of particular ions and assignment of possible structures or by comparison to a library of prerecorded spectra. If the spectra are recorded at high resolution, the precise mass values of the ions provide additional information in that their elemental compositions are limited to only a few possibilities.

2.3.2. Quantitative

Two methods of quantitation by mass spectrometry are used. Spectra may be recorded as the sample, perhaps in conjunction with an internal standard, is evaporated in the ion source. The time profiles of several ions specific to the compound may then be reconstructed from sequential spectra, and an estimate computed of the amount of unknown from their intensities by comparison with the intensities of the ions of an internal standard or a known amount of the compound under investigation admitted subsequently. The advantage of this method is that complete spectra are available for identification. Since all ions in the spectrum are recorded, however, the time spent on each is short, and thus, the sensitivity is relatively low.

An alternative and preferred method is that of selected ion monitoring (SIM).[6] This technique possesses many neologisms such as mass fragmentography, integrated ion current procedure, multiple-ion monitoring, single-ion recording, etc. In this procedure, the MS is adjusted so that it can record the signal from only a few selected masses, perhaps one to eight, as a function of

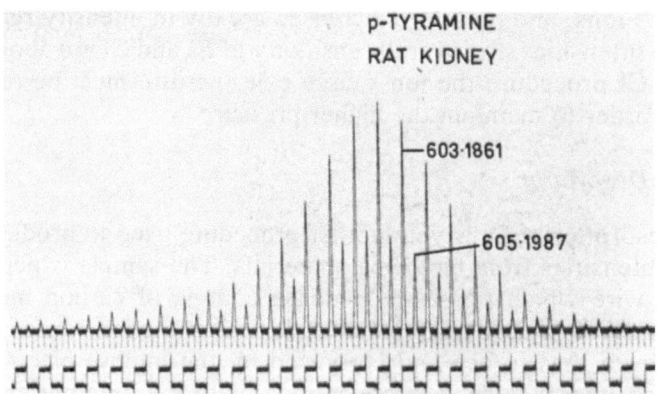

Fig. 4.   High-resolution ($R = 7000$) SIM profile of the molecular ions of di-(1-dimethylaminon-aphthalene-5-sulfonyl) derivatives of *p*-tryamine and the internal standard dideutero-*p*-tyramine as they are evaporated from the direct probe after isolation by thin-layer chromatography.

time. The sensitivity is increased, since only a part of the spectrum is detected, and the time spent on each ion is increased, thus increasing the signal-to-noise ratio. The resolution of the mass spectrometer may be either low (i.e., mass-fragmentography) or high (i.e., integrated ion current procedure). Quantitation then occurs by comparing the signals from an internal standard of different mass mixed with the sample or from an external standard which may be the same compound admitted separately. If the selected ions are the major fragments in the spectrum of the sample, detected at low resolution using a quadrupole or magnetic MS with magnetic peak selection, the related intensities of the ions can be used to verify the identity of the compound being quantitated, and this may be done over the whole mass range. Under high-resolution conditions, the mass range is more limited (within 20% of the mass of the lowest-mass ion), but a precise mass value and hence elemental composition can be selected in order to provide the necessary specificity. A typical high-resolution SIM profile is illustrated in Fig. 4.

2.4. Internal Standards for Quantitative Analysis

In order to be able to quantitate reliably the low levels of neurotransmitters in physiological extracts and fluids, it is usually necessary to use internal standards.

The ideal internal standard is a stable isotope-labeled analogue with identical reactive and physical chemical properties to the substance being analyzed. Deuterated analogues are the best because they are relatively easy to synthesize at high isotopic purity. The deuterium atoms must be located in nonexchangeable (or very slowly exchangeable) positions in the molecule so that they are not lost either during derivatization and isolation or in the neutral fragment if a fragment ion is to be monitored. Labeling with more than one deuterium atom avoids interference from the [^{13}C]species of the unlabeled compound. With respect to most compounds, the properties of the deuterated and proteo species are essentially identical, although some chromatographic separation has been observed.[7,8]

Other atoms, ^{13}C, ^{15}N, or ^{18}O, have also been used to label internal standards, with [^{15}N]histamine[9] being a notable example.

The second choice for an internal standard is a homologue or structural isomer of the compound to be analyzed. These are generally acceptable when a GC separation is used, but they are not suitable for TLC separations since they separate from the unknown.

In some situations, external standards have been employed. For example, the sample may be divided into aliquot parts one or more of which are then supplemented with the compound of interest. In other cases, a standard may only be incorporated in part of the anlysis, e.g., the mass spectrometric step. In all of these "nonideal" cases, the sample to be quantitated is exposed to different conditions than the sample containing the standard. These nonideal procedures are only acceptable when relatively large (microgram) quantities of the neurotransmitter or metabolite are to be quantitated.

3. CHROMATOGRAPHIC–MASS SPECTROMETRIC METHODS

3.1. Thin-Layer Chromatography–Mass Spectrometry

The TLC–MS method has been used primarily for the analysis of amines as their 1-dimethylaminonapthalene-5-sulphonyl (dansyl) (DNS) derivatives. The amines are isolated from the fluid or tissue homogenate by extraction or cation exchange after the appropriate deuterated standard has been added. After derivatization with DNS chloride, they are separated unidimensionally on silica gel thin layers in two or more different solvent systems. They are quantitated by SIM[10,11] of the signals arising from the molecular ions of the deutero and proteo amines at high resolution (7,000 to 10,000).

3.2. Gas Chromatography–Mass Spectrometry

The interconnecting of GC and MS has produced the greatest number of analytical procedures because of the variety of GC techniques such as packed columns, support (SCOT) or wall (WCOT) coated high-resolution capillary columns and low- or high-resolution MS (LRMS or HRMS).

Packed columns GC–LRMS (conventional GC–MS) was the first method adopted. Its advantages are relatively high sensitivity since large aliquots (up to 3 μl) of the sample may be injected, and the MS may be operated at maximum sensitivity and short analysis time. The disadvantages are that the specificity may be less than that of other methods because a packed column does not provide good separation of some compounds, and there may be ambiguity in the identity of the source of the ion if a fragment ion common to many compounds is being detected. To reduce these possible ambiguities, considerable "preseparation" of the sample prior to GC–MS analysis may be required. In addition, the monitoring of several fragment ions should be undertaken in order to increase the specificity. The method appears to yield excellent results for neurotransmitters and related substances if they are present in relatively high concentrations. For compounds present in low concentrations, the signal-to-noise ratio may be quite low because of the extended GC peak elution time (i.e., large peak width), in which case other GC–MS procedures become preferable. Compounds must invariably be derivatized in order to increase their volatilities and reduce the activities of their functional groups. Elution of column packing materials (i.e., column bleed) causes an increased background with concurrent obscuration of the desired signal; as a consequence, only low-bleed liquid phases such as the polysiloxanes are recommended for GC-MS use. The use of GC–HRMS reduces this background contribution by resolving the sample ions from the ions arising from column bleed.[12]

Further increases in specificity can be obtained by combining high-resolution (WCOT) capillary columns with LRMS (i.e., HRGC–MS). Because the GC peak width is reduced and the peak height increased, the signal-to-noise ratio is improved. This offsets somewhat the expected lower sensitivity imposed by the low capacity of the column.

Maximum specificity is obtained when high-resolution capillary columns are combined with HRMS (i.e., HRGC–HRMS), since in this situation the elemental composition of the ions being monitored can be predetermined. It is usually necessary, however, to clean up the sample prior to GC–MS analysis.

Finally, it may be concluded that the present ideal compromise with respect to specificity and sensitivity is a combination of SCOT capillary columns with their relatively high permissible loading and lower column bleed and HRMS with its precise mass analysis.

4. SUMMARIES OF METHODS

4.1. Amines

4.1.1. Catecholamines

The catecholamines, dopamine (DA), norepinephrine (NA), and epinephrine (adrenaline, AD), and their methylated metabolites, 3-methoxytyramine (3-methoxy-4-hydroxyphenyl ethylamine, 3-MT), metanephrine (MN), normetanephrine (NMN), and epinine (EP), were among the first amines to be analyzed by MS. In all cases, derivatives were prepared and were separated either by TLC or by GC. Several groups have used the DNS derivatives[13–17] which were separated by TLC and introduced to the MS by the direct probe (DP). Most of the analyses were qualitative, but AD was quantitated with a deuterated internal standard.[17] A similar derivative, BNS, was utilized with FD ion formation,[18] but the masses of the molecular ions of the amine and its associated internal standard become very high, and the sensitivity with FD is probably too low for tissue analyses. Consequently, GC–MS procedures using different derivatives have become preferred.

The GC–MS analyses first focused on determining DA and NA in tissues using perfluorinated acyl derivatives[4,19] (i.e., PFP derivatives with α-MDA or α-MNA as internal standards). These derivatives produce virtually no molecular ions under EI, and so the fragment ions m/z 428 and m/z 442 for DA and α-MDA, and m/z 176 and m/z 190 for NA and α-MNA, respectively had to be used. Following this successful methodology, which has been extended to include AD,[20] the optimum reaction conditions for formation of PFP derivatives and selection of the appropriate polysiloxane stationary phases for GC[21,22] were established. A further major improvement was the introduction of suitable tri- or tetradeuterated analogues (DA-d_3, NA-d_3, or DA-d_4) to serve as internal standards[23–29] (see Table I). In addition, other preseparation steps such as alumina adsorption[28] or boric acid gel separation[29] have been incorporated to increase specificity and sensitivity, and more recently the use of CI instead of EI[30,32] resulted in an increase in sensitivity of about tenfold.[30] Introduction of a PFP-derivatized extract by DP in conjunction with scanning over a narrow mass range of the CI spectrum[33] are a simpler but somewhat less sensitive procedure than GC–MS.

Other fluorinated acyls, sometimes in conjunction with trimethylsilylation of the phenolic groups, have also now been used. The TFA derivative was

Table I

Mass Spectrometric Methods for the Quantitative Analysis of Amines

Compound	Method	Ionization	Preseparation	Stationary phase	Derivative	Internal standard	Ions monitored	Sensitivity	Comments	Reference
Catecholamines										
Dopamine (DA)	GC–MS	EI	None	OV-17	PFP	Ring d₃ or alkyl-d₄	F 281/284 F 428/431	N3 ng	Acceptable for large (> 10 ng) quantities	23–27
	GC–MS	CI	Alumina	OV-17	PFP	Alkyl-d₂	F 590/592	—	CI produces a larger fragment ion as base peak	30–33
	GC–MS	EI	None or alumina	SE-54	TFA	Alkyl-d₄	F 328/331	10 pg	TFA is more sensitive but less stable than PFP derivative	34,35
	HRGC–MS	CI	Alumina	WCOT-SE-30	PFB-TMS	[¹⁴C]DA	MH 476/478	100 pg	Deuterated standard preferred, since ¹⁴C may be a hazard	37
Norepinephrine (NA)	GC–HRMS	CI	Alumina	OV-17	TFA-TMS	Isoproterenol	F 355.1568	—	Deuterated standard would be preferable	12
	GC–MS	CI	GPC-10	OV-101	PFP	NA-d₆	F 590/596	—	Preseparation and CI increase specificity	30,31
	GC–MS	EI	Boric acid gel	OV-1	PFP	Alkyl-d₃	F 590/592	25 pg	This appears to be the most sensitive procedure	29
	GC–MS	EI	Alumina	OV-1	PFP-Et	Epinine	F 431/428	40 pg	Benzylic OH is derivatized to give larger fragment ion as base peak	28
Epinephrine (AD)	HRGC–MS	CI	Alumina	WCOT-SE-30	PFB-TMS	[¹⁴C]NA	MH 564/566	100 pg	Internal standard is radioactive	37
	TLC–HRMS	EI	Direct derivatization	Silica gel	DNS	Ring-d₃	M 882.2427 M 885.2614	<1 ng	Precise mass measurement at 7000 resolution was used	17
	GC–HRMS	EI	Alumina	OV-17	TFA-TMS	Isoproterenol	F 355.1568	—	Deuterated standard would be preferable	12
3-Methoxytyramine (3-MT)	GC–MS	EI	—	OV-17	PFP	Ring-d₃	F 283/286 F 296/299	~1 ng	Determined simultaneously with DA	24,25
Metanephrine (MN)	GC–MS	EI	XAD-2	Poly I-110 OV-1	PFP	Alkyl-d₃	F 458/460	10 ng	This method for urine could be applied to tissues	39,40
Normetanephrine (NMN)	GC–MS	EI	XAD-2	Poly I-110 OV-1	PFP	Alkyl-d₃	F 445/446	10 ng	This method for urine could be applied to tissues	39,40
Epinine (EP)	GC–MS	EI	—	OV-17 or OV-1	PFP	—	F 428	—	Epinine must be separated from dopamine	42
Phenylalkylamines										
Phenylethylamine (PE)	TLC–HRMS	EI	None or Dowex® 50-X2	Silica gel	DNS	Alkyl-d₂ or d₄	M 354, 356, 358	200 pg	High specificity and sensitivity	43
	GC–MS	CI	n-Butanol heptane	OV-101 SP 2250	DNP-TMS	Alkyl-d₄	MH 288	<1 ng	DNP derivative and CI are more specific than PFP derivative and EI	44

Compound	Technique	Ionization	Extraction/Preparation	Column	Derivative	Internal standard	Ions	Sensitivity	Comments	Ref.
Phenylethanolamine	TLC–HRMS	EI	Direct derivatization	Silica gel	DNS	Alkyl-d3	M 370/373	200 pg	High specificity and sensitivity	10
	GC–MS	CI	n-Butanol heptane	SP 2250	DNS-acetyl DNP-TMS	Alkyl-d3	M 412/415 MH 376/379	<1 ng	GC-MS may not be as specific as TLC–HRMS	44
m-Tyramine (m-TA)	TLC–HRMS	EI	Direct derivatization	Silica gel	DNS	Alkyl-d4	M 603/607	200 pg	High specificity and sensitivity	10,50
p-Tyramine (p-TA)	TLC–HRMS	EI	Direct derivatization	Silica gel	DNS	Alkyl-d4	M 603/607	200 pg	High specificity and sensitivity	10,48
m-, p-Octopamine (m-OCT, p-OCT)	TLC–HRMS	EI	Direct derivatization	Silica gel	DNS-acetyl	Alkyl-d3	M 661/664	200 pg	High specificity and sensitivity	54
	GC–MS	EI	Dowex® 50	OV-17	PFP	Alkyl-d3	F 428-430	—	May not be as specific as TLC–HRMS	51
m-, p-Synephrine (m-SYN, p-SYN)	TLC–HRMS	EI	Direct derivatization	Silica gel	DNS-acetyl	Alkyl-d3	M 675/678	200 pg	High specificity and sensitivity	54,55
Indolylalkylamines Tryptamine (TRYPT)	TLC–HRMS	EI	None or Dowex® 50-X2	Silica gel	DNS	Alkyl-d2, d4	M 393/395 M 393/397	500 pg 200 pg	Sensitivity is improved with the d4 standard	56
	GC–MS	EI	CG-50	OV-17	N-Ac-N-TFA	Alkyl-d2	M 280/282 F 183/185	<1 ng	Blank is higher because of presence of (M-2)$^+$ ions; a d4 standard may decrease it	57
5-Hydroxytryptamine (5-HT)	TLC–MS	EI	Acetone extraction	Silica gel	DNS	Synephrine	M 642/633	~1 ng	Internal standard only used for MS step; a deuterated standard is preferable	60
	GC–MS	EI	n-Butanol extraction	SP-2100/QF-1	PFP	Alkyl-d4	F 451/454 F 438/440	1-2 ng	Tris-PFP derivative	62,63
	GC–MS	CI	n-heptane extraction	OV-17	PFP	Alkyl-d4	MH 469/473	<1 ng	Di-PFP derivative, methane reagent gas	64
Melatonin (MEL)	GC–MS	NI-CI	CHCl3 extraction	OV-225	PFP	Alkyl-d4	F 320/323	1 pg	Negative ion CI gives excellent sensitivity	65
	GC–MS	EI	CHCl3 extraction	OV-17	TMS	ω-N-n-hexanoyl-5-MT	F 245, 232	10 pg	High-resolution MS was used to confirm fragment ion compositions	66
6-Hydroxymelatonin (6-HMEL)	GC–MS	NI-CI	EtOAc extraction	OV-1	TBDMS-PFP	6-HMEL-d4	F 470/473 F 450/453	—	Negative ion CI	67,68
5-Methoxytryptamine (5-MT)	HRGC–MS	EI	CHCl3 extraction	WCOT-OV-1	TMS	ω-N-n-Hexanoyl-5-MT	F 174/232	1 pg	m/z 174 is an ion common to amine TMS derivatives	72
N,N-Dimethyltryptamine (DMT)	HRGC–MS	EI	CHCl3 extraction	WCOT-SE-30	TMS	5-MeO-DMT	F 58	100 pg	m/z 58 is an ion common to dimethyl amines	76
DMT and bufotonin (BUF)	GC–MS	EI	XAD-2, silica gel	OV-101	TMS	DMT-d5, BUF-d4	F 202/207 F 290/294	~1 ng ~1 ng	m/z 202 gives greater precision but lower sensitivity	74,75

(Continued)

Table I. (Continued)

Compound	Method	Ionization	Preseparation	Stationary phase	Derivative	Internal standard	Ions monitored	Sensitivity	Comments	Reference
	TLC–MS	EI	Silica gel	Silica gel	DNS	—	M 437	—	Derivative with the most intense molecular ion	77
Imidazolyl amines										
Histamine (Hist)	GC–MS	EI	Extraction + gel chromatography	SE-30	HFB-EtO	[^{15}N$_2$]Hist	M 379/381, F 306/308	1 ng	It appears that an (M-2)$^+$ ion causes a high blank of 2%	9,78
t-Methylhistamine (t-MH)	GC–MS	EI	Extraction + gel chromatography	SE-30	HFB	Me-d$_3$-t-MH	M 324/327	200 pg	Blank is better, since N is methylated	78
Quaternary amines										
Choline (Ch) and acetylcholine (ACh)	GC–MS	EI	Extraction, demethylation	OV-101	O-Propionylation	Ch-d$_4$, d$_9$, ACh-d$_4$, d$_9$	F 58/60/64, F 58/60/64	1 ng	The amines are demethylated to improve their GC characteristics	3,80
	DP–MS	FD	Extraction	—	—	Ch-d$_2$-d$_9$, ACh-d$_9$	M 104/106/113, M 146/155	10 pg	Quaternary amines are desorbed from the FD probe	82
Polyamines										
Putrescine (Put)	TLC–MS	EI	Dowex® 50-X8	Silica gel	DNS	DNS hexamethylene diamine	M 554, M 582	—	DNS hexamethylene diamine was used for the MS step only; a deuterated standard would improve these procedures	84,87
Cadaverine (Cad)	TLC–MS	EI	Direct derivatization	Silica gel	DNS	DNS hexamethylene diamine	M 568, M 582	—	DNS hexamethylene diamine was used for the MS step only; a deuterated standard would improve these procedures	85,86
All polyamines	GC–MS	EI	None (no tissues)	OV-101	TFA	d$_4$, d$_6$, or d$_8$ (see ref.)	(M-CF$_3$), (M-CF$_3$CO)	100 pg	All four polyamines can be determined simultaneously	88
	GC–MS	NI-CI	Several extractions	OV-17	TFA	d$_4$ or d$_6$ (see ref.)	(M-HF)$^-$	<1 pg	Negative ion CI gives excellent sensitivity	90
Cycloalkylamines										
Piperidine (PIP)	TLC–MS	EI	Direct derivatization	Silica gel	DNS	DNS-pyrrolidine	M 318, M 304	—	Molecular ion gives good specificity, but DNS γ-butyrolactam (GABA) must be isolated	91
	TLC–MS	EI	Direct derivatization	Silica gel	BNS	BNS-pyrrolidine	F 359, F 345	—	M-43 ion gives a higher sensitivity; a deuterated standard would improve the method	92
	GC–MS	EI	Steam distillation	OV-17	DNB	Pip-d$_{11}$	F 234/243	—	Probably the most sensitive method	93

Ions monitored: F = Fragment ion, M = molecular ion, MH = pseudo molecular ion.

found to have a lower chemical stability than the PFP analogue but exhibited a greater sensitivity[34,35] and has been used in conjunction with TMS[12]; and HRMS for increased specificity. The benzylimine derivative PFB-TMS was observed to give excellent sensitivity using either EI[36] or CI and HRGC–MS.[37]

The PFP derivatives (along with suitable internal standards) have also now been used to analyze the methylated CA metabolites 3-MT in tissue[4,24,25,38] and MN and NMN in urine.[39,40] To determine the three amines in plasma, Wang *et al.*[41] chose to use the TFA derivatives. Epinine has recently been detected in adrenal medulla using perfluoroacyl derivatives,[42] but high GC resolution is required, as the spectra of EP and DA are almost identical.[21] More complete details about selected methods for these compounds are given in Table I which includes the substance to be identified and quantitated, the MS combination, the ionization process, details on preseparation, the GC station-ary phase, the derivatives used, the ions monitored, the internal standards used, the sensitivity achieved (when applied to tissue samples, i.e., not standard solutions), other general comments on the suitability or otherwise of the pro-cedure, and the reference sources.

4.1.2. Phenylalkylamines

The phenylalkylamines also referred to as the trace amines: phenylethyl-amine (PE), phenylethanolamine (PEOH), *meta*-tyramine (*m*-TA), *para*-ty-ramine (*p*-TA), *meta*- and *para*-octopamine (*m*-OCT, *p*-OCT) and *meta*- and *para*-synephrine (*m*-SYN, *p*-SYN) have been identified in tissues and physi-ological fluids using both TLC–MS and GC–MS.

Durden *et al.*[43] were the first to report a procedure for measuring trace quantities of PE in brain and other tissues. They used preparations of DNS derivatives on silica gel TLC followed by quantitation using HRMS (i.e., TLC–HRMS) in which, in the example of PE, its molecular ion and that of DNS-dideutero-PE as internal standard were monitored by high-resolution SIM. More recently, Edwards *et al.*[44] have been able to confirm the values using the N-dinitrophenyl (DNP) derivative and GC–MS with CI. The PFP derivative has been used with low-resolution GC–MS and tetradeutero-PE internal standard,[45,46] but the low-mass ions used for quantitation may not be specific.

The TLC–HRMS method using the DNS derivative has been applied to the other trace amines PEOH,[10] *p*-TA,[47–49] and *m*-TA.[49,50] It is worthy of note that this procedure results in lower values than those obtained by GC–MS[44–46] with excellent consistency and reproducibility.

Buck *et al.*[51] have used packed-column LRGC–MS of the PFP derivative for the analysis of *p*-OCT, whereas Couch and Williams used di-O-TMS-N-TFA derivatives for *o*- and *m*-OCT and the N,O-TFA derivative for *m*-SYN.[52,53] The use of packed-column GC and perfluorinated acetyl derivatives does not appear to be the method of choice for these compounds, as the derivatives produce mainly low-mass ions that are also present in the spectra of many other compounds, and HRGC would be required to completely sep-arate the compounds and verify identities.

The TLC–HRMS method using trideuterointernal standards has recently been extended to enable quantitation of β-hydroxyphenylalkylamines (i.e., *m*- and *p*-OCT, *m*- and *p*-SYN) by preparation of the DNS-acetyl derivatives which produce more stable and intense molecular ions than the DNS derivatives.[54,55]

4.1.3. Indolylalkylamines

Both TLC-MS and GC-MS methods have been applied to the analysis of the indolylalkylamines, tryptamine (TRYPT), 5-hydroxytryptamine (5-HT), 5-methoxytryptamine (5-MT), melatonin (MEL), 6-hydroxymelatonin (6-HMEL), N,N-dimethyltryptamine (DMT), and bufotenin (BUF). The TLC–HRMS analysis of tryptamine as the DNS derivative by Philips *et al.*[56] appears to be the most sensitive method and is capable of determining less than 500 pg from brain and other tissues. Warsh *et al.*[57] chose the $^{\omega}$N-acetyl-N-TFA derivative for GC–MS analysis. Acetylation of the amino group results in a cyclic compound on perfluoroacylation of the indole nitrogen.[58] This compound produces an intense molecular ion which improves the specificity, but an $(M-2)^+$ ion increases the blank. Donike *et al.*[59] have suggested the use of the $^{\omega}$N-TFA-N-TMS derivative which also exhibits an intense molecular ion.

As shown in Table I, 5-HT has been assessed by TLC–MS of its DNS derivative using N,O-bis-DNS-synephrine as internal standard for the MS step[60] and by GC–MS using the tris PFP derivative on a packed column (OV-17)[61] and α,α,β,β-tetradeutero-5-HT as internal standard instead of α-methyl-5-HT.[62,63] It appears that 5-HT may not react uniformly with pentafluoropropionic anhydride to produce the tris derivative,[22] and therefore, Liuzzi *et al.*[64] quantitated 5-HT (with 5-HT-d_2 internal standard) as its di-PFP derivative using CI–GC–MS for greater sensitivity.

Analysis of MEL as its PFP derivative[61] with EI–GC–MS was not particularly sensitive, but this was improved dramatically (to less than 1 pg/ml plasma) by using negative-ion CI with tetradeutero-MEL as internal standard.[65] More recently, EI–GC–MS of the N-TMS derivative with $^{\omega}$N-*n*-hexanoyl-5-MT internal standard[66] has demonstrated a similar sensitivity.

The major MEL metabolite, 6-HMEL, has been identified and quantitated in normal human urine using a tetradeutero analogue and negative CI–GC–MS of the *t*-butyldimethylsilylpentafluoropropionyl (TBDMS-PFP) derivative.[67,68]

The claimed GC–MS identification of 5-MT in tissue and physiological fluids was based originally on detection of the N-PFP derivative at the fragment ion mass *m/z* 306 (and *m/z* 308 of the deuterated analogue).[61,69,70] Narasimhachari *et al.*,[71] however, have shown that melatonin is transacylated to 5-MT-PFP by pentafluoropropionic anhydride, and so the validity of this procedure is in question. This transacylation is variable, as Beck and Bosin[70] were not able to confirm previous findings[61] for rat brain 5-MT. A recent alternative method which should have been capable of avoiding artifactual derivatization products utilized the TMS derivative,[72] but unfortunately, a nonspecific ion of *m/z* 174 resulting from $N(TMS)_2^+$ was monitored. For 5-MT, a new unambiguous procedure such as TLC–HRMS of the DNS derivative is required.

Other methylated indoleamines such as DMT and BUF are particularly difficult to determine mass spectrometrically, since the major peak in the spectra of the free bases and of most derivatives is m/z 58, $N(CH_3)_2{}^+$ which is not a specific fragment ion. The trimethylsilyl derivative and GC–MS[73–76] have been used in the main, and either m/z 58 or m/z 202 used for quantitation. Extensive prepurification using XAD resin and silica gel TLC[75] or glass capillary HRGC[76] are required to verify the source of the monitored ion. The use of TLC–MS of the DNA derivative of BUF, which exhibits an intense molecular ion m/z 437 in its EI spectrum, is an alternate procedure[77] to GC–MS.

4.1.4. Imidazolylamines

A notable exception to the use of deuterated internal standards is the analysis of histamine in which a [^{15}N]-labeled analogue is usually used.[9,78] The N^α-heptafluorobutyryl-N'-ethoxycarbonyl (HFB-EtO) derivative was prepared and monitored as its molecular ion with LRGC–MS. The methylated metabolite *tele*-methylhistamine has also been determined by GC–MS as the HFB derivative with an internal standard that was either the deuterated analogue[78] or the structural isomer *pros*-methylhistamine.[79]

4.1.5. Quaternary Amines

The quaternary ammonium neurotransmitters acetylcholine (ACh) and choline (Ch) have been identified by GC–MS and DP–FD–MS. For GC–MS analysis, an extract containing Ch and ACh (and their deuterated internal standards) is first esterified so that Ch is O-propionylated, and then both compounds are demethylated to form the tertiary amines propionyldeanol and acetyldeanol which can be chromatographed.[3,80,81] The base peak ions caused by $C_3H_8N^+$ (i.e., m/z 58, 60, and 64 of the proteo, d_4, and d_9 standards) are monitored as the tertiary amines elute from the GC.[80,81] This procedure is satisfactory because of the relatively high concentrations of these amines in tissue samples.

Field desorption is an ideal method for quarternary ammonium compounds, since the onium ions themselves are desorbed, thus eliminating the necessity of demethylation. The procedure reported by Lehmann and Schulten[82] used a deuterium standard for Ch.

4.1.6. Polyamines

The term "polyamines" is usually taken to include the diamines putrescine (Put) and cadaverine (Cad), the triamine spermidine (Sper), and the tetraamine spermine (Sp). Early methods[13,83–86] for the diamines used TLC–MS of the DNS derivative which were quantitated either by fluorimetry, with the MS being used essentially only for identification purposes, or with a DNS homologue used as the internal standard in the MS step,[87] whereas the tri- and tetraamines were only identified by MS.[13,87] The inclusion of stable isotope-labeled standards for the complete analysis would have made the DNS method an excellent procedure.

Smith and Daves[88] have reported a GC–MS method in which all four polyamines can be determined at the picomole level (with deuterated analogues as internal standards) as their trifluoroacetyl derivatives after a single GC injection. The procedure has been extended to include the metabolite N-acetylspermidine.[89] Negative ion CI–GC–MS increases the sensitivity of analysis of Put, Sper, and Sp as their TFA derivatives.[90] The highly electrophilic derivatives of Sper and Sp exhibit a considerable increase in sensitivity over the results obtained from positive CI ion formation.

4.1.7. Cycloalkylamines

Piperidine (Pip) has been identified and quantitated in tissues as its DNS derivative by TLC–MS.[91] An increase in sensitivity was observed when the BNS derivative, which fragments to give an intense $(M-43)^+$ ion in a cleaner part of the spectrum, was chosen instead.[92] DNS- or BNS-pyrrolidine was added to the derivatized extracts to act as an internal standard for the MS step only. The use of a deuterated standard in conjunction with the BNS derivatives would undoubtably improve the sensitivity of the procedure. A deuterium-labeled internal standard has been used by Miyata *et al.*[93] in a GC–MS procedure utilizing DNP derivatives.

4.2. Acids

The acid metabolites of the biogenic amines have been analyzed almost exclusively by GC–MS procedures, with most being packed-column, low-resolution methods. As is the case in the analysis of amines, acids usually require a preseparation step followed by the preparation of volatile derivatives for gas chromatography. The most popular derivatives are methyl esters (Me), trimethylsilyl esters (TMS), and fluorinated ethyl (TFE) and propyl (HFIP, PFnP) esters for the carboxyl group and fluorinated acyls (TFA, PFP, HFB), or TMS derivatives for other (phenolic) functional groups.

4.2.1. Catecholamine Metabolites

The catecholamine acid metabolites include the direct metabolites of the catecholamines: 3,4-dihydroxyphenylacetic acid (DOPAC), 4-hydroxy-3-methoxyphenylacetic acid (HVA, homovanillic acid), 4-hydroxy-3-methoxylphenylacetic acid (VMA, vanillyl mandelic acid), and their 3-hydroxy-4-methoxy isomers (iso-HVA and iso-VMA).

The primary metabolite of DA, DOPAC, has been identified and quantified by LRGC–MS using either the PFnP-PFP derivative[94,95] or the Me-PFP derivative[96,97] with pentadeutero-DOPAC internal standard in all cases. The PFnP derivative appears to be more suitable, as its spectrum exhibits an intense molecular ion and the ion used for SIM, *m/z* 415 (see Table II), is the base peak.

Homovanillic acid appears to have received the greatest analytical attention of all the biogenic acids and was determined originally as methylfluoroacyl derivatives[4,98-101] with di-, tri-, or pentadeutero-HVA internal standards. It was observed, however, that diazomethane, chosen to esterify the carboxyl group, methylated a phenolic group of DOPAC and artificially enhanced HVA concentrations.[101] This is obviated by the choice of the PFnP ester for the carboxyl group.[94,95] The resulting PFnP-PFP derivative again exhibits an intense molecular ion m/z 460, which is also the base peak.[102] Vogt *et al.*[103] have increased the specificity and sensitivity of HVA determinations by using HRGC–HRMS of the O-dimethylthiophosphatemethyl ester. The 4-methoxy isomer, iso-HVA, was separated from HVA as either the HFIP-TFA[104] or the Me-TFA[105] derivative with packed column GC–MS.

Vanillyl mandelic acid has been quantified by LRGC–MS as the methyl fluoroacyl derivatives Me-TFA[101] and Me-PFP[96,97,106,107] and as the HFIP-TFA derivative.[104] Separation of iso-HVA has also been reported.[105]

4.2.2. Phenylalkyl Acids

Methods for the identification and quantitation of the phenylalkyl acids, phenylacetic (PAA), *o*-, *m*-, and *p*-hydroxyphenylacetic (*o*-HPA, *m*-HPA, and *p*-HPA), and *o*-, *m*- and *p*-hydroxymandelic (*o*-HMA, *m*-MHA, and *p*-HMA) acids, have only recently been described.

Phenylacetic acid has been determined in CSF as the pentafluorobenzyl ester (PFB) with a PAA-d_7 internal standard and packed-column GC–MS[108] and in urine as its PFnP ester.[109] The PFB derivative has the higher molecular weight but also requires a higher GC temperature for elution, with a concomitant increase in signal arising from column bleed.

m-Hydroxyphenylacetic acid (*m*-HPA) in rat brain has been determined using a SCOT column and accurate mass measurement of the molecular ion (HRGC–HRMS) of the Me-HFB derivative by Durden and Boulton.[8] The internal standard was pentadeutero-*m*-HPA. The same procedure was also used for *p*-HPA with tetradeutero internal standard, and the high-resolution SCOT column allows base-line resolution of the two isomers. Karoum *et al.*[96,97] have used the Me-PFP derivative, *p*-HPA-d_4, as internal standard and a LRGC–MS procedure. They did not, however, report on the presence or separation of *m*-HPA. A less sensitive procedure for *o*-HPA, *m*-HPA, and *p*-HPA in urine using a common fragment ion of the trimethylsilyl derivative has been reported by Narasimhachari *et al.*[110]

The acid metabolites of octopamine, *o*-HMA, *m*-HMA, and *p*-HMA, were originally determined in human urine using LRGC–MS[111] as the Me-TFA derivatives and later by HRGC–HRMS[112] with Me-HFB derivatives.

4.2.3. Indolylalkyl Acids

The acidic metabolites of the indolylamines that have received most attention to date are indole-3-acetic acid (IAA), 5-hydroxyindole-3-acetic acid (5-HIAA), and 5-methoxyindole-3-acetic acid (5-MeOIAA).

Indole-3-acetic acid has been derivatized only as its O-methyl-N-fluoroacetyl derivative. Bertilsson and Palmer[113] originally used the Me-HFB de-

Table II

Mass Spectrometric Methods for the Quantitative Analysis of Acids

Compound	Method	Ionization	Preparation	Stationary phase	Derivative	Internal standard	Ions monitored	Sensitivity	Comments	Reference
Catecholamine Metabolites										
3,4-Dihydroxyphenylacetic acid (DOPAC)	GC–MS	EI	EtOAc extraction	OV-17	PFnP-PFP	DOPAC-d5	M 592/595 F 415/420	2 ng	The molecular ion provides greater specificity and m/z 415 greater sensitivity than the following procedure	94
	GC–MS	EI	EtOAc extraction	SE-54	Me-PFP	DOPAC-d5	F 387/392 F 415/420	—	p-HMA acid exhibits the same fragment ions at 387 and 415	96,97
Homovanillic acid (HVA)	GC–MS	EI	EtOAc or XAD-2	SE-30 or OV-17	Me-HFB	HVA-d5 HVA-d2	M 392/397 M 392/394	—	iso-HVA may make a small contribution using packed-column GC	99,100
	GC–MS	EI	Alumina	OV-17	Me-TFA	HVA Methyl-d3	F 292/295	1 ng	Alumina was used to remove catechols which could be methylated by diazomethane	101
	GC–MS	EI	EtOAc extraction	OV-17	PFnP-PFP	HVA-d2	M 460/462	—	The intense molecular ion makes this a preferred derivative	94,95,102
	HRGC–HRMS	EI	EtOAc extraction	WCOT SE-30	DMTP-Me	HVA-d3	M 288.0585 M 291.0773	500 pg	Useful for small volumes, 50 to 100 µl, of CSF	103
iso-Homovanillic acid (iso-HVA)	GC–MS	EI	EtOAc extraction	OV-225	Me	iso-HVA-d5	M 196/201	—	Method for urine; the catecholic OH is not derivatized	105
Vanillylmandelic acid (VMA)	GC–MS	EI	Alumina	OV-17	Me-TFA	HVA Methyl-d3	F 345/348	1 ng	Alumina was used to remove catecholic acids	101
	GC–MS	EI	EtOAc extraction	SE-54	Me-PFP	VMA-d3	F 445/448, 417/420	—		96,97
	GC–MS	EI	XAD-2	OV-17	Me-PFP	VMA-d3	F 445/448 M 504/507	1–2 ng	Use of the molecular ion increases the specificity	106,107
iso-Vanillylmandelic acid (iso-VMA)	GC–MS	EI	Solvent extraction	OV-225	Me-TFA	iso-VMA-d5	F 345/348	—	Method for urine	105

Phenylalkyl

Compound	Technique	Ionization	Sample prep	Column	Derivative	Internal standard	Ions	Sensitivity	Comments	Ref.
Phenylacetic acid (PAA)	GC–MS	EI	EtOAc extraction	OV-1	PFB	PAA-d$_7$	M 316/323	—	Method for CSF and plasma	108
m-Hydroxyphenyl-acetic acid (m-HPA)	GC–MS	EI	None	SE-54	PFnP	PAA-d$_7$	M 268/275	—	Method for urine	109
	HRGC–HRMS	EI	EtOAc extraction	SCOT-SP-2250	Me-HFB	m-HPA-d$_5$	M 362/367	300 pg	m and p isomers are well separated by the SCOT column	8
p-Hydroxyphenyl-acetic acid (p-HPA)	HRGC–HRMS	EI	EtOAc extraction	SCOT-SP-2250	Me-HFB	p-HPA-d$_4$	M 362/366	350 pg		8
	GC–MS	EI	EtOAc extraction	SE-54	Me-PFP	p-HPA-d$_4$	F 253/257	—	m-HPA and p-HPA may not be separated by the packed column	96,97
Hydroxymandelic acid (o-HMA, m-HMA, p-HMA)	GC–MS	EI	EtOAc extraction	OV-101	Me-TFA	o-HMA-d$_2$, m-HMA-d$_3$, p-HMA-d$_2$	F 315/317/318, F 374/376/377	—	The isomers were not well separated	111
	HRGC–HRMS	EI	EtOAc extraction	SCOT-SP-2250	Me-HFB	m-HMA-d$_3$	F 514.9964, F 518.0152	—	Capillary column and high resolution allow complete separation of the isomers	112

Indolylalkyl

Compound	Technique	Ionization	Sample prep	Column	Derivative	Internal standard	Ions	Sensitivity	Comments	Ref.
Indoleacetic acid (IAA)	GC–MS	EI	Ether extraction of XAD-2	OV-17	Me-PFP	IAA-d$_2$	M 335/337, F 276/278	<1 ng		114,115
5-Hydroxyindole-3-acetic acid (5-HIAA)	GC–MS	EI	Solvent extraction	XE-60	Me-HFB	5-HIAA-d$_2$	F 538/540	—	The deuterated standard used by these workers was 20% undeuterated	116
	GC–MS	EI	XAD-2	OV-17	Me-PFP	5-HIAA-d$_2$	F 438/440, M 497, M 497/499	<1 ng	The increased isotopic purity of the standard increases sensitivity	115
	GC–MS	EI	EtOAC extraction	OV-17	PFnP-PFP	5-HIAA-d$_2$	M 615/617, F 438/440	<1 ng	Recovery of 5-HIAA was improved by addition of 5-MeOIAA	62,102,116
5-Methoxyindole-3-acetic acid (5-MeOIAA)	GC–MS	EI	Dowex® 50W-X4 TLC	OV-17	TMS	5-MeOIAA-d$_2$	M 349/351, F 232/234	10 ng/ml	This method is not very sensitive	120

Table III
Mass Spectrometric Methods for the Quantitative Analysis of Alcohols

Compound	Method	Ionization	Preseparation	Stationary phase	Derivative	Internal standard	Ions monitored	Sensitivity	Comments	Reference
Catecholic										
3-Hydroxy-4-methoxyphenyl-ethylene glycol (HMPG)	GC–MS	EI	EtOAC extraction	SE-54	PFP	HMPG-d₃ or HMPG-d₅	F 311/314 F 458/461	1 ng	m/z 445 and 448 may also be used for the d_5 internal standard	96,121,122
	GC–MS	EI	EtOAC extraction	XE-60 or SE-30 or OV-17	TFA	HMPG-d₂ or HMPG-d₃	M 472/474 M 472/475	1 ng	The molecular ion increases specificity	101,124, 125
	GC–MS	EI	Acetylation and solvent extraction	OV-17	Ac-TFA	HMPG-d₃	F 376/379 F 249/252	1 ng	Acetylation improves HMPG extraction but the molecular ion intensity is lower	107
3-Hydroxy-4-methoxyphenyl-ethanol (HMPE)	GC–MS	EI	EtOAc extraction	SE-54	PFP	HMPE-d₂	M 460/462 F 296/297	1 ng	PFP gives a more intense M ion than does TMS derivative	96
3,4-Dihydroxy-phenylethylene glycol (DHPG)	GC–MS	EI	EtOAC extraction	OV-1	PFP	DHPG-d₂	F 590/592	>2–5 ng	Urinary analysis with a sensitivity of 0.5 μg/g creatinine	126
3,4-Dihydroxy-phenylethanol (DHPE)	GC–MS	EI	EtOAC extraction	OV-1	TMS	DHPE-d₅	M 370/375	<2–5 ng	Urinary analysis	127

Phenylalkyl

Compound	Method	Ionization	Extraction	Column	Derivative	Internal standard	Ions	Sensitivity	Comments	Ref.
Phenylethylene glycol (PEG)	GC–MS	CI	EtOAc extraction	OV-1, QF-1 or SP-2250	PFP	—	F 267	—	Major fragment ion (MG-164) is formed by loss of C_2F_2COOH from the pseudomolecular ion	128,129
p-Hydroxyphenyl-ethanol (p-HPE)	GC–MS	CI	EtOAc extraction	OV-1, QF-1 or SP-2250	PFP	—	F 267	—	Deuterated standards would improve this method	128
p-Hydroxyphenyl-ethylene glycol (p-HPG)	GC–MS	CI	EtOAc	OV-1, QF-1 or SP-2250	PFP	—	F 249	—		128

Indolylalkyl

Compound	Method	Ionization	Extraction	Column	Derivative	Internal standard	Ions	Sensitivity	Comments	Ref.
5-Hydroxytryptophol (5-HTOL)	GC–MS	EI	EtOAc extraction	OV-17	di-PFP	5-HTOL-d4	M 469/473 F 305/308	1 ng, 0.2 ng	Sensitivity is increased by monitoring the fragment ions	131
	HRGC–MS	EI	CHCl3 extraction	WCOT-SE-52	tri-PFP	5-HTOL-d4	M 615/619 F 438/454 F 451/454	—		132
	HRGC–HRMS	EI	Ether extraction	WCOT-OV-101	HFB	—	M 765.0067 F 551.0202	—	10,000 resolution accurate mass; a deuterated standard would improve quantitation	133
5-Methoxytrypto-phol (5-MTOL)	GC–MS	EI	Ether extraction	SE-54	PFP	5-αMTr	F 319, F 321	—	Internal standard was not a tryptophol	130
O-Acetyl-5-methoxytrypto-phol (AMT)	GC–MS	EI	CHCl3 extraction	OV-1	TMS	(See text)	F 232	—		134
	GC–MS	EI	CHCl3 extraction	OV-1	TMS	—	—	—	The spectrum was taken for identification only	135

rivative and 5-methylindole-3-acetic acid as an internal standard to demonstrate the presence of IAA in human CSF. This procedure has been improved recently by incorporating α-dideutero-IAA as the internal standard and Me-PFP as the derivative.[114,115]

5-Hydroxyindole-3-acetic acid, as might be expected, has received much more attention. Although methyl ester perfluoroacyl derivatives were originally used,[115,116] the PFnP-PFP derivative has been found to be more useful, as it gives an intense molecular ion, m/z 615, which is also the base peak.[62,102,116–118] The TMS derivative has also been used,[119] but the sensitivity appears lower than when the above derivatives are used.

Urinary 5-MeOIAA has been assessed by Hoskins *et al.*[120] using the TMS derivative. Again, the molecular ion intensity is low, and the PFnP-PFP derivative with its intense molecular ion[116] would undoubtably afford increased sensitivity. A procedure similar to that of Faull *et al.*[116] for 5-HIAA with inclusion of a deuterated standard could quite possibly provide an adequate method for the assessment of 5-MeOIAA.

4.3. Alcohols

4.3.1. Catecholamine Metabolites

The major neutral metabolite of NA (and AD) is 4-hydroxy-3-methoxyphenylethylene glycol (HMPG, also abbreviated MHPG or MOPEG), which is readily determined by GC–MS, frequently in conjunction with the acidic metabolites VMA and HVA. A minor metabolite, less frequently determined, is 3,4-dihydroxyphenylethylene glycol (DHPG or DOPEG). Two other minor neutral metabolites of DA have also been reported, 4-hydroxy-3-methoxyphenylethanol (HMPE, MHPE, or MOPET) and 3,4-dihydroxyphenylethanol (DHPE or DOPET) (Table III).

4-Hydroxy-3-methoxyphenylethylene glycol is frequently determined simultaneously with the catecholic acids HVA or VMA, since the reagents used to deactivate the catechol hydroxyl groups also react with the alcohol hydroxyl groups. Following the work of Anggard and Sedvall[4] who demonstrated the usefulness of the fluorinated acyl derivatives TFA, PFP, and HFB, many workers chose to derivatize HMPG with PFP because of its greater stability in comparison with the TFA derivative; Karoum *et al.*,[96,97] Swahn *et al.*,[121] and Muskiet *et al.*[105,122] have all reported HRGC–MS procedures using deuterated internal standards for HMPG in brain tissue, *CSF*, amniotic fluid, and urine. The PFP derivative has also been used to determine the ratio between the 3-hydroxy-4-methoxy isomer (i.e., isoHMPG) and HMPG.[105] Recently, an increased specificity in the procedure has been obtained by using a 25-m SE-54 WCOT capillary column to improve the separation of iso-HMPG from HMPG.[123]

Although the TFA derivative of HMPG has been reported to be less stable than either the PFP or HFB derivatives, the spectrum of tris-TFA-HMPG exhibits a strong molecular ion intensity at m/z 472 which could be used for quantitation.[124] Procedures using GC–MS for HMPG levels in CSF, tissues, and urine have been developed by Bertilsson,[124] Gordon *et al.*,[101] and Sjoquist *et al.*[125] using the TFA derivative and deuterated standards. Takahashi *et al.*[107] acetylated the phenolic function prior to solvent extraction and so increased the selectivity of extraction of HMPG-Ac which was then further derivatized to produce HMPG-Ac-TFA$_2$. It was quantitated using fragment ions because its molecular ion exhibits a lower relative intensity than does that of the tris-TFA derivative.

The minor metabolites HMPE, DHPG, and DHPE have received less attention but have been quantitated in CSF, rat brain, and urine by Karoum *et al.*[96] and by Muskiet *et al.* as either the PFP[126] or TMS[127] derivatives. Considering the small concentrations of those metabolites found in tissue or CSF, the relatively intense molecular ion of the PFP derivative[96] would appear to make it the derivative of choice.

Edwards *et al.*[128] have described a procedure using CI–GC–MS of the PFP derivatives of all the neutral alcohol and glycol metabolites. Although the MH$^+$ pseudomolecular ion intensity was low for all compounds, each spectrum exhibited a single prominent fragment ion, and because of this reduced fragmentation, CI may enhance the specificity of GC–MS analyses of these metabolites. Unfortunately, deuterated standards were not used, and recoveries were estimated from standard curves.

4.3.2. Phenylalkyl Alcohols

The phenylalkylamine metabolites phenylethylene glycol (PEG), *p*-hydroxyphenylethanol (*p*-HPE), and *p*-hydroxyphenylethylene glycol (*p*-HPG) have been detected in urine by Edwards *et al.*[128,129] using the CI method described above.

4.3.3. Indolylalkyl Alcohols

Mass spectrometric procedures have been developed for the indolylalkyl alcohols 5-hydroxytryptophol (5-HTOL), 5-methoxytryptophol (5-MTOL), and O-acetyl-5-methoxytryptophol (AMT), primarily as their fluorinated acyl derivatives.

Curtius *et al.*[130] first reported the use of LRGC–MS of the di-PFP derivative to determine 5-HTOL in CSF. They used 5-fluoro-α-methyltryptamine (5-α-MFTr) as internal standard, since a deuterated analogue was not available. The synthesis of tetradeutero-5-HTOL increased the specificity and sensitivity of the analysis and is now used as one of the preferred procedures developed by Takahashi *et al.*[131] Beck *et al.*[132] have utilized the tri-PFP-5-HTOL derivative for an analysis in which the GC separation took place on a 12-m SE-52 WCOT capillary column (HRGC–MS). The ultimate analytical combination, HRGC–HRMS, has been used by Diggory *et al.*[133] to detect 5-HTOL in mouse brain as its HFB derivative (tri-HFB-5-HTOL) using accurate mass analysis of the molecular ion m/z 765·0067 at 10,000 resolution. Unfortunately, the amounts could only be estimated, since an internal standard was not incorporated.

The methoxy metabolite 5-MTOL was detected along with 5-HTOL by Curtius et al.,[130] again as a di-PFP derivative. As a way of validating the specificity of assays for 5-MTOL in plasma, Leone et al.[134] chose to use two different derivatives and different internal standards. In the first assay, the internal standard was a homologue of 5-MTOL, 3-(butan-3'-butyrate)-5-methoxyindole; it was used with TMS derivatization, and a common ion m/z 232 was monitored. The results were then verified using PFP derivatization and another homologue, 1,3-(butan-3-one)-5-methoxyindole, as the alternate internal standard. In none of the methods described above was a deuterated standard used. O-Acetyl-5-methoxytryptophol (AMT) has recently been reported by these workers to be present in pineal gland.[135]

4.4. Amino Acids

4.4.1. Phenylalkyl Amino Acids

The procedures reported for the analysis of the two major neurotransmitter precursor amino acids phenylalanine and p-tyrosine all utilize GC–MS (see Table IV).

Zagalak et al.[136] have quantitated phenylalanine (Phe) and p-tyrosine (p-Tyr) in plasma using various combinations of TMS, HFB, TFA, and Me and isobutyl ester derivatives, none of which exhibit usable molecular ion intensities in their EI spectra. They chose the N(O)-trifluoroacetyl-O-methyl esters (Me-TFA) for SIM ion quantitation using deuterated internal standards. The n-butyl-PFP ester was selected by Sjoquist,[137] and the diacetylphenylthiohydantoin derivative by Trefz et al.[138] This latter derivative appears to be preferable since it produces the most intense molecular ion of all derivatives tried.

4.4.2. Indolylalkyl Amino Acids

A GC–MS procedure for tryptophan (Trp) and N-acetyl-tryptophan (N-Ac-Trp) has been described by Wegmann et al.[139] Tryptophan was analyzed as its N-PFP-Me ester and N-Ac-Trp as its N-TMS-Me ester; each was quantitated using the molecular ion, and the fully deuterated substances tryptophan-d_8 and N-acetyltryptophan-d_{11} were included as internal standards. Reaction conditions for the formation of mixed PFP–TMS derivatives of Trp have recently been reported by Martinez and Gelpi.[140] Several isomeric products are possible, with the TMS ester of 5-O-PFP-N^1-TMS, N^{ω}-PFP-hydroxytryptophan, being the major product. It is detectable in picogram quantities using the base peak m/z 364.

Table IV

Mass Spectrometric Methods for the Quantitative Analysis of Amino Acids

Compound	Method	Ionization	Preseparation	Stationary phase	Derivative	Internal standard	Ions monitored	Sensitivity	Comments	Reference
Phenylalkyl										
Phenylalanine (Phe)	GC–MS	EI	Dowex® 50W-X8	SE-30	Me-TFA	Phe-d_1, Phe-d_5	F 162/163/167	2.5 ng/ml	Other derivatives were also used (see text)	136
p-Tyrosine (p-Tyr)	GC–MS	EI	Dowex® 50W-X8	SE-30	Me-TFA	p-Tyr-d_7	F 274/280	2.5 ng/ml		136
	HRGC–MS	EI	Amberlite® IR-120	WCOT-OV-17	n-Butyl-PFP	p-Tyr-d_3, α-Me-Tyr	F 428/430/442, F 336/369/380	—	p-Tyr-d_3 was quantitated using α-Me-Tyr internal standard. p-Tyr-d_6 was quantitated after loading with Phe-d_7	137
	HRGC–MS	EI	HPLC	WCOT-SE-30	di-Ac-PTH	p-Tyr-d_2, p-Tyr-d_6	M 382/384, M 382/388	—		138
Indolylalkyl										
Tryptophan (Trp)	GC–MS	EI	Dowex® 50W-X2	QF-1	Me-PFP	Trp-d_8	M 364/372	—	PFP was more stable than TFA derivative	139
N-Acetyltryptophan (N-Ac-Trp)	GC–MS	EI	Dowex® 50W-X2	SE-30	Me-TMS	N-Ac-Trp-d_4, N-Ac-Trp-d_{11}	M 332/343	—		139
Alkyl										
γ-Aminobutyric acid (GABA)	TLC–HRMS	EI	Direct derivatization	Silica gel	DNS	GABA-d_2	M 318.1038, M 320.1164	1 ng	Accurate mass at 7,000 resolution; molecular ion is the base peak	149
	GC–MS	EI	Amberlite®, CG-120, CM-sepharose	OV-17	HFIP-PFP	GABA-d_2	F 204/206	—	Prepurification is necessary for physiological fluids	141,142
	GC–MS	EI	EtOAc, CH₂Cl₂ extraction	OV-101	DNP-Et	GABA-d_2	F 252/254	—	Claim that resin prepurification is not required	143
	GC–MS	EI	Dowex® 50	OV-17	DMF	GABA-d_2 or AVA	M 172/176/186	—	GABA and its lactam, 2-pyrrolidinone, are separated by Dowex® 50	146
	GC–MS	CI	Dowex 50W-X8	ASI	Me-PFP	GABA-d_2, -d_1, -d_5	MH 264/266, F 232/236	—	Methane CI causes some fragmentation	148
Glutamic acid (Glu)	GC–MS	EI	—	QF-1 or OV-17	HFIP-PFP	None	F 426	—	These procedures could be combined i.e., monitor 202, 204, 426, 428, with d_2 std.	150
		EI	Direct derivatization		HFIP-PFP	Glu-d_2	F 202/204	—		141
Glutamine (Gln)	GC–MS	CI		QF-1	HFIP-PFP	—	F 426	—	No internal standard	150
Glycine (Gly)	GC–MS	EI	Direct derivatization	OV-17/QF-1	HFIP-HFB	Gly-$^{13}C_2^{15}N$	F 224/228	—	HFB derivative has higher molecular weight than PFP or TFA	152

4.4.3. Alkyl Amino Acids

The alkyl amino acids γ-aminobutyric acid (GABA), glutamic acid (Glu), glutamine (Gln), and glycine (Gly) have been identified and quantitated in brain and physiological fluids by mass spectrometry.

GABA has received the greatest attention, with most procedures utilizing LRGC–MS. Bertilsson and Costa[141] were the first to incorporate a deuterated internal standard and the HFIP-PFP derivative to quantitate GABA in cerebellar nuclei and sympathetic ganglia. When the procedure was applied to human CSF, however, it was found that an ion-exchange prepurification step was required.[142] A pitfall in the use of the HFIP-PFP derivative is the possible formation of 2-pyrrolidinone (γ-butyrolactam) by acid-catalyzed ring closure of the ester. In order to reduce this possibility and to be able to eliminate the prepurification step, Colby and McCaman[143] developed a GC–MS method using the N-dinitrophenyl ethyl ester which they claim possesses a specificity adequate to permit the isolation of GABA from a complex mixture of CSF amino acids. Other GABA derivatives that have been used include HFIP-TFA,[144] TMS[145] (without a deuterated internal standard), dimethylformamide (DMF),[146] and Me-PFP[147,148] (with a deuterated internal standard). In this last mentioned procedure, Ferkany et al.[148] used CI (methane) to achieve a sensitivity of 10 pmol (1 ng) when the pseudomolecular ion (NH$^+$, m/z 264) was monitored.

A non-GC–MS procedure for GABA has been described by Wu et al.[149] GABA was cyclized into γ-butyrolactam, derivatized with DNS chloride, and monitored by HRMS after isolation on TLC. The major advantage of this derivative is that under EI, the spectrum contains an intense molecular ion that may be used for identification and quantitation. GABA-d_2 was used as the internal standard.

Glutamine and Glu have been quantitated simultaneously by CI–GC–MS of the HFIP-PFP derivatives without internal standards,[150] and Glu has been analyzed simultaneously with GABA by EI–GC–MS with a dideutero internal standard.[141] As is the case with GABA, Glu can be cyclized to pyroglutamic acid by the acidic conditions that occur in preparation of the derivative,[151] thus limiting the usefulness of this procedure.

The shortest amino acid of this series, Gly, has also been determined using the HFIP ester. The procedure reported by Lapin and Karobath[152] used the HFB amide derivative and [1,2-$^{13}C_2$,^{15}N] glycine as internal standard.

Many other derivatives have been investigated in the MS analysis of amino acids. For example, Leimer et al.[153] have described N-TFA-n-butyl esters (TAB derivatives), Liardon et al.[154] oxazolidones of α-amino acids, Seiler et al.[155] DNS derivatives, and Iwase et al.[156] TMS spectra. Schulman and Abramson[157] have reported a procedure to quantitate plasma amino acids (a deuterated internal standard for each acid is added to the plasma) in which the MS is scanned over a wide mass range as the TAB derivatives elute from the GC. This procedure offers the clear advantage of detecting simultaneously many amino acids, but the sensitivity is lower than is the case if SIM is used.

ACKNOWLEDGMENT. We thank Saskatchewan Health and the Medical Research Council of Canada for continuing financial support.

APPENDIX

Techniques

CI	Chemical ionization
DP	Direct probe
EI	Electron impact (ionization)
F	Fragment ion
FD	Field desorption
GC	Gas chromatography
HR	High resolution
HRGC	High-resolution gas chromatography (capillary column)
HRMS	High-resolution mass spectrometry
LR	Low resolution
LRGC	Low-resolution gas chromatography (packed column)
LRMS	Low-resolution mass spectrometry
M	Molecular ion
MH	Protonated pseudomolecular ion
MS	Mass spectrometry
m/z	Mass-to-charge ratio
NI-CI	Negative ion chemical ionization
SCOT	Support-coated open tube
SIM	Selected ion monitoring
TLC	Thin-layer chromatography
WCOT	Wall-coated open tube

Compounds

ACh	Acetylcholine
AD	Epinephrine (Adrenaline)
AMT	O-Acetyl-5-methoxytryptophol
AVA	5-Amino-*n*-valeric acid
BUF	Bufotenin
Cad	Cadaverine
Ch	Choline
DA	Dopamine
DHPE	3,4-Dihydroxyphenylethanol
DHPG	3,4-Dihydroxyphenylethylene glycol
DMT	N,N-Dimethyltryptamine
DOPAC	3,4-Dihydroxyphenylacetic acid
EP	Epinine
EtOAc	Ethyl acetate
GABA	γ-Aminobutyric acid
Gln	Glutamine
Glu	Glutamic acid
Gly	Glycine
Hist	Histamine
5-HIAA	5-Hydroxyindoleacetic acid
m-HMA	*meta*-Hydroxymandelic acid
o-HMA	*ortho*-Hydroxymandelic acid
p-HMA	*para*-Hydroxymandelic acid
6-HMEL	6-Hydroxymelatonin
HMPE	4-Hydroxy-3-methoxyphenylethanol
HMPG	4-Hydroxy-3-methoxylphenylethylene glycol
m-HPA	*meta*-Hydroxyphenylacetic acid
p-HPA	*para*-Hydroxyphenylacetic acid
p-HPE	*para*-Hydroxyphenylethanol
p-HPG	*para*-Hydroxyphenylethylene glycol
5-HT	5-Hydroxytryptamine, serotonin
5-HTOL	5-Hydroxytryptophol
HVA	Homovanillic (4-hydroxy-3-methoxyphenylacetic) acid
IAA	Indoleacetic acid
α-MDA	α-Methyldopamine
MEL	Melatonin
5-MeOIAA	5-Methoxyindoleacetic acid
5-α-MFTr	5-Fluoro-α-methyltryptamine
t-MH	*t*-Methylhistamine
α-MNA	α-Methylnorepinephrine
MN	Metanephrine
3-MT	3-Methoxy-4-hydroxyphenylethylamine
5-MT	5-Methoxytryptamine

Compounds (Cont.)

		Derivatives	
5-MTOL	5-Methoxytryptophol		
N-Ac-Trp	N-Acetyltryptophan	Ac	Acetyl
NA	Norepinephrine	BNS	1-Di-*n*-butylaminonap-
NMN	Normetanephrine		thalene-5-sulfonyl (ban-
m-OCT	*meta*-Octopamine		syl)
p-OCT	*para*-Octopamine	DMTP	O-Dimethylthiophos-
PAA	Phenylacetic acid		phate
PE	Phenylethylamine	DNP	Dinitrophenyl
PEG	Phenylethylene glycol	DNS	1-Dimethylaminonap-
PEOH	Phenylethanolamine		thalene-5-sulfonyl (dan-
Phe	Phenylalanine		syl)
Pip	Piperidine	Et	Ethyl
Put	Putrescine	EtO	Ethoxycarbonyl
Sp	Spermine	HFB	Heptafluorobutyryl
Sper	Spermidine	HFIP	Hexafluoroisopropyl
m-SYN	*meta*-Synephrine	Me	Methyl
p-SYN	*para*-Synephrine	MeO	Methoxy
m-TA	*meta*-Tyramine	PFnP	Pentafluoro-*n*-propyl
p-TA	*para*-Tyramine	PFP	Pentafluoropropionyl
Trp	Tryptophan	PTH	Phenylthiohydantoin
TRYPT	Tryptamine	TBDMS	*t*-Butyldimethylsilyl
p-Tyr	*para*-Tyrosine	TFA	Trifluoroacetyl
VMA	Vanillylmandelic acid	TFE	Trifluoroethyl
		TMS	Trimethylsilyl

REFERENCES

1. Ryhage, R., 1964, *Anal. Chem.* **36**:759–764.
2. Watson, J. T., and Biemann, K., 1964, *Anal. Chem.* **36**:1135–1137.
3. Hammar, C. G., Hanin, I., Holmstedt, B., Kitz, R. J., Jenden, D. J., and Karlen, B., 1968, *Nature* **220**:915–917.
4. Angaard, E., and Sedvall, G., 1969, *Anal. Chem.* **41**:1250–1256.
5. Boulton, A. A., Pollit, R. J., and Majer, J. M., 1967, *Nature* **215**:132–134.
6. Watson, J. T., Falkner, F. C., and Sweetman, B. J., 1974, *Biomed. Mass Spectrom.* **1**:156–157.
7. Sweeley, C. C., Elliott, W. H., Fries, I., and Ryhage, R., 1966, *Anal. Chem.* **38**:1549–1552.
8. Durden, D. A., and Boulton, A. A., 1981, *J. Neurochem.* **36**:129–135.
9. Mita, H., Yasueda, H., and Shida, T., 1980, *J. Chromatogr.* **181**:153–159.
10. Durden, D. A., 1978, *Res. Methods Neurochem.* **4**:205–250.
11. Durden, D. A., and Boulton, A. A., 1979, *Tech. Life Sci.* **B214**:1–25.
12. Jacob, K., Vogt, W., Knedel, M., and Schwertfeger, G., 1978, *J. Chromatogr.* **146**:221–226.
13. Creveling, C. R., Kondo, K., and Daly, J. W., 1968, *Clin. Chem.* **14**:302–309.
14. Seiler, N., Schneider, H., and Sonnenberg, K.-D., 1970, *Z. Anal. Chem.* **252**:127–136.
15. Chapman, D. I., Chapman, J. R., and Clark, J., 1972, *Int. J. Biochem.* **3**:66–72.
16. Knoche, H., Alfes, H., Mollmann, H., and Reisch, J., 1969, *Experientia* **25**:515–516.
17. Juorio, A. V., and Durden, D. A., 1977, *Can. J. Biochem.* **55**:761–765.
18. Lehmann, W. D., Beckey, H. D., and Schulten, H.-R., 1976, *Anal. Chem.* **48**:1572–1575.
19. Koslow, S. H., Cattabeni, F., and Costa, E., 1972, *Science* **176**:177–180.
20. Koslow, S. H., and Schlumpf, M., 1974, *Nature* **251**:530–531.
21. Karoum, F., Cattabeni, F., Costa, E., Ruthven, C. R. J., and Sandler, M., 1972, *Anal. Biochem.* **47**:550–561.

22. Gelpi, S., Peralta, E., and Segura, J., 1974, *J. Chromatogr. Sci.* **12**:701–709.
23. Ko, H., Lahti, R. A., Duchamp, D. J., and Royer, M. E., 1974, *Anal. Lett.* **7**:243–255.
24. Wiesel, F.-A., 1976, *Adv. Mass Spectrom. Biochem. Med.* **1**:171–180.
25. Kilts, C. D., Vrbanac, J. J., Rickert, D. E., and Rech, R. H., 1977, *J. Neurochem.* **28**:465–467.
26. Warsh, J. J., Chiu, A., Li, P. P., and Godse, D. D., 1980, *J. Chromatogr.* **183**:483–486.
27. Holdiness, M. R., Rosen, M. T., Justice, J. B., and Neill, D. B., 1980, *J. Chromatogr.* **198**:329–336.
28. Erhardt, J.-D., and Schwartz, J., 1978, *Clin. Chim. Acta* **88**:71–79.
29. Yoshida, J.-I., Yoshino, K., Matsunaga, T., Higa, S., Suzuki, T., Hayashi, A., and Yamamura, Y., 1980, *Biomed. Mass Spectrom.* **7**:396–398.
30. Miyazaki, H., Hashimoto, Y., Iwanaga, M., and Kubodera, T., 1974, *J. Chromatogr.* **99**:575–586.
31. Hashimoto, Y., and Miyazaki, H., 1979, *J. Chromatogr.* **168**:59–68.
32. Mizuno, Y., and Ariga, T., 1979, *Clin. Chim. Acta* **98**:217–224.
33. Freed, C. R., Weinkam, R. J., Melmon, K. L., and Castagnoli, N., 1977, *Anal. Biochem.* **78**:319–332.
34. Curtius, H. C., Wolfensberger, M., Steinmann, M., Redweik, U., and Siegfried, J., 1974, *J. Chromatogr.* **99**:529–540.
35. Wang, M.-T., Imai, K., Yoshioka, M., and Tamura, Z., 1975, *Clin. Chim. Acta* **63**:13–19.
36. Maume, B. F., Bournot, P., Lhuguenot, J. C., Baron, C., Barbier, F., Maume, G., Prost, M., and Padieu, P., 1973, *Anal. Chem.* **45**:1073–1082.
37. Lhuguenot, J. C., and Maume, B. F., 1980, *Biomed. Mass Spectrom.* **7**:529–532.
38. Galli, C. L., Cattabeni, F., Eros, T., Spano, P. F., Algeri, S., DiGuilio, A., and Gropetti, A., 1976, *J. Neurochem.* **27**:795–798.
39. Robertson, D., Heath, E. C., Falkner, F. C., Hill, R. E., Brilis, G. M., and Watson, J. T., 1978, *Biomed. Mass Spectrom.* **5**:704–708.
40. Muskiet, F. A. J., Thomasson, C. G., Gerding, A. M., Fremouw-Ottevangers, D. C., Nagel, G. T., and Wolthers, B. G., 1979, *Clin. Chem.* **25**:453–460.
41. Wang, M.-T., Yoshioka, M., Imai, K., and Tamura, Z., 1975, *Clin. Chim. Acta* **63**:21–27.
42. Claeys, M., Verzele, M., Vandenwalle, M., Leysen, J., and Laduron, P., 1974, *Biomed. Mass Spectrom.* **1**:103–108.
43. Durden, D. A., Philips, S. R., and Boulton, A. A., 1973, *Can. J. Biochem.* **51**:995–1002.
44. Edwards, D. J., Doshi, P. S., and Hanin, I., 1979, *Anal. Biochem.* **96**:308–316.
45. Willner, J., LeFevre, H. F., and Costa, E., 1974, *J. Neurochem.* **23**:857–859.
46. Karoum, F., Nasrallah, H., Potkin, S., Chuang, L., Moyer-Schwing, J., Phillips, I., and Wyatt, R. J., 1979, *J. Neurochem.* **33**:201–212.
47. Axelsson, S., Bjorklund, A., and Seiler, N., 1973, *Life Sci.* **13**:1411–1419.
48. Philips, S. R., Durden, D. A., and Boulton, A. A., 1974, *Can. J. Biochem.* **52**:366–373.
49. Philips, S. R., Rozdilski, B., and Boulton, A. A., 1978, *Biol. Psychiatry* **13**:51–57.
50. Philips, S. R., Davis, B. A., Durden, D. A., and Boulton, A. A., 1975, *Can. J. Biochem.* **53**:65–69.
51. Buck, S. H., Murphy, R. C., and Mollinoff, P. B., 1977, *Brain Res.* **122**:281–297.
52. Williams, C. M., and Couch, M. W., 1978, *Life Sci.* **22**:2113–2120.
53. Midgley, J. M., Couch, M. W., Crowley, J. R., and Williams, C. M., 1980, *J. Neurochem.* **34**:1225–1230.
54. Durden, D. A., Juorio, A. V., and Davis, B. A., 1980, *Anal. Chem.* **52**:1815–1820.
55. Durden, D. A., Juorio, A. V., and Davis, B. A., 1978, *Quant. Mass Spectrom. Life Sci.* **2**:389–397.
56. Philips, S. R., Durden, D. A., and Boulton, A. A., 1974, *Can. J. Biochem.* **52**:447–451.
57. Warsh, J. J., Godse, D. D., Stancer, H. C., Chan, P. W., and Coscina, D. V., 1977, *Biochem. Med.* **18**:10–20.
58. Blau, K., King, G. S., and Sandler, M., 1977, *Biomed. Mass Spectrom.* **4**:232–236.
59. Donike, M., Gola, R., and Jaenicke, L., 1977, *J. Chromatogr.* **134**:385–395.
60. Seiler, N., and Bruder, K., 1975, *J. Chromatogr.* **106**:159–173.
61. Cattabeni, F., Koslow, S. H., and Costa, E., 1972, *Science* **178**:166–168.
62. Beck, O., Wiesel, F.-A., and Sedvall, G., 1977, *J. Chromatogr.* **134**:407–414.

63. Curtius, H.-C., Farner, H., and Rey, F., 1980, *J. Chromatogr.* **199:**171–179.
64. Liuzzi, A., Foppen, F. H., Saavedra, J. M., Levi-Montalcini, R., and Kopin, I. J., 1977, *Brain Res.* **133:**345–357.
65. Lewy, A. J., and Markey, S. P., 1978, *Science* **201:**741–743.
66. Wilson, B. W., Snedden, W., Silman, R. E., Smith, I., and Mullen, P., 1977, *Anal. Biochem.* **81:**283–291.
67. Sisak, M. E., Markey, S. P., Colburn, R. W., Zavadil, A. P., III, ⟨ ⟩ Kopin, I. J., 1979, *Life Sci.* **25:**803–806.
68. Tetsuo, M., Markey, S. P., and Kopin, I. J., 1980, *Life Sci.* **27:**105–109.
69. Green, A. R., Koslow, S. H., and Costa, E., 1973, *Brain Res.* **51:**371–374.
70. Beck, O., and Bosin, T. R., 1979, *Biomed. Mass Spectrom.* **6:**19–22.
71. Narasimhachari, N., Kempster, E., and Anbar, M., 1980, *Biomed. Mass Spectrom.* **7:**231–235.
72. Wilson, B. W., and Snedden, W., 1979, *J. Neurochem.* **33:**939–941.
73. Narasimhachari, N., and Himwich, H. E., 1973, *Biochem. Biophys. Res. Commun.* **55:**1064–1071.
74. Raisanen, M., and Karkkainen, J., 1978, *Biomed. Mass Spectrom.* **5:**596–600.
75. Raisanen, M., and Karkkainen, J., 1979, *J. Chromatogr.* **162:**579–584.
76. Walker, R. W., Mandel, L. R., Kleinman, J. E., Gillin, J. C., Wyatt, R. J., and Vandenheuvel, W. J. A., 1979, *J. Chromatogr.* **162:**539–546.
77. Axelsson, S., Bjorklund, A., and Seiler, N., 1971, *Life Sci.* **10:**745–749.
78. Mita, H., Yasueda, H., and Shida, T., 1980, *J. Chromatogr.* **221:**1–7.
79. Hough, L. B., Stetson, P. L., and Domino, E. F., 1979, *Anal. Biochem.* **96:**56–63.
80. Jenden, D. J., Roch, M., and Booth, R. A., 1973, *Anal. Biochem.* **55:**438–448.
81. Jenden, D. J., Roch, M., and Fainman, F., 1978, *Life Sci.* **23:**291–300.
82. Lehmann, W. D., and Schulten, H. R., 1978, *Biomed. Mass Spectrom.* **5:**591–595.
83. Seiler, N., and Lamberty, U., 1973, *J. Neurochem.* **20:**709–717.
84. Seiler, N., and Askar, A., 1971, *J. Chromatogr.* **62:**121–127.
85. Dolezalova, H., Stepita-Klauco, M., and Fairweather, R., 1974, *Brain Res.* **77:**166–168.
86. Stepita-Klauco, M., and Dolezalova, H., 1974, *Nature* **252:**158–159.
87. Seiler, N., 1975, *Res. Methods Neurochem.* **3:**409–441.
88. Smith, R. G., and Daves, G. D., Jr., 1977, *Biomed. Mass Spectrom.* **4:**146–151.
89. Smith, R. G., Bartos, D., Bartos, F., Grettie, D. P., Frick, W., Campbell, R. A., and Daves, G. D., Jr., 1978, *Biomed. Mass Spectrom.* **5:**515–517.
90. Shipe, J. R., Jr., Hunt, D. F., and Savoy, J., 1979, *Clin. Chem.* **25:**1564–1571.
91. Stepita-Klauco, M., Dolezalova, H., and Fairweather, R., 1974, *Science* **183:**536–537.
92. Seiler, N., and Schneider, H. H., 1974, *Biomed. Mass Spectrom.* **1:**381–385.
93. Miyata, T., Okano, Y., Murao, K., Fukunaga, K., Takahama, K., and Kasé, Y., 1979, *Life Sci.* **25:**1731–1738.
94. Wiesel, F.-A., Fri, C.-G., and Sedvall, G., 1974, *J. Neural Transm.* **35:**319–326.
95. Gordon, E. K., Markey, S. P., Sherman, R. L., and Kopin, I. J., 1976, *Life Sci.* **18:**1285–1292.
96. Karoum, F., Gillin, J. C., Wyatt, R. J., and Costa, E., 1975, *Biomed. Mass Spectrom.* **2:**183–189.
97. Karoum, F., Gillin, J. C., and Wyatt, R. J., 1975, *J. Neurochem.* **25:**653–658.
98. Sjoquist, B., and Anggard, E., 1972, *Anal. Chem.* **44:**2297–2301.
99. Sjoquist, B., Lindstrom, B., and Anggard, E., 1973, *Life Sci.* **13:**1655–1664.
100. Fri, C.-G., Wiesel, F.-A., and Sedvall, G., 1974, *Psychopharmacologia* **35:**295–305.
101. Gordon, E. K., Oliver, J., Black, K., and Kopin, I. J., 1974, *Biochem. Med.* **11:**32–40.
102. Fri, C.-G., Wiesel, F.-A., and Sedvall, G., 1974, *Life Sci.* **14:**2469–2480.
103. Vogt, W., Jacob, K., Ohnesorge, A.-B., and Schwertfeger, G., 1980, *J. Chromatogr.* **199:**191–197.
104. Takahashi, S., Yoshioka, M., Yoshiue, S., and Tamura, Z., 1978, *J. Chromatogr.* **145:**1–9.
105. Muskiet, F. A. J., Fremouw-Ottevangers, D. C., Nagel, G. T., and Wolthers, B. G., 1979, *Clin. Chem.* **25:**1708–1713.
106. Sjoquist, B., 1975, *J. Neurochem.* **24:**199–201.

107. Takahashi, S., Godse, D. D., Warsh, J. J., and Stancer, H. C., 1977, *Clin. Chem. Acta* **81**:183–192.
108. Fellows, L. E., King, G. S., Pettit, B. R., Goodwin, B. L., Ruthven, C. R. J., and Sandler, M., 1978, *Biomed. Mass Spectrom.* **5**:508–11.
109. Martin, M. E., Karoum, F., and Wyatt, R. J., 1979, *Anal. Biochem.* **99**:283–287.
110. Narasimhachari, N., Prakash, U., Helgeson, E., and Davis, J. M., 1978, *J. Chromatogr. Sci.* **16**:263–267.
111. Midgley, J. M., Couch, M. W., Crowley, J. R., and Williams, C. M., 1979, *Biomed. Mass Spectrom.* **6**:485–490.
112. Davis, B. A., and Boulton, A. A., 1981, *J. Chromatogr.* **222**:271–275.
113. Bertilsson, L., and Palmer, L., 1972, *Science* **177**:74–76.
114. Warsh, J. J., Chan, P. W., Godse, D. D., Coscina, D. V., and Stancer, H. C., 1977, *J. Neurochem.* **29**:955–958.
115. Artigas, F., and Gelpi, E., 1979, *Anal. Biochem.* **92**:233–242.
116. Faull, K. F., Anderson, P. J., Barchas, J. D., and Berger, P. A., 1979, *J. Chromatogr.* **163**:337–349.
117. Bertilsson, L., Atkinson, A. J., Jr., Althaus, J. R., Harfast, A., Lindgren, J.-E., and Holmstedt, B., 1972, *Anal. Chem.* **44**:1434–1438.
118. Godse, D. D., Warsh, J. J., and Stancer, H. C., 1977, *Anal. Chem.* **49**:915–918.
119. Domino, E. F., Mathews, B. N., and Tait, S. K., 1979, *Biomed. Mass Spectrom.* **6**:331–334.
120. Hoskins, J. A., Pollit, R. J., and Evans, S., 1978, *J. Chromatogr.* **145**:285–289.
121. Swahn, C.-G., Sandgarde, B., Wiesel, F.-A., and Sedval, G., 1976, *Psychopharmacology* **48**:147–152.
122. Muskiet, F. A. J., Jeuring, H. A., Nagel, G. T., deBruyn, H. W. A., and Wolthers, B. G., 1978, *Clin. Chem.* **24**:1899–1902.
123. Muskiet, F. A. J., Nagel, G. T., and Wolthers, B. G., 1980, *Anal. Biochem.* **109**:130–136.
124. Bertilsson, L., 1973, *J. Chromatogr.* **87**:147–153.
125. Sjoquist, B., Lindstrom, B., and Anggard, E., 1975, *J. Chromatogr.* **105**:309–316.
126. Muskiet, F. A. J., Fremouw-Ottevangers, D. C., Nagel, G. T., Wolthers, B. G., and deVries, J. A., 1978, *Clin. Chem.* **24**:2001–2008.
127. Muskiet, F. A. J., Fremouw-Ottevangers, D. C., van der Meulen, J., Wolthers, B. G., and de Vries, J. A., 1978, *Clin. Chem.* **24**:122–127.
128. Edwards, D. J., Rizk, M., and Neil, J., 1979, *J. Chromatogr.* **164**:407–416.
129. Edwards, D. J., and Rizk, M., 1979, *Clin. Chem. Acta* **95**:1–10.
130. Curtius, H. C., Wolfensberger, M., Redweik, U., Leimbacher, W., Maibach, R. A., and Isler, W., 1975, *J. Chromatogr.* **112**:523–531.
131. Takahashi, S., Godse, D. D., Naqvi, A., Warsh, J. J., and Stancer, H. C., 1978, *Clin. Chim. Acta* **84**:55–62.
132. Beck, O., Borg, S., Holmstedt, B., and Stibler, H., 1980, *Biochem. Pharmacol.* **29**:693–696.
133. Diggory, G. L., Caesar, P. M., Hazelby, D., and Taylor, K. T., 1979, *J. Neurochem.* **32**:1323–1325.
134. Leone, R. M., Silman, R. E., Hooper, R. J. L., Finnie, M. D. A., Carter, S. J., Edwards, R., Smith, I., Towell, P., and Mullen, P. E., 1979, *J. Endocrinol.* **82**:243–251.
135. Smith, I., Francis, P., Leone, R. M., and Mullen, P. E., 1980, *Biochem. J.* **185**:537–540.
136. Zagalak, M.-J., Curtius, H.-C., Leimbacher, W., and Redweik, U., 1977, *J. Chromatogr.* **142**:523–531.
137. Sjoquist, B., 1979, *Biomed. Mass Spectrom.* **6**:392–395.
138. Trefz, F. K., Erlenmaier, T., Hunneman, D. H., Bartholomé, K., and Lutz, P., 1979, *Clin. Chim. Acta* **99**:211–220.
139. Wegmann, H., Curtius, H.-C., and Redweik, U., 1978, *J. Chromatogr.* **158**:305–312.
140. Martinez, E., and Gelpi, E., 1978, *J. Chromatogr.* **167**:77–90.
141. Bertilsson, L., and Costa, E., 1976, *J. Chromatogr.* **118**:395–402.
142. Huizinga, J. D., Teelken, A. W., Muskiet, F. A. J., Jeuring, H. J., and Wolthers, B. G., 1978, *J. Neurochem.* **30**:911–913.
143. Colby, B. N., and McCaman, M. W., 1978, *Biomed. Mass Spectrom.* **5**:215–219.

144. Schmid, R., and Karobath, M., 1977, *J. Chromatogr.* **139**:101–109.
145. Cattabeni, F., Galli, C. L., and Eros, T., 1976, *Anal. Biochem.* **72**:1–7.
146. Callery, P. S., Stogniew, M., and Geelhaar, L. A., 1979, *Biomed. Mass Spectrom.* **6**:23–26.
147. Faull, K. F., DoAmaral, J. R., Berger, P. A., and Barchas, J. D., 1978, *J. Neurochem.* **31**:1119–1122.
148. Ferkany, J. W., Smith, L. A., Seifert, W. E., Jr., Caprioli, R. M., and Enna, S. J., 1978, *Life Sci.* **22**:2121–2128.
149. Wu, P. H., Durden, D. A., and Hertz, L., 1979, *J. Neurochem.* **32**:379–390.
150. Wolfensberger, M., Redweik, U., and Curtius, H.-C., 1979, *J. Chromatogr.* **172**:471–475.
151. Collins, F. S., and Summer, G. K., 1978, *J. Chromatogr.* **145**:456–463.
152. Lapin, A., and Karobath, M., 1980, *J. Chromatogr.* **193**:95–99.
153. Leimer, K. R., Rice, R. H., and Gehrke, C. W., 1977, *J. Chromatogr.* **141**:121–144.
154. Liardon, R., Ott-Kuhn, U., and Husek, P., 1979, *Biomed. Mass Spectrom.* **6**:381–391.
155. Seiler, W., Schneider, H. H., and Sonnenberg, K.-D., 1971, *Anal. Biochem.* **44**:451–457.
156. Iwase, H., Takeuchi, Y., and Muria, A., 1979, *Chem. Pharm. Bull.* **27**:1307–1315.
157. Schulman, M. F., and Abramson, F. P., 1975, *Biomed. Mass Spectrom.* **2**:9–14.

Gas Chromatography

Ronald T. Coutts and Glen B. Baker

1. INTRODUCTION

The application of a number of quantitative procedures to studies of nervous tissue and body fluids has done much to advance our knowledge of neurochemistry. In this chapter, we discuss one of the analytical tools in wide use, gas chromatography (GC). The purpose of this brief discussion of the principles involved and the use of selected examples of how the GC has been employed in analysis of endogenous components of the nervous system and in analysis of psychotropic drugs is to give the reader an appreciation of the versatility of GC in neurochemical investigations.

2. BASIC PRINCIPLES OF GAS CHROMATOGRAPHY

Gas chromatography, also termed gas–liquid chromatography (GLC), is an analytical procedure used to separate mixtures of organic compounds prior to their identification and/or quantitation. Gas chromatographic separations are usually conducted on solutions of compounds in inert solvents and are accomplished on glass, metal, and occasionally Teflon® columns containing a nonvolatile liquid (the stationary phase), usually coated onto an inert solid support material with a large surface area. The GC column is contained in the oven of a gas chromatograph; the oven temperature can be adjusted at will, but normally over a 20–300° range, depending on the volatility of the solutes and the stability of the stationary phase. The components of a mixture (including the solvent) are carried through the column by an inert carrier gas and separated from one another according to their partition coefficients between the carrier gas and the stationary phase. As each component elutes from the column, it is detected in such a manner that it can be displayed as a peak on

Ronald T. Coutts • Neurochemical Research Unit, Faculty of Pharmacy and Pharmaceutical Sciences, University of Alberta, Edmonton, Alberta T6G 2N8, Canada. *Glen B. Baker* • Neurochemical Research Unit, Department of Psychiatry, University of Alberta, Edmonton, Alberta T6G 2N8, Canada.

a recorder. The time interval between the point of injection of the solution and the apex of the recorded peak is termed the retention time of the peak and is characteristic of, but not unique to, the compound giving rise to it under the GC conditions employed.

Chromatographic separations can be achieved isothermally (constant column temperature) or by temperature programming in which the column temperature is altered at preset rates during the analysis. Temperature programming permits a great reduction in analysis time when solutions containing a wide range of components are being investigated.

2.1. Gas Chromatographic Columns

Gas chromatographic columns made of glass, stainless steel, copper, aluminum, or Teflon® and glass-lined metal columns are available commercially. Glass columns are popular in biological studies because of their inertness; undesirable degradations occur much more frequently on heated metal columns. Both packed and capillary columns are utilized. The former are typically 1–2 m in length and 2–4 mm in internal diameter and are packed with an inert solid support that has previously been coated with the stationary phase. Capillary columns are usually made of glass and are generally 10–75 m in length and 0.25–0.50 mm in internal diameter and coated with a thin layer of the stationary phase. Two types of glass capillary columns are in general use. Wall-coated open tubular (WCOT) columns have a thin liquid-phase film coated directly onto the inside surface of the capillary. With support-coated open tubular (SCOT) columns, the inside surface of the capillary is coated with a thin layer of support material which is coated with a thin film of liquid phase. The SCOT columns generally have a higher sample loading capacity than WCOT columns. In general, the use of capillary columns permits much better resolution of components of a mixture than can be achieved on packed columns. However, packed columns are relatively inexpensive and robust, and the analyst can prepare a virtually unlimited number of stationary phases on various solid supports for insertion into empty columns. Glass capillary columns are quite fragile and easily broken during installation, but fused silica columns with a high degree of flexibility (and of inertness) are now commercially available.

Literally hundreds of stationary phases are used in GC analyses, and the amount of stationary phase used to coat the solid support can vary from the amount that will produce a monomolecular surface layer to as much as 30% w/w. In practice, however, a small number of stationary phases are used routinely. One to 5% w/w of a stationary phase on a solid support is the commonly used concentration. There are relatively few support materials in popular use. Those most commonly encountered are prepared from diatomaceous earth (kieselguhr) which is very porous and has a high surface area. They are available in different densities; the low-density supports (e.g., Chromosorb® W) can be loaded with up to 30% stationary phase, whereas high-density supports (e.g., Chromosorb® G) can accommodate much lower amounts of stationary phase. Chromosorb® 750 is of medium density and is composed of hard particles that generate no undesirable fine particles during coating or packing. Other solid supports commonly encountered include Gas Chrom Q,® Haloport F,® and glass beads.

2.2. Detectors

Compounds eluting from a gas chromatograph are detected by a variety of techniques. The thermal conductivity detector (TCD) has a limited use in neurochemistry; the flame ionization detector (FID) and the electron-capture detector (ECD) are in routine use; and the nitrogen–phosphorous detector (NPD) is being used increasingly. The mass spectrometer is also used as a sophisticated detector in the GC analysis of neurochemicals.

2.2.1. Thermal Conductivity Detector

The TCD is one of the oldest GC detection devices. It is durable, easy to operate, and is capable of detecting a wide range of organic compounds because of its lack of selectivity. Thermal conductivity detection is nondestructive; eluted compounds can be collected for further study. Another advantage of the TCD is that it produces a linear response over a wide range of sample amounts (*ca.* 10^5). However, the TCD is rarely used in neurochemical studies because of its poor sensitivity relative to that of other detectors. The minimum detectable limit for a TCD is around 1 μg.

2.2.2. Flame Ionization Detector

The FID has been the most widely used detector in biological studies. It operates on a simple principle. The effluent from the GC column is mixed with hydrogen, passed through a metal jet, and burned in an atmosphere of air. On combustion, positive and negative ions are produced, and one of these species is collected on a charged collector located above the jet. An electric current is produced, the strength of which is directly proportional to the amount of compound combusted. The current is amplified and displayed as a function of time on a recorder. Only compounds that combust with ionization in a hydrogen/air flame produce an FID response. Virtually all organic compounds do; notable exceptions are carbon dioxide and carbon disulfide as well as water, ammonia, inert gases, nitrogen, and other simple gases. Detection sensitivity of an FID is approximately proportional to the number of carbon atoms in the molecule being detected and is linear over a wide range of sample amounts (*ca.* 10^7). Sample quantities as low as 1 ng can usually be detected. To ensure maximum sensitivity and stable detector operation, the flow rates of the three gases utilized (carrier gas, air, hydrogen) require critical adjustment.

2.2.3. Electron-Capture Detector

The ECD is a selective detector that can detect as little as 1 pg of an organic compound that contains an electrophoric substituent. An ECD usually contains a ^{63}Ni (or ^3H) source which emits relatively high-energy β particles.

These particles collide with carrier gas molecules (normally 95% argon/5% methane) and produce a large number of low-energy secondary electrons. A small current (the standing current) is produced in the detector. Some of these electrons are absorbed by the sample eluting from the GC column, thus reducing the strength of the standing current which returns to its original level when the sample has left the detector. The change in current is amplified and recorded as a peak on the recorder. The sensitivity of the recorder varies greatly according to the ability of a compound to absorb electrons. Compounds that contain halogen atoms, ketone or nitro groups, or other electrophoric groups are particularly suitable for ECD. Bromobenzene, for example, is 7500 times more sensitive than benzene to electron-capture detection.

2.2.4. Nitrogen–Phosphorous Detector

The NPD, a selective detector, is, as the name suggests, extremely sensitive to most compounds that possess nitrogen- and/or phosphorous-containing functions. The effluent from the GC column is mixed with a smaller volume of hydrogen, and the mixture enters the electrically heated detector chamber which contains an alkali source (often a rubidium salt). A low-temperature plasma rather than a discrete flame is formed. By a mechanism that is not completely understood, this treatment produces a minute electric current of a magnitude proportional to the amount of compound reaching the detector. The current produced is amplified and recorded. The NPD is about 30,000 times more sensitive towards nitrogen and 60,000 times more sensitive towards phosphorous than it is towards carbon. Low picogram quantities of nitrogen- and phosphorous-containing compounds can be detected.

2.2.5. Mass Spectrometric Detection

A mass spectrometer can be employed as a sensitive, specific detector of compounds eluting from a GC and is now widely used in neurochemical studies. The topic of GC–MS analysis is dealt with in detail in Chapter 17, and the present chapter is concerned with GC analysis using the other detectors just mentioned above.

3. DERIVATIZATION FOR ANALYSIS BY GAS CHROMATOGRAPHY

Generally, derivatization of substances for GC analysis is done for one or more of the following reasons: (1) to increase volatility; (2) to increase stability; (3) to reduce the polarity of the substance, since polar compounds such as amines, alcohols, or phenols generally chromatograph poorly; (4) to improve extraction efficiency from aqueous solution (e.g., acetylation of phenolic amines); and (5) to introduce a functional group that is particularly sensitive to selective detectors such as ECD or NPD.

a   R-OH $\xrightarrow[\text{base}]{R^1X}$ R-O-R^1 + HX

b   R-NH$_2$ $\xrightarrow{(CF_3CO)_2O}$ R-NHCOCF$_3$ + CF$_3$COOH

c   R-OH $\xrightarrow{(CH_3)_3SiX}$ R-OSi(CH$_3$)$_3$ + HX

d   R-NH$_2$ $\xrightarrow{O=CR^1R^2}$ R-N=CR^{1}R^2 + H$_2$O

Fig. 1. Examples of the principal types of derivatizations used for GC analysis: (a) alkylation; (b) acylation; (c) silylation; and (d) condensation.

Most derivatizations involve the replacement of active hydrogen in polar groups (e.g., NH, OH, SH), and the main types of derivatization are alkylation, acylation, silylation, and condensation.[1] General reactions for each type of derivatization are shown in Fig. 1. Alkylation of OH (including COOH), SH, and NH (including CONH) groups results in replacement of active hydrogens with alkyl groups, yielding ethers, esters, thioethers, thiolesters, N-alkyl amines, or N-alkyl amides. Replacement of an active hydrogen on OH, SH, or NH by a trimethylsilyl group is the most common type of silylation. The various trimethylsilylating reagents available and their silyl donor strength and reactivity are discussed by Knapp.[1] Acylation of OH, SH, and NH groups is usually performed with acyl anhydrides, acyl halides, or activated acyl amide reagents. Condensations are reactions in which two molecules are joined with loss of water.[1] Derivatives formed for GC analysis by condensation reactions include boronates, oximes, and hydrazones. The reactions described above account for the majority of derivatizations used for GC work, but other products that have been utilized include phosphoryl and sulfonyl derivatives, ketals, ureas, oxazolidines, and oxazolidinones.[1] Some excellent texts are available describing derivatization procedures in considerable detail.[1,2]

4. ANALYSIS OF SPECIFIC TYPES OF COMPOUNDS BY GAS CHROMATOGRAPHY

4.1. Endogenous Arylalkylamines

Norepinephrine (NA), dopamine (DA), 5-hydroxytryptamine (5-HT), histamine (HIST), and the "trace" amines [2-phenylethylamine (PEA), tyramine (TA), octopamine (OA), phenylethanolamine (PEOH), and tryptamine (T)] are of particular interest to neurochemists because of their possible role as neurotransmitters or neuromodulators and their proposed involvement in the etiology of a number of neurological and psychiatric disorders.

Quantitative neurochemical GC analysis of these endogenous monoamines has largely been by GC with an electron-capture detector (GC–ECD) because of the high inherent detector sensitivity and because of the ease with which most amines can be derivatized by reagents containing groups with a high affinity for electrons. Many of these biogenic amines contain other functional groups, such as phenols and alcohols, in their structure which may also be derivatized readily, resulting in increased sensitivity. Most commonly used of the halogenated derivatizing reagents are trifluoroacetic, pentafluoropropionic, and heptafluorobutyric anhydrides, but examples of other reagents that may be used with success include trifluoroacetyl-, pentafluoropropionyl-, and heptafluorobutyrylimidazole, pentafluoropropionyl chloride, pentafluoropropionyltriazole, N-methyl-bis-trifluoroacetamide, 3-trifluoromethylphenylisocynate, acetic anhydride substituted with chlorine or combinations of chlorine and fluorine, halogenated chloroformates, and pentafluorobenzaldehyde.[2,3] Amines with another polar group present on the molecule (e.g., a phenol or alcohol) at appropriate positions can react with an alkylboronic acid to yield boronate derivatives that are sensitive to ECD. A number of reagents containing nitro groups will react with amines to give derivatives with good ECD sensitivity.[4] Mixed derivatives of phenolic or alcoholic amines may also be prepared; i.e., two or more reagents may be used simultaneously or sequentially to form different derivatives of different functional groups on the molecule.[5–10] For obtaining optimum reaction and separation conditions as well as information regarding sensitivities and stabilities of the derivatives mentioned above, a number of excellent articles and reviews are available.[1,2,11–16]

For the catecholamines, GC–ECD analysis frequently involves adsorption on alumina, elution with acid, removal of the liquid under vacuum, and reaction of the residue with a perfluoroacylating agent.[7,17–21] Bertani *et al.*[22] and Nelson *et al.*[23] used GC–ECD to quantitate perfluoroacylated derivatives of 3-O-methylated amine metabolites of the catecholamines after these amines had been extracted from urine. Coutts *et al.*[24] utilized acetylation under aqueous conditions and extraction with ethyl acetate to isolate 3-methoxytyramine (3-MT) and normetanephrine (NME). The acetylated phenolic moieties were then specifically hydrolyzed with NH_4OH, and the resultant compounds were reacted with trifluoroacetic anhydride (TFAA). This technique has now been modified for analysis of these 3-O-methylated amines in brain (D. F. LeGatt, G. B. Baker, and R. T. Coutts, unpublished data). Martin and Ansell[18] developed a GC–ECD technique for 5-HT in brain in which the TFA derivative of this amine was formed after extraction from the tissue. Baker *et al.*[25] employed extraction with a liquid ion exchanger, acetylation under aqueous conditions, organic extraction, and reaction with pentafluoropropionic anhydride (PFPA) to analyze 5-HT in brain.

Traces amines have been detected and/or quantitated in brain tissue and body fluids by GC–ECD using techniques such as reaction with nitro-containing reagents,[26,27] reaction with PFPA,[28,29] reaction with pentafluorobenzoyl chloride,[30,31] and acetylation followed by reaction with TFAA or PFPA.[32–35] By

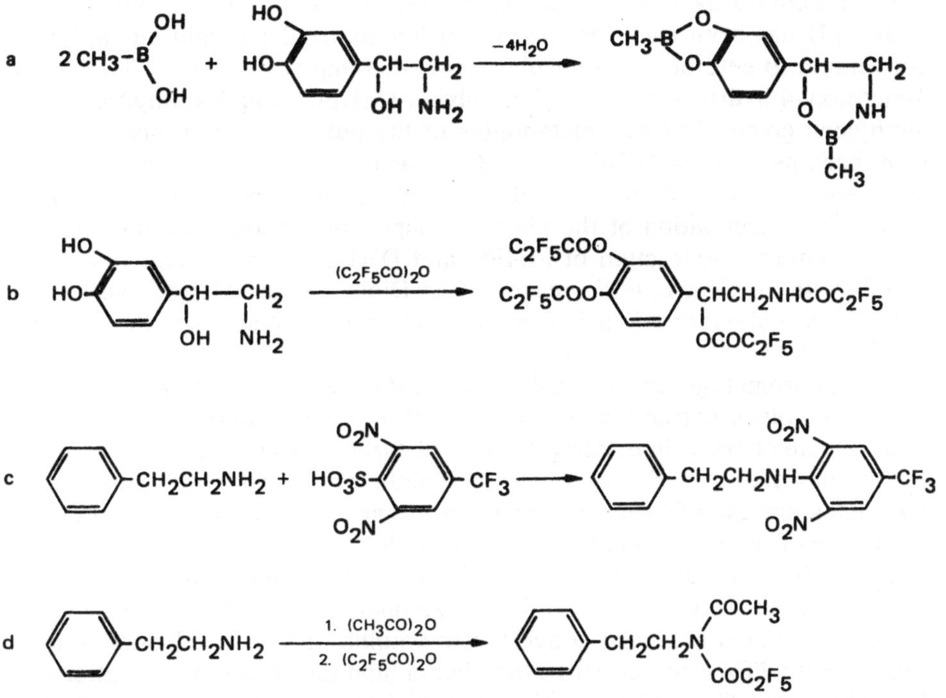

Fig. 2. Examples of derivatizations of some arylalkylamines of neurochemical interest: (a) reaction of norepinephrine with methylboronic acid; (b) reaction of norepinephrine with pentafluoropropionic anhydride; (c) reaction of 2-phenylethylamine with 2,6-dinitro-4-trifluoromethylbenzenesulfonic acid; (d) reaction of 2-phenylethylamine with acetic anhydride and pentafluoropropionic anhydride.

utilizing acetylation, specific hydrolysis of acetylated phenolic groups, and then reaction with a perfluoroacylating agent, it has been possible to assay simultaneously PEA, *m*- and *p*-TA, OA, NME, 3-MT, T, and 5-HT.[3,36,37]

Histamine has been a difficult amine to assay by means of GC, but recent reports utilizing reaction with heptafluorobutyric anhydride (HFBA) and ethyl chloroformate,[38] with PFPA,[39] and with 2,6-dinitro-4-trifluoromethylbenzenesulfonic acid[40] look promising for future experimentation with GC–ECD. Navert and Wollin[41] recently reported an assay that involves reaction with HFBA and acetic anhydride and GC analysis using NPD.

Examples of derivatization of amines with some of the reagents mentioned above are shown in Fig. 2.

4.2. Acidic and Neutral Metabolites of Endogenous Arylalkylamines

Alcoholic (and phenolic in some cases) and/or carboxylic acid groups are present on these substances, and these moieties are often derivatized for GC analysis. In addition to derivatization with the halogenated acylating agents listed in the previous section on amines, bromomethyldimethylsilyl, iodomethyldimethylsilyl, and pentafluorophenyldimethylsilyl (flophemesyl) derivatives of alcohols have been prepared for GC–ECD.[42,43] Gas chromatography with NPD has been used for analysis of low-molecular-weight alcohols after conversion to heterocyclic phosphorous-containing derivatives.[44] The glycols 3-methoxy-4-hydroxyphenylethylene glycol (MHPG) and 3,4-dihydroxyphenylethylene glycol (DHPG), metabolites of the putative neurotransmitter NA, have been assayed by GC–ECD: MHPG has been extracted from body fluids and brain tissue and derivatized for analysis with perfluoroacylating reagents.[45–47] Acetylation of the phenol group(s) in aqueous medium has been used to improve extraction of MHPG and DHPG before derivatizing the alcohols with an electrophoric reagent.[48–52] Diols such as MHPG can be derivatized with alkylboronic acids to yield boronate derivatives with sensitivity on ECD.[53]

Gas chromatographic analysis of the acidic metabolites of biogenic amines usually involves extraction of the acids from an acidified sample of tissue homogenate or body fluid, derivatization of carboxylic acid moieties, and preparation of different derivatives of any phenolic or alcoholic functions present. Examples are esterification of the carboxyl group with an alcohol and derivatization of the phenols and alcohols (or indole nitrogens) with reagents such as acyl halides or acyl anhydrides.[54–58] Davis[59] has employed pentafluorobenzyl bromide in a crown ether to derivatize phenolic acids for GC–ECD. Vanillylmandelic acid has been converted to vanillin or vanillyl alcohol and then reacted with TFAA to derivatize the phenol (and the alcohol in the latter case) for subsequent GC–ECD quantitation.[60,61] Gas chromatographic analysis of acidic metabolites of phenylethyl- and indolealkylamines in urine and brain has also been performed using TMS derivatives.[62,63]

4.3. Polyamines

Gas chromatography has not been used extensively in the analysis of these alkylamines (e.g., putrescine, spermidine, spermine). Problems have been encountered in attempts to trifluoroacetylate polyamines,[64] but a successful analysis of picomole amounts of polyamines by GC–ECD using pentafluorobenzoyl chloride as a derivatizing reagent has been reported.[65]

4.4. Amino Acids

Amino acids are not volatile enough to be gas chromatographed satisfactorily without prior derivatization. Because of their amphoteric nature, it is difficult to extract them directly into organic solvents from aqueous solutions in high yield. The use of ion-exchange chromatography is a commonly used method of removing amino acids, even in trace amounts, from aqueous solutions. Extraction methodology has been described in detail by Gehrke and colleagues[66]; the analysis of amino acids by gas chromatography has been reviewed by Husek and Macek,[67] and derivatization procedures have been reviewed by Knapp.[1] Generally, the carboxylic acid group of the amino acid is esterified, and the amino group acylated, most often with a perfluoroacylating agent. Methyl, ethyl, n-propyl, n-butyl, isobutyl, isoamyl, and other esters

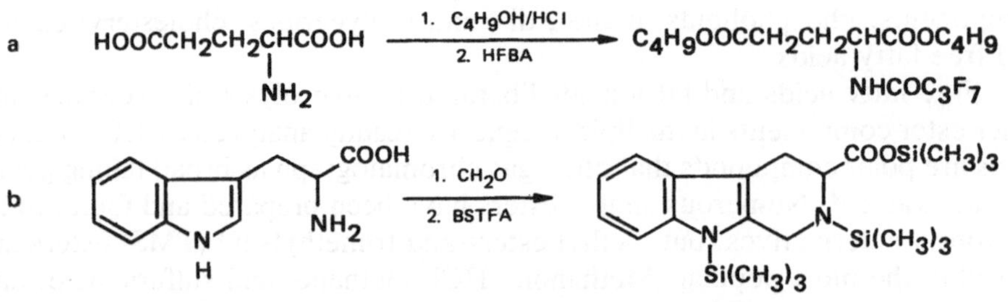

Fig. 3. (a) Derivatization of glutamic acid by alkylation and heptafluorobutyrylation. (b) Product formed by reaction of tryptophan with BSTFA in presence of formalin.[1]

have been described. Trifluoroacetylation, pentafluoropropionylation, and heptafluorobutyrylation are most frequently used for derivatization of the amine group. The reaction depicted for glutamic acid is typical (Fig. 3).

The sensitivity of the reaction is normally in the 5 to 100-ng range, depending on the identity of the amino acid. Claims have been made that *n*-butylated, N-trifluoroacetylated amino acids can be detected to the picogram level using GC with ECD.[68]

Another common method of derivatizing amino acids for GC analysis[69] is to react them with a trimethylsilylating reagent that will simultaneously esterify the acid group and trimethylsilylate the amino group and any alcoholic and phenolic hydroxyl groups that are present in the molecule. Silylating agents that are used for this purpose include bis(trimethylsilyl)trifluoroacetamide (BSTFA), trimethylsilyldiethylamine (TMSDEA), N-methylacetamide (MSA), and mixtures of bis(trimethylsilyl)acetamide (BSA), trimethylsilylimidazole (TMSI), and trimethylchlorosilane (TMCS). Various solvents are used, depending on the nature of the reagent. Numerous other derivatizing reagents have been described and reviewed.[1]

The reaction of tryptophan with BSTFA in the presence of formalin is of interest. A trimethylsilylated β-carboline was formed which had good GC properties (Fig. 3).[1,69] The reaction is also applicable to tryptamines.

4.5. Acetylcholine and Choline

A major problem in the GC analysis of acetylcholine (Ach) and choline (Ch) has been the lack of volatility of these compounds. Demethylation, either by chemical means[70,71] or by pyrolysis,[72] is usually employed to increase volatility. This may be combined with acylation of choline to propionyl choline.[73] Detection of the derivatives just mentioned is generally by FID, although NPD[74] has also been utilized. These methods have now been used widely for analysis of these substances in brain and other tissues.

4.6. Lipids

Lipids can be extracted from a biological sample into an organic solvent and can be separated by column or thin-layer chromatography into their major

components: phospholipids, mono-, di-, and triglycerides, cholesteryl esters, and free fatty acids.

Free fatty acids and fatty acids liberated by hydrolysis of glycerides and other ester components in the lipid sample are readily analyzed by GC. Organic acids are polar compounds that often gas chromatograph as broad tailing peaks if underivatized. Numerous simple esters have been prepared and found to be appropriate derivatives, but methyl esters and trimethylsilyl (TMS) esters are probably the most popular. Methanolic HCl, methanol and sulfuric acid, and diazomethane are suitable reagents for methylation, and BSTFA, trimethyl-silylimidazole (TMSI), and t-butyldimethylsilyl chloride (TBDMS-Cl) are suitable silylating agents.

A typical GC analysis of fatty acids in a liquid sample would be to reflux 1–10 mg of the sample with 5% HCl in dry methanol (4 ml)/benzene (0.5 ml) for 2 hr, cool, extract with petroleum ether, and concentrate the petroleum ether solution for GC analysis. This procedure will hydrolyze the triglycerides and other esters and convert the resulting acids to their methyl esters.

Fatty acids can also be converted to esters which are particularly sensitive to ECD. Halogen-containing alcohols or alkyl halides are generally employed in ester formation. 2-Chloroethanol and 2,2,2-trichloroethanol are typical alcohol reagents in which the acid is dissolved and either boron trifluoride or trichloride is added. On refluxing, esterification occurs. Pentafluorobenzyl bromide is the most commonly used alkyl halide in the esterification of fatty acids for ECD. It reacts readily at room temperature with acids in basic solution. Numerous other halogenated groups can be introduced on the fatty acid. Generally, 1–100 ng of a fatty acid or less can be readily detected by GC–ECD.[1,2]

4.7. Steroids

Gas chromatographic methods have been developed for various steroids including cortisol and its metabolites.[1] The presence of polar substituents usually makes derivatization a mandatory procedure prior to GC analysis. Most steroids contain an alcohol group or groups; other substituents encountered are phenolic hydroxyl, carbonyl, and carboxylic acid.

Hydroxysteroids (steroids with alcoholic groups) are often acylated prior to GC analysis. Perfluoroacylation is common, with a perfluoroacyl anhydride or chloride used as the derivatizing reagent. Other anhydrides, especially chloroacetic anhydride, have also been used as esterifying agents.

Trimethylsilyl derivatives of sterols have been studied extensively. Both alcohol and phenolic hydroxyl groups react readily with various silylating agents. Reaction conditions vary with each reagent, and not all silylating reagents react in the same manner with steroids containing alcoholic groups.[75] The 11β-hydroxyl group in cortisol, for example, resists derivatization with BSTFA. In contrast, with TMSI, the 11β-hydroxyl group is trimethylsily-lated.[76]

Ketone groups in steroids can also be derivatized. Usually, the ketosteroid is reacted with an alkyloxyamine ($RONH_2$; $R = CH_3$, nC_4H_9, $C_6H_5CH_2$, $C_6F_5CH_2$) and structurally related reagents. Reagents that introduce a halogen-

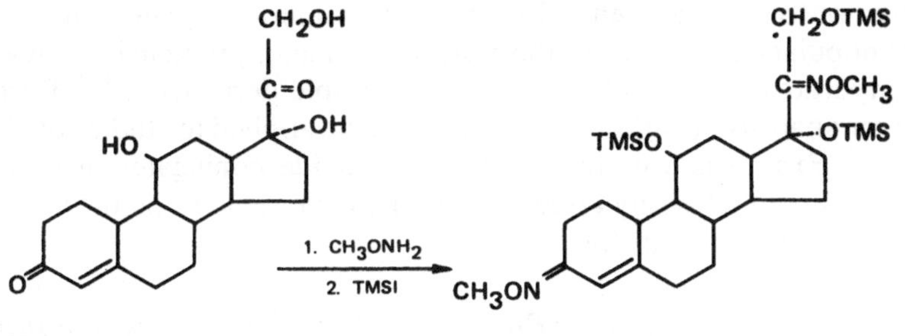

Fig. 4.   Formation of a derivative of a ketosteroid suitable for GC analysis.

containing group are ECD sensitive. A combination of ketone group derivatization and trimethylsilylation of alcoholic substituents gives products with good GC properties[77] (Fig. 4).

Steroids that contain carboxylic acid groups, e.g., bile acids, are esterified prior to GC analysis.

4.8. Prostaglandins

The primary prostaglandins are unsaturated acids containing 20 carbons, including a cyclopentane ring with hydroxyl and/or carboxyl substituents. For this reason, GC analysis usually involves esterification of the carboxyl group along with additional derivatization procedures for the ring substituent(s). Derivatives of prostaglandins have been prepared for GC–FID analysis of these substances in lung,[78] menstrual fluid,[79] and semen[80] and for studies on their metabolism.[81] Gas chromatography with ECD provides the opportunity for increased sensitivity. Sensitive derivatives of prostaglandins $F_{1\alpha}$ and $F_{2\alpha}$ have been produced for GC–ECD analysis by utilizing diazomethane for methylation of the carboxyl moiety and HFBA for derivatization of phenolic groups.[82] Bromomethyldimethylsilyldiethylamine has been reacted with F prostaglandins following esterification of the carboxyl group to produce derivatives suitable for GC–ECD analysis in tissues and amniotic fluid.[83] Carbonyl-containing prostaglandins are often derivatized by formation of an oxime derivative of this moiety; for example, a combination of esterifying and trimethylsilylating agents and pentafluorobenzylhydroxylamine yields compounds with good GC–ECD properties.[84,85] Wickramasinghe and Shaw[86] esterified the carboxyl groups of prostaglandin $F_{2\alpha}$ and formed the TMS derivatives of the phenols. Detector sensitivity down to 12.5 pg was reported with such derivatives. For those prostaglandins that have suitably positioned hydroxyl groups, boronate derivatives may be formed for GC–ECD analysis,[87] but these derivatives do not seem to have been employed widely.

4.9. Carbohydrates

Because of the presence of multiple hydroxyl groups in the structure of carbohydrates, numerous derivatives of these substances have been prepared in an attempt to analyze them. However, the trimethylsilyl derivatives remain the most popular derivatives for this purpose. A major problem in derivatization of carbohydrates is the tendency to yield multiple derivatives.[1,88] Derivatives suitable for analysis of carbohydrates can also be applied to studies of glycoside conjugates. Examples are analyses of glucuronide conjugates of drugs using methylation or trimethylsilylation. Comprehensive reviews on the GC analysis of carbohydrates are available.[89-91]

4.10. Purine and Pyrimidine Bases, Nucleosides, and Nucleotides

These substances have been derivatized by permethylation, silylation, and acylation or a combination of these three procedures. Excellent reviews describing derivatization of these compounds have been published.[1,92,93]

4.11. Antidepressants

Although GC–FID has been employed for analysis of tricyclic antidepressants in body fluids, GC–ECD and GC–NPD are now popular for quantitation of these drugs. Secondary amines such as desmethylimipramine may be readily derivatized with electrophoric acylating agents such as TFAA, but tertiary amines (e.g., imipramine) are much more difficult to derivatize. However, halogenated carbamate derivatives and anthraquinone derivatives have been prepared for analysis by GC–ECD.[94-97]

Gas chromatography with NPD is proving very popular for analysis of tricyclics, since it is not necessary to derivatize the tertiary amines. Simultaneous analysis of the parent drug and the demethylated metabolite is accomplished conveniently,[98] although the demethylated drug may be derivatized[99,100] to improve resolution and improve peak shape. Maprotoline, a tetracyclic antidepressant, has a secondary amine group in its aliphatic side chain which may be derivatized with reagents such as HFBA for GC–ECD analysis.[101]

Phenelzine (phenylethylhydrazine) and tranylcypromine (*trans*-phenylcyclopropylamine) are monoamine oxidase-inhibiting antidepressants that are still widely prescribed. One procedure for analysis of phenelzine involves oxidation of the drug to ethylbenzene and styrene, which are quantitated using GC–FID.[102] Cooper *et al.*[103] have measured phenelzine by GC–ECD after extraction from plasma and derivatization with HFBA. Plasma levels of tranylcypromine (TCP) have been determined by GC–NPD after perfluoroacylation of this drug.[104] In the procedure of Calverley *et al.*,[105] TCP was acetylated and subsequently reacted with TFAA or PFPA for GC–ECD analysis.

4.12. Neuroleptics

Gas chromatography has played an important role in analysis of neuroleptics (antipsychotics, major tranquillizers) in body fluids and tissues. Gas chromatography with FID is adequate for analysis of phenothiazines and other neuroleptics in urine and in the upper range of therapeutic levels in blood samples.[106] Much of the analysis of chlorpromazine (CPZ) and its metabolites at therapeutic levels has been done using GC–ECD,[107,108] and Curry[109] has discussed some of the problems involved in the analysis of CPZ and related neuroleptics using this analytical tool. High oven temperatures are often necessary for GLC analysis of phenothiazines and thioxanthenes, and there is difficulty in eluting some of the higher-molecular-weight compounds such as the sulfoxide metabolites. The metabolites of the phenothiazines may be derivatized by the use of techniques such as acetylation and trifluoroacetylation[107,110] to improve sensitivity, separation, and/or peak shape. Gas chromatography with FID has been used for quantitation of thioridazine and some of its metabolites in blood samples.[111] Analysis of phenothiazines and some of their metabolites by GC–NPD is now becoming popular.[112,113] Fluphenazine has been notoriously difficult to analyze in biological fluids at low levels[107]; however, it has now been reported that adequate sensitivity for measurement of this drug at low levels in blood may be obtained by derivatization with N,O-bis(trimethylsilyl)trifluoroacetamide for GC–ECD analysis[114] or by using GC–NPD analysis.[115,116]

Gas chromatographic analysis of therapeutic levels of butyrophenones in blood has been largely by ECD,[117-120] although measurement of haloperidol using GC–NPD[121] has also been reported.

4.13. Benzodiazepines

Most of the clinically efficacious benzodiazepines (BZDs) have in their structure several electrophoric substituents (halogen or nitro group in the 7 position, a halogen in the 2' position of the phenyl ring, and a carbonyl group in position 2 of the 1,4-benzodiazepine ring), making them particularly amenable to analysis by GC–ECD. Benzophenone derivatives with good ECD sensitivity can be prepared from several BZDs.[106]

Many BZDs are analyzed intact, although simple derivatizations may be performed to improve volatility, peak shape, and detector response.[122-124] Methyl derivatives of the N-desalkyl-1,4-benzodiazepin-2-ones are prepared to provide chemical stability, to enhance sensitivity to ECD, and to minimize adsorption losses. Procedures for extraction of a number of BZDs from blood have been summarized by De Silva.[106]

4.14. CNS Stimulants and Related Compounds

Amphetamines and structurally related compounds (e.g., phentermines, fenfluramine) can be analyzed without derivatization, but they are usually converted to products that are particularly sensitive to ECD. Additional advantages of derivatization are that the products have superior chromatographic properties and give more characteristic mass spectra. Reaction with halogenated acid chlorides or anhydrides is most common. The choice of derivatizing reagent is important, because ECD responses to different derivatives of amphetamines vary considerably.[2]

A difficulty associated with the analysis of trace quantities of amphetamine and amphetaminelike substances is loss of drug by volatilization and adsorption

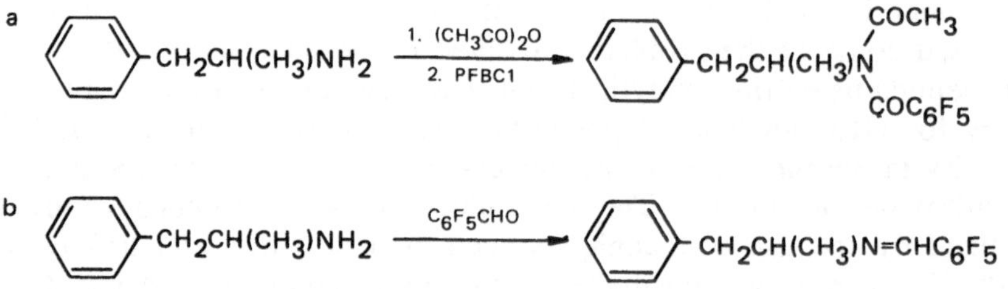

Fig. 5. Derivative formed by reaction of amphetamine with (a) acetic anhydride in aqueous solution and then with pentafluorobenzoyl chloride under anhydrous conditions and (b) pentafluorobenzaldehyde.

onto glassware during solvent extraction and evaporation. In the case of amphetamine, loss was minimal when HCl gas and diethylamine were added to a benzene extract of amphetamine prior to concentration.[125] In preliminary studies in our laboratory, it has been found that aqueous solutions of amphetamine can be acetylated with acetic anhydride, and the N-acetylated derivative efficiently extracted. Further derivatization with pentafluorobenzoyl chloride gives a product (Fig. 5) that has good sensitivity on GC–ECD.[126] Primary amine drugs (amphetamine, phentermine, chlorphentermine, and others) also react with simple ketones or aromatic aldehydes to give Schiff bases (Fig. 5). If halogenated aldehydes or ketones or nitrated aldehydes are employed, the products are sensitive to GC–ECD analysis, and low picogram quantities of amines can be quantitated.[127,128]

The reactions just described can be applied to other basic CNS stimulants. The secondary amine methylphenidate, for example, has been analyzed by GC–ECD as its trichloroacetate.[129]

Reagents are also available that permit the separation and quantitative analysis of (+)- and (−)-isomers in enantiomeric mixtures of amphetamine or related compounds.[130]

4.15. Barbiturates, Hydantoins, and Related Compounds

The CNS-depressant barbiturates and structurally related compounds (e.g., the hydantoin antiepileptics, some xanthines) are routinely analyzed by GC. They can often be chromatographed without derivatization, but in most cases, they are derivatized to improve peak shape and separation. Generally, the NH groups are alkylated, and various reagents are used for this purpose.

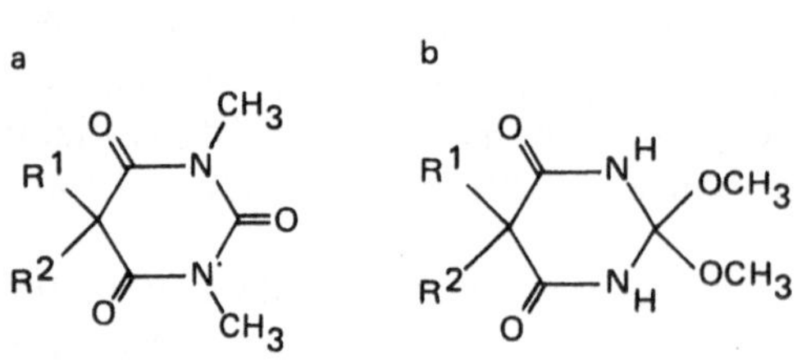

Fig. 6. (a) A dialkyl derivative and (b) a dimethylketal derivative of a barbiturate.

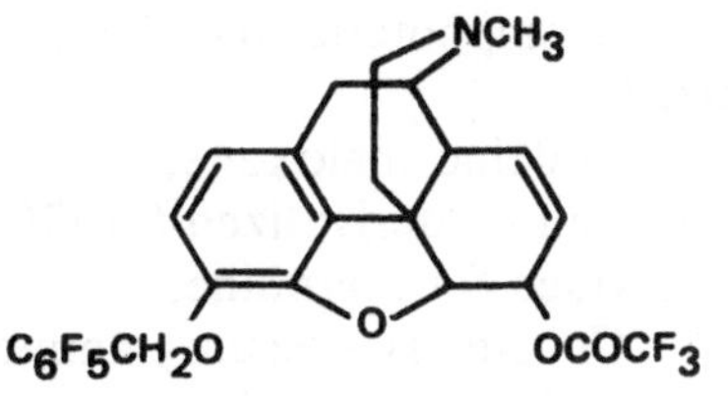

Fig. 7. Derivative formed by pentafluorbenzoylation of morphine under aqueous conditions and trifluoroacetylation under anhydrous conditions.[142]

Dimethyl sulfate, methyl iodide, benzyl bromide, benzyl iodide, and other alkyl or aralkyl halides in an alkaline medium have all been employed. The use of diazomethane is usually avoided, because mixtures of products may be obtained.

Tetraalkylammonium hydroxides (TAAHs) are also used extensively to derivatize barbiturates, hydantoins, xanthines, and analogous compounds.[131,132] The drug is extracted from acidified solution, and the dried extract is dissolved in methanol to which the TAAH is added. Reaction occurs in the injection port of the gas chromatograph, and dialkylated products (Fig. 6) are formed. Barbiturates and related compounds may also be converted to dimethylketals (Fig. 6) for GC analysis by reaction with dimethylformamide dimethylacetal.[133]

4.16. Alkaloids

As well as a tertiary amine group, morphine contains phenolic and alcoholic groups, and GC analytical procedures are conducted on products in which both hydroxyl groups are derivatized. Acylation, silylation, and alkylation reactions have been utilized to prepare suitable derivatives. Perfluoroacylated products are popular and permit analysis of morphine down to 100 pg/30 mg of brain tissue or 500 pg/ml of plasma.[134] Reaction of morphine with a mixture of BSTFA and TMCS or other silylating reagents produces an analogous derivative in which both hydroxyl groups are trimethylsilylated [OH → OSi $(CH_3)_3$]. When analyzed by GC, 25 ng of the alkaloid as the trimethylsilylated derivative could be detected.[135]

Morphine has also been analyzed by GC–ECD as its pentafluorobenzyl, trifluoroacetyl derivative (Fig. 7).[136] An interesting aspect of this assay procedure was that the pentafluorobenzylation of the phenolic hydroxyl group was accomplished prior to extraction of the plasma sample. The product was no

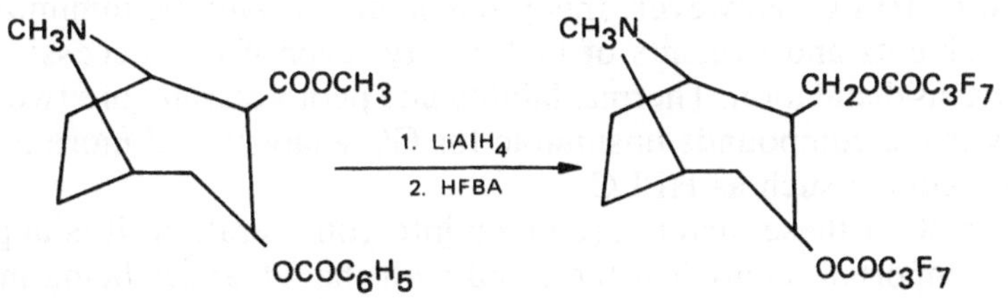

Fig. 8. Derivatization of cocaine for GC–ECD analysis.[143]

longer amphoteric and therefore was readily extracted, then further derivatized.

Codeine, naloxone, naltrexone, norcodeine, normorphine, and pentazocine can be derivatized for GC analysis using techniques similar to those just described for morphine.

Cocaine is a tertiary amine that contains two ester groups (Fig. 8). It can be gas chromatographed underivatized, but enhanced GC sensitivity is obtained by reducing the ester groups to alcohol functions, followed by perfluoroacylation to the compound shown in Fig. 8.[137]

5. ADVANTAGES AND DISADVANTAGES OF GAS CHROMATOGRAPHY

As can be seen from the above discussion, GC is applicable to the analysis of a wide variety of substances of neurochemical interest. The large number of derivatizing reagents and column types and advances in instrument technology (automatic samplers, increased programming capability) make this method of analysis a very versatile one. The use of detectors with increased sensitivity and selectivity has permitted routine analysis of endogenous components and drugs present in low concentrations in tissue and body fluids. Gas chromatography is relatively inexpensive, particularly when compared to mass spectrometric techniques.

On the negative side, practical sensitivity is generally lower than that obtained with radioenzymic procedures, and there is decreased certainty of specificity compared to mass spectrometric analysis. For the latter reason, we employ GC–MS to confirm structures of derivatives and to monitor the course of reaction sequences while assays are being developed for application to GC. When development of the assay is completed, GC is used for routine analysis. We have found this procedure to be very satisfactory, and the results we have obtained for concentrations of amines and related compounds in brain using GC–ECD are in good agreement with values reported using mass spectrometric methods, e.g., mean values obtained for whole-brain concentrations of PEA, p-TA, and 5-HT are 1.1,[33] 1.8,[34] and 581[35] ng/g, respectively.

Depending on the assay involved, interference with the analysis of final derivatives by impurities acquired during the preparative stages of the assay may be more of a problem with GC than with mass spectrometry, radioenzymic methods, or HPLC. However, this problem can usually be minimized by employing solvents and reagents of high purity, even if this necessitates procedures such as distillation. Thermal lability and poor volatility are two properties that may make compounds unsuitable for GC analysis and more amenable to other procedures such as HPLC.

When all of these factors are taken into consideration, it is apparent that GC is an important analytical tool, and with the advances being made in the field of GC technology, it should continue to play an important role in neurochemical research.

REFERENCES

1. Knapp, D. R., 1979, *Handbook of Analytical Derivatization Reactions*, John Wiley & Sons, New York.
2. Blau, K., and King, G. S., 1978, *Handbook of Derivatives for Chromatography*, (K. Blau and G. S. King, eds.), Heyden, London, pp. 104–151.
3. Baker, G. B., Coutts, R. T., and Martin, I. L., 1981, *Prog. Neurobiol.* **17**:1–24.
4. Edwards, D. J., 1978, *Handbook of Derivatives for Chromatography*, (K. Blau and G. S. King, eds.), Heyden, London, pp. 391–410.
5. Horning, M. G., Moss, A. M., Boucher, E. A., and Horning, E. C., 1968, *Anal. Lett.* **1**:311–321.
6. Cancalon, P., and Klingman, J. D., 1972, *J. Chromatogr. Sci.* **10**:253–256.
7. Arnold, E. L., and Ford, R., 1973, *Anal. Chem.* **45**:85–89.
8. Lhuguenot, J.-C., and Maume, B. F., 1974, *J. Chromatogr. Sci.* **12**:411–418.
9. Donike, M., Gola, R., and Jaenicke, L., 1977, *J. Chromatogr.* **134**:385–395.
10. Freed, C. R., Weinkam, R. J., Melmon, K. L., and Castagnoli, N., 1977, *Anal. Biochem.* **78**:319–332.
11. Änggård, E., and Sedvall, G., 1969, *Anal. Chem.* **41**:1250–1256.
12. Karoum, F., Cattabeni, F., Costa, E., Ruthven, C. R. J., and Sandler, M., 1972, *Anal. Biochem.* **47**:550–561.
13. Moffat, A. C., Horning, E. C., Matin, S. B., and Rowland, M., 1972, *J. Chromatogr.* **66**:255–260.
14. Gelpí, E., Paralta, E., and Segura, J., 1974, *J. Chromatogr. Sci.* **12**:701–709.
15. Ahuja, S., 1976, *J. Pharm. Sci.* **65**:163–182.
16. Applied Science Labs, 1978, Technical Bulletin No. 25, Applied Science Labs, State College, Pennsylvania.
17. Imai, K., Sugiura, M., and Tamura, Z., 1971, *Chem. Pharm. Bull.* **19**:409–411.
18. Martin, I. L., and Ansell, G. B., 1973, *Biochem. Pharmacol.* **22**:521–533.
19. Bigdeli, M. G. and Collins, M. A., 1975, *Biochem. Med.* **12**:55–65.
20. Kilts, C. D., Vrbanac, J. J., Rickert, D. E., and Rech, R. H., 1977, *J. Neurochem.* **28**:465–467.
21. Kawano, T., Niwa, M., Fujita, Y., Ozaki, M., and Mori, K., 1978, *Jpn. J. Pharmacol.* **28**:168–171.
22. Bertani, L. M., Dziedzic, S. W., Clarke, D. D., and Gitlow, S. E., 1970, *Clin. Chim. Acta* **30**:227–233.
23. Nelson, L. M., Bubb, F. A., Lax, P. M., Weg, M. W., and Sandler, M., 1979, *Clin. Chim. Acta* **92**:235–240.
24. Coutts, R. T., Baker, G. B., LeGatt, D. F., and Pasutto, F., 1980, *Proceedings World Conference on Clinical Pharmacology and Therapeutics*, Macmillan, London.
25. Baker, G. B., Martin, I. L., Coutts, R. T., and Benderly, A., 1980, *J. Pharmacol. Methods* **3**:173–179.
26. Edwards, D. J., and Blau, K., 1972a, *Anal. Biochem.* **45**:387–402.
27. Edwards, D. J., and Blau, K., 1973, *Biochem. J.* **132**:95–100.
28. Martin, W. R., Sloan, J. W., Buchwald, W. F., and Bridges, S. R., 1974, *Psychopharmacologia (Berl.)* **37**:189–198.
29. Sloan, J. W., Martin, W. R., Clements, T. H., Buchwald, W. F., and Bridges, S. R., 1975, *J. Neurochem.* **24**:523–532.
30. Reynolds, G. P., Sandler, M., Hardy, J., and Bradford, H., 1980, *J. Neurochem.* **34**:1123–1125.
31. Reynolds, G. P., Ceasar, P. M., Ruthven, C. R. J., and Sandler, M., 1978, *Clin. Chim. Acta* **84**:225–231.
32. Martin, I. L., and Baker, G. B., 1976, *J. Chromatogr.* **123**:45–50.
33. Martin, I. L., and Baker, G. B., 1977, *Biochem. Pharmacol.* **26**:1513–1516.
34. Baker, G. B., Coutts, R. T., and LeGatt, D. F., 1980, *Can. J. Neurol. Sci.* **7**:235.
35. Calverley, D. G., Baker, G. B., McKim, H. R., and Dewhurst, W. G., 1980, *Can. J. Neurol. Sci.* **7**:237.
36. LeGatt, D. F., Baker, G. B., and Coutts, R. T., 1981, *J. Chromatogr. Biomed. Appl.* **225**:301–308.

37. Coutts, R. T., Baker, G. B., LeGatt, D. F., McIntosh, G. J., Hopkinson, G., and Dewhurst, W. G., 1981, *Prog. Neuropsychopharmacol.* **5:**565–568.
38. Mita, H., Yasueda, H., and Shida, T., 1980, *J. Chromatogr.* **181:**153–159.
39. Mahy, N., and Gelpí, N., 1978, *Chromatographia* **11:**573–577.
40. Doshi, P. S., and Edwards, D. J., 1979, *J. Chromatogr.* **176:**359–366.
41. Navert, H., and Wollin, A., 1980, *Union Med. Can.* **109:**1507.
42. Brooks, J. B., Liddle, J. A., and Alley, C. C., 1975, *Anal. Chem.* **47:**1960–1965.
43. Burkinshaw, P. M., and Morgan, E. D., 1977, *J. Chromatogr.* **132:**548–551.
44. Vilceanu, R., and Schulz, P., 1973, *J. Chromatogr.* **82:**279–284.
45. Wilk, S., Gitlow, S. E., Clarke, D. D., and Paley, D. H., 1967, *Clin. Chim. Acta* **16:**403–408.
46. Wilk, S., Davis, K. L., and Thacker, S. B., 1970, *Anal. Biochem.* **39:**498–504.
47. Dekirmenjian, H., and Maas, J. W., 1970, *Anal. Biochem.* **35:**113–122.
48. Sharman, D. F., 1969, *Br. J. Pharmacol.* **36:**523–534.
49. Kahane, Z., Jindal, S. P., and Vestergaard, P., 1976, *Clin. Chim. Acta* **73:**203–206.
50. Tang, S. W., Helmeste, D. M., and Stancer, H. C., 1978, *Naunyn Schmiedeberg's Arch. Pharmacol.* **305:**207–211.
51. Warsh, J. J., Godse, D. D., Li, P., and Cheung, S., 1980, *Can. J. Neurol. Sci.* **7:**230.
52. Baker, G. B., Coutts, R. T., Hopkinson, G., Hay, J. L., and Liu, S.-F., 1980, *Proceedings 27th Canadian Conference on Pharmaceutical Research*, Association of Faculties of Pharmacy of Canada, Calgary, p. 17.
53. Cagnasso, M., and Biondi, P. A., 1976, *Anal. Biochem.* **71:**597–600.
54. Sjöquist, B., and Änggård, E., 1972, *Anal. Chem.* **44:**2297–2301.
55. Wiesel, F.-A., Fri, C.-G., and Sedvall. G., 1974, *J. Neural. Transm.* **35:**319–326.
56. Watson, E., Travis, B., and Wilk, S., 1974, *Life Sci.* **15:**2167–2178.
57. Dziedzic, S. W., Bertani, L. M., Dziedzic, L., and Gitlow, S. E., 1973, *J. Lab. Clin. Med.* **82:**829–835.
58. Davis, B. A., Durden, D. A., Pun-Li, P., and Boulton, A. A., 1977, *J. Chromatogr.* **142:**517–522.
59. Davis, B., 1977, *Anal. Chem.* **49:**832–834.
60. Wilk, S., Gitlow, S. E., Mendlowitz, M., Franklin, M. J., Carr, H. E., and Clarke, D. D., 1965, *Anal. Biochem.* **13:**544–551.
61. Dekirmenjian, H., and Maas, J. W., 1971, *Clin. Chim. Acta* **32:**310–312.
62. Sprinkle, T. J., Porter, A. H., Greer, M., and Williams, C. M., 1969, *Clin. Chim. Acta* **25:**409–411.
63. Addanki, S., Kinnenkamp, E. R., and Sotos, J. F., 1976, *Clin. Chem.* **22:**310–314.
64. Seiler, N., 1980, *Polyamines in Biomedical Research* (J. M. Gaugas ed.), John Wiley & Sons, New York, pp. 435–461.
65. Makita, M., Yamamoto, S., and Kono, M., 1975, *Clin. Chim. Acta* **61:**403–405.
66. Gehrke, C. W., Roach, D., Zumwalt, R. W., Stalling, D. L., and Wall, L. L., 1968, *Quantitative Gas–Liquid Chromatograph of Amino Acids in Proteins and Biological Substances*, Analytical Biochemistry Labs, Columbia, Missouri.
67. Husek, P., and Macek, K., 1975, *J. Chromatogr.* **113:**139–230.
68. Zumwalt, R. W., Kuo, K., and Gehrke, C. W., 1971, *J. Chromatogr.* **57:**193–208.
69. Middleditch, B. S., 1976, *J. Chromatogr.* **126:**581–589.
70. Hanin, I., and Jenden, D. J., 1969, *Biochem. Pharmac.* **18:**837–845.
71. Jenden, D. J., and Hanin, I., 1974, *Choline and Acetylcholine: Handbook of Chemical Assay Methods* (I. Hanin, ed.), Raven Press, New York, pp. 135–150.
72. Szilagyi, P. I. A., Green, J. P., Brown, O. M., and Margolis, S., 1972, *J. Neurochem.* **19:**2555–2566.
73. Stavinoha, W. B., and Weintrub, S. T., 1974, *Science* **183:**964–965.
74. Kilbinger, H., 1973, *J. Neurochem.* **21:**421–429.
75. Gleispach, H., 1974, *J. Chromatogr.* **91:**407–412.
76. Sakauchi, N., and Horning, E. C., 1971, *Anal. Lett.* **4:**41–52.
77. Thenot, J.-P., and Horning, E. C., 1972, *Anal. Lett.* **5:**21–33.
78. Samuelsson, B., 1964, *Biochim. Biophys. Acta* **84:**707–713.
79. Eglinton, G., Raphael, R. A., and Smith, G. N., 1963, *Nature* **200:**960–995.

80. Taylor, P. L., and Kelly, R. W., 1974, *Nature* **250**:665–667.
81. Morozowich, W., Oesterling, T. O., and Brown, L. W., 1979, *GLC and HPLC Determination of Therapeutic Agents, Part 3* (K. Tsuji, ed.), Marcel Dekker, New York, pp. 975–1014.
82. Levitt, M. J., 1973, *Anal. Chem.* **45**:618–620.
83. Jouvenaz, G. H., Nugteren, D. H., and Van Dorp, D. A., 1973, *Prostaglandins* **3**:175–187.
84. Fitzpatrick, F. A., Wynalda, M. A., and Kaiser, D. G., 1977, *Anal. Chem.* **49**:1032–1035.
85. Wickramasinghe, J. A. F., Morozowich, W., Hamlin, W. E., and Shaw, S. R., 1973, *J. Pharm. Sci.* **62**:1428–1431.
86. Wickramasinghe, A. J. F., and Shaw, R. S., 1974, *Biochem. J.* **141**:179–187.
87. Pace-Asciak, C., and Wolfe, L. S., 1971, *J. Chromatogr.* **56**:129–133.
88. Palmer, J. K., 1979, *GLC and HPLC Determination of Therapeutic Agents* (K. Tsuji, ed.), Marcel Dekker, New York, pp. 1317–1340.
89. Clamp, J. R., Bhatti, T., and Chambers, R. E., 1971, *Methods of Biochemical Analysis,* Volume 19 (D. Glick, ed.), Interscience Publishers, New York, pp. 229–344.
90. Dutton, G. G. S., 1973, *Advances in Carbohydrate Chemistry and Biochemistry,* Volume 28 (R. S. Tipson, and D. Horton, eds.), Academic Press, New York, pp. 11–160.
91. Dutton, G. G. S., 1974, *Advances in Carbohydrate Chemistry and Biochemistry,* Volume 30 (R. S. Tipson, and D. Horton, eds.), Academic Press, New York, pp. 9–110.
92. Schram, K. H., and McCloskey, J. A., 1979, *GLC and HPLC Determination of Therapeutic Agents, Part 3.* (K. Tsuji, ed.), Marcel Dekker, New York, pp. 1149–1190.
93. Poole, C. F., 1978, *Handbook of Derivatives for Chromatography* (K. Blau and G. King, eds.), Heyden, London, pp. 152–200.
94. Hartvig, P., Handl, W., Vessman, J., and Svahn, C. M., 1976, *Anal. Chem.* **48**:390–396.
95. Hartvig, P., and Näslund, B., 1977, *J. Chromatogr.* **133**:367–371.
96. Hamilton, H. E., Wallace, J. E., and Blum, K., 1975, *Anal. Chem.* **47**:1139–1143.
97. Wallace, J. E., Hamilton, H. E., Goggin, L. K., and Blum, K., 1975, *Anal. Chem.* **47**:1516–1519.
98. Cooper, T. B., Allen, D., and Simpson, G. M., 1976, *Psychopharmacol. Commun.* **2**:105–116.
99. Vasiliades, J., and Bush, K. C., 1976, *Anal. Chem.* **48**:1708–1711.
100. Jorgensen, A., 1975, *Acta Pharmacol. Toxicol. (Kbh.)* **36**:79–90.
101. Geiger, U. P., Rajagopalan, T. G., and Reiss, W., 1975, *J. Chromatogr.* **114**:167–173.
102. Gelbicova-Ruzicková, J., Novak, J., and Chundela, B., 1971, *Biochem. Med.* **5**:537–547.
103. Cooper, T. B., Robinson, D. S., and Nies, A., 1978, *Commun. Psychopharmacol.* **2**:505–512.
104. Bailey, E., and Barron, E. J., 1980, *J. Chromatogr. Biomed. Appl.* **183**:25–31.
105. Calverley, D. G., Baker, G. B., Coutts, R. T., and Dewhurst, W. G., 1981, *Biochem. Pharmacol.* **30**:861–867.
106. De Silva, J. A. F., 1978, *GLC and HPLC Determination of Therapeutic Agents,* Part 3. (K. Tsuji, ed.), Marcel Dekker, New York, pp. 581–636.
107. Fairbrother, J. E., 1979, *Pharm. J.* **218**:271–275.
108. Curry, S. H., and Mould, G. P., 1969, *J. Pharm. Pharmacol.* **21**:674–677.
109. Curry, S. H., 1978, *Blood, Drugs and Other Analytical Challenges* (E. Reid, ed.), John Wiley & Sons, New York, pp. 109–117.
110. Hammar, C. G., and Holmstedt, B., 1968, *Experientia* **24**:98–99.
111. Dinovo, E. C., Gottschalk, L. A., Nandi, B. R., and Geddes, P. G., 1976, *J. Pharm. Sci.* **65**:667–669.
112. Riedmann, M., 1974, *J. Chromatogr.* **92**:55–59.
113. Linnoila, M., and Dorrity, F., 1978, *Acta Pharmacol. Toxicol. (Kbh.)* **42**:264–270.
114. Rivera-Calimlim, L., and Sircusa, A., 1977, *Commun. Psychopharmacol.* **1**:233–236.
115. Franklin, M., Wiles, D., and Harvey, D., 1978, *Clin. Chem.* **24**:41–44.
116. Dekirmenjian, H., Javaid, J. I., Duslak, B., and Davis, J. M., 1978, *J. Chromatogr.* **160**:291–296.
117. Marcucci, F., Airoldi, L., Mussini, E., and Garattini, S., 1971, *J. Chromatogr.* **59**:174–177.
118. Zingales, I. A., 1971, *J. Chromatogr.* **54**:15–24.
119. Marcucci, F., Mussini, E., Airoldi, L., Fanelli, R., Frigerio, A., de Nadai, F., Bizzi, A., Rizzo, M., Morselli, P. L., and Garattini, S., 1971, *Clin. Chim. Acta* **34**:321–326.
120. Forsman, F. A., Mårtensson, E., Nyberg, G., and Öhman, R., 1974, *Naunyn Schmiedeberg's Arch. Pharmacol.* **286**:113–124.

121. Javaid, J. I., Dekirmenjian, H., Dysken, M., and Davis, J. M., 1980, *Adv. Biochem. Psychopharmacol.* **24:**585–589.

122. Ehrsson, M., and Tilly, A., 1973, *Anal. Lett.* **6:**197–210.

123. De Silva, J. A. F., and Bekersky, I., 1974, *J. Chromatogr.* **99:**447–460.

124. Belvedere, G., Tognoni, G., Frigerio, A., and Morselli, P. L., 1972, *Anal. Lett.* **5:**531–541.

125. O'Brien, J. E., Zazulak, W., Abbey, V., and Hinsvark, O., 1972, *J. Chromatogr. Sci.* **10:**336–341.

126. Cristofoli, W. A., Baker, G. B., and Coutts, R. T., 1981, *Proc. Can. Fed. Biol. Soc.* **24:**242.

127. Moffat, A. C., and Horning, E. C., 1970, *Anal. Lett.* **3:**205–216.

128. Brandenberger, R. H., and Brandenberger, H., 1978, *Handbook of Derivatives for Chromatography* (K. Blau, and G. King, eds.), Heyden, London, pp. 234–261.

129. Ray, R. S., Noonan, J. S., Murdick, P. W., and Tharp, V. L., 1972, *Am. J. Vet. Res.* **33:**27–31.

130. Gal, J., 1978, *Biomed. Mass Spectrom.* **5:**32–37.

131. MacGee, J., 1971, *Clin. Chem.* **17:**587–591.

132. Giovanniello, T. J., and Pecci, J., 1977, *Clin. Chem.* **23:**2154–2155.

133. Venturella, V. S., Gualario, V. M., and Lang, R. E., 1973, *J. Pharm. Sci.* **62:**662–668.

134. Dahlström, B., and Paalzow, L., 1975, *J. Pharm. Pharmacol.* **27:**172–176.

135. Sine, H. E., Kubasik, N. P., and Woytash, J., 1973, *Clin. Chem.* **19:**340–341.

136. Cole, W. J., Parkhouse, J., and Yousef, Y. Y., 1977, *J. Chromatogr.* **136:**409–416.

137. Blake, J. W., Ray, R. S., Noonan, J. S., and Murdick, P. W., 1974, *Anal. Chem.* **46:**288–289.

High-Performance Liquid Chromatography

Stanley Stein

1. INTRODUCTION

High-performance liquid chromatography (HPLC) has become a standard technique in the laboratory. The intent of this chapter is to present an overview of the technical aspects of HPLC rather than an in-depth discussion. For more detailed information on HPLC, the reader is referred to the recently revised text by Synder and Kirkland.[1] There is an extensive literature in the scientific journals covering the HPLC of virtually every class of compound, and it should be consulted by the reader for his own applications. Furthermore, manufacturers of HPLC equipment can be helpful in obtaining information about specific applications.

In order to illustrate the scope and utility of HPLC, a variety of applications are presented. Some of them are based on studies from the author's laboratory and deal with the isolation and analysis of peptides and proteins of neurochemical interest. A unique aspect of these studies is the use of fluorescence detection for continuously monitoring amino acids, peptides, and proteins in column effluents at nanogram to microgram levels. Other applications presented illustrate the use of ultraviolet absorption, electrochemical detection, and radioactivity for monitoring column effluents.

2. PRINCIPLES AND PRACTICES

2.1. HPLC Supports

The technique of HPLC is based on the use of small (10 μm or less), uniformly sized, rigid, porous particles which are tightly packed into steel columns at high pressure. Interaction with the solid support causes the mol-

Stanley Stein • Roche Institute of Molecular Biology, Nutley, New Jersey 07110.

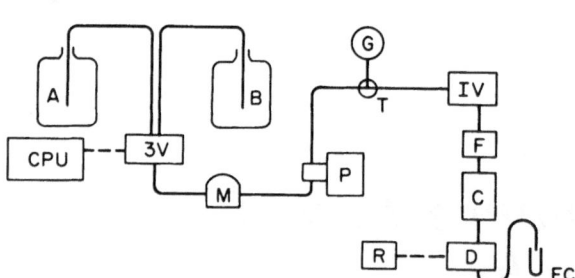

Fig. 1. A typical HPLC system. The gradient, which is controlled by a central processing unit (CPU), is formed by switching the three-way valve (3V) intake between reservoirs A and B. A single pump (P) is used. The eluents are mixed (M) before the pump. The pressure gauge (G), connected via a tee-fitting (T), has the additional function of damping the pulses when a pulsatile pump is used. Either an in-line filter or a precolumn (F) is placed between the sample injection valve (IV) and the column (C). The column effluent is monitored by UV absorption, and the output of the detector (D) is plotted on a chart recorder (R). For preparative applications, a fraction collector (FC) is used.

ecules of each substance to migrate at a particular rate with the mobile liquid phase. Each substance stays in a narrow band and elutes from the column as a sharp peak (see Section 2.2). Therefore, separations can be achieved with substances that differ only slightly with respect to their interaction with the support. The porous particles have an extremely large surface area and, therefore, can handle large sample loads (several milligrams per gram of support).

The silica matrix has been utilized directly or derivatized with a variety of functional groups. Underivatized silica is used for adsorption chromatography. This mode is generally used for uncharged substances and is carried out with nonaqueous eluents. Adsorption chromatography is unique in its usefulness for separating isomers (e.g., *cis/trans*) of a compound. This property is based on the fact that the solute is binding to a surface, and the interaction of its functional groups depends on their spatial orientation.

In the reverse-phase support, a hydrophobic moiety is covalently coupled to the matrix. The name for this technique is derived from the fact that nonpolar solvents are stronger eluents than are polar solvents (the reverse situation is true in adsorption chromatography). At the moment, reverse-phase chromatography is the most popular mode of separation.

A weakly hydrophobic coating will mask the strong adsorptive properties of the matrix. This type of support (called diol or glycophase) is used for permeation chromatography (separation by size) of polar molecules such as proteins. Aqueous buffers are typically used for elution. The same support can also be used for normal-phase chromatography of polar molecules. In the normal-phase mode, the strength of the eluent increases with its polarity. Both cation and anion exchange supports based on the silica matrix are available. A combination of the hydrophilic coating of the silica and an ion-exchange group is especially useful for the chromatography of proteins.[2]

2.2. Efficiency and Selectivity

HPLC provides a dramatic improvement in resolving power and speed of analysis over conventional liquid chromatography. The resolution or separation of components in a sample depends on both the efficiency and the selectivity of the column. Efficiency (N) is a measure of the sharpness of the peaks and is defined as

$$N = 16(t_r)/W_t$$

where t_r is the elution time corresponding to the apex of the peak, and W_t is the width of the peak at the base in terms of time. From the equation, it is seen that the higher the plate number is, the sharper the peaks will be.

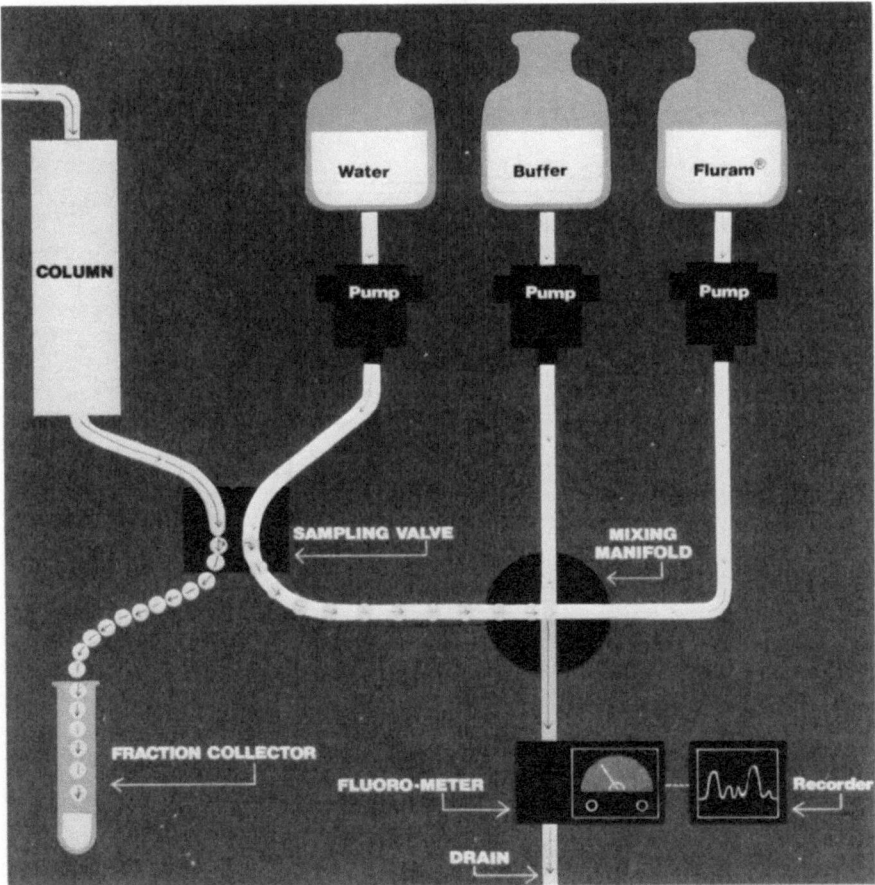

Fig. 2. Diagram of the automatic fluorescent column monitoring system. Aliquots of column effluent are taken at regular intervals for detection. The aliquots are picked up in a stream of water and then mixed sequentially with borate buffer (pH 9.3) and fluorescamine in acetone (0.2 mg/ml). After a delay of 30 sec, the mixture passes through a filter fluorometer and then to waste. Fluorescence is plotted on a chart recorder.

Many factors, including the quality of the support, the method used to pack the column, the chromatography conditions, the dimensions of the column, and the properties of the sample, influence the efficiency. A standard 25 × 0.46 cm column of 10 μm support will typically give an efficiency of about 5000 theoretical plates under test conditions. The same column packed with a 5-μm or 3-μm support can give severalfold more theoretical plates.

HPLC columns have such high plate numbers principally for two reasons. First, the mobile phase moves down the column in a uniform fashion. Therefore, the molecules of each component do not tend to move ahead or lag behind at any point across the column bed. To minimize "wall effects," HPLC columns are highly polished inside and are of large diameter (usually 4.6 mm).

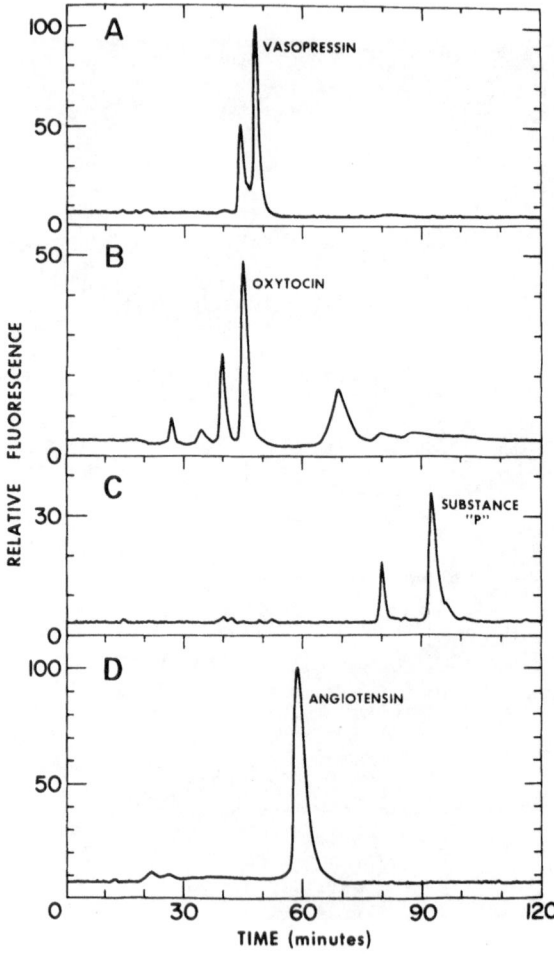

Fig. 3. Chromatography of synthetic peptides. Four commercial preparations of purportedly pure peptides were chromatographed on an ion-exchange HPLC column. Individual peaks were collected, and rechromatography gave single peaks for each (from Radhakrishnan et al.[13]).

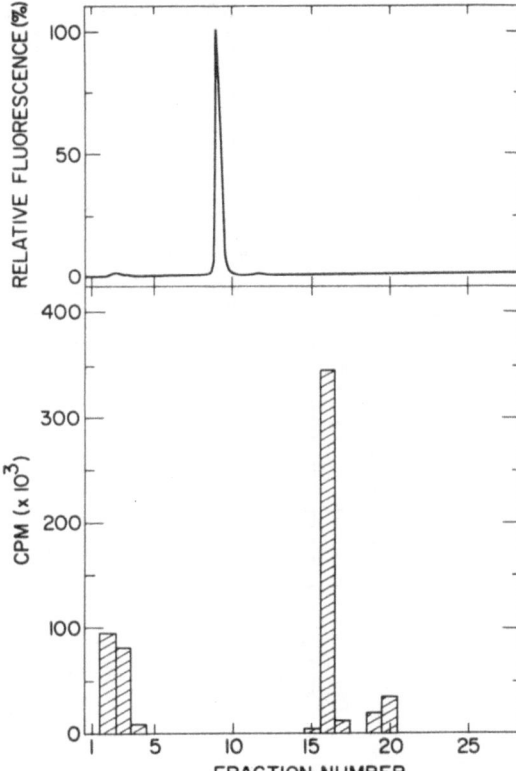

Fig. 4. Chromatography of a radiolabeled peptide. A synthetic analogue of Leu-enkephalin was iodinated and cleared of excess iodine by passage through Sephadex G-25. Chromatography on a reverse-phase HPLC column indicated that the major component (fractions 15–17) was the monoiodinated peptide. The radiolabeled material that eluted immediately from the column is free iodine, whereas the material in fractions 19 and 20 is most likely the disubstituted peptide. Chromatography of the native peptide is shown in the top panel.

Second, the solute molecules can rapidly reach a state of equilibrium between the stationary and mobile phases and will therefore distribute between the phases in a uniform manner. This rapid equilibration allows the use of a high mobile-phase velocity.

The ability to resolve different substances also depends on the selectivity of the column packing. Selectivity is based on the differential interaction of the functional groups of the solutes with the HPLC support. Indeed, the type of column chosen may be a more important consideration than the efficiency of the column. Even the same type of column from different manufacturers can give different separations. Reverse-phase columns with different ligands (e.g., cyanopropyl vs. octyldecyl) can display differences in selectivity.

2.3. The Sample

For rapid analyses, the sample should be injected into the column in a small volume (e.g., 20 μl). For preparative work there is no limit to the volume applied to the column as long as the substances of interest are retained during sample application. Therefore, in preparative work, the column also serves as a means of concentrating the sample. The capacity of the column, which is at least several milligrams for standard columns, should not be exceeded.

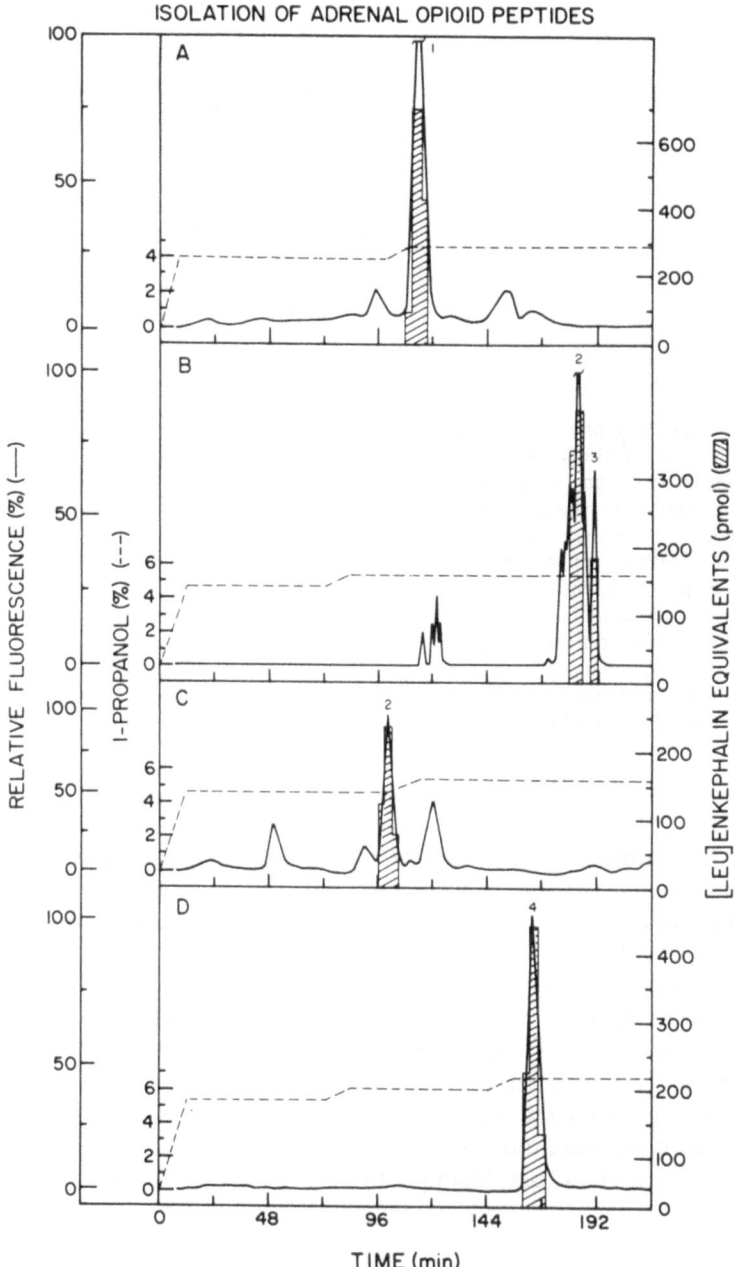

ISOLATION OF ADRENAL OPIOID PEPTIDES

Fig. 5. Purification of endogenous enkaphalin congeners. Chromaffin granules were prepared from bovine adrenal glands and extracted with dilute acid containing protease inhibitors. Proteins were precipitated by treatment with trichloroacetic acid, and the supernatant was chromatographed on a Lichrosorb® RP-18 column (not shown). Three separate cuts were taken from that column to purify four peptides, as shown in this figure. (A) After rechromatography on a Lichrosorb® RP-18 column (not shown), peptide 1 was purified on a diphenyl (10 μm) reverse-phase column using

The sample should not be deleterious to the column. It should be free of particulate matter and be at a pH within the limits of the stability of the support. To protect the column, an in-line particle filter (0.2 μm) or a precolumn may be used (see Fig. 1). The in-line filter will remove particulate matter from the sample or the eluents. The precolumn will remove substances in the sample that may be irreversibly retained. Replacement of the precolumn is considerably less expensive than replacement of the HPLC column. The support of the precolumn has the same functional group as the HPLC column and may be a surface-coated nonporous matrix (pellicular) or a larger bead size porous matrix (e.g., 50 μm). The efficiency of the HPLC column is not noticeably affected by a precolumn.

2.4. Chromatography Conditions

The eluents chosen should be compatible with the sample, the column packing, and the detector. For example, pyridine buffers are unacceptable for use with UV detectors but are quite useful with the fluorometric detection described later. For reverse-phase chromatography, a change in pH or the use of an ion-pairing reagent or other organic modifier can markedly change the separation. In preparative work, it is convenient to use completely volatile eluents. A UV-transparent, volatile elution system based on trifluoroacetic and acetonitrile has been devised for peptides.[3]

Columns are generally run at room temperature. Elevated temperature may improve efficiency slightly and also affect the selectivity. Low temperature may be employed for certain sensitive substances, such as enzymes. Close control of temperature can be critical for the reproducibility of some analytical applications.

Eluent flow rate is an important factor for macromolecules. Low-molecular-weight substances show a small increase in peak width with increasing flow rate, and their speed of separation is often limited only by the pressure capabilities of the pump. However, there is a marked loss in apparent column efficiency with increasing flow rate for macromolecules such as proteins because of their slow rates of diffusion. Protein chromatography is, therefore, carried out at a flow rate of 30 ml/hr or less when using a 25 × 0.46 cm column.

Preparative separations often require the sequential use of different columns or the same column run under different elution conditions. A single

a step gradient of 4.0–4.6% *n*-propanol. (B) Peptides 2 and 3 were rechromatographed on an Ultraspheres® ODS (5 μm) column with a 4.6–5.3% gradient. Peptide 3 was thereby purified. (C) purification of peptide 2 was on a diphenyl column with a 4.6–5.3% gradient. (D) After rechromatography on a Lichrosorb® RP-18 column (not shown), peptide 4 was purified on the diphenyl column with a 5.3–6.0% gradient. All columns were eluted at 10 ml/hr with 0.5 M formic acid/0.4 M pyridine, pH 4.0, and a step gradient of propanol, as indicated. Ten percent of the column effluent was diverted to the fluorescamine detection system, and aliquots of the collected fractions were tested or opiate receptor binding. The peptides were identified as (1) oxidized Metenkephalin, (2) Metenkephalin-Arg⁶Arg⁷, (3) Met-enkephalin-Lys⁶, and (4) Met-enkephalin-Arg⁶ (from Stern *et al.*[20]).

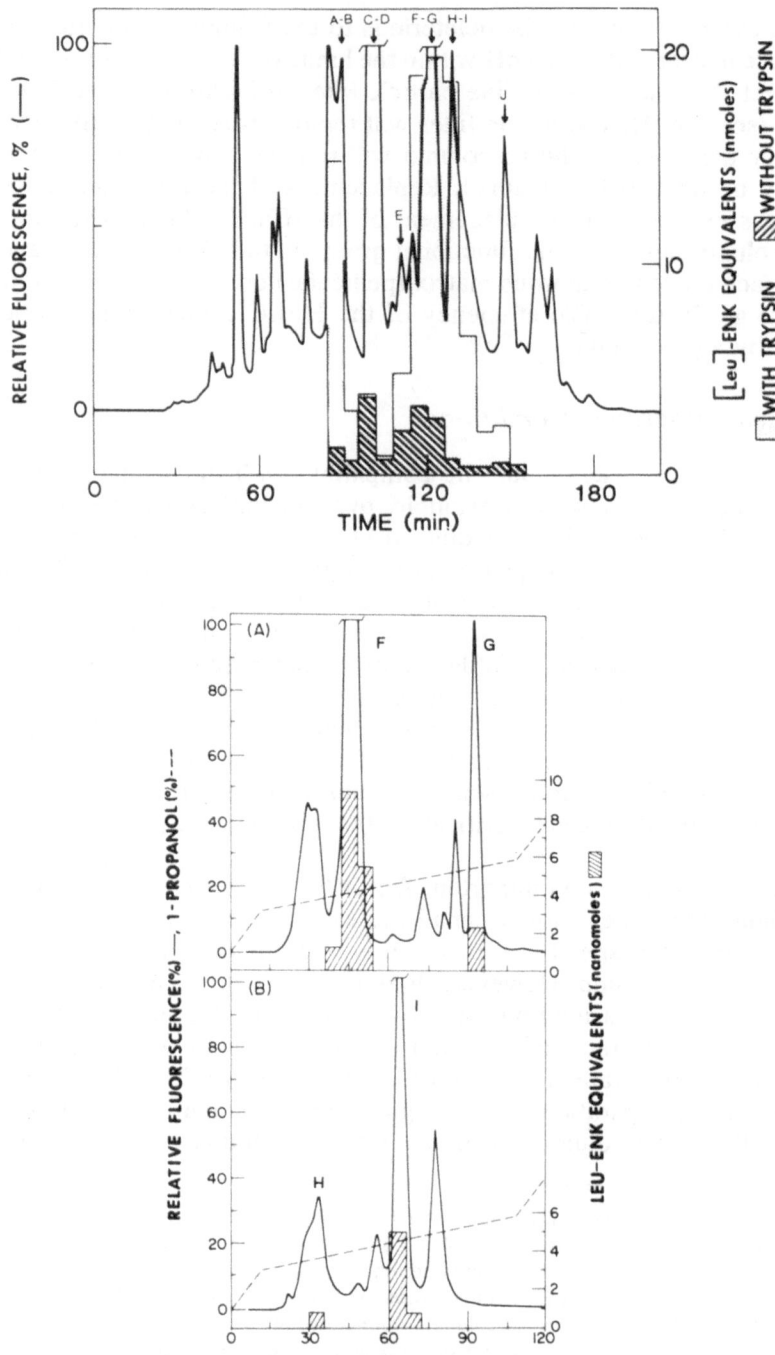

Fig. 6.   Purification of endogenous enkephalin-containing polypeptides. Bovine adrenal chromaffin granules were extracted with dilute acid containing protease inhibitors. A cut representing 2000–5000 daltons was taken from a Sephadex G-75 chromatograph, pumped onto a Lichrosorb®

HPLC run is desirable for analytical applications. In the latter case, extensive sample processing may be required before proceeding to the HPLC column.

3. EQUIPMENT

3.1. Pumps

High-pressure pumps are available at prices ranging from several hundred to several thousand dollars. The higher-priced pumps are relatively pulse-free and can usually achieve higher pressures. With the higher-priced pumps, the delivery rate is proportional to the speed of a stepping motor and is controllable by a microprocessor. The flow rate of the inexpensive pumps is proportional to the distance of the piston stroke, and these pumps deliver the eluent in a pulsatile manner.

The degree of pulsation tolerable depends on the type of detector as well as on the sensitivity range used. Generally, fluorometers and UV detectors are relatively insensitive to flow rate fluctuation. In contrast, electrochemical detectors require extremely pulse-free flow in order to produce a stable base line. It should be noted that the column and pressure gauge (see Fig. 1) dampen the pulses (in an analogous fashion to an electrical RC circuit). In the author's experience, pulsatile flow has not been found to be deleterious to the HPLC column or to cause irreproducible elution profiles.

3.2. Gradient Systems

Isocratic elution is sufficient for many applications. However, the capability for gradient elution is highly recommended for applications in which components in the sample have large differences in their affinity for the support. Gradient elution is also useful for screening uncharacterized samples. A typical chromatography system is shown in Fig. 1. In this configuration, the gradient is formed by opening and closing valves to different reservoirs; a single pump is required. The alternative approach for forming a gradient employs two pumps, each of which delivers a different eluent. In the former case, the proportions of the eluents are controlled by the duration of opening of each valve, whereas in the latter case, the flow rates of the two pumps are varied. Both types of gradient systems have been used in the author's laboratory, and comparable performance has been observed.

RP-18 column, and eluted at pH 4.0 with a gradient of 0 to 20% *n*-propanol (top panel). Two separate cuts were taken for rechromatography. Peptide F (middle panel) and peptide I (bottom panel) were each purified on a Spherisorb® CN column (5 μm). Columns were eluted and monitored in a similar fashion to that described in Fig. 5. Peptide F was identified as being composed of 34 amino acids and having a Met-enkaphalin sequence at both the N-terminal and C-terminal positions. Peptide I was identified as being composed of 39 amino acids and having an internal Met-enkephalin sequence and C-terminal Leu-enkephalin sequence (from Kimura *et al.*[21]).

3.3. Detectors

The most common detector is the UV monitor. Many types of biochemicals absorb ultraviolet light (e.g., proteins at 280 nm and nucleic acids at 260 nm). Most substances absorb strongly at short wavelengths (*ca.* 210 nm). For example, whereas only those peptides with aromatic residues are detectable at 280 nm, all peptides can be detected at 210 nm at the microgram level.[4]

The electrochemical detector with a thin-layer flow cell can also be used to monitor HPLC columns. In this detector, a potential difference is established between the working electrode and the mobile phase passing through the electrochemical flow cell. The potential is constantly measured with a reference electrode and is maintained at a constant value by feedback control of an auxiliary electrode. In the oxidation mode, the working electrode has a positive charge. Any substance capable of being oxidized will transfer electrons to the working electrode, and this flow of current is measured. The current is proportional to the concentration of the particular substance. Selectivity is achieved by adjusting the potential difference to a value just above the oxidation potential of the component of interest. Any components with higher oxidation potentials will not be oxidized and, therefore, will not be detected. Likewise, the electrochemical detector may be operated in the reduction mode.

Compounds of interest to the neurochemist that may be detected by this method include catecholamines and their metabolites as well as tryptophan and other indolamines. The technique is applicable at the nanogram level. The reader is referred to a recent minireview of this methodology.[5] A typical application is presented later in this chapter (Section 4.6).

It is also possible to add reagents to the column effluent and then to monitor the products of a particular reaction. This technique has been applied to the detection of groups of isoenzymes separated by HPLC; the reaction product is monitored by absorption at the appropriate wavelength.[6] In a similar fashion, amino acid analyzers employ postcolumn reaction with the fluorometric reagents *o*-phthaldialdehyde[7] and fluorescamine.[8] The application of precolumn fluorescence derivatization to compounds of interest to the neurochemist is presented in a recent minireview.[9]

Fluorescence detection is used almost exclusively in the author's laboratory for detecting amino acids, peptides, and proteins. Amino acid analysis is based on postcolumn reaction with fluorescamine[10] or precolumn reaction with *o*-phthaldialdehyde.[11] For the preparative isolation of peptides and proteins, the column effluent must be stream-split.[12] By the use of a stream-sampling valve, a portion of the column effluent is automatically and accurately transferred into the detection system, and the remainder of the column effluent goes to a fraction collector (Fig. 2). The transferred column effluent is picked up in a stream of water and mixed with alkaline borate buffer and fluorescamine solution. The reaction mixture is monitored by a flow-through filter fluorometer and then is discarded. This approach offers both high sensitivity (picomole level) and specificity of detection.

Radioisotopically labeled substances can also be monitored in column effluents in a continuous fashion. For β emitters, the scintillation fluid is added to the column effluent with a separate reagent pump, and the mixture passes

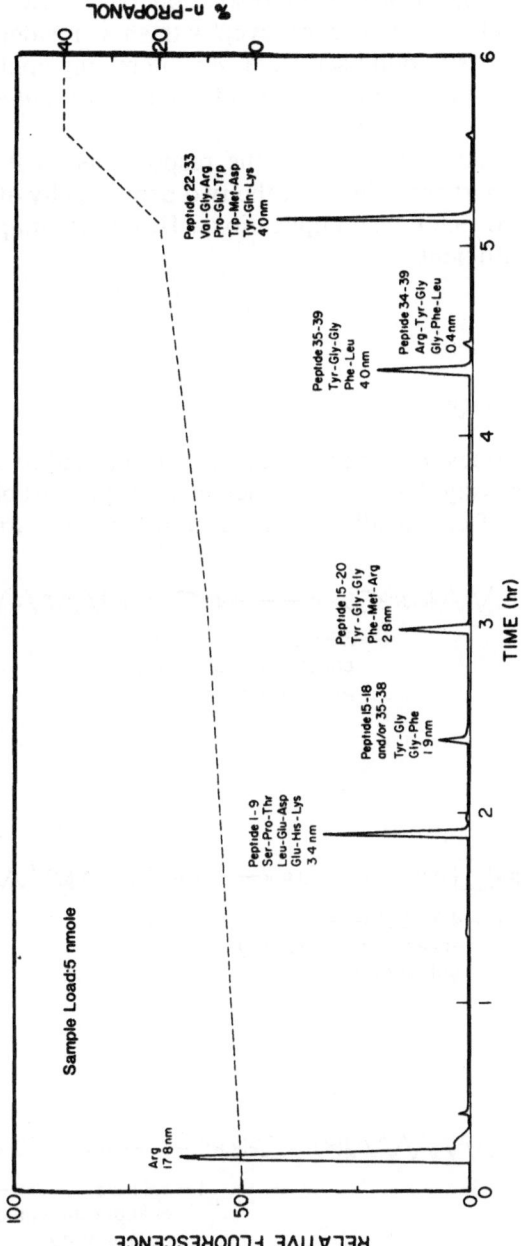

Fig. 7. Separation of tryptic peptides. Twenty micrograms of peptide I (see Fig. 6) were digested with trypsin (0.5 μg) and applied to a Lichrosorb® RP-18 column. Elution was at 20 ml/hr with 0.5 M pyridine/0.4 M formic acid, pH 4.0, using a linear gradient of 0–20% *n*-propanol. Five percent of the column effluent was diverted to the fluorescamine detection system. The collected peptides were characterized by amino acid analysis and microsequence analysis by automated Edman degradation.

through a flow cell positioned between the photomultiplier tubes. To increase the sensitivity of detection, the residence time in the flow cell would be increased. This would result in a loss of resolution, and hence, the optimal compromise should be chosen. However, even with short residence times, the radiochemical detector is still quite sensitive. An interesting application of the simultaneous use of UV and radiochemical detection is presented later in this chapter (Section 4.7).

Whatever type of detection is used, the response is most conveniently plotted on a dual-pen chart recorder with the two pens usually at a 10:1 ratio. An integrator is used for analytical applications. However, simply measuring peak height is often sufficient.

4. APPLICATIONS

4.1. Evaluation of Purity

Homogeneous peptides (or other substances) are needed to serve as standards for radioimmunoassay, bioassay, or chemical assay. Most of the peptides used as standards are of chemically synthetic origin and at various levels of

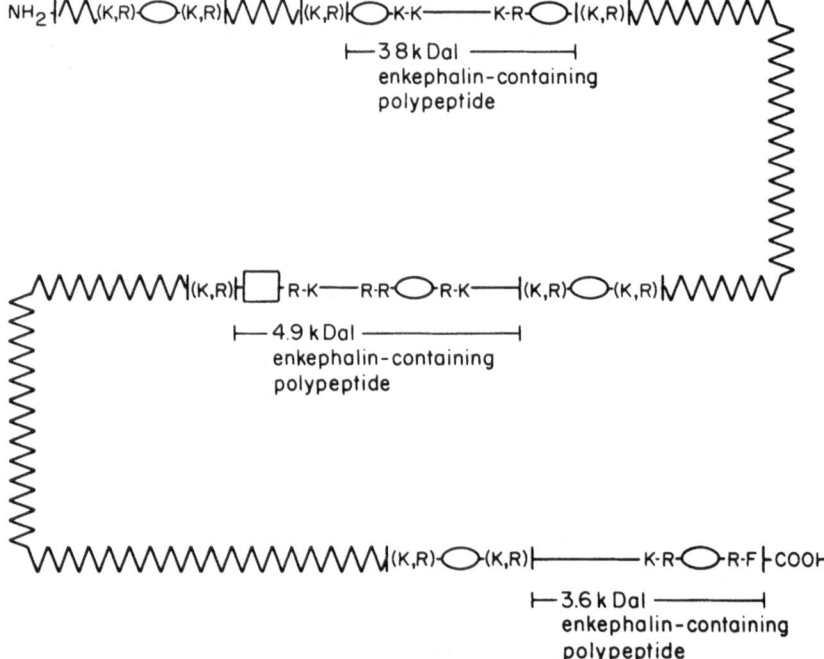

Fig. 8. Proposed partial structure of proenkephalin. Endogenous peptides representing fragments of proenkephalin were isolated from bovine adrenal medulla and sequenced. Proenkephalin contains several copies of the MET-enkephalin sequence (ovals) and at least one copy of the Leu-enkephalin sequence (rectangle) within its primary structure. A general feature is that each enkephalin sequence is both preceded and followed by pairs of basic amino acids (K is lysine, and R is arginine).

purity. Chromatography of several commercial preparations[13] is shown in Fig. 3. These determinations (Fig. 3) were made several years ago, and the quality of commercial preparations is usually, but not always, satisfactory now.

Another illustration of this application is in the preparation of radioiodinated peptides. An enkephalin analogue (Tyr-D-Ala-Gly-Phe-D-Leu-Lys) was labeled at the tyrosine residue by the chloramine T method and purified on Sephadex G-25. The HPLC run indicated that the monoiodinated product is the major species, with a small amount of diiodinated product and some free iodine being present (Fig. 4).

4.2. Purification of Peptides and Proteins

In the examples from the previous section, it is clear that chemically synthesized peptides and their derivatives can be purified by HPLC. This section, however, deals with the purification of peptides and proteins from tissue extracts. This is illustrated by the studies on the precursor, proenkephalin, and the intermediates in the biosynthetic pathway of Met-enkephalin and Leu-enkephalin.

The first endogenous peptides possessing opioid activity were isolated from brain extracts.[14] They were named Met-enkephalin (Tyr-Gly-Gly-Phe-Met) and Leu-enkephalin (Tyr-Gly-Gly-Phe-Leu). It was originally postulated that Met-enkephalin is derived biosynthetically from β-endorphin which had been isolated from pituitary extracts.[15] β-Endorphin is derived through the intermediate β-lipotropin from the precursor, proopiocortin, which is also the precursor of corticotropin.[16] The precursor for Leu-enkephalin had not been identified, although two Leu-enkephalin-containing peptides had been characterized.[17,18] Furthermore, the actual production of Met-enkephalin from β-endorphin had not been established.

Our studies, presented in this and the following section, with bovine adrenal medulla clarified the biosynthetic pathway of the enkephalins. The purification procedures involved preparation of chromaffin granules, extraction of the granules, and column chromatography. Enkephalins and enkephalin-containing peptides and proteins were detected by a radioreceptor assay or by radioimmunoassay. Details of the work may be found in the publications cited below.

The initial chromatographic separation of the chromaffin granule extract on Sephadex G-75 revealed a series of enkephalin-containing peptides of different molecular weights.[19] Five arbitrary molecular weight cuts were made. From the low-molecular-weight region (<1000 daltons), several opioid peptides were purified by HPLC[20] (some are shown in Fig. 5). They were identified by sequence analysis as Met-enkephalin, Leu-enkephalin, Met-enkephalin-Lys[6], Met-enkephalin-Arg[6], Met-enkephalin-Arg[6]-Arg[7], Met-enkephalin-Arg[6]-Phe[7], and Leu-enkephalin-Arg[6]. Other peptides were detected but were not purified.

From the fractions of the Sephadex column corresponding to the molecular weight range of 2000 to 5000, several distinct enkephalin-containing peptides were discerned by HPLC.[21] The purification of two of these peptides is shown

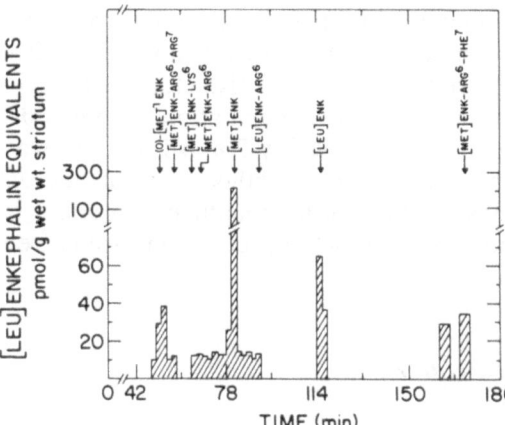

Fig. 9. Determination of enkaphalin analogues in brain. Bovine striatum was homogenized in dilute acid containing protease inhibitors. The extract was treated with trichloroacetic acid to precipitate proteins. The supernatant was resolved on a reverse-phase HPLC column that previously had been calibrated with the enkephalin congeners isolated from the adrenal gland. Collected fractions were assayed for opiate receptor binding. The pattern and relative proportions of the opioid peptides are similar to those observed in the chromaffin granule extract (from Stern *et al.*[20]).

in Fig. 6. Among the other enkephalin-containing peptides purified by HPLC were those of 8,000, 11,000, 14,000, and 22,000 daltons.[22] The precursor, proenkephalin (*ca.* 50,000 daltons) has been identified and partially purified by HPLC.

4.3. Elucidation of Peptide Structure

Fragmentation of large peptides or proteins by enzymatic or chemical treatment is a standard approach used in sequence analysis. Separation by HPLC of the fragments produced by trypsin digestion of one of the enkephalin-containing peptides is shown in Fig. 7. Fragments produced by cyanogen bromide cleavage were also purified by HPLC (not shown).

The original enkephalin-containing peptides as well as their cleavage products were sequenced by special high-sensitivity techniques.[23] One sequencing method used automatic Edman degradation and identification of the resultant phenylthiohydantoin derivatives of the released amino acids by HPLC and UV detection.[24] A second sequencing method was based on the time course release of amino acids by either aminopeptidase or carboxypeptidase; the amino acids were analyzed by HPLC and precolumn flourometric derivatization with *o*-phthaldialdehyde.[11]

A partial structure of proenkephalin has been deduced,[22] and it is shown in a schematic fashion in Fig. 8. The pentapeptide sequence of Met-enkephalin is repeated several times in the proenkephalin primary structure, and the Leu-enkephalin pentapeptide sequence is represented at least once. Each enkephalin sequence is preceded by a pair of basic amino acids. The presence of a pair of basic amino acids immediately before a peptide hormone in the structure of its precursor is a common occurrence. Furthermore, with but one known exception, each enkephalin sequence is also followed by a pair of basic amino acids.

4.4. HPLC-Coupled Assay

HPLC can be used to rapidly partially purify samples prior to a bioassay, radioimmunoassay, or any other specific type of assay. This partial purification may be useful for removing interfering substances or for concentrating the component of interest. Furthermore, HPLC can provide the qualitative identification, whereas the subsequent assay provides the quantitation.

The importance of this approach is illustrated by the assay of enkephalin in brain.[20] A low-molecular-weight fraction from a brain extract was resolved

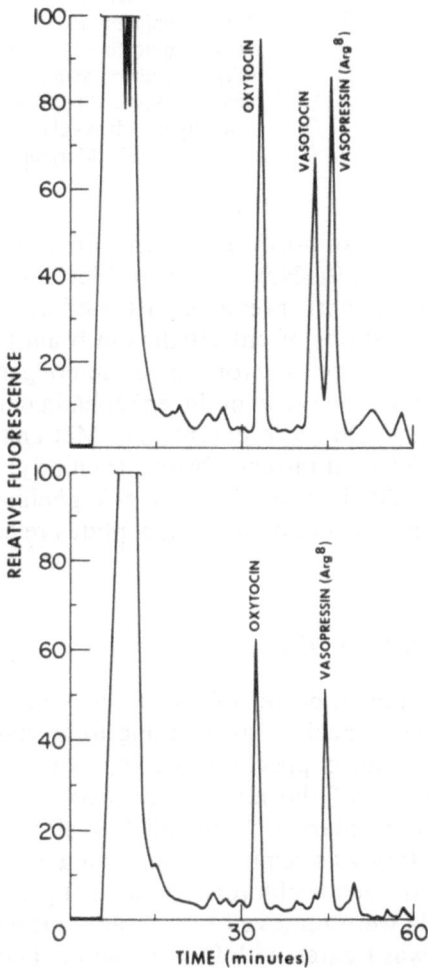

Fig. 10. Assay of oxytocin and vasopressin in rat pituitary. Individual rat posterior pituitaries were homogenized in acid. A peptide fraction was prepared from each extract and was then reacted with fluorescamine. The resulting peptide fluorophors were resolved on a reverse-phase HPLC column (bottom panel). Vasotocin was added to the pituitary homogenate as an internal standard (top panel) (from Gruber et al.[26]).

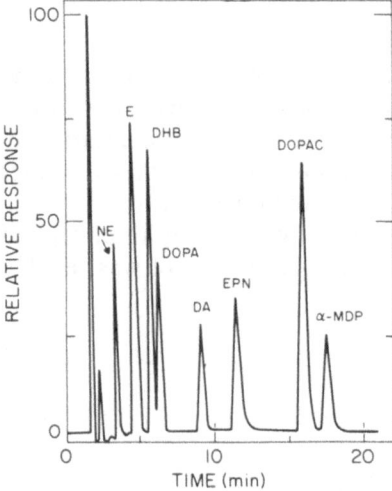

Fig. 11. Chromatography of a standard mixture of catechols. The catechols (100 pmol each) were eluted from a reverse-phase column with 5% methanol in sodium phosphate buffer at pH 2.8 with the ion-pairing reagent 1-octanesulfonic acid (30 mg/liter). The electrochemical detector was set at +0.72 V (vs. Ag/AgCl). NE, norepinephine; E, epinephrine; DHB, dihydroxybenzylamine (used as an internal standard); DOPA, dihydroxyphenylalanine; DA, dopamine; EPN, epinine; DOPAC, dihydroxyphenylacetic acid; α-MDP, α-methyldihydroxyphenylalanine. Courtesy of Drs. T. Kawano and F. Margolis.

on an HPLC column that had previously been calibrated with the enkephalins and known congeners (Fig. 9). Not only were Met-enkephalin and Leu-enkephalin found, but the congeners, previously isolated from adrenal glands, were also observed. Determinations of enkephalins in brain that do not use HPLC for sample preparation can be in error, since the congeners do cross react in various assays. Of course, the assay of the enkephalin congeners, one of which was shown to be severalfold more potent than Met-enkephalin as an antinociceptive agent when injected intracerebroventricularly,[25] should not be overlooked. Analogously, radioimmunoassay of enkephalins in blood should account for each of the enkephalin-containing peptides released from the adrenal gland.

4.5. Quantitative Analysis of Peptides

Using the high resolving power of HPLC, it is possible to detect by absorption or fluorescence a peak corresponding to an individual peptide (or a protein or any other substance) present in a complex mixture. This is illustrated by the fluorometric assay of the nonapeptides oxytocin and vasopressin in individual rat posterior pituitaries.[26] Immediately after decapitation, the posterior lobe of the pituitary was removed and homogenized in dilute acid. Proteins were precipitated with trichloroacetic acid, lipids were extracted with ether, and amino acids were removed by passage through copper–Sephadex. The resultant sample was treated with fluorescamine, and the derivatives were resolved on an HPLC column using a gradient optimized for the resolution of the oxytocin and vasopressin fluorophors (Fig. 10). The structurally related nonapeptide vasotocin, which is not present in pituitaries, was added as an internal standard.

The approach of direct assay has certain advantages over radioimmunoassay or bioassay. For example, it was found that oxytocin and vasopressin are present in equimolar amounts in each pituitary and that water deprivation causes a parallel decrease in the pituitary levels of both peptides. Since these two peptides have a certain degree of cross reactivity in the bioassays available at the time these studies were undertaken, the more accurate results of the direct assay implied a coordinated, if not common, biosynthetic mechanism. Furthermore, the direct assay allowed for confirmation of the accuracy of the method. The peaks corresponding to the oxytocin and vasopressin fluorophors were collected, hydrolyzed in acid, and subjected to amino acid analysis. The amino acid analyses were in agreement with the compositions and amounts of each peptide fluorophor.

A similar procedure was developed for the assay of carnosine in mouse olfactory bulb.[27] In these last two instances, the peptides were present at relatively high concentrations. The sensitivity of fluorescence or short-wave

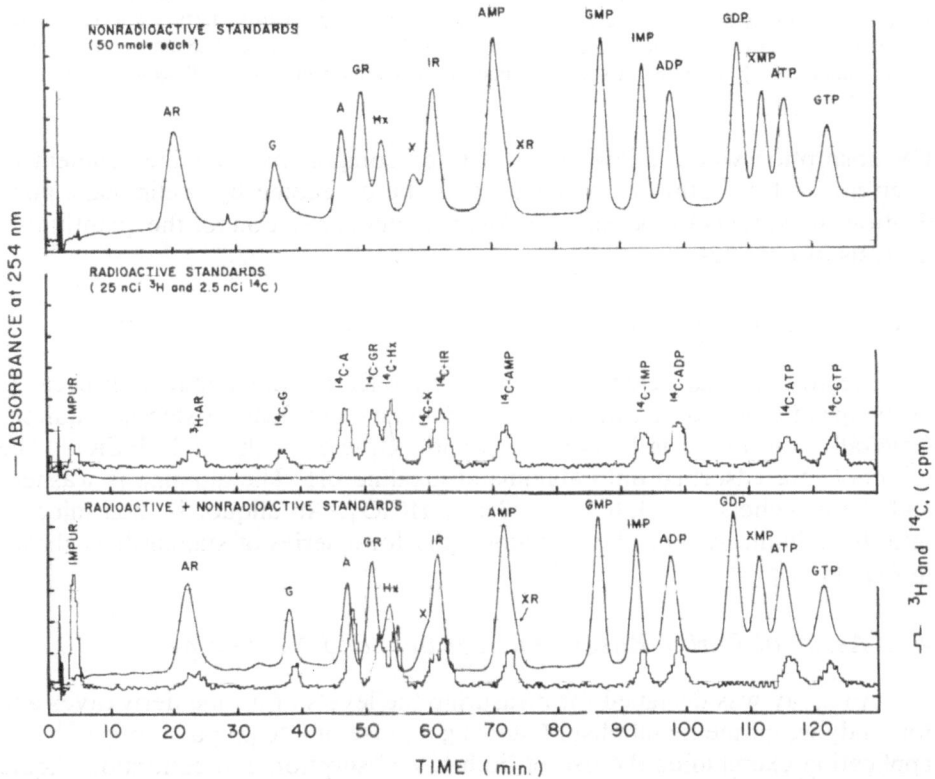

Fig. 12. Chromatography of mixtures of radioactive and nonradioactive purine standards (from Bakay et al.[28]). Elution from a polystyrene anion exchange column was with a gradient of increasing NH₄Cl at pH 9. Top: UV detection of a mixture of 16 standards. Middle: Liquid scintillation monitoring of a sample containing 5500 dpm and 55,000 dpm of [14]-labeled and [3H]-labeled standards, respectively. Bottom: Chromatography of a mixture of the above radioactive and nonradioactive standards.

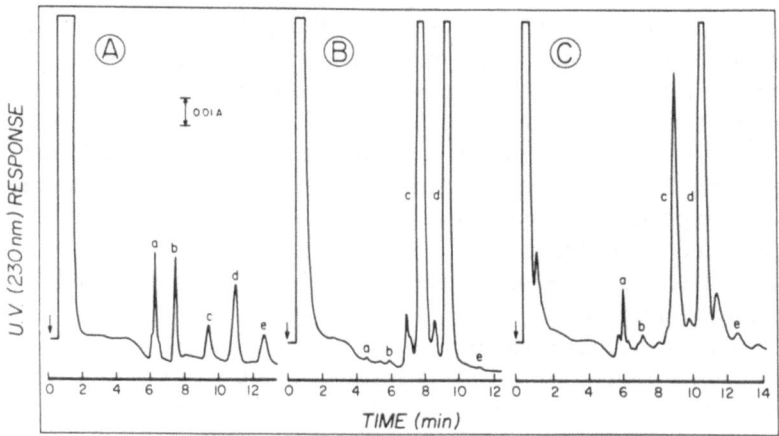

Fig. 13. Chromatography of benzoylated neutral glycosphingolipids and ceramides. Elution from an HPLC silica column was with a gradient of increasing dioxane in benzene. A: a mixture of standards. They are: a, nonhydroxy fatty acid-ceramide (0.55 nmol); b, hydroxy fatty acid-ceramide (0.5 nmol); c, nonhydroxy fatty acid-cerebrosides (0.12 nmol); d, hydroxy fatty acid-cerebrosides (0.22 nmol); e, lactosylceramide (0.22 nmol). B: Normal trigeminal nerve (0.25 mg wet tissue, 32.5 μg total lipids). C: Neurinoma of the trigeminal nerve (4.7 mg wet tissue, 110 μg total lipids).

UV absorption would allow assay of trace components, but the problem of interference from other components must be overcome by preliminary purification steps. It is also possible to use postcolumn reaction for the quantitative analysis of peptides.

4.6. Catecholamine Assay

Electrochemical detection is an ideal choice for this application because of its specificity and sensitivity.[5] A preliminary purification step is required. Generally, the tissue is homogenized and extracted with 0.1 N $HClO_4$. The catechols are adsorbed onto alumina at alkaline pH. The alumina is washed, and the catechols are eluted with 0.1 N $HClO_4$. An aliquot is then injected into the column. A typical chromatography for a series of standards is shown in Fig. 11.

4.7. Assay of Purine Bases, Nucleosides, and Nucleotides

An assay was designed for measuring the levels of purine derivatives and for studying purine metabolism following a pulse of [^{14}C]hypoxanthine.[28] This application exemplifies the use of both UV absorption and continuous liquid scintillation counting for monitoring column effluents. The column effluent first passed through the UV detector, was then mixed with the scintillation fluids, and then passed through the scintillation flow monitor. Chromatograms for nonradioactive standards alone, radioactive standards alone, and a combination of both are shown in Fig. 12.

In this particular application, skin fibroblasts were pulsed for 2 hr with the radiolabeled precursor and extracted with cold 0.4 M $HClO_4$. The extracts were neutralized and analyzed on the column. The data obtained from the analyses in this study were used to help establish the molecular causes of abnormal purine metabolism in individuals with uric acid overproduction.

4.8. Assay of Neutral Glycosphingolipids and Ceramides

This assay[29] illustrates the covalent introduction of a chromophoric moiety into a group of related substances. After extraction from the tissue, the lipids were fractionated on a silicic acid column. The appropriate fraction was then esterified with benzoyl chloride, and the resultant derivatives were chromatographed on an HPLC silica column using a gradient of increasing dioxane in hexane (Fig. 13). This method provides a rapid, sensitive, and quantitative procedure and has been used to study minor components in small amounts of tissue samples such as neural tumors.

ACKNOWLEDGMENTS. The author wishes to thank Dr. Frank Margolis for useful suggestions during the preparation of this manuscript and for providing Fig. 11. The cooperation of Dr. Bohdan Bakay for providing Fig. 12 and of Dr. Firoze B. Jungalwala for providing Fig. 13 is greatly appreciated.

REFERENCES

1. Snyder, L. R., and Kirkland, J. J., 1979, *Introduction to Modern Liquid Chromatography*, 2nd ed., John Wiley & Sons, New York.
2. Regnier, F. E., and Gooding, K. M., 1980, *Anal. Biochem.* **103**:1–25.
3. Bennet, H. P. J., Browne, C. A., Goltzman, D., and Solomon S., 1979, *Peptides: Structure and Biological Function* (E. Gross and J. Meienhofer, eds.), Pierce Chemical Co., Rockford, Illinois, pp. 121–124.
4. Rivier, J. E., 1978, *J. Liquid Chromatogr.* **1**:343–366.
5. Kissinger, P. T., Bruntlett, C. S., and Shoup, R. A., 1981, *Life Sci.* **28**:455–466.
6. Chang, S. H., Noel, R., and Regnier, F. E., 1976, *Anal. Chem* **48**:1839–1845.
7. Benson, J. R., and Hare, P. E., 1975, *Proc. Natl. Acad. Sci. U.S.A.* **72**:619–622.
8. Udenfriend, S., Stein, S. Böhlen, P., Dairman, W., Leimgruber, W., and Weigele, M., 1972, *Science* **178**:871–872.
9. Anderson, G. M., and Young, J. G., 1981, *Life Sci.* **28**:507–517.
10. Stein, S., Böhlen, P., Stone, J., Dairman, W., and Udenfriend, S., 1973, *Arch. Biochem. Biophys.* **155**:202–212.
11. Jones, B. N., Pääbo, S., and Stein, S., 1981, *J. Liquid Chromatogr.* **4**:565–586.
12. Böhlen, P., Stein, S., Stone, J., and Udenfriend, S., 1975, *Anal. Biochem* **67**:438–445.
13. Radhakrishnan, A. N., Stein, S., Licht, A., Gruber, K. A., and Udenfriend, S., 1977 *J. Chromatogr.* **132**:552–555.
14. Hughes, J., Smith, T. W., Kosterlitz, H., Fothergill, N. L., Morgan, B., and Morris, H. R., 1975, *Nature* **258**:577–579.
15. Li, C. H., and Chung, D., 1976, *Proc. Natl. Acad. Sci. U.S.A.* **73**:1145–1148.
16. Mains, R. E., Eipper, B. A., and Ling, N., 1977, *Proc. Natl. Acad. Sci. U.S.A.* **74**:3014–3018.
17. Kangawa, K., Matsuo, H., and Igarashi, M., 1979, *Biochem. Biophys. Res. Commun.* **86**:153–160.

18. Goldstein, A., Tachibani, S., Lowney, L. I., Hunkapiller, M. W., and Hood, L. E., 1979, *Proc. Natl. Acad. Sci. U.S.A.* **76:**6666–6670.
19. Lewis, R. V., Stern, A. S., Rossier, J., Stein, S., and Udenfriend, S., 1979, *Biochem. Biophys. Res. Commun.* **89:**822–829.
20. Stern, A. S., Lewis, R. V., Kimura, S., Rossier, J., Stein, S., and Udenfriend, S., 1980, *Arch. Biochem. Biophys.* **205:**606–613.
21. Kimura, S., Lewis, R. V., Stern, A. S., Rossier, J., Stein, S., and Udenfriend, S., 1980, *Proc. Natl. Acad. Sci. U.S.A.* **77:**1681–1685.
22. Lewis, R. V., Stern, A. S., Kimura, S., Rossier, J., Stein, S., and Udenfriend, S., 1980, *Science* **208:**1459–1461.
23. Jones, B. N., Stern, A. S., Lewis, R. V., Kimura, S., Stein, S., Udenfriend, S., and Shivley, J. E., 1980, *Arch. Biochem. Biophys.* **204:**392–395.
24. Hunkapiller, M. W., and Hood, L. E. 1980, *Science* **207:**523–525.
25. Inturrisi, C. E., Umas, J. G., Wolff, D., Stern, A. S., Lewis, R. V., Stein, S., and Udenfriend, S., 1980, *Proc. Natl. Acad. Sci. U.S.A.* **77:**7474–7475.
26. Gruber, K. A., Stein, S., Radhakrishnan, A. N., Brink, L., and Udenfriend, S., 1976, *Proc. Natl. Acad. Sci. U.S.A.* **73:**1314–1318.
27. Wideman, J., Brink, L., and Stein, S., 1978, *Anal. Biochem.* **86:**670–678.
28. Bakay, B., Nissinen, R., and Sweetman, L., 1978, *Anal. Biochem.* **86:**65–77.
29. Chou, K. H., and Jungalwala, F. B., 1981, *J. Neurochem.* **36:**394–401.

Index